Sustainable Environmental Waste Management Strategies

Series Editors

Alok Prasad Das, Department of Life Sciences, Rama Devi Women's University, Bhubaneswar, Odisha, India

Eric van Hullebusch, Équipe ACE, Institut de Physique du Globe de Paris, PARIS CEDEX 05, France

Sustainable waste management is an absolute imperative in a world that is already dealing with the effects of climate change. This book series seeks to publish an extensive collection of scientific books targeting researchers, scientists, professionals, academics, students, and anyone else involved in the wide, diverse field of sustainable waste management. The series accepts monographs, edited volumes, textbooks, and conference proceedings. Topics covered include but are not limited to environmental pollution and remediation, sources of environmental waste, environmental and human health, sustainable waste management, circular economy, nature-inspired environmental solutions, pollution prevention, and energy-efficient technologies for waste management. The series puts a particular emphasis on developing countries/the majority world but accepts proposals globally as long as the topic is in scope.

Alok Prasad Das • Sailaja Priyadarsini
Editors

Electronic Waste and Environmental Pollution

Impacts and Sustainable Management through Circular Economy

Springer

Editors
Alok Prasad Das
Department of Life Sciences
Rama Devi Women's University
Bhubaneswar, Odisha, India

Sailaja Priyadarsini
Department of Life Sciences
Rama Devi Women's University
Bhubaneswar, Odisha, India

ISSN 3005-1622 ISSN 3005-1630 (electronic)
Sustainable Environmental Waste Management Strategies
ISBN 978-3-031-90286-4 ISBN 978-3-031-90287-1 (eBook)
https://doi.org/10.1007/978-3-031-90287-1

This Springer imprint is published by the registered company Springer Nature Switzerland AG
The registered company address is: Gewerbestrasse 11, 6330 Cham, Switzerland

Preface

E-waste or electronic waste is the fastest growing waste, generated from major electrical and electronic devices that contain many hazardous materials, including heavy metals such as lead, mercury, and chromium. The improper disposal and management of e-waste lead to soil and water pollution affecting human health and the environment. Heavy metal and other pollutants can also leach deep into the soil and water, creating significant environmental threats and leading to the loss of many precious metals. To tackle the major challenges, urgent innovations in recycling technology are needed to align with the circular economy principles. Informal recycling methods, such as pyrometallurgy, hydrometallurgy, and direct recycling, require a high energy supply and yield a low recovery rate. In contrast, bioleaching is a biotechnological-based bio-recovery method to extract metals by leveraging the natural potential of microorganisms. Recycling of valuable metals from e-waste using circular economy approaches maximizes resource recovery efficiency compared to traditional linear economy. This book, *Electronic Waste and Environmental Pollution: Impacts and Sustainable Management Through Circular Economy*, delves into the critical aspects of the sources, transport, fate, and impact of e-waste on the environment and human health. This book will also describe the various methods for the sustainable management of e-waste, aligned with circular economy approaches. Additionally, it will focus on many biotechnological-based, cutting-edge recycling technologies for the recovery of many precious metals from e-waste. This book is a valuable resource for scholars, legislators, and industry experts looking for long-term solutions to e-waste that link technical progress with environmental issues. It promotes a shift from the traditional linear economy to a circular economy model that emphasizes sustainability, resource efficiency, and climate change mitigation. We gratefully acknowledge the contributions of all authors who offered informative chapters to this book.

Bhubaneswar, Odisha, India — Alok Prasad Das
Bhubaneswar, Odisha, India — Sailaja Priyadarsini

Contents

An Introduction to Electronic Waste Pollution and Its Sustainable Management

Sailaja Priyadarsini and Alok Prasad Das

Abstract Considering a sharp increase in the consumption of electronic devices, pollution from electronic waste has arisen as an enormous environmental and public health hazard. The global growth in e-waste creation, fueled by technology improvements and short product life cycles, has had major environmental and socioeconomic consequences. Poor e-waste disposal procedures and informal recycling practices offer significant environmental and health risks by releasing dangerous compounds into the environment, including heavy metals, brominated flame retardants, and persistent organic pollutants. Successful management initiatives, such as efficient recycling, eco-design, and bio-based recovery strategies, are required to offset these impacts. This chapter investigates the current condition of e-waste pollution, focusing on its environmental and human health impacts, as well as its long-term management. Sustainable e-waste management can reduce resource consumption and environmental damage by incorporating regulatory frameworks, innovative recycling technology, and green chemistry principles, opening the path for a more circular and sustainable electronics industry. This book seeks to provide a comprehensive understanding of e-waste, including its environmental and health consequences, as well as the urgent need for long-term solutions. It investigates multiple options for responsible e-waste management, ranging from technological advancements in recycling to legislative initiatives and circular economy strategies, such as bioleaching processes for extracting precious metals from e-waste. This book aims to educate researchers, policymakers, industry experts, and students about the intricacies of e-waste management while also encouraging innovative and sustainable approaches.

Keywords E-waste · Pollution · Toxic · Management · Bioleaching

---

S. Priyadarsini · A. P. Das (✉)
Department of Life Sciences, Rama Devi Women's University, Bhubaneswar, Odisha, India

A. P. Das, S. Priyadarsini (eds.), *Electronic Waste and Environmental Pollution*, Sustainable Environmental Waste Management Strategies,
https://doi.org/10.1007/978-3-031-90287-1_1

1 Introduction

Electronic waste (e-waste) pollution has arisen as one of the most obstinate environmental challenges in the technological advancement era (Priyadarsini & Das, 2024). The electronics sector is the world's major and fastest-growing industrial sector, due to the substantial dependence on modern technologies purely to live a luxurious existence (Benya et al., 2023; Das & Ghosh, 2022). E-waste is commonly demarcated as electrical or electronic devices that are waste, with all components, parts, and consumables that are integral to the product at the time of disposal. With rapid technological advances and an ever-increasing demand for electronic products, the accumulation of discarded gadgets, appliances, and batteries has reached frightening levels (Prabhakar et al., 2019; Raj & Das, 2023). Considering the growing adoption of new technologies, there is a significant amount of e-waste in the environment (Das et al., 2011; Das & Ghosh, 2023a; Priyadarsini & Das, 2025). According to a referendum, there are around seven billion individuals on the earth, with just five billion having access to and able to obtain their basic needs. More than 6.3 billion people use mobile phones, which are today considered an elementary need (Krishnamoorthy et al., 2018). Recent assessments of comprehensive worldwide electronic waste estimates have indicated a global generation of 54.5 Mt e-waste in 2019, up from 42.6 Mt in 2015 and 45.2 Mt in 2018, with additional projections to climb to approximately 75 Mt by 2030 (Ahirwar & Tripathi, 2021). E-waste discarded in landfills frequently pollutes the environment by leaking into soil and groundwater contaminating the nearby soil and surface water. Many valued resources are also wasted during disposal under poor conditions (Priyadarsini & Das, 2025). The global e-waste generated in 2023 is predicted to be 19.5 Mt of minor equipment, 14.5 Mt of bulky equipment, 11.9 Mt of temperature conversation equipment, and 5.2 Mt of tiny IT and telecommunication equipment (Augusto et al., 2020). India, Japan, Indonesia, and China contributed maximum of the e-waste among Asian large countries (Ahirwar & Tripathi, 2021). Reprocessing electronic waste without proper control measures for pollution can introduce wide range pollutants into the environment, including toxic heavy metals (Das et al., 2012a; Ghosh et al., 2016b, 2023; Mishra et al., 2022), polycyclic aromatic hydrocarbons (PAHs) (Behera et al., 2020; Dey et al., 2023; Mohanta et al., 2024), plastics (Das et al., 2024; Jagatee et al., 2024; Mishra et al., 2021; Singh Jadaun et al., 2022), and other pollutants. To strike an appropriate equilibrium between the adverse consequences of e-waste reprocessing on the health of humans and the environment, as well as resource compensation, a multi-directional method is required to comprehensively address the e-waste problem while also aiding the global economy and resource conservation.

This book offers a thorough investigation of e-waste sources, pollution, transport, and impact on the ecosystem and human health. It also emphasizes its sustainable management strategies, bringing together renowned specialists in the industry to investigate new recovery approaches. It addresses an extensive variety of sources, including e-waste origins, health consequences, circular economy solutions, bioleaching, and advanced recycling methods. Together, the chapters highlight the necessity for regulatory structures, advanced biotechnological tools, and resource recovery ways to reduce pollution and maximize sustainability. The book will be useful to researchers, policymakers, and industry experts since it combines interdisciplinary research with real-world case studies. It promotes proper e-waste management to protect ecosystems and human health.

2 Electronic Waste Pollution and Its Health Effect

Inadequate e-waste management has numerous undesirable consequences on the environment, including soil deterioration and pollution, water contamination, and the production of poisonous vapors (from e-waste burning), all of which have a direct impact on living organisms' health. Though e-waste makes up only 4%–7% of the entire solid volume, it accounts for more than 60% of toxicity. Heavy metals like chromium (Cr), gallium (Ga), beryllium (Be), lead (Pb), palladium (Pd), and others pollute our food chain by interacting with air, water, and soil (Liu et al., 2021). When e-waste gets disposed of inappropriately, chemicals such as dioxins get contaminated into the environment, contributing to air pollution, and damaging respiratory health (Mundada et al., 2004). Cancers and tumors, particularly chronic diseases, are at severe risk in people because e-waste emits extremely small particles that are problematic to handle, causing health problems in humans and other organisms. The substantial dangers are predominantly for humans who deal with this type of trash (Jain et al., 2023). Heavy elements from e-waste penetrate the ground, ultimately a way to groundwater, which ultimately ends up in rivers, ponds, and lakes. These processes cause acidification and toxic effects in the water, showing damaging effects on animals, plants, and organisms (Thakur et al., 2020). E-waste can damage humans in several ways, including variations in thyroid function, unfavorable newborn outcomes, alterations in cellular function, psychological changes in behavior and temperament, and a failure in lung function (Ankit et al., 2021). E-waste contaminant elements such as dioxins were detected in increased proportions in hair, placentas, and human milk, entering the human body via food and air in sufficient proportions to cause serious health risks (Chan et al., 2007). The pollutants are introduced to children from birth, beginning with the mother's breastfeeding. Lithium from mobile, photography, video, and batteries can be transferred to an infant through the mother's milk and have undesirable adverse effects on health (Priyadarsini & Das, 2025). Numerous studies have indicated that children possess much greater quantities of Pb and Cd, leading to worse cognitive abilities as related to unaffected e-waste sites, and causing respiratory difficulties such as coughing (Li et al., 2017). A significant quantity of lead was discovered in children's blood in the reprocessing of e-waste, relating to the creation of immunoglobulin E (IgE) and serves as the cause of asthma in humans, and it also leads to a weakened immunological response, such as the body's inability to fight hepatitis (Zeng et al., 2016). The damaging effect of e-waste on plants has not been fully studied, but its components, particularly in heavy metals present in leachate, are known toxicants (Das & Mishra, 2009; Paul et al., 2023).

3 Sustainable Management of E-Waste Pollutants

The removal and management of e-waste is a major undertaking, especially as the world's population and industrialization increase people's reliance on electronic devices and equipment. Waste management solutions are critical witnessing the tremendous spike in the volume of e-waste (Mishra et al., 2023). Several nations have currently resorted toward successful e-waste management policies and advanced recycling to manage these wastes responsibly and extract all the potential useful components (Yong et al., 2019). Waste recycling refers to the creation, assembly, purchase, or promotion of new electronic products made from waste. A completely useful electronic gadget quickly becomes waste due to expanding market diffusion in emerging nations, a quick spare market in wealthy nations, and an ever-increasing rate (Ankit et al., 2021). Disposing and managing e-waste is a massive undertaking, especially as the world's population and industrialization increase people's reliance on electronic equipment. E-waste has now become an indispensable component of solid garbage. Still, only a few options, such as landfilling and incineration, are available to deal with trash; yet, each has advantages and downsides. Currently, landfilling is one of the most widely used methods for disposing of e-waste (Shirodkar & Terkar, 2017). This approach involves digging channels in the ground and burying waste in them, which is then enclosed with a bushy layer of dirt to avoid exposure to air. However, it has several shortcomings, as landfilling techniques do not solve many problems due to the leaking of chemicals, uncontrolled fires, and so on. Furthermore, landfilling is not an appropriate treatment strategy for compounds that are volatile and nonbiologically degradable (Annamalai, 2015). Incineration is a complete combustion-regulated method in which waste material is burned at extremely high temperatures, often 900–1000 °C, in specially built incinerators. Advantages include reduced e-waste and the utilization of combustible materials' energy content. The key disadvantages of e-waste incineration are the emission of absconding flue gases, the vast amounts of leftovers from the combustion process, and gas cleaning (Fu et al., 2016). There are three major methods for recycling most e-waste, depending on the content and composition of the waste materials: pyrometallurgical method, hydrometallurgical method, and bioleaching. Hydrometallurgy-based e-waste recycling entails extracting soluble metals from a solid to a liquid media using particular solvents. Cyanide leaching is the most popular and traditional method, particularly for gold leaching. The extraction of metal via the hydrometallurgical technique is predictable and simple to manage when compared to pyrometallurgy; nevertheless, this technique also uses poisonous solvents such as cyanides and chlorides. After removing plastic and refractory oxides, crumpled e-waste scrap is added to the molten bath to produce metal-containing slag. This process produces harmful byproducts such as polybrominated dibenzofurans and metal recovery is only partial. Pyrometallurgy and hydrometallurgy both have disadvantages: they are expensive, need greater temperatures, emit harmful gases such as dioxins and furans, and require hazardous solvents for leaching, as well as a larger energy input (Das et al., 2012b; Das & Ghosh, 2023e; Ghosh et al., 2016a). Recently, a biological technique employing microorganisms for the extraction of metals from e-waste has been proposed; it is the most ecologically friendly and cost-effective of the previously discussed methods (Das & Ghosh, 2023b, 2023d; Ghosh & Das, 2017; Mohanty et al., 2017). Microorganisms absorb and bioaccumulate metals as part of their essential functions; bioabsorption occurs via metabolic oxidation and reduction. Acidophilic bacteria, chemolithoautotrophic bacteria, cyanogenic bacteria, and heterotrophic fungi are all employed in e-waste bleach (Bala et al., 2024; Das & Ghosh, 2018, 2023c; Ghosh & Das, 2018). Bioleaching is the process of converting metals from their insoluble solid form to water-soluble states using microorganisms (Das et al., 2015a, 2015b; Ghosh et al., 2020; Mohanty et al., 2016). Their metabolism byproducts and enzyme activities play a specific role in this process. Microorganisms for bioleaching are commonly found in nature and can be recovered from leachate and acid mine drainage. Iron-oxidizing bacteria from sulfidic concentrates have been thoroughly studied for the bioleaching process. In addition various remediation of e-waste could be solved through many approaches, with good laws and regulations serving as a vital part of the strategic framework for tackling the problem legally and universally across the country.

4 Conclusion

E-waste has evolved into an important environmental issue as a result of rapid technological advances and the growing consumption of electricity. Soil pollution is a key worry with e-waste because toxic metals leak into the earth from discarded devices, harming soil and groundwater quality. E-waste includes a complex mixture of valuable and harmful components, making recycling and disposing of it both an environmental issue and an economic opportunity. Electronic waste management involves an integrated strategy that combines technical improvements, regulatory frameworks, and circular economy ideas. Furthermore, developing biotechnological and eco-friendly recycling methods, such as microbiological and hydrometallurgical processes, presents great opportunities for resource recovery and pollution reduction. In the future, international coordination between politicians, companies, and researchers is required to build standardized e-waste management methods that promote sustainability. Embracing a circular economy strategy and investing in green technologies allow communities to move forward with more responsible and effective electronic waste solutions, ultimately promoting environmental protection and resource conservation.

References

Ahirwar, R., & Tripathi, A. K. (2021). E-waste management: A review of recycling process, environmental and occupational health hazards, and potential solutions. *Environmental Nanotechnology, Monitoring & Management, 15*, 100409. https://doi.org/10.1016/J.ENMM.2020.100409

Ankit, Saha, L., Kumar, V., Tiwari, J., Sweta, Rawat, S., Singh, J., & Bauddh, K. (2021). Electronic waste and their leachates impact on human health and environment: Global ecological threat and management. *Environmental Technology & Innovation, 24*, 102049. https://doi.org/10.1016/J.ETI.2021.102049

Annamalai, J. (2015). Occupational health hazards related to informal recycling of E-waste in India: An overview. *Indian Journal Of Occupational And Environmental Medicine, 19*(1), 61–65. https://doi.org/10.4103/0019-5278.157013

Augusto, C., Forti, B., & Freitas, K. S. (2020). Relação entre os dividendos e a classificação de rating de empresas brasileiras. *Revista Contemporânea de Contabilidade, 17*(42), 120–137. https://doi.org/10.5007/2175-8069.2020V17N42P120

Bala, K., Dogra, A., Lata, P., & Das, A. P. (2024). Discussing sources and biological-chemical remediation approaches for microplastic pollution. In *Sustainable microbial technology for synthetic and cellulosic microfiber bioremediation* (pp. 37–59). https://doi.org/10.1007/978-3-031-62600-5_3

Behera, I. D., Basak, G., Kumar, R. R., Sen, R., & Meikap, B. C. (2020). Treatment of petroleum refinery sludge by petroleum degrading bacterium Stenotrophomonas pavanii IRB19 as an efficient novel technology. *Journal of Environmental Science and Health, Part A, 56*(2), 226–239. https://doi.org/10.1080/10934529.2020.1866924

Benya, A., Mohanty, S., Hota, S., Das, A. P., Rath, C. C., Achary, K. G., & Singh, S. (2023). Endangered Curcuma caesia Roxb.: Qualitative and quantitative analysis for identification of industrially important elite genotypes. *Industrial Crops and Products, 195*, 116363. https://doi.org/10.1016/J.INDCROP.2023.116363

Chan, J. K. Y., Guan, H. X., Xu, Y., Liang, Y., Ling, X. C., Sheng, C. W., Wong, C. K. C., Leung, C. K. M., & Wong, M. H. (2007). Body loadings and health risk assessment of polychlorinated dibenzo-p-dioxins and dibenzofurans at an intensive electronic waste recycling site in China. *Environmental Science and Technology, 41*(22), 7668–7674. https://doi.org/10.1021/ES071492J/SUPPL_FILE/ES071492J-FILE002.PDF

Das, A. P., & Ghosh, S. (2018). Bioleaching of manganese from mining waste materials. *Materials Today Proceedings, 5*(1), 2381–2390. https://doi.org/10.1016/J.MATPR.2017.11.459

Das, A., & Ghosh, S. (2022). Role of microorganisms in extenuation of mining and industrial wastes. *Geomicrobiology Journal, 39*, 173–175. https://doi.org/10.1080/01490451.2022.2038953

Das, A. P., & Ghosh, S. (2023a). Evolution of biomining technology. In *Manganese mining microorganisms* (pp. 5–19). https://doi.org/10.1016/B978-0-12-822147-1.00012-1

Das, A. P., & Ghosh, S. (2023b). Mechanism of manganese bioleaching. In *Manganese mining microorganisms* (pp. 87–98). https://doi.org/10.1016/B978-0-12-822147-1.00014-5

Das, A. P., & Ghosh, S. (2023c). Microorganisms in manganese biomining processes. In *Manganese mining microorganisms* (pp. 55–61). https://doi.org/10.1016/B978-0-12-822147-1.00013-3

Das, A. P., & Ghosh, S. (2023d). Mixed culture bioleaching: An insight into manganese biomining process and efficacy. In *Manganese mining microorganisms* (pp. 129–141). https://doi.org/10.1016/B978-0-12-822147-1.00001-7

Das, A. P., & Ghosh, S. (2023e). Role of metabolites and metabolomics for understanding of manganese bioleaching. In *Manganese Mining Microorganisms* (pp. 117–127). https://doi.org/10.1016/B978-0-12-822147-1.00011-X

Das, A. P., & Mishra, S. (2009). Hexavalent chromium [Cr (VI)]: Yellow water pollution and its remediation. *Sarovar Saurabh the ENVIS Newsletter on Wetland Ecosystems, 5*(2), 1.

Das, A. P., Sukla, L. B., Pradhan, N., & Nayak, S. (2011). Manganese biomining: A review. *Bioresource Technology, 102*(16), 7381–7387. https://doi.org/10.1016/J.BIORTECH.2011.05.018

Das, A. P., Sukla, L. B., & Pradhan, N. (2012a). Microbial recovery of manganese using *Staphylococcus epidermidis*. *International Journal of Nonferrous Metallurgy, 2012*(02), 9–12. https://doi.org/10.4236/IJNM.2012.12002

Das, A. P., Swain, S., Panda, S., Pradhan, N., & Sukla, L. B. (2012b). Reductive acid leaching of low grade manganese ores. *Geomaterials, 2012*(04), 70–72. https://doi.org/10.4236/GM.2012.24011

Das, A. P., Ghosh, S., Mohanty, S., & Sukla, L. B. (2015a). Advances in manganese pollution and its bioremediation. In L. B. Sukla, N. Pradhan, S. Panda, & B. K. Mishra (Eds.), *Environmental microbe biotechnology* (1st ed., pp. 313–328). Springer. https://doi.org/10.1007/978-3-319-19018-1_16

Das, A. P., Ghosh, S., Mohanty, S., & Sukla, L. B. (2015b). Consequences of manganese compounds: A review. *Toxicological & Environmental Chemistry, 96*(7), 981–997. https://doi.org/10.1080/02772248.2015.1005428

Das, A., Dey, S., & Das, A. P. (2024). Ecotoxicological impacts of synthetic microfiber pollutants and development of sustainable mitigation strategies. *Environmental Chemistry and Ecotoxicology, 7*, 201. https://doi.org/10.1016/J.ENCECO.2024.12.003

Dey, S., Das, A., Mallick, K., Sahu, A., & Das, A. P. (2023). Environmental petroleum waste: Pollution, toxicity, sustainable remediation. In *Impact of petroleum waste on environmental pollution and its sustainable management through circular economy* (pp. 159–175). https://doi.org/10.1007/978-3-031-48220-5_7

Fu, Z., Wang, Y., Chen, J., Wang, Z., & Wang, X. (2016). How PBDEs are transformed into dihydroxylated and dioxin metabolites catalyzed by the active center of cytochrome P450s: A DFT study. *Environmental Science and Technology, 50*(15), 8155–8163. https://doi.org/10.1021/ACS.EST.6B00524/SUPPL_FILE/ES6B00524_SI_001.PDF

Ghosh, S., & Das, A. P. (2017). Bioleaching of manganese from mining waste residues using Acinetobacter sp. *Geology, Ecology, and Landscapes, 1*(2), 77–83. https://doi.org/10.1080/24749508.2017.1332847

Ghosh, S., & Das, A. P. (2018). Metagenomic insights into the microbial diversity in manganese-contaminated mine tailings and their role in biogeochemical cycling of manganese. *Scientific Reports, 8*(1), 1–12. https://doi.org/10.1038/s41598-018-26311-w

Ghosh, S., Mohanty, S., Akcil, A., Sukla, L. B., & Das, A. P. (2016a). A greener approach for resource recycling: Manganese bioleaching. *Chemosphere, 154*, 628–639. https://doi.org/10.1016/J.CHEMOSPHERE.2016.04.028

Ghosh, S., Mohanty, S., Nayak, S., Sukla, L. B., & Das, A. P. (2016b). Molecular identification of indigenous manganese solubilising bacterial biodiversity from manganese mining deposits. *Journal of Basic Microbiology, 56*(3), 254–262. https://doi.org/10.1002/JOBM.201500477

Ghosh, S., Gandhi, M., van Hullebusch, E. D., & Das, A. P. (2020). Proteomic insights into Lysinibacillus sp.-mediated biosolubilization of manganese. *Environmental Science and Pollution Research, 28*(30), 40249–40263. https://doi.org/10.1007/S11356-020-10863-4

Ghosh, S., Tripathy, B., Dey, S., & Das, A. P. (2023). Comparative investigation of fungal and bacterial manganese biomining mechanisms. In *Manganese mining microorganisms* (pp. 99–115). https://doi.org/10.1016/b978-0-12-822147-1.00016-9

Jagatee, S., Priyadarshini, S., Rath, C. C., & Das, A. P. (2024). Synthetic microfiber: An enduring environmental problem linked to sustainable development. In *Renewable energy generation and value addition from environmental microfiber pollution through advanced Greener solution* (pp. 93–112). https://doi.org/10.1007/978-3-031-51792-1_6

Jain, M., Kumar, D., Chaudhary, J., Kumar, S., Sharma, S., & Singh Verma, A. (2023). Review on E-waste management and its impact on the environment and society. *Waste Management Bulletin, 1*(3), 34–44. https://doi.org/10.1016/J.WMB.2023.06.004

Krishnamoorthy, Y., Ma, V., Sakthivel, M., & Sarveswaran, G. (2018). Emerging public health threat of e-waste management: Global and Indian perspective. *Reviews on Environmental Health, 33*(4), 321–329. https://doi.org/10.1515/REVEH-2018-0021/MACHINEREADABLECITATION/RIS

Li, X., Tian, Y., Zhang, Y., Ben, Y., & Lv, Q. (2017). Accumulation of polybrominated diphenyl ethers in breast milk of women from an e-waste recycling center in China. *Journal of Environmental Sciences, 52*, 305–313. https://doi.org/10.1016/J.JES.2016.10.008

Liu, Y., Zhang, J., Yang, X., Yang, W., Chen, Y., & Wang, C. (2021). Efficient recovery of valuable metals from waste printed circuit boards by microwave pyrolysis. *Chinese Journal of Chemical Engineering, 40*, 262–268. https://doi.org/10.1016/J.CJCHE.2020.11.008

Mishra, S., Rout, P. K., & Das, A. P. (2021). Emerging microfiber pollution and its remediation. In R. Prasad (Ed.), *Environmental and microbial biotechnology* (pp. 247–266). Springer. https://doi.org/10.1007/978-981-15-5499-5_9

Mishra, S., Dash, D., Al-Tawaha, A. R. M. S., & Das, A. P. (2022). A review on heavy metal ion adsorption on synthetic microfiber surface in aquatic environments. *Applied Biochemistry and Biotechnology, 194*(10), 4639–4654. https://doi.org/10.1007/S12010-022-04029-W/METRICS

Mishra, S., Ghosh, S., van Hullebusch, E. D., Singh, S., & Das, A. P. (2023). A critical review on the recovery of base and critical elements from electronic waste-contaminated streams using microbial biotechnology. *Applied Biochemistry and Biotechnology, 2023*, 1–30. https://doi.org/10.1007/S12010-023-04440-X

Mohanta, S., Pradhan, B., & Behera, I. D. (2024). Impact and remediation of petroleum hydrocarbon pollutants on agricultural land: A review. *Geomicrobiology Journal, 41*(4), 345–359. https://doi.org/10.1080/01490451.2023.2243925

Mohanty, S., Ghosh, S., Nayak, S., & Das, A. P. (2016). Isolation, identification and screening of manganese solubilizing fungi from low-grade manganese ore deposits. *Geomicrobiological Journal, 34*(4), 309–316. https://doi.org/10.1080/01490451.2016.1189016

Mohanty, S., Ghosh, S., Nayak, S., & Das, A. P. (2017). Bioleaching of manganese by aspergillus sp. isolated from mining deposits. *Chemosphere, 172*, 302–309. https://doi.org/10.1016/J.CHEMOSPHERE.2016.12.136

Mundada, M. N., Kumar, S., & Shekdar, A. V. (2004). E-waste: A new challenge for waste management in India. *International Journal of Environmental Studies, 61*(3), 265–279. https://doi.org/10.1080/0020723042000176060

Paul, A., Dey, S., Ram, D. K., & Das, A. P. (2023). Hexavalent chromium pollution and its sustainable management through bioremediation. *Geomicrobiology Journal, 41*, 1. https://doi.org/10.1080/01490451.2023.2218377

Prabhakar, A., Mishra, S., & Das, A. P. (2019). Isolation and identification of Lead (Pb) solubilizing bacteria from automobile waste and its potential for recovery of Lead from end of life waste batteries. *Geomicrobiological Journal, 36*(10), 894–903. https://doi.org/10.1080/01490451.2019.1654044

Priyadarsini, S., & Das, A. P. (2024). Analytical and structural characterization of waste lithium-ion batteries for their effective recycling strategy. *Environmental Chemistry and Ecotoxicology, 7*, 182. https://doi.org/10.1016/j.enceco.2024.12.004

Priyadarsini, S., & Das, A. P. (2025). Lithium bioleaching: A review on microbial-assisted sustainable technology for lithium bio-circularity. *Journal of Water Process Engineering, 69*, 106744. https://doi.org/10.1016/j.jwpe.2024.106744

Priyadarsini, S., & Das, A. P. (2025). Isolation and Identification of Native Lithium-Resistant Bacteria from Electronic Waste Contaminated sites: A focused Investigation on Tolerance Assessment with Potential Lithium Bio-recovery Application. *Geomicrobiology Journal*, 1–15. https://doi.org/10.1080/01490451.2025.2539509

Raj, K., & Das, A. P. (2023). Lead pollution: Impact on environment and human health and approach for a sustainable solution. *Environmental Chemistry and Ecotoxicology, 5*, 79–85. https://doi.org/10.1016/J.ENCECO.2023.02.001

Shirodkar, N., & Terkar, R. (2017). Stepped recycling: The solution for E-waste management and sustainable manufacturing in India. *Materials Today Proceedings, 4*(8), 8911–8917. https://doi.org/10.1016/J.MATPR.2017.07.242

Singh Jadaun, J., Bansal, S., Sonthalia, A., Rai, A. K., & Singh, S. P. (2022). Biodegradation of plastics for sustainable environment. *Bioresource Technology, 347*, 126697. https://doi.org/10.1016/J.BIORTECH.2022.126697

Thakur, D., Ganguly, R., Kumar Gupta, A., & Ghali, V. (2020). Evaluation of existing solid waste management system in Una town, India. In *Sustainable waste management: Policies and case studies: 7th IconSWM—ISWMAW 2017: Volume 1* (pp. 367–381). https://doi.org/10.1007/978-981-13-7071-7_33

Yong, Y. S., Lim, Y. A., & Ilankoon, I. M. S. K. (2019). An analysis of electronic waste management strategies and recycling operations in Malaysia: Challenges and future prospects. *Journal of Cleaner Production, 224*, 151–166. https://doi.org/10.1016/J.JCLEPRO.2019.03.205

Zeng, X., Xu, X., Zheng, X., Reponen, T., Chen, A., & Huo, X. (2016). Heavy metals in PM2.5 and in blood, and children's respiratory symptoms and asthma from an e-waste recycling area. *Environmental Pollution, 210*, 346–353. https://doi.org/10.1016/J.ENVPOL.2016.01.025

Source, Transport, and Accumulation of E-Waste in the Aquatic and Terrestrial Ecosystem

Ravi Kumar Goswami, Priti Giri, Dhitri Borah, Kiran Kumar Salam, Prachi, Sudeshna Mandal, and Bronson Kumar Khangembam

Abstract Electronic waste (e-waste) is the world's fastest-growing solid waste source, expanding threefold faster than the global population. E-waste is a rapidly growing environmental problem caused by the increased usage of electronic gadgets and the short lifecycle span of modern technology. The improper disposal of e-waste, which includes consumer electronics, and industrial and home appliances, exposes the environment to hazardous compounds such as heavy metals (lead, manganese, cadmium), persistent organic pollutants (POPs), and flame retardants. These hazardous compounds contaminate aquatic and terrestrial ecosystems, creating an imminent risk for both human health and the environment. In terrestrial ecosystems, e-waste is often disposed of in landfills or by unlawful dumping, where it decomposes and releases hazardous substances into the soil and groundwater. Surface runoff from landfills and informal dumping sites introduce toxins into aquatic ecosystems, resulting in bioaccumulation and biomagnification throughout the food chain, with major implications for aquatic biodiversity and human health. The e-waste pollutants occur through multiple pathways, including wind dispersal, water runoff, and inappropriate recycling practices. In many developing nations

R. K. Goswami · K. K. Salam · Prachi
Department of Zoology, Hindu College, University of Delhi, Delhi, India

P. Giri
Department of Botany, University of Delhi, Delhi, India

D. Borah
Department of Zoology, Biswanath College, Gauhati University, Biswanath Chariali, Assam, India

S. Mandal
Department of Zoology, Vivekananda College, University of Calcutta, Kolkata, West Bengal, India

B. K. Khangembam (✉)
Department of Zoology, Tripura University (A Central University), Suryamaninagar, Tripura, India
e-mail: bronsonkhangembam@tripurauniv.ac.in

A. P. Das, S. Priyadarsini (eds.), *Electronic Waste and Environmental Pollution*, Sustainable Environmental Waste Management Strategies,
https://doi.org/10.1007/978-3-031-90287-1_2

weak or poorly enforced environmental regulations that are informal e-waste recycling practices contribute to the spread of pollutants across wide areas, harming both terrestrial and aquatic ecosystems. The accumulation of e-waste pollutants in ecosystems can cause long-term ecological harm, disrupt ecosystem services, and pose health risks to populations exposed to these hazardous substances. This chapter addresses the environmental challenges posed by e-waste, which requires improved global e-waste management practices, stronger regulatory frameworks, and innovative solutions to reduce e-waste generation and promote sustainable recycling. Through the integration of newly developed techniques with numerous existing methodologies related to the sources, transportation, and accumulation of e-waste, it provides ample opportunities to evaluate e-waste generation and offers a comprehensive strategy.

Keywords E-waste · Aquatic ecosystem · Terrestrial ecosystem · Human health · Sustainable recycling

1 Introduction

One of the major concerns directly affecting human health and the environment in the twenty-first century is the rapidly increasing accumulation of electronic waste or e-waste. Rapid advancement in Information and Communication Technology has driven the increasing use of electronic devices like computers, laptops, tablets, mobile phones, and various other small and large household equipment worldwide (Ramadan et al., 2024). E-waste includes all discarded household, consumer, and industrial electrical and electronic equipment (EEEs) that has become obsolete or irreparable and hence discarded as Waste Electrical Electronic Equipment or WEEE. It is reported that approximately 11.2 billion tonnes of e-waste is accumulated annually (UNEP, 2020). The issue of e-waste in developing countries is more alarming, due to ineffective environmental policies and legal loopholes (Kalamaras et al., 2021) along with inadequate waste management systems. Its management is most challenging due to the constant evolution of electronic devices and their specific material compositions (Abalansa et al., 2021).

Electronic items include a variety of dangerous and harmful components along with nonhazardous components. E-waste is generated when EEEs are not properly discarded or recycled. The primary reason for discarding EEEs is the end of its functional life, mechanical damage, malfunctions, outdated software, or complete breakdown (Mishra et al., 2023). Furthermore, with the rapid advancement of modern technology, the upgradation of existing EEEs by adding new features to the product coupled with increasing consumer purchasing power may also encourage consumers to buy new EEEs and discard the existing ones, thereby increasing the generation of e-waste. Aggressive public advertisement and social networks may also be common reasons to discard EEEs in many cases (Ongondo et al., 2011).

Every year, there is almost 10% increase in the production of e-waste (Shamim et al., 2015), of which a relatively small portion of about 20% is reported to undergo recycling (Baldé et al., 2017). Nevertheless, electronic waste may have a significant detrimental effect on humans through its penetration into the environment, thereby causing environmental problems and economic losses. Without a proper disposal system, electronic waste can release harmful substances like heavy metals such as mercury, lead, and cadmium into water, soil, and air, causing severe environmental pollution (Raj & Das, 2023). These toxins can contaminate ecosystems, harm wildlife, and leach into groundwater, posing long-term ecological threats. Not surprisingly, lead and cadmium in blood samples were found higher in children living near e-waste recycling zone (Zheng et al., 2008), and polychlorinated biphenyls in breast milk samples of females exposed to an electronic waste recycling site in China (Xing et al., 2009). These toxic substances may enter the food chain or directly impact the health of the workers involved in their recycling process (Ha et al., 2009). The concentrations of copper in the hair samples of workers in India were found to be higher than their concentration in the site itself (Ha et al., 2009). Direct inhalation of heavy metals could lead to an increased risk of negative effects on the thyroid, neonatal health, and lung health (Huang et al., 2016). Geeraerts et al. (2015) observed neurological disorders in 81,300 children after exposure to sites of e-waste recycling. Thus, e-waste pollutants can cause serious health issues in the neurological, respiratory, and developmental process, especially in vulnerable populations like children and workers in informal recycling sectors. Economically, on the other hand, precious metals (gold, silver, and rare earth metals) are lost when e-waste is discarded rather than recycled. Additionally, inefficient management increases waste treatment costs and burdens public health systems. Addressing these impacts requires sustainable e-waste management practices, robust recycling systems, and global cooperation to mitigate its far-reaching consequences. This chapter discusses the characteristics of e-waste, its various sources, and the current status. The major transport routes and accumulation in aquatic and terrestrial ecosystems are described in detail. Finally, the chapter ends with a brief description of various management and mitigation strategies, and current regulations in various countries to fight the menace of e-waste.

2 Current Trends of E-Waste Production

The UN Global E-Waste Monitor (2024) states that global e-waste production is a rapidly growing environmental challenge since only a small amount of all e-waste generated is formally recycled after proper collection (in 2022 just 22.3% of the 62 million tonnes of e-waste generated was recycled). The total e-waste production is expected to reach 82 million tonnes by the year 2030, considering the increasing consumer demand for electronics, rapid technological obsolescence, and shorter product lifecycles. Europe has the highest per capita production of e-waste, while Asia leads in the total e-waste generation (Forti et al., 2020). Unfortunately, less than a quarter of this undergoes recycling after proper collection, resulting in the majority of it being improperly disposed of, posing significant environmental and health risks. The largest market for electronic devices is Asia, and because of poor management, there is an illegal e-waste recycling industry that typically deals with and gets rid of it (Andeobu et al., 2021). Understanding the various degrees of e-waste production is crucial for addressing this worldwide problem since it allows for the appropriate management methods. The amount of e-waste generated by developed nations like the United States, Germany, and the UK is far higher than that of developing nations. Asian countries like Taiwan, Hong Kong, and South Korea generate relatively less e-waste overall, but their per capita production is on par with that of the United States, Canada, and the EU (Baldé et al., 2017; StEP, 2018).

With a population of 1.35 billion people, India is facing an immediate threat because of the increased amount of e-waste generated (Mundada et al., 2004). Weak legislation for e-waste management may be responsible for the import of used and scrapped devices such as projectors, CPUs, monitors, printers, computer accessories, PVC wires, mobile phones, etc., from Western nations (Agoramoorthy & Chakraborty, 2012; Borthakur & Govind, 2017). The corporate, IT, and government associations account for nearly 70% of all e-waste in India (Ahmed et al., 2014), while households account for 15% through a wide range of electrical products and other white goods (Rajya Sabha, 2011). With 19.8% of the total, Maharashtra leads the generation, followed by Tamil Nadu with 13%. About 20–80% of all illegally dumped e-waste in India and partially obsolete electronic items in India originate from the United States and EU (Khattar et al., 2007). According to IBEF (2019), India is the second largest manufacturer of mobile phones globally, which makes it one of the top 4 countries with a potential for generation of e-waste (ASSOCHAM-NEC, 2018). Nonetheless, innovative strategies and robust policies for the effective recovery of metals and the development of commercially viable and environmentally beneficial management systems may help overcome this issue.

3 Characteristics and Composition of E-Waste

Electrical or electronic devices that have ceased to perform their function or do not further meet or satisfy their current owners for their original function and/or whose useful life has reached its terminal end are discarded and collectively referred to as "electronic waste" or "e-waste" (OECD, 2001; EU, 2002; Sinha, 2004). These are also commonly known as waste electrical and electronic equipment or WEEE (Kahhat et al., 2008). However, it might also include electrical and electronic devices that use energy and release carbon dioxide, therefore contributing to the

Table 1 Major categories of electrical devices and e-waste generated from them (Garlapati, 2016; Carroll 2018)

Electronic and electrical devices	Waste produced
Computer and its accessories (monitors, circuit boards, Mouse, keyboards, etc.)	Heavy metals (Cd, Cr, Hg), beryllium-brominated flame retardants, plastic, screws, lead, metals, fiberglass, and precious metals
Electric bulb	Contact wires, insulation, tungsten filament, low-pressure inert gas, outline glass, highly concentrated lead and phosphorous, and tiny metal components
Electric motor	Metals (ferrous and nonferrous), copper coil, wires, Hg, Pb, plastic
LCD screens	Heavy metals (Hg, Pb)
Laptop and its parts	Li, Ni, Cd, Hg, plastics
Mobile phones	Heavy metal such as Pb, Pd, As, Cd
Optical fibers	Pb, Cu, Y, F, Zr
Rechargeable batteries	Cd, Li, Ni
Television	Glass, Pb, Hg, circuit boards, plastic, and small metal components

source of e-waste (Carroll, 2018). Table 1 lists the different types of waste generated by various electrical and electronic devices.

The general composition of e-waste consists of Al, Cu, Ni, Sn, Pb, Fe, and Mn (metals) and various other components such as various types of glass fiber, plastics, ceramics, and metal oxides (Das & Ghosh, 2023; Prabhakar et al., 2019). Nevertheless, significant differences may exist in its composition based on the manufacturing company, origin, type, or age of the EEEs (Bigum et al., 2017). Plastics account for 20–40% in many obsolete home appliances, making it the most significant and heaviest component (Mallick et al., 2024; Mishra et al., 2019; Sahoo et al., 2022). For instance, 25–55% of waste mobile phones are composed of plastics (Nnorom & Osibanjo, 2009; Tan et al., 2017). Interestingly, smaller devices like mobile cellular handsets, tablets, and laptops contain higher potential contaminants than the heavy electrical equipment like TVs and refrigerators (Suja et al., 2014). The heavy metals like Hg, Pb, Cd, persistent organic compounds (BFRs), phthalates, and other toxic and hazardous substances are among its constituents, which can leak into the environment and cause health problems in humans, if not properly processed, recycled, or safely disposed (UNEP, 2005). Metal is the major constituent (60.2%), while the other components comprise of plastic materials (15.21%), polychlorinated biphenyls (1.7%), metal-plastic mixtures (4.9%), CRT and LCD screens (11.9%), cables (2.7%), etc. (Widmer et al., 2005). On the other hand, some metals like Li, Ni, Co, and Mn may be extracted from e-waste generated from some electronic devices like mobile phones (Priyadarsini & Das, 2025).

Although e-waste consists of different materials, the major components are steel (50%), polymers (21%), and metals like Cu and Al (13%). Toxic and harmful metals can make up to 1% of the entire weight, whereas precious metals such as silver

(0.2%), gold (0.1%), and palladium (0.005%) are also important recoverable components. Although 95% of the funding for the recycling infrastructure comes from recoverable precious metals, other components like Pb, Ni, and polymers may still be valuable after being recovered. The heavy metals present in e-waste can impede the biodegradation process. These become hazardous when they leach out from the e-waste, accumulate over the threshold levels, and pollute different ecosystems. For instance, dibutyl phthalate and diethyl hexyl phthalate become toxic at concentrations above 0.5% (CPCB 2008 Guidelines). The components listed in Annex II of the WEEE Directives include hazardous compounds such as capacitors that contain polychlorinated biphenyls (PCBs), parts that contain mercury, including switches, lamps, batteries, mobile phones, and PC boards, which must be removed from the e-waste devices immediately (European Commission, 2006).

Since short-lived electronic devices like computers and smartphones are becoming more common, the production of e-waste scrap and associated chemical pollutants is rising. Additionally, a greater variety of metals may be found other than those previously linked to PCBs such as Pb, Sn, and Cu (Dalrymple et al., 2007). Heavy metals, many of which may be harmful, are constituents in most of the 500–1000 different parts of normal mobile phones (Song & Li, 2014). TBBPA (tetra-

Table 2 List of the different types of hazardous substances present in e-waste and their different sources

Hazardous substance	Main sources	References
Barium (Ba)	Fluorescent lamps, spark plugs, CRTs	Oskarsson (2015), Peana et al. (2021)
Beryllium (Be)	Some electrical components, power supply boxes, computers, circuit boards, x-ray equipment, motherboards, etc.	Adanu et al. (2020)
Chromium (Cr)	Data tapes, anticorrosion coatings, floppy disks	Shanker and Venkateswarlu (2011), Kuntawee et al. (2020)
Cadmium (Cd)	CRTs, mobile phones, semi-conductor chips, PCBs, rechargeable computer batteries, contacts and switches, Ni-Cd batteries, photocopying machines	Li et al. (2008), Ebrahimi et al. (2020)
Lead (Pb)	CRTs (4–22% of Pb), television sets, PC monitors, batteries, PCBs light bulbs	Li et al. (2008), Sharma et al. (2020)
Lithium (Li)	Li-batteries, mobile phones	Hendricks et al. (2015)
Mercury (Hg)	Monitors, CRTs, PCBs, thermostats, sensors, fluorescent lamps	Kim et al. (2016)
Nickel (Ni)	Ni-Cd batteries, CRTs electron guns	Kuntawee et al. (2020)
PBDE, PBB, TBBA	Fire retardants for electronic equipment, circuit boards, plastic castings as insulation for cables, computers, etc. components of capacitors, transformers, mobile phones	Jarema et al. (2015), Arvaniti and Kalantzi (2021)
PVC	Insulation on wires and cables	Boyle et al. (2020)

bromobisphenol-A), a persistent organic pollutant, is an important fire retardant used in the manufacture of mobile phones, and it may finally contribute as e-waste when discarded (Cobbing, 2008). A summary of the primary harmful materials found in e-waste is listed in Table 2. It should however be emphasized that a large number of metals are not dangerous in their metallic forms and only turn dangerous when consumed in liquid, gaseous, or dust/soot form at quantities higher than those listed.

4 Sources, Transport, and Accumulation of E-Waste

When e-waste is not properly handled or processed, it may cause contamination of the air, water, and soil by releasing various harmful toxicants and ultimately reaching the food chain (Zhao et al., 2008). E-waste originates from various sources which include both public and private sectors, homes and manufacturing businesses, academic and research institutions (Fig. 1). Many sectors are expanding rapidly and when any equipment gets outdated, their contribution to e-waste reaches its maximum. However, residential garbage is freely disposed of by local rag pickers in the absence of proper management, resulting in multiple negative health impacts on the recyclers, even if large institutions and government agencies are somewhat required

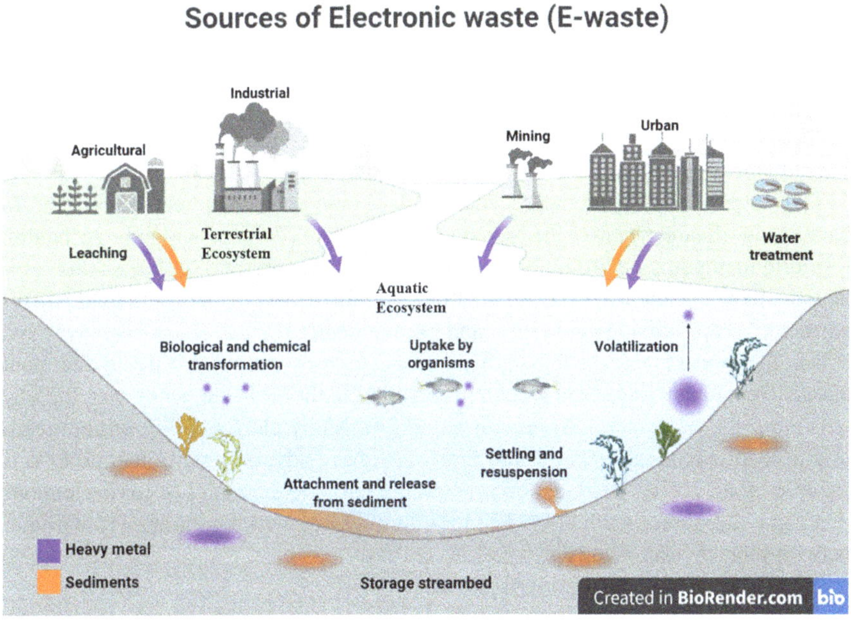

Fig. 1 Sources of e-waste in aquatic and terrestrial ecosystems

to dispose of their waste to the proper dealer. Some noteworthy sources include big and small household goods, hospital and private sector medical equipment, government-owned structures, telephone and information technology devices, hospital and private sector healthcare devices, and machinery and other industrial goods. About 30% of e-waste includes washing devices, drying equipment, cleaners for vacuums, air conditioners, coffee makers, irons, and other appliances which are followed by refrigerators (20%), mobile phones, printing equipment, CDs, DVDs, faxes, high-fidelity sound collections, etc. constituting 30%, TVs 10%, and monitors alone 10% (Manikandan et al., 2023). During 2019, the small equipment comprised the highest global e-waste quantity which is followed by large temperature exchange equipment (Gupta & Nath, 2020).

Environmental pollution from inappropriate e-waste recycling and unlawful disposal practices can harm the biota zone by causing direct and indirect exposures through the soil, air, dust, and water surrounding e-waste recycling operations. Environmental outflows of biogeochemical are fueled by processes like complexation (aggregations of heavy metals and organic matter), leaching, adsorption–desorption, volatilization (in air or soil), chemical and biological degradation, and plant uptake. The environment's ecological, physico-chemical, geomorphological, and pedological characteristics frequently affect the future uses of the dangerous substances released by electronic waste (Robinson, 2009; Grant et al., 2013). Hazardous substances leak into waters from the landfilling of electronic waste, damaging aquatic and terrestrial environments (Fig. 1). Health of individuals in countries that are developing, where most digital debris gets disposed of, may be affected by this. Nevertheless, if the wasted goods can be reused following appropriate waste disposal, it becomes feasible to retrieve the components such as copper, aluminum, gold, and silver (Manikandan et al., 2023). Despite a variety of contact pathways, improper disposal or handling of these substances can result in negative consequences on individuals, wildlife, and plants.

Due to poor management procedures, electronic waste has been discovered to have a significant negative impact on both the environment and human health, resulting in ongoing environmental contamination. The presence of hazardous components in electronic trash has led to a major contribution to environmental endangerment, compromising ecosystem and human health (Lee et al., 2024). Over and above, the modest increase in health hazards caused by improper e-waste treatment has also resulted in long-term hazardous residues in the earth, air, and water, thereby poisoning the ecosystem (Apprey et al., 2024). Many elements like lithium, cadmium, lead, zinc, and nickel found in e-waste have adverse environmental effects (Kumar et al., 2024). Lead, cadmium, and mercury are examples of toxic elements that can either directly affect the health of individuals working during extraction or leak into the environment (Perkins et al., 2014).

4.1 E-Waste in Aquatic Ecosystem

Groundwater, one of the most valuable resources of earth, is gradually getting contaminated over the years, and electronic waste has become a major contributor to its contamination. One of the main reasons for this threat is thought to be e-waste. Electronic waste disposal sites nearby can have significant adverse effects on the physicochemical properties of the groundwater. Major environmental and health problems arise in poor countries like India due to a lack of knowledge on the proper disposal of e-waste (Gupta & Nath, 2020). The presence of heavy metals like mercury and cyanide, as well as supplementary wastes, have a high potential to seep into the soil and create organic as well as inorganic compounds there (Sepúlveda et al., 2010). Harmful substances in e-waste can react with groundwater, thereby altering the turbidity, pH, conductivity, total dissolved solids (TDS), and chloride levels, which may be detrimental to human health (Dharini et al., 2017). Hazardous e-waste seeps into the earth and contaminates water found in underground waterways, and the majority of enterprises are situated close to water sources. Both drinking and other approved uses of this water are rendered dangerous. Pollutant concentrations have been shown to be much greater near the landfills and gradually become low as the separation between the groundwater and the waste site increases (Gupta & Nath, 2020). Aquatic systems may also become contaminated by the breakdown or deposition of airborne pollutants, the discharge of acid into rivers or onto soils after hydrometallurgical processes, or both. Nearby a reprocessing factory in Guiyu, China, Wang and Guo (2006) discovered a level of 0.4 mg/L lead in river water, which was eight times greater than the regional water for drinking standard (0.05 mg/L). Carp from the Nanyang River, close to Guiyu, China, were absorbing polybrominated diphenyl ethers (PBDEs) to levels of up to 766 ng/g (fresh weight), according to Luo et al. (2007a). Additionally, they noted that the same river's sLiDE (up to 16,000 ng/g) (Luo et al., 2007b). Wu et al. (2008) found that the water snake in an aquatic system near e-waste reprocessing plant showed an average 1091 ng/g PBDEs and 16,512 ng/g polychlorinated biphenyls (PCBs) on a wet weight basis. Prawns, crucian carp, and mud carp also showed an increased level of these contaminants (Shi et al., 2009).

Hazardous chemicals from e-waste released into the environment can significantly contribute to food contamination, bioaccumulation, and extensive ecological exposure, which can result in widespread health hazards (Sepúlveda et al., 2010). There are reports of metal-contaminated sediments and high levels of dissolved metals in rivers located in and around e-waste recycling towns. Landfill disposal can lead to the transfer of toxic materials into groundwater. Leachates from e-waste dumping sites, processing plants, landfills, etc., which may be produced by precipitation, water infiltration, etc., are a source of various harmful or toxic chemical pollutants, that may harm any living organism (Ankit et al., 2021). Some of these are persistent organic pollutants (POPs) and polychlorinated biphenyls (PCBs) which are harmful to the environment (Omondi et al., 2015). Leachates from e-waste dumping sites contain harmful aneugenic and clastogenic substances (Bakare et al., 2009; Purchase et al., 2020) and higher amounts of Al, Ba, Be, Cd, Cr, Cu, Co, Pb, Ni, and V than those generated from the other types of waste (Kiddee et al., 2013b, 2014). Exposure to metals such as Pb and Cd can lead to reduced erythrocyte immunity, osteoporosis, and liver metabolism disruption (Chen et al., 2019), whereas chromium compounds cause DNA damage and respiratory problems in humans (Das & Mishra, 2008, 2009; Paul et al., 2023). Mercury results in skin and respiratory disorders causing chronic damage to the brain. Moreover, leachates from landfills where e-waste gets mixed with municipal or industrial waste become more toxic and hazardous as they increase the complexity through the dissolution of various pollutants in the e-waste (Ankit et al., 2021). By-products of recycling of e-waste and disposal of acid post-hydrometallurgy can also introduce toxic substances (chromium, copper, mercury, lead, nickel, silver, dioxin, polycyclic aromatic hydrocarbons, and polychlorinated biphenyls) into the environment (Puckett & Smith, 2002; Rawat et al., 2020; Chakraborty et al., 2023; Hossain & Sahajwalla, 2024).

4.2 E-Waste in Terrestrial Ecosystems

The most common method for getting rid of electronic waste in terrestrial ecosystems is landfilling (Shirodkar & Terkar, 2017), in which electronic debris goes underground in ditches dug out on soil and filled with a thick covering of dirt to keep it from being exposed to the air. When improperly disposed of, e-waste often ends up in landfills, where it leaches harmful chemicals such as mercury, lead, cadmium, and brominated flame retardants into the soil. These toxic substances can alter the chemical composition of the soil, reducing its fertility and disrupting microbial ecosystems essential for plant growth. Over time, the contamination may spread, affecting nearby agricultural land and water sources. Harmful chemicals in the dust and smoke emitted by open-air burning operations of e-waste also readily spread throughout the production region due to wind patterns and the principle of diffusion (Robinson, 2009). Furthermore, heavy metals and other pollutants from e-waste are persistent in the environment, making remediation difficult and expensive. This contamination not only threatens biodiversity but also poses health risks to humans through the food chain, as plants and animals absorb these toxic substances from the soil. Soils samples from an area used for recovery of valuable metals by the method of acid leaching were found to be contaminated with PBDEs by up to a concentration of 4250 ng/g (Leung et al., 2007). There are reports of elevated concentrations of PCBs, PBDEs, and polycyclic aromatic hydrocarbons or PAHs (Shen et al., 2009) in agricultural lands near e-waste recycling sites in China.

E-waste significantly affects the food system by contaminating soil and water resources which are vital for agriculture. It is a rapidly growing environmental challenge that poses risks to food safety and security. Improper disposal and informal recycling practices release harmful components into the environment, directly and indirectly affecting the food system. Increased levels of PCBs and PBDEs in snails, plants, and soils in Guiyu, China, and its nearby vicinities were reported by Liu et al. (2008). Two kilometers away from an electronic waste disposal workshop, Luo et al. (2009) found that farming soils had PBDE concentrations ranging from 191 to 9156 ng/g (dry weight). Shen et al. (2009) also reported the presence of up to 100, 330, and 20,000 ng/g, respectively, of polychloronated dibenzo-p-dioxins and dibenzofurans (PCDD/Fs), PCBs, and PAHs in soils from this area. PBDEs' content has been found (on a dry matter basis) in the leaves of spider fern (116 ng/g), sorghum (162 ng/g), Japanese dock (278 ng/g), Eastern daisy fleabane (326 ng/g), and 25,479 ng/g in the soil (Yang et al., 2008). Plant absorption may help these pollutants enter food systems even when their bioaccumulation coefficients are low (<0.01). Ha et al. (2009) reported contamination of up to 180, 4.6, 49, 957, 2850, 39, and 2.7 mg/kg, respectively by Sb, In, Hg, Sn, Pb, Cd, and Bi in soil samples from an e-waste reprocessing slum in Bangalore. Pb and Cd concentrations in polished rice were found to be 2–4 times higher than their maximum permitted quantities in Chinese foods in an e-waste manufacturing town of Eastern China (Fu et al., 2008). Liang et al. (2008) found increased levels of PBDEs in chicken tissues in the same municipality, which could be harmful to both the public and ecosystems. In Zhejiang province, it has been reported that rice paddy soils next to e-waste recycling facilities lower the rate of rice germination (Jun-Hui & Hang, 2009). Notable correlations between lead and chromium, and copper and cadmium in soil samples were reported by Panwar and Ahmed (2018), which indicate their interrelationship. Similar associations were also reported for nickel and chromium, and copper and cadmium in water samples (Das & Bissoyi, 2011; Mishra et al., 2009). Similarly, livestock that graze on polluted land or consume contaminated water are exposed to these hazardous substances, which can accumulate in animal products like milk, meat, and eggs. This contamination not only reduces the nutritional quality of food but also poses severe health risks to humans, including neurological, developmental, and reproductive disorders.

5 Movement of E-Waste and Factors Influencing It

The bulk of e-consumption takes place in developed nations, but the residual waste is often transported to developing Asian and African countries to dodge the high costs, strict labor and environmental regulations, and labor-intensive recycling processes in these countries of origin. Annually, 20–50 million tons of waste from electronic devices is generated globally, 75–80% of which lands in Asia and Africa for recycling and disposal (Caravanos et al., 2011). Increased consumerism, coupled with the weak enforcement of regulations in these countries, makes them attractive dumping grounds for e-waste from wealthier nations (Ilankoon et al.,

2018). The importing nations, many of which lack the necessary infrastructure for managing WEEE in an environmentally friendly manner, may find these shipments to be problematic. Hopson and Pucket (2016) verified the existence of transboundary WEEE movement, mostly from the United States and EU to developing countries in Asia, using Global Positioning System (GPS) tracking devices. Currently, E-waste exports originate not only from North America and Western Europe but increasingly from countries like Australia, Japan, Hong Kong, and South Korea (Kahhat & Williams, 2009; Lau et al., 2013).

Only a quarter of the 50 Mt (metric tonnes) globally generated e-waste is accumulated and treated legally (Baldé et al., 2017), which suggests that a huge portion is left to be dealt illegally. Remarkably, 80% of e-waste transported for the purpose of recycling results in illegal shipping to developing nations, where it is processed by vast numbers of informal workers (Julander et al., 2014). Often disguised as functional, these are shipped to various African and East Asian countries from Europe. For instance, Nigeria received 100 kilo metric tons (kt) of nonfunctional e-waste which was 30% of the entire "used items" imported (Ogungbuyi et al., 2012). Lack of efficient management and processing of e-waste may result in human and environmental health-related issues and waste of recoverable useful resources (Geeraerts et al., 2015).

Understanding the mechanisms and processes contributing to the accumulation of e-waste is crucial for designing appropriate waste management strategies. One major contributor is the short lifecycle of electronic equipment, which has significantly decreased over the years (Rabani & Thakur, 2020). Coupled with low recycling rates and the relentless demand for continuous technological upgrades, especially in affluent societies, the production of e-waste has reached alarming levels. This phenomenon is particularly prevalent in developed countries, where the appetite for cutting-edge technology is insatiable (Wang et al., 2019). Consumers often replace devices, not because they are nonfunctional but because they lack the latest features or compatibility (Robinson, 2009). Planned obsolescence (products with limited lifespans) of EEEs by manufacturers may also drive repeated purchases, making devices difficult to repair or upgrade, and encouraging consumers to dispose of old devices prematurely (Ongondo et al., 2011). The low cost of disposal or recycling in poorer developing countries also attracts e-waste exports to these countries. Moreover, lower labor and operating expenses have made the extraction of precious elements from electronic gadgets, such as copper, platinum, and gold, a lucrative industry in these countries. However, lack of advanced technology, infrastructure, and robust waste management organizations has led to large-scale accumulation of e-waste in many developing countries (Amankwah-Amoah, 2016).

6 Management of E-Waste

The management of the rapidly increasing pollution due to electronic waste is a major challenge across the world today. For the reduction of electronic waste and improving the environment, developed countries have come forward with a EU Directive 2002/96/EC, also known as WEEE (waste electric and electronic equipment) Directive (EU, 2002) which requires the separation, recycling, and recovery of precious rare metals. A 3-D, multi-latitude, regulatory framework (Barapatre & Rastogi, 2022) for e-waste management requires governmental cooperation to minimize the hazardous ecological impacts and maximize the economic interests of labor and producer responsibility (Chen et al., 2022). Some of the current strategies available for e-waste management are discussed as follows.

6.1 *Material Flow Analysis (MFA)*

MFA is a decision support system to examine the route of e-waste going to disposal areas, recycling sites, or stocks of materials in space and time by linking sources, pathways, and the final destinations of the material. It is used for environmental and waste management taking into consideration the flow of e-waste and assessing its social, environmental, and economic values (Islam & Huda, 2019). The MFA has been used to investigate the movement of e-waste (Ismail & Hanafiah, 2021), which can be beneficial when combined with economic evaluation (Streicher-Porte et al., 2007). Countries such China, India, and Nigeria, use MFA to examine sinks into which exported e-waste is dumped.

6.2 *Life Cycle Assessment (LCA)*

It is used for designing eco-friendly electronic equipment with an aim to reduce the problems due to e-waste (Goh et al., 2022). Since the 1990s, significant research work regarding LCA has been done on various electronic devices (Pollock & Coulon, 1996; Prek, 2004; Muñoz et al., 2009). The LCA has been employed in countries such as the United Kingdom (Mayers et al., 2005), United States (Andeobu et al., 2021), and Australia (Islam & Iyer-Raniga, 2023).

6.3 *Multi-criteria Analysis (MCA)*

It is a decision-making tool (El-Gawhari et al., 2024) which can be applied to various environmental issues including hazardous or management of e-waste (Hatami-Marbini et al., 2013). It can also be used for selecting suitable sites for establishing recycling units for electronic waste (Queiruga et al., 2008) or for examining alternative systems for the management of e-waste (Rousis et al., 2008).

6.4 Recycling of E-Waste

Only 22.3% of the global e-waste material was officially collected and recycled (Balde et al., 2024). In this regard, the efforts of the developed nations, who are the main producers of e-waste, remain insufficient (Popescu, 2015). The recycling process involves the destruction and disassembly of electronic waste equipment to recover new materials and has a lower ecological impact than landfilling even after incineration (Hischier et al., 2005). Most electronic items contain sufficient quantities of recyclable parts, although their quantity may vary, such as 95% in computers and up to 45% in CRT monitors (Kiddee et al., 2013a). Recycling has a lesser ecological impact as compared to landfilling of e-waste even after incineration. However, the inclusion of marketing sectors, manufacturers, and recycling companies is necessary for successful recycling efforts. Recycling is done through different stakeholders in many countries like the United States (Petridis et al., 2020), Japan (Chaudhary & Vrat, 2018), and China (Wang et al., 2020).

Nonmetallic components of e-waste include rubber, plastic, and glass (Yousef et al., 2017) of which plastic has the highest recycling value, second only to metal. The amount of waste plastics can be decreased by using techniques including pyrolysis, incineration, gasification, and hydrothermal liquefaction (Das et al., 2023; Mishra et al., 2022). Technological improvements have also led to the development of complete recycling of nonmetals and waste plastics (Mohanty et al., 2024; Ray et al., 2024; Sahu et al., 2024). At present efforts are on for the development of full-component recycling methods, as a highly complex structure of e-products makes it difficult to recover all valuable materials. As a result, a huge number of valuable resources get wasted during the process of recycling (Liu et al., 2023). Pyrolysis and electrochemical extraction have been adopted for the fine separation of rare metals. These processes can also be used to convert low-value components into useful building materials (Murthy & Ramakrishna, 2022). The government can also fortify the recycling enterprises via fund incentives and intervention of external capital (Thakur & Kumar, 2022).

6.5 Recovery of Valuable Metals

This involves physical and mechanical techniques, biometallurgy, pyrometallurgy, and hydrometallurgy (Xiu et al., 2022). Electrodeposition can be used for the extraction of copper, silver, gold, and other precious metals through selective leaching (Li et al., 2018). New emerging technologies like molten salt, supercritical fluid, mechanochemistry, and electrochemistry have been developed for extracting valuable metals in an eco-friendlier manner (Li & Xu, 2019; Ashiq et al., 2020; Kaya, 2022). However, due to a lack of technical maturity, hydrometallurgy and pyrometallurgy continue to dominate the recycling of e-waste metals (Priyadarsini & Das, 2024).

6.6 *Disposal Processes*

This process involves steps such as dismantling, segregation and enrichment, regeneration, recovery of energy, and treatment of pollutants (Li et al., 2017). Harmful and toxic parts are separated and the waste is subjected to enrichment techniques like separation by gravity, magnetic force, eddy current, and airflow (Huynh et al., 2020). Post the required processing, segregated components can be regenerated into new functional products (Sahajwalla & Gaikwad, 2018). For energy recovery, left-out waste materials can be used as fuel (Chandrasekaran et al., 2018).

6.7 *Advanced E-Waste Collection and Management Systems*

Creating an online e-waste recycling network may be a solution to the problem of collecting and recycling e-waste (Sun et al., 2022). This system allows makers to codify e-products, registration of these products with identity information by the sellers, and submission of information about old e-products by buyers to earn money funded by the government.

6.8 *Green Product Design: Reducing E-Waste Production*

The important elements of green designing e-products include principles of design, the raw materials utilized, and techniques of construction. Factors such as recyclability, disassembly, reusability, and maintainability are taken into consideration under the principles of design. Under the material utilized category, recyclable eco-friendly materials are encouraged, while hazardous materials are prohibited. To avoid the emission of heavy metals and harmful compounds, green recycling techniques are used in buildings. The use of new eco-friendly and high-powered materials should be encouraged for designing electronic products. The producers should be responsible for manufacturing energy-saving, nonhazardous, reliable, and easily recyclable products for safe disposal and the reduction of electronic waste (Li et al. 2020; Kirschner, 2022).

6.9 Extended Producer Responsibility (EPR)

EPR, based on "polluter-pays" principles, is a policy in which the producers or manufacturers are attributed with the responsibility to take back their products after use (OECD, 2001). EPR is in use in many countries such as Japan, Germany, Switzerland, Netherlands, Canada, United States, UK, India, and Thailand (Faibil et al., 2023; Patra & Mishra, 2024).

6.10 Laws and Legislation for E-Waste Management

Proper regulations are essential for the protection of the environment and humans from hazardous materials spreading and contaminating ecosystems by leaching into groundwater, polluting food, and affecting air quality (Chen et al., 2012). Many countries have enforced various laws and published several guidelines for the proper management of electronic waste. However, the absence of appropriate e-waste management laws and regulations in Asia and South Asia has affected the ecosystem of these regions (Priyashantha et al., 2022). According to Herat (2021), China continues to have low adherence to safety regulations and unlawful e-waste imports that continue to damage the environment even after regulations have been put in place. Therefore, countries need to regulate stringent laws and legislation to control e-waste disposal by big corporations and prevent the transboundary movement of toxic wastes, especially from developed countries.

7 Conclusions and Future Directions

As the culture of consumerism continues to grow, the volume of electronic waste generated is expected to increase with the growing market for electronic devices. The growing menace of e-waste represents a global challenge with far-reaching consequences if not properly managed. Numerous hazardous components of e-waste such as halogenated compounds like heavy metals (Hg, Cr, As, Cd, and Cu), PCBs, TBBPA, and polybrominated biphenyl (PBB) can adversely impact plants, microbes, and human beings. These toxicants reach air, soil, and water through various routes and accumulate in the food chain. Hence, sustainable management of e-waste is the need of the hour as it is a threat to both aquatic and terrestrial ecosystems. Moreover, the majority of e-waste is routed to developing countries from developed nations, where recycling is often conducted through rudimentary and dangerous informal means. This increases the risk of environmental pollution from e-waste manifold. While developed nations must take greater responsibility for recycling their own electronic waste, there is an urgent need to provide support to developing countries to help them establish proper waste management systems. Strengthening the Basel

Convention, enhancing national e-waste management legislation, and encouraging ethical recycling methods are some of the measures being taken to address the e-waste challenge. Campaigns for public awareness can also motivate customers to support eco-friendly producers and recycle their used devices in an appropriate manner. International collaboration, stricter regulations, and the promotion of sustainable practices are essential to address this growing crisis and ensure a safer, more equitable future for all.

References

Abalansa, S., El Mahrad, B., Icely, J., & Newton, A. (2021). Electronic waste, an environmental problem exported to developing countries: The GOOD, the BAD and the UGLY. *Sustainability, 13*(9), 5302.

Adanu, S. K., Gbedemah, S. F., & Attah, M. K. (2020). Challenges of adopting sustainable technologies in e-waste management at Agbogbloshie, Ghana. *Heliyon, 6*(8).

Agoramoorthy, G., & Chakraborty, C. (2012). Control electronic waste in India. *Nature, 485*(7398), 309–309.

Ahmed, S., Panwar, R. M., & Sharma, A. (2014). Forecasting e-waste amounts in India. *International Journal of Engineering Research and General Science, 2*(6), 2091–2730.

Amankwah-Amoah, J. (2016). Global business and emerging economies: Towards a new perspective on the effects of e-waste. *Technological Forecasting and Social Change, 105*, 20–26.

Andeobu, L., Wibowo, S., & Grandhi, S. (2021). An assessment of e-waste generation and environmental management of selected countries in Africa, Europe and North America: A systematic review. *Science of the Total Environment, 792*, 148078.

Ankit, Kumar, A., Jain, V., Deovanshi, A., Lepcha, A., Das, C., et al. (2021). Environmental impact of COVID-19 pandemic: More negatives than positives. *Environmental Sustainability, 4*(3), 447–454.

Apprey, M. W., Dzah, C., Agbevanu, K. T., Agyapong, J. O., & Selase, G. S. (2024). E-waste management from electronic repair workshops: Societal implications and environmental consequences. *Societal Impacts, 4*, 100077.

Arvaniti, O. S., & Kalantzi, O. I. (2021). Determinants of flame retardants in non-occupationally exposed individuals–A review. *Chemosphere, 263*, 127923.

Ashiq, A., Cooray, A., Srivatsa, S. C., & Vithanage, M. (2020). Electrochemical enhanced metal extraction from E-waste. In *Handbook of electronic waste management* (pp. 119–139). Butterworth-Heinemann.

ASSOCHAM-NEC. (2018). *India among the top five countries in e-waste generation: ASSOCHAM-NEC study*. Retrieved September 25, 2025, from https://www.business-standard.com/article/news-cm/india-among-the-top-five-countries-in-e-waste-generation-assocham-necstudy118060500207_1.html

Bakare, A. A., Okunola, A. A., Adetunji, O. A., & Jenmi, H. B. (2009). Genotoxicity assessment of a pharmaceutical effluent using four bioassays. *Genetics and Molecular Biology, 32*, 373–381.

Baldé, C. P., Forti, V., Gray, V., Kuehr, R., & Stegmann, P. (2017). The global e-waste monitor. United Nations University (UNU), International Telecommunication Union (ITU) & International Solid Waste Association (ISWA), Bonn/Geneva/Vienna, 1–109.

Balde, C. P., Kuehr, R., Yamamoto, T., McDonald, R., D'Angelo, E., Althaf, S., et al. (2024). *The global e-waste monitor 2024*. International Telecommunication Union (ITU) and United Nations Institute for Training and Research (UNITAR).

Barapatre, S., & Rastogi, M. (2022). E-waste management: A transition towards a circular economy. In *Handbook of solid waste management: Sustainability through circular economy* (pp. 1499–1521). Springer Nature Singapore.

Bigum, M., Damgaard, A., Scheutz, C., & Christensen, T. H. (2017). Environmental impacts and resource losses of incinerating misplaced household special wastes (WEEE, batteries, ink cartridges and cables). *Resources, Conservation and Recycling, 122*, 251–260.

Borthakur, A., & Govind, M. (2017). Emerging trends in consumers' E-waste disposal behaviour and awareness: A worldwide overview with special focus on India. *Resources, Conservation and Recycling, 117*, 102–113.

Boyle, D., Catarino, A. I., Clark, N. J., & Henry, T. B. (2020). Polyvinyl chloride (PVC) plastic fragments release Pb additives that are bioavailable in zebrafish. *Environmental Pollution, 263*, 114422.

Caravanos, J., Clark, E., Fuller, R., & Lambertson, C. (2011). Assessing worker and environmental chemical exposure risks at an e-waste recycling and disposal site in Accra, Ghana. *Journal of Health and Pollution, 1*(1), 16–25.

Carroll, A. (2018). *Solid waste management: A comparative carbon footprint and cost analysis* (Master's thesis, Colorado State University).

Chakraborty, D., Chatterjee, S., Althuri, A., Palani, S. G., & Mohan, S. V. (2023). Sustainable enzymatic treatment of organic waste in a framework of circular economy. *Bioresource Technology, 370*, 128487.

Chandrasekaran, S. R., Avasarala, S., Murali, D., Rajagopalan, N., & Sharma, B. K. (2018). Materials and energy recovery from e-waste plastics. *ACS Sustainable Chemistry & Engineering, 6*(4), 4594–4602.

Chaudhary, K., & Vrat, P. (2018). Case study analysis of e-waste management systems in Germany, Switzerland, Japan and India: A RADAR chart approach. *Benchmarking: An International Journal, 25*(9), 3519–3540.

Chen, Y., Li, J., Liu, L., & Zhao, N. (2012). Polybrominated diphenyl ethers fate in China: A review with an emphasis on environmental contamination levels, human exposure and regulation. *Journal of Environmental Management, 113*, 22–30.

Chen, S., Chen, M., Shu, J., & Deng, Y. (2022). Comparative analysis of recycling modes of power batteries based on extended producer-responsibility principle. *Circular Economy, 1*(2), 100013.

Chen, Y., Xu, X., Zeng, Z., Lin, X., Qin, Q., & Huo, X. (2019). Blood lead and cadmium levels associated with hematological and hepatic functions in patients from an e-waste-polluted area. *Chemosphere, 220*, 531–538.

Cobbing, M. (2008). Toxic tech: not in our backyard. uncovering the hidden flows of e-waste. Report from Greenpeace International, Amsterdam, pp 1–12. http://www.greenpeace.org/raw/content/Belgium/fr/press/reports/toxictech.Pdf

Dalrymple, I., Wright, N., Kellner, R., Bains, N., Geraghty, K., Goosey, M., & Lightfoot, L. (2007). An integrated approach to electronic waste (WEEE) recycling. *Circuit World, 33*(2), 52–58.

Das, A. P., & Bissoyi, A. (2011). Computational approach for comparative phylogenetic analysis of isolated chromium resistant strain Brevibacterium casei. *Journal of Engineering and Technology Research, 3*(3), 82–87. https://doi.org/10.5897/JETR.9000020

Das, A. P., & Ghosh, S. (2023). *Manganese mining microorganisms.*

Das, A. P., & Mishra, S. (2008). Hexavalent Chromium (VI): Health hazards and environmental pollutant. *The Journal of Environment & Development, 2*(3), 386–392.

Das, A. P., & Mishra, S. (2009). Hexavalent chromium [Cr (VI)]: Yellow water pollution and its remediation. *Sarovar Saurabh ENVIS Newsletter Wetland Ecosystem, 5*(2).

Das, A. P., Mohanty, A. K., van Hullebusch, E. D., Figueiredo, G., Das, A. P., Mohanty, A. K., van Hullebusch, E. D., & Figueiredo, G. (2023). Bioremediation as an emerging technology for the removal of synthetic microplastic pollutants from marine ecosystem. *Marine Pollution Bulletin, 194*, 115297. https://doi.org/10.1016/J.MARPOLBUL.2023.115297

Dharini, K., Cynthia, J. B., Kamalambikai, B., Celestina, J. A. S., & Muthu, D. (2017). Hazardous E-waste and its impact on soil structure. In *IOP conference series: Earth and environmental science* (Vol. 80, No. 1, p. 012057). IOP Publishing.

Ebrahimi, M., Khalili, N., Razi, S., Keshavarz-Fathi, M., Khalili, N., & Rezaei, N. (2020). Effects of lead and cadmium on the immune system and cancer progression. *Journal of Environmental Health Science and Engineering, 18*, 335–343.
El-Gawhari, I. S., Abdel Hameed, I. M., & Sami, F. M. (2024). Strategies to manage electronic-wastes in Malaysia: A review study. *Zagazig Journal of Agricultural Research, 51*(2), 157–169.
EU. (2002). Directive 2002/96/EC of the European parliament and of the council of 27 January 2003 on waste electrical and electronic equipment (WEEE). *Official Journal of the European Union, L037*, 0024–39.
European Commission. (2006). Commission Regulation (EC) No 1881/2006 of 19 December 2006 setting maximum levels for certain contaminants in foodstuffs. Retrieved 14 June 2010 from: http://ec.europa.eu/food/food/chemicalsafety/contaminants/dio xins_en.htm
Faibil, D., Asante, R., Agyemang, M., Addaney, M., & Baah, C. (2023). Extended producer responsibility in developing economies: Assessment of promoting factors through retail electronic firms for sustainable e-waste management. *Waste Management & Research, 41*(1), 117–142.
Forti, V., Baldé, C. P., Kuehr, R., & Bel, G. (2020). The global e-waste monitor 2020. United Nations University (UNU), International Telecommunication Union (ITU) & International Solid Waste Association (ISWA), Bonn/Geneva/Rotterdam, 120.
Fu, J., Zhou, Q., Liu, J., Liu, W., Wang, T., Zhang, Q., & Jiang, G. (2008). High levels of heavy metals in rice (Oryzasativa L.) from a typical E-waste recycling area in Southeast China and its potential risk to human health. *Chemosphere, 71*(7), 1269–1275.
Garlapati, V. K. (2016). E-waste in India and developed countries: Management, recycling, business and biotechnological initiatives. *Renewable and Sustainable Energy Reviews, 54*, 874–881.
Geeraerts, K., Mutafoglu, K., & Illes, A. (2015). Illegal e-waste shipments from the EU to China. Quantitative andmonetary analysis of illegal shipments and its environmental, social and economic impacts. A study compiled as part of the EFFACE project. Available online: https://collections.unu.edu/eserv/UNU, 1624.
Goh, P. G., Maghfouri, M., Onn, C. C., & Loo, S. C. (2022). Life cycle assessment on recycled e-waste concrete. *Case Studies in Construction Materials, 17*, e01412.
Grant, K., Goldizen, F. C., Sly, P. D., Brune, M. N., Neira, M., van den Berg, M., & Norman, R. E. (2013). Health consequences of exposure to e-waste: A systematic review. *The Lancet Global Health, 1*(6), e350–e361.
Gupta, N., & Nath, M. (2020). Groundwater contamination by E-waste and its remedial measure-a literature review. *Journal of Physics: Conference Series, 1531*(1), 012023. IOP Publishing.
Ha, N. N., Agusa, T., Ramu, K., Tu, N. P. C., Murata, S., Bulbule, K. A., et al. (2009). Contamination by trace elements at e-waste recycling sites in Bangalore, India. *Chemosphere, 76*(1), 9–15.
Hatami-Marbini, A., Tavana, M., Moradi, M., & Kangi, F. (2013). A fuzzy group Electre method for safety and health assessment in hazardous waste recycling facilities. *Safety Science, 51*(1), 414–426.
Hendricks, C., Williard, N., Mathew, S., & Pecht, M. (2015). A failure modes, mechanisms, and effects analysis (FMMEA) of lithium-ion batteries. *Journal of Power Sources, 297*, 113–120.
Herat, S. (2021). E-waste management in Asia Pacific region: Review of issues, challenges and solutions. *Nature Environment and Pollution Technology, 20*(1), 45–53.
Hischier, R., Wäger, P., & Gauglhofer, J. (2005). Does WEEE recycling make sense from an environmental perspective?: The environmental impacts of the Swiss take-back and recycling systems for waste electrical and electronic equipment (WEEE). *Environmental Impact Assessment Review, 25*(5), 525–539.
Hopson, E., & Puckett, J. (2016). *Scam Recycling: E-Dumping on Asia by US Recyclers*. Basel Action Network. www.ban.org
Hossain, R., & Sahajwalla, V. (2024). Current recycling innovations to utilize e-waste in sustainable green metal manufacturing. *Philosophical Transactions A, 382*(2284), 20230239.
Huang, C. L., Bao, L. J., Luo, P., Wang, Z. Y., Li, S. M., & Zeng, E. Y. (2016). Potential health risk for residents around a typical e-waste recycling zone via inhalation of size-fractionated particle-bound heavy metals. *Journal of Hazardous Materials, 317*, 449–456.

Huynh, G. H., Chen, T. L., Hsu, C. H., Chen, Y. H., & Chiang, P. C. (2020). Process integration of E-waste carbonization and high-gravity rotating packed bed for optimal gold recovery and the fine particles reduction. *Separation and Purification Technology, 241*, 116686.

IBEF. (2019). *Mobile phone manufacturing in India: Towards a brighter future*! Retrieved September 25, 2025, from https://ibef.org/blogs/mobile-phonemanufacturing-in-india-towards-a-brighter-future

Ilankoon, I. M. S. K., Ghorbani, Y., Chong, M. N., Herath, G., Moyo, T., & Petersen, J. (2018). E-waste in the international context—A review of trade flows, regulations, hazards, waste management strategies and technologies for value recovery. *Waste Management, 82*, 258–275. https://doi.org/10.1016/j.wasman.2018.10.018

Islam, M. T., & Huda, N. (2019). Material flow analysis (MFA) as a strategic tool in E-waste management: Applications, trends and future directions. *Journal of Environmental Management, 244*, 344–361.

Islam, M. T., & Iyer-Raniga, U. (2023). Life cycle assessment of e-waste management system in Australia: Case of waste printed circuit board (PCB). *Journal of Cleaner Production, 418*, 138082.

Ismail, H., & Hanafiah, M. M. (2021). Evaluation of e-waste management systems in Malaysia using life cycle assessment and material flow analysis. *Journal of Cleaner Production, 308*, 127358.

Jarema, K. A., Hunter, D. L., Shaffer, R. M., Behl, M., & Padilla, S. (2015). Acute and developmental behavioral effects of flame retardants and related chemicals in zebrafish. *Neurotoxicology and Teratology, 52*, 194–209.

Julander, A., Lundgren, L., Skare, L., Grandér, M., Palm, B., Vahter, M., & Lidén, C. (2014). Formal recycling of e-waste leads to increased exposure to toxic metals: An occupational exposure study from Sweden. *Environment International, 73*, 243–251.

Jun-Hui, Z., & Hang, M. I. N. (2009). Eco-toxicity and metal contamination of paddy soil in an e-wastes recycling area. *Journal of Hazardous Materials, 165*(1–3), 744–750.

Kahhat, R., Kim, J., Xu, M., Allenby, B., Williams, E., & Zhang, P. (2008). Exploring e-waste management systems in the United States. *Resources, Conservation and Recycling, 52*(7), 955–964.

Kahhat, R., & Williams, E. (2009). Product or waste? Importation and end-of-life processing of computers in Peru. *Environmental Science & Technology, 43*(15), 6010–6016.

Kalamaras, G., Kloukinioti, M., Antonopoulou, M., Ntaikou, I., Vlastos, D., Eleftherianos, A., & Dailianis, S. (2021). The potential risk of electronic waste disposal into aquatic media: The case of personal computer motherboards. *Toxics, 9*(7), 166.

Kaya, M. (2022). State-of-the-art lithium-ion battery recycling technologies. *Circular Economy, 1*(2), 100015.

Khattar, V., Kaur, J., Chaturvedi, A., & Arora, R. (2007). E-waste assessment in India: Specific 1222 focus on Delhi: A quantitative understanding of generation, disposal & recycling of 1223 electronic waste. *BIRD & GTZ*, 1–66.

Kiddee, P., Naidu, R., & Wong, M. H. (2013a). Electronic waste management approaches: An overview. *Waste Management, 33*(5), 1237–1250.

Kiddee, P., Naidu, R., & Wong, M. H. (2013b). Metals and polybrominated diphenyl ethers leaching from electronic waste in simulated landfills. *Journal of Hazardous Materials, 252*, 243–249.

Kiddee, P., Naidu, R., Wong, M. H., Hearn, L., & Muller, J. F. (2014). Field investigation of the quality of fresh and aged leachates from selected landfills receiving e-waste in an arid climate. *Waste Management, 34*(11), 2292–2304.

Kim, K. H., Kabir, E., & Jahan, S. A. (2016). A review on the distribution of Hg in the environment and its human health impacts. *Journal of Hazardous Materials, 306*, 376–385.

Kirschner, M. (2022). Why the circular economy will drive green and sustainable chemistry in electronics. *Advanced Sustainable Systems, 6*(2), 2100046.

Kumar, A., Rathore, P. S., Dubey, A. K., Srivastav, A. L., Dutt, V., & Kumar, T. A. (Eds.). (2024). *Sustainable management of electronic waste*. John Wiley & Sons.

Kuntawee, C., Tantrakarnapa, K., Limpanont, Y., Lawpoolsri, S., Phetrak, A., Mingkhwan, R., & Worakhunpiset, S. (2020). Exposure to heavy metals in electronic waste recycling in Thailand. *International Journal of Environmental Research and Public Health, 17*(9), 2996.

Lau, W. K. Y., Chung, S. S., & Zhang, C. (2013). A material flow analysis on current electrical and electronic waste disposal from Hong Kong households. *Waste Management, 33*(3), 714–721.

Lee, J., Choi, H., & Kim, J. (2024). Environmental and economic impacts of e-waste recycling: A systematic review. *Chemical Engineering Journal*, 152917.

Leung, A. O., Luksemburg, W. J., Wong, A. S., & Wong, M. H. (2007). Spatial distribution of polybrominated diphenyl ethers and polychlorinated dibenzo-p-dioxins and dibenzofurans in soil and combusted residue at Guiyu, an electronic waste recycling site in Southeast China. *Environmental Science & Technology, 41*(8), 2730–2737.

Li, K., & Xu, Z. (2019). A review of current progress of supercritical fluid technologies for e-waste treatment. *Journal of Cleaner Production, 227*, 794–809.

Li, J., He, X., & Zeng, X. (2017). Designing and examining e-waste recycling process: Methodology and case studies. *Environmental Technology, 38*(6), 652–660.

Li, H., Eksteen, J., & Oraby, E. (2018). Hydrometallurgical recovery of metals from waste printed circuit boards (WPCBs): Current status and perspectives–A review. *Resources, Conservation and Recycling, 139*, 122–139.

Li, Y. L., Liu, Y. G., Liu, J. L., Zeng, G. M., & Li, X. (2008). Effects of EDTA on lead uptake by Typha orientalis Presl: A new lead-accumulating species in southern China. *Bulletin of Environmental Contamination and Toxicology, 81*(1), 36–41.

Li, Y., Chang, Q., Luo, Z., Zhang, J., Liu, Y., Duan, H., & Li, J. (2020). Transfer of POP-BFRs within e-waste plastics in recycling streams in China. *Science of the Total Environment, 717*, 135003.

Liang, S. X., Zhao, Q., Qin, Z. F., Zhao, X. R., Yang, Z. Z., & Xu, X. B. (2008). Levels and distribution of polybrominated diphenyl ethers in various tissues of foraging hens from an electronic waste recycling area in South China. *Environmental Toxicology and Chemistry: An International Journal, 27*(6), 1279–1283.

Liu, H., Zhou, Q., Wang, Y., Zhang, Q., Cai, Z., & Jiang, G. (2008). E-waste recycling induced polybrominated diphenyl ethers, polychlorinated biphenyls, polychlorinated dibenzo-p-dioxins and dibenzo-furans pollution in the ambient environment. *Environment International, 34*(1), 67–72.

Liu, K., Tan, Q., Yu, J., & Wang, M. (2023). A global perspective on e-waste recycling. *Circular Economy, 2*(1), 100028.

Luo, Q., Wong, M., & Cai, Z. (2007a). Determination of polybrominated diphenyl ethers in freshwater fishes from a river polluted by e-wastes. *Talanta, 72*(5), 1644–1649.

Luo, Q., Cai, Z. W., & Wong, M. H. (2007b). Polybrominated diphenyl ethers in fish and sediment from river polluted by electronic waste. *Science of the Total Environment, 383*(1–3), 115–127.

Luo, Y., Luo, X. J., Lin, Z., Chen, S. J., Liu, J., Mai, B. X., & Yang, Z. Y. (2009). Polybrominated diphenyl ethers in road and farmland soils from an e-waste recycling region in Southern China: Concentrations, source profiles, and potential dispersion and deposition. *Science of the Total Environment, 407*(3), 1105–1113.

Mallick, K., Sahu, S., Sahu, A., Dey, S., & Das, A. P. (2024). A critical review of marine microfiber pollution routes, toxicity, and its sustainable remediation. *Environmental Science and Engineering, Part, F2333*, 189–211. https://doi.org/10.1007/978-3-031-51792-1_10

Manikandan, S., Inbakandan, D., Nachiyar, C. V., & Namasivayam, S. K. R. (2023). Towards sustainable metal recovery from e-waste: A mini review. *Sustainable Chemistry for the Environment, 2*, 100001.

Mayers, C. K., France, C. M., & Cowell, S. J. (2005). Extended producer responsibility for waste electronics: An example of printer recycling in the United Kingdom. *Journal of Industrial Ecology, 9*(3), 169–189.

Mishra, S., Das, A. P., & Seragadam, P. (2009). Microbial remediation of hexavalent chromium from chromite contaminated mines of Sukinda Valley, Orissa (India). *The Journal of Environment & Development*, 1122–1127.

Mishra, S., Rath, C. c., & Das, A. P. (2019). Marine microfiber pollution: A review on present status and future challenges. *Marine Pollution Bulletin, 140*, 188–197. https://doi.org/10.1016/J.MARPOLBUL.2019.01.039

Mishra, S., Dash, D., & Das, A. P. (2022). Detection, characterization and possible biofragmentation of synthetic microfibers released from domestic laundering wastewater as an emerging source of marine pollution. *Marine Pollution Bulletin, 185*, 114254. https://doi.org/10.1016/J.MARPOLBUL.2022.114254

Mishra, S., Ghosh, S., van Hullebusch, E. D., Singh, S., & Das, A. P. (2023). A critical review on the recovery of base and critical elements from electronic waste-contaminated streams using microbial biotechnology. *Applied Biochemistry and Biotechnology, 2023*, 1–30. https://doi.org/10.1007/S12010-023-04440-X

Mohanty, J., Dey, S., Bala, K., & Das, A. P. (2024). Recent advances in microplastic pollution for its sustainable management. In *Sustainable management of urban plastic waste through circular economic approaches* (1st ed., pp. 27–48). CRC Press.

Mundada, M. N., Kumar, S., & Shekdar, A. V. (2004). E-waste: A new challenge for waste management in India. *International Journal of Environmental Studies, 61*(3), 265–279.

Muñoz, I., Gazulla, C., Bala, A., Puig, R., & Fullana, P. (2009). LCA and ecodesign in the toy industry: Case study of a teddy bear incorporating electric and electronic components. *The International Journal of Life Cycle Assessment, 14*(1), 64–72.

Murthy, V., & Ramakrishna, S. (2022). A review on global E-waste management: Urban mining towards a sustainable future and circular economy. *Sustainability, 14*(2), 647.

Nnorom, I. C., & Osibanjo, O. (2009). Toxicity characterization of waste mobile phone plastics. *Journal of Hazardous Materials, 161*(1), 183–188.

OECD. (2001). *Extended producer responsibility: A guidance manual for governments*. OECD.

Ogungbuyi, O., Nnorom, I. C., Osibanjo, O., & Schluep, M. (2012). E-waste country assessment Nigeria. In *E waste Africa Project of the Secretariat of the Basel Convention* (p. 94).

Omondi, E. A., Ndiba, P. K., & Njuru, P. G. (2015). Phytoremediation of polychlorobiphenyls (PCB's) in landfill e-waste leachate with water hyacinth (E. crassipes). *International Journal of Scientific and Technology Research, 4*, 147–156.

Ongondo, F. O., Williams, I. D., & Cherrett, T. J. (2011). How are WEEE doing? A global review of the management of electrical and electronic wastes. *Waste Management, 31*(4), 714–730.

Oskarsson, A. (2015). *Barium handbook on the toxicology of metals* (Vol. 2, 4th ed., pp. 625–634).

Panwar, R. M., & Ahmed, S. (2018). Assessment of contamination of soil and groundwater due to e-waste handling. *Current Science*, 166–173.

Patra, B. C., & Mishra, D. N. (2024). Extended producer responsibility (EPR) and E-waste management: A roadmap for entrepreneurs. In *Sustainable solutions for E-waste and development* (pp. 136–145). IGI Global.

Paul, A., Dey, S., Ram, D. K., & Das, A. P. (2023). Hexavalent chromium pollution and its sustainable management through bioremediation. *Geomicrobiology Journal*, 1–11. https://doi.org/10.1080/01490451.2023.2218377

Peana, M., Medici, S., Dadar, M., Zoroddu, M. A., Pelucelli, A., Chasapis, C. T., & Bjørklund, G. (2021). Environmental barium: Potential exposure and health-hazards. *Archives of Toxicology, 95*(8), 2605–2612.

Perkins, D. N., Drisse, M. N. B., Nxele, T., & Sly, P. D. (2014). E-waste: A global hazard. *Annals of Global Health, 80*(4), 286–295.

Petridis, N. E., Petridis, K., & Stiakakis, E. (2020). Global e-waste trade network analysis. *Resources, Conservation and Recycling, 158*, 104742.

Pollock, D., & Coulon, R. (1996). Life cycle assessment: Of an inkjet print cartridge. In *Proceedings of the 1996 IEEE International Symposium on Electronics and the Environment. ISEE-1996* (pp. 154–160). IEEE.

Popescu, M. L. (2015). Waste electrical and electronic equipment management in Romania. Harmonizing national environmental law with the UE legislation. *Procedia-Social and Behavioral Sciences, 188*, 264–269.

Prabhakar, A., Mishra, S., & Das, A. P. (2019). Isolation and identification of Lead (Pb) solubilizing bacteria from automobile waste and its potential for recovery of Lead from end of life waste batteries. *Geomicrobiological Journal, 36*(10), 894–903. https://doi.org/10.1080/01490451.2019.1654044

Prek, M. (2004). Environmental impact and life cycle assessment of heating and air conditioning systems, a simplified case study. *Energy and Buildings, 36*(10), 1021–1027.

Priyadarsini, S., & Das, A. P. (2024). Analytical and structural characterization of waste lithium-ion batteries for their effective recycling strategy. *Environmental Chemistry and Ecotoxicology*. https://doi.org/10.1016/j.enceco.2024.12.004

Priyadarsini, S., & Das, A. P. (2025). Lithium bioleaching: A review on microbial-assisted sustainable technology for lithium bio-circularity. *Journal of Water Process Engineering, 69*, 106744. https://doi.org/10.1016/j.jwpe.2024.106744

Priyashantha, A. K. H., Pratheesh, N., & Pretheeba, P. (2022). E-waste scenario in South-Asia: An emerging risk to environment and public health. *Environmental Analysis, Health and Toxicology, 37*(3).

Puckett, J., & Smith, T. (2002). Exporting harm: the high-tech trashing of Asia. The Basel Action Network. Seattle7 Silicon Valley Toxics Coalition.

Purchase, D., Abbasi, G., Bisschop, L., Chatterjee, D., Ekberg, C., Ermolin, M., et al. (2020). Global occurrence, chemical properties, and ecological impacts of e-wastes (IUPAC Technical Report). *Pure and Applied Chemistry, 92*(11), 1733–1767.

Queiruga, D., Walther, G., Gonzalez-Benito, J., & Spengler, T. (2008). Evaluation of sites for the location of WEEE recycling plants in Spain. *Waste Management, 28*(1), 181–190.

Rabani, B., & Thakur, B. (2020). Recycling potential of e-waste for Jammu City. *International Journal of Progressive Research in Science and Engineering, 1*, 29–32.

Raj, K., & Das, A. P. (2023). Lead pollution: Impact on environment and human health and approach for a sustainable solution. *Environmental Chemistry and Ecotoxicology, 5*, 79–85. https://doi.org/10.1016/J.ENCECO.2023.02.001

Rajya Sabha Secretariat. (2011). E-waste in India. http://greene.gov.in/wp-content/uploads/2018/01/e-waste_in_india-Document.pdf

Ramadan Elbaioumi Shaddad, A., & Jember, B. (2024). A step toward effective language learning: An insight into the impacts of feedback-supported tasks and peer-work activities on learners' engagement, self-esteem, and language growth. *Asian-Pacific Journal of Second and Foreign Language Education, 9*(1), 39.

Rawat, S., Verma, L., & Singh, J. (2020). Environmental hazards and management of E-waste. In *Environmental concerns and sustainable development: Volume 2: Biodiversity, soil and waste management* (pp. 381–398).

Ray, S. K., Dey, S., Das, A. P., & John, A. (2024). Microplastic waste management through greener technologies: A review. In *Sustainable management of urban plastic waste through circular economic approaches* (pp. 237–261). CRC Press.

Robinson, B. H. (2009). E-waste: An assessment of global production and environmental impacts. *Science of the Total Environment, 408*(2), 183–191.

Rousis, K., Moustakas, K., Malamis, S., Papadopoulos, A., & Loizidou, M. (2008). Multi-criteria analysis for the determination of the best WEEE management scenario in Cyprus. *Waste Management, 28*(10), 1941–1954.

Sahu, A., Dey, S., Mallick, K., Priyadarsini, S., Behera, I. D., & Das, A. P. (2024). A review on advanced technology for sustainable management of synthetic microplastic waste. In *Sustainable management of urban plastic waste through circular economic approaches* (pp. 262–279). CRC Press.

Sahajwalla, V., & Gaikwad, V. (2018). The present and future of e-waste plastics recycling. *Current Opinion in Green and Sustainable Chemistry, 13*, 102–107.

Sahoo, P. P., Singh, S., Rout, P. K., Mishra, S., & Das, A. P. (2022). Microbial remediation of plastic pollutants generated from discarded and abandoned marine fishing nets. *Biotechnology and Genetic Engineering Reviews*. https://doi.org/10.1080/02648725.2022.2152629

Sepúlveda, A., Schluep, M., Renaud, F. G., Streicher, M., Kuehr, R., Hagelüken, C., & Gerecke, A. C. (2010). A review of the environmental fate and effects of hazardous substances released from electrical and electronic equipments during recycling: Examples from China and India. *Environmental Impact Assessment Review, 30*(1), 28–41.

Shamim, A., Mursheda, A. K., & Rafiq, I. (2015). E-waste trading impact on public health and ecosystem services in developing countries. *International Journal of Waste Resources, 5*(4), 1–18.

Shanker, A. K., & Venkateswarlu, B. (2011). Chromium: Environmental pollution, health effects and mode of action (pp. 650–659). Reference module in earth systems and environmental sciences, from Encyclopedia of Environmental Health.

Sharma, S., Wakode, S., Sharma, A., Nair, N., Dhobi, M., Wani, M. A., & Pottoo, F. H. (2020). Effect of environmental toxicants on neuronal functions. *Environmental Science and Pollution Research, 27*, 44906–44921.

Shen, C., Chen, Y., Huang, S., Wang, Z., Yu, C., Qiao, M., et al. (2009). Dioxin-like compounds in agricultural soils near e-waste recycling sites from Taizhou area, China: Chemical and bioanalytical characterization. *Environment International, 35*(1), 50–55.

Shi, T., Chen, S. J., Luo, X. J., Zhang, X. L., Tang, C. M., Luo, Y., et al. (2009). Occurrence of brominated flame retardants other than polybrominated diphenyl ethers in environmental and biota samples from southern China. *Chemosphere, 74*(7), 910–916.

Shirodkar, N., & Terkar, R. (2017). Stepped recycling: The solution for E-waste management and sustainable manufacturing in India. *Materials Today: Proceedings, 4*(8), 8911–8917.

Sinha, D. (2004). *The management of electronic waste: A comparative study on India and Switzerland*. University of St. Gallen.

Song, Q., & Li, J. (2014). Environmental effects of heavy metals derived from the e-waste recycling activities in China: A systematic review. *Waste Management, 34*(12), 2587–2594.

StEP (Solving the e-waste problem). (2018). http://www.step-initiative.org/. Accessed 1 Apr 2018.

Streicher-Porte, M., Bader, H. P., Scheidegger, R., & Kytzia, S. (2007). Material flow and economic analysis as a suitable tool for system analysis under the constraints of poor data availability and quality in emerging economies. *Clean Technologies and Environmental Policy, 9*, 325–345.

Suja, F., Abdul Rahman, R., Yusof, A., & Masdar, M. S. (2014). E-waste management scenarios in Malaysia. *Journal of Waste Management, 2014*(1), 609169.

Sun, B., Schnoor, J. L., & Zeng, E. Y. (2022). Decadal journey of e-waste recycling: What has it achieved? *Environmental Science & Technology, 56*(18), 12785–12792.

Tan, Q., Dong, Q., Liu, L., Song, Q., Liang, Y., & Li, J. (2017). Potential recycling availability and capacity assessment on typical metals in waste mobile phones: A current research study in China. *Journal of Cleaner Production, 148*, 509–517.

Thakur, P., & Kumar, S. (2022). Evaluation of e-waste status, management strategies, and legislations. *International journal of Environmental Science and Technology, 19*(7), 6957–6966.

UNEP. (2005). E-waste the hidden side of IT equipment's manufacturing and use. Environment Alert Bulletin 5, January 2005, United Nation Environmental Program, Nairobi, Kenya.

UNEP. (2020). *Solid waste management | UNEP—U.N. Environment Programme*. United Nations Environment Programme. Available at: https://www.unep.org/explore-topics/resource-efficiency/what-we-do/cities/solid-waste-management. Accessed 20 Oct 2021.

Wang, J. P., & Guo, X. K. (2006). Impact of electronic wastes recycling on environmental quality. *Biomedical and Environmental Sciences, 19*(2), 137.

Wang, B., Ren, C., Dong, X., Zhang, B., & Wang, Z. (2019). Determinants shaping willingness towards on-line recycling behaviour: An empirical study of household e-waste recycling in China. *Resources, Conservation and Recycling, 143*, 218–225.

Wang, Q., Kong, L., Li, J., Li, B., & Wang, F. (2020). Behavioral evolutionary analysis between the government and uncertified recycler in China's E-waste recycling management. *International Journal of Environmental Research and Public Health, 17*(19), 7221.

Widmer, R., Oswald-Krapf, H., Sinha-Khetriwal, D., Schnellmann, M., & Böni, H. (2005). Global perspectives on e-waste. *Environmental Impact Assessment Review, 25*(5), 436–458.

Wu, J. P., Luo, X. J., Zhang, Y., Luo, Y., Chen, S. J., Mai, B. X., & Yang, Z. Y. (2008). Bioaccumulation of polybrominated diphenyl ethers (PBDEs) and polychlorinated biphenyls (PCBs) in wild aquatic species from an electronic waste (e-waste) recycling site in South China. *Environment International, 34*(8), 1109–1113.

Xing, G. H., Chan, J. K. Y., Leung, A. O. W., Wu, S. C., & Wong, M. H. (2009). Environmental impact and human exposure to PCBs in Guiyu, an electronic waste recycling site in China. *Environment International, 35*(1), 76–82.

Xiu, F. R., Song, Z., Lu, Y., Qi, Y., & Wang, M. (2022). A novel conversion strategy for organic compounds in waste liquid crystal displays based on the near/supercritical methanol process. *Waste Management, 153*, 397–404.

Yang, Z. Z., Zhao, X. R., Zhao, Q., Qin, Z. F., Qin, X. F., Xu, X. B., et al. (2008). Polybrominated diphenyl ethers in leaves and soil from typical electronic waste polluted area in South China. *Bulletin of Environmental Contamination and Toxicology, 80*, 340–344.

Yousef, S., Tatariants, M., Bendikiene, R., & Denafas, G. (2017). Mechanical and thermal characterizations of non-metallic components recycled from waste printed circuit boards. *Journal of Cleaner Production, 167*, 271–280.

Zhao, G., Wang, Z., Dong, M. H., Rao, K., Luo, J., Wang, D., et al. (2008). PBBs, PBDEs, and PCBs levels in hair of residents around e-waste disassembly sites in Zhejiang Province, China, and their potential sources. *Science of the Total Environment, 397*(1–3), 46–57.

Zheng, L., Wu, K., Li, Y., Qi, Z., Han, D., Zhang, B., et al. (2008). Blood lead and cadmium levels and relevant factors among children from an e-waste recycling town in China. *Environmental Research, 108*(1), 15–20.

In-Depth Coverage of E-Waste Sources, Characteristics, and Environmental Impact

N. Aarthi

Abstract Electronic waste (e-waste) has grown in recent years due to the increase in electronic gadget usage. Wherever it proliferates, there is a detrimental effect on the ecosystem, plants, animals, and human beings. The regulation of e-waste abandonment has been lacklustre and hence has polluted the ecosystem. The pollutants contributing to e-waste include lithium, lead, cadmium, mercury, chromium, arsenic, and nickel, which are heavy metal pollutants in the environment that create various kinds of diseases. These hazardous chemicals metamorphose the ecosystem into an unsustainable environment for the survival of flora and fauna. Apart from the flora and fauna, this also coerces humans to orbit the environment without being aware of the potential threat, which ruins life at a snail's pace. To put a stop to the contamination of these pollutants against the adverse effect on human beings and the environment, it is essential to be aware of the essential quality or attribute of various unpredictable materials in e-waste so that people will recognize the detrimental effect of e-waste treated in uncontrolled conditions. This chapter will focus on the sources, characteristics, burden, environmental impacts, and health impacts of e-waste.

Keywords Electronic waste · Pollutants · Hazardous · Impact · Environment

1 Introduction

Technological and industrial development have paved a new way for a sophisticated and luxurious lifestyle. Humans are dependent on advanced technologies merely to acclimate themselves and lead a luxurious existence. Reliance upon these gadgets has increased as the electronic industry has expanded at a breakneck speed in the

N. Aarthi (✉)
Department of Zoology, Sri Sarada College for Women (Autonomous), Salem, Tamil Nadu, India

A. P. Das, S. Priyadarsini (eds.), *Electronic Waste and Environmental Pollution*, Sustainable Environmental Waste Management Strategies,
https://doi.org/10.1007/978-3-031-90287-1_3

industrial sector (Prabhakar et al., 2019; Raj & Das, 2023). The utilization of electronic gadgets and appliances by the public is remarkably prevalent in today's culture, aiding the global economy in a massive expansion of inventions and applications of new technologies (Priyadarsini & Das, 2025). Due to the spread of advanced gadgets, there is a substantial portion of end-of-life (EoL) electronics on the planet. Initially, the production, demand, and management of end-of-life electronic (EoL) appliances and gadgets have made leaps and bounds globally (Priyadarsini & Das, 2024). According to a survey, more than 8 billion people live on Earth, of whom only 6 billion have their basic needs met to sustain their lives, whereas more than 6 billion people have mobile phones, which have become a fundamental necessity. The term *obsolete* is more aptly applied in the field of electronic gadget manufacturing than anywhere else, since whenever it is introduced, new features are upgraded, especially in mobile development. This enhances the functionality and efficiency of the operating system and speeds up the process, so people rely on the new version, and hence, the older version becomes obsolete. The obsolete electronic items, broken, and batteries in mobiles and other gadgets like laptops, tablets also a major source of heavy metal waste. In India, most electronic waste gets dumped in the houses, as people are not aware of how to discard it. Different categories of e-waste are discussed in Fig. 1.

The term *e-waste* refers to outdated and nonworking electronics like televisions, laptops, generators, freezers, computers, Oven-toaster-griller (OTG), and other

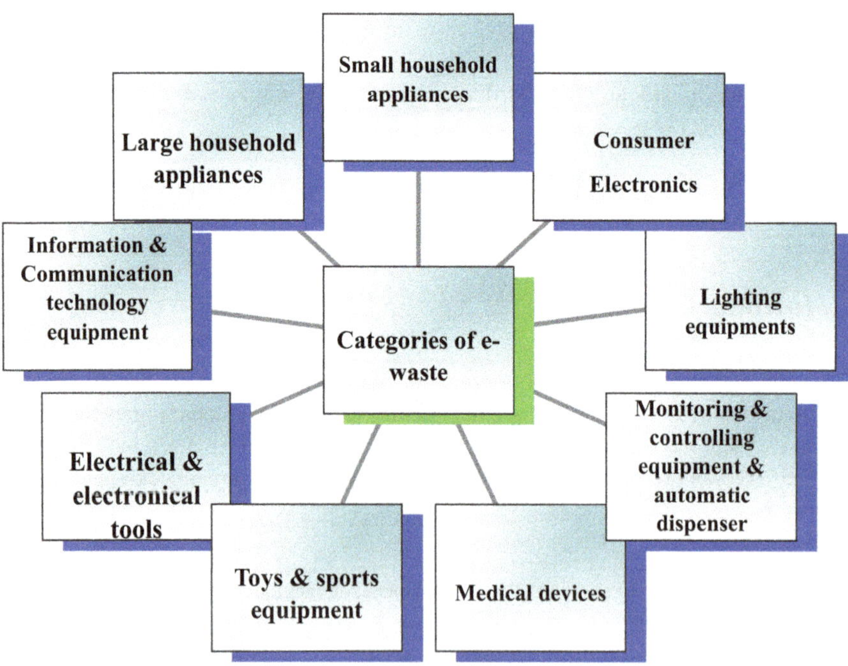

Fig. 1 Categories of e-waste

electronic gadgets tossed out by their owners. Today, computers and mobile phones are the most used tools for all types of activities—in schools, homes, offices, hospitals, and the manufacturing industry. Computer components are highly hazardous, and they possess heavy metals such as brominated flame retardants for printed circuits, cadmium and lead in printed circuit boards, lead oxide and barium in cathode ray tubes, cadmium batteries, mercury in transformers, flat screens, printed circuit boards, capacitors, and old transformers. A survey estimates that 500 million computers worldwide contain 1.3 kilograms of plastic, 716.7 million kilograms of lead, and 286,700 kg of mercury. The global collection of standard 14-in. vacuum tube screens accounts for 2.5 to 4 kg of lead. Lead from these waste sources can leach into the water level of a landfill and contaminate it. According to the World Health Organization (WHO), lead poisoning can affect both physical and mental development in children as well as other age groups.

The enormous quantity of E-waste generated by the human community presents life-threatening effects. The categories of e-waste are:

- Large household appliances.
- Small household appliances.
- Information and communication technology equipment.
- Consumer electronics.
- Lighting equipment.
- Monitoring & controlling equipment & automatic dispenser.
- Medical devices.
- Toys and sports equipment.
- Electrical and electronic tools.

2 Examples of E-Waste

E-waste, also known as Waste Electrical and Electronic Equipment (WEEE), presents a concern that continues to evolve as electronic consumption grows, and recycling becomes essential for sustainable practices. Recent studies emphasize categorizing e-waste based on material composition, environmental hazards, and recyclability (Mishra et al., 2023). One study classifies e-waste into three groups based on the threat to the environment. This classification approach focuses on alleviating environmental risk and improving the environment (Buchka, 2024). Another research emphasis is on isolating e-waste into metallic and nonmetallic fragments using thermal imaging, and reutilization rates for components such as plastics, printed circuit boards (PCBs), and metals (Gundupalli et al., 2018; Dey et al., 2024; Jagatee et al., 2024; Mishra et al., 2020). Universally, discussions of the E-waste topic address household appliances, IT devices, and entertainment systems. Newer categorization considers size and redeemable value of obsolete or expired products (Chatterjee & Abraham, 2017). As e-waste is traded and administered internationally, a structural or comprehensive allocation or arrangement is needed to reconcile resource extraction with public health protection (Ilankoon et al., 2018). Emerging technical proposals recommend a blending of synthetic and organic innovations to recover precious metals like copper, gold, and silver from e-waste streams (Patil et al., 2014). A system of classifications that aims to enhance recycling efficiency can also address ecological concerns and bridge gaps in regulatory practices across nations.

Large household appliances include refrigerators, freezers, washing machines, and air conditioners. Computers, laptops, mobile phones, computer accessories, printers, and copy machines are the best examples of Information and Communication Technology (ICT) equipment. Toasters, coffee machines, clocks, watches, hair dryers, TVs, radios, video cameras, amplifiers, and electric razors are examples of consumer electronics. Vacuum cleaners, grinders, ovens, toasters, and grills are examples of small household appliances. Drills, sewing machines, and saws are examples of electrical and electronic tools. CFLs (compact fluorescent lights), mercury vapor lamps, fans, switches, and wires are examples of lighting equipment. Phones, video games, electric trains, and toys are examples of leisure and sports equipment. Radiotherapy and cardiology devices like ECG (electrocardiogram) devices, neurology devices, and dialysis equipment are examples of medical devices. Smoke detectors, thermostats, ATMs (automated teller machines), and coffee and juice vending machines are examples of monitoring and controlling equipment and automatic dispensers.

Items from these categories of appliances and other equipment are usually disposed of after use. Waste may be classified as either hazardous or nonhazardous, based on its nature. These wastes are creating hazardous effects. Various types of waste are discarded into our environment, but only nonbiodegradable wastes, which cannot be degraded by our ecosystem, pose significant and concerning threats to the organisms in the environment. Let us review the sources of waste material.

3 Sources of E-Waste

Electronic waste is generated by various sources. These substances are classified into hazardous and nonhazardous groups. Different types of e-waste streams encompass computers, mobile phones, large household appliances, and medical equipment. A large quantity of e-waste has been recycled using unreliable activities, but e-waste has also been dumped in domestic areas, stored in warehouses, and illegally exported. When people lack knowledge about the hazards of certain types of waste, it may be dumped in the trash instead of being recycled. Public knowledge about electronic trash is very poor, and so it accumulates in the ecosystem, which causes many unidentified problems. E-waste accumulation is the result of the widespread use of electronic/electrical gadgets, and the tendency of users to constantly update to newer models.

Different varieties of components comprise the dumped products. The percentages of the hazardous nature of the compounds vary. All e-waste contains three

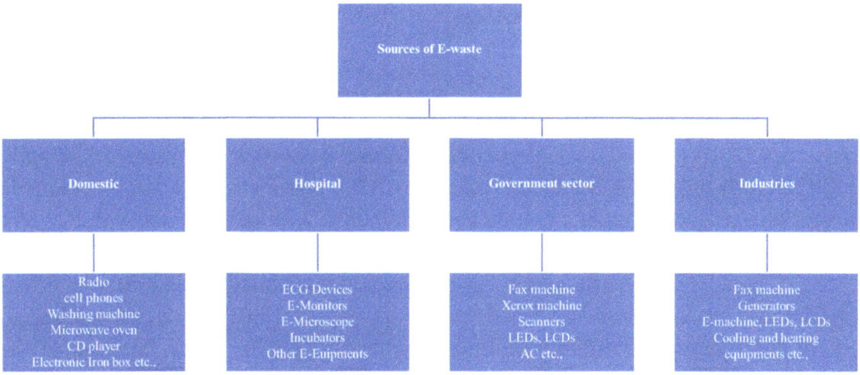

Fig. 2 Depicts the different sources of e-waste

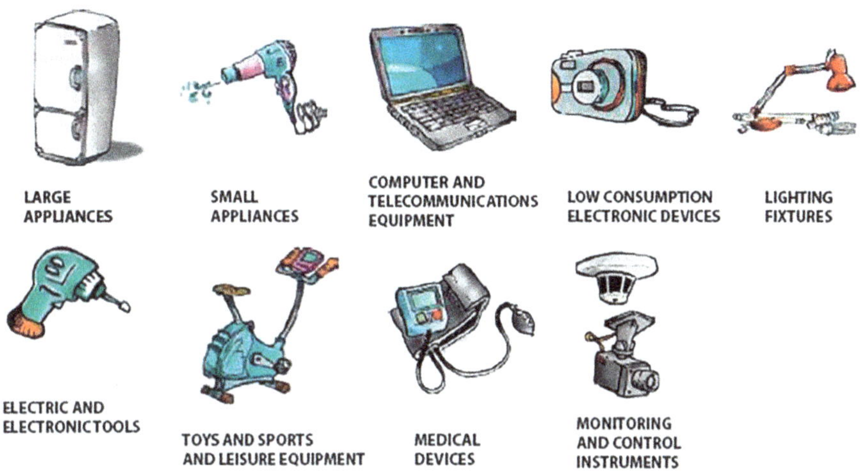

Fig. 3 Depiction of sources of e-waste

main constituents—glass, metal, and plastics (Bala et al., 2024; Sahoo et al., 2022; Singh Jadaun et al., 2022). Among the metals, some expensive and rare metals such as copper, nickel, palladium, silver, and gold are included. It is possible to recover and reuse the valuable metals found in e-waste as raw materials for new products. However, new product development requires a myriad of discrete components and substances. Recovering and reusing these metals is a very tedious and time-consuming process, and hence it is often not attempted (Das et al., 2011, 2015a, 2015b; Das & Ghosh, 2018). Figures 2 and 3 represent the different sources of e-waste. Among these, a faster network, the rapid expansion of cellular and broadband networks, and digital marketing paved the way for the increased consumption of electrical gadgets. Access to the internet and smartphones in all places is another source of waste.

4 Characteristics of E-Waste

E-waste, encompassing junk electronic devices, is a rapidly flourishing waste stream with critical environmental and economic implications. The *Global E-Waste Monitor* (2024) reported a staggering 62 million tons of e-waste generated globally in 2022, with only 22% effectively recycled. This waste stream contains valuable metals such as gold, copper, and rare earth elements, with an estimated value of $91 billion annually. However, the recycling rates are insufficient, particularly in the Global South, where insufficient infrastructure and inadequate policies hinder effective e-waste management. Informal processing prevails in these areas, leading to severe environmental adulteration and a menace to public health because of the unhealthy effects of toxic materials like lead and mercury.

Studies in 2023 have also highlighted the socioeconomic and legislative dynamics of e-waste control. In developing nations, factors such as consumer behavior, lack of insight, and weak enforcement of regulations contribute to the burgeoning problem. Emerging strategies include urban mining initiatives to reclaim valuable materials, along with the utilization of extended producer responsibility (EPR) schemes that mandate manufacturers to manage end-of-life recycling for their products. Pilot programs utilizing artificial intelligence have shown promise in enhancing recycling efficiency and consumer engagement (EWCRA, 2023).

Illegal transnational trafficking of e-waste remains a challenge, especially the transport of e-waste from developed to developing countries. Networks exploit regulatory loopholes, with significant environmental and social costs in recipient regions. Advanced regulatory frameworks and global standards are essential for mitigating these flows (IMPEL, 2023).

Sustainable solutions like circular economy practices, ecodesign, and innovative recycling technologies have been adopted. These include bioleaching and pyrolysis, although challenges such as high costs and pollutant emissions persist. Awareness campaigns and incentivized programs are increasingly recognized as key to promoting responsible consumer behavior and proper disposal practices (Fawole et al., 2023; EWCRA, 2023).

5 Burden of E-Waste

E-waste poses a significant global burden due to its rapid growth and complex environmental and health implications. The mishandling and improper recycling of e-waste release hazardous chemicals such as lead, mercury, and cadmium, which contaminate the soil, water, and air, particularly in developing countries where regulations are less stringent (Wirtu & Tucho, 2022). These hazardous conditions harm the environment and cause health issues, such as bronchitis, neurological damage, and malignancy, particularly for informal laborers who are often untrained, inexperienced, and lacking in protective measures (Lama et al., 2022). Recently, in Africa,

Fig. 4 Illustrates the burden of e-waste as a danger to the environment

the unofficial reprocessing of e-waste has exposed people to heavy metals and brominated inflammables, underscoring a dreadful community health catastrophe (Orisakwe et al., 2019).

The growing quantity of e-waste has been galvanized by rapidly advancing technologies and the improper dumping or disposal of the accumulated e-waste. An underdeveloped nation faces intricate challenges as it readily accepts obsolete products from a first-world country, which exacerbates the degradation of waste in a complicated way (Deng, 2019). Weak legal frameworks worsen these issues, as many countries adopt weak general methodologies for handling e-waste disposal rather than adhering to strict regulations. Renewable solutions include a closed-loop system that reinforces extended producer responsibility (EPR), creating awareness among the general public regarding the threat and suggesting ways to handle e-waste. International cooperation is critical for establishing equitable policies and reducing the burden on vulnerable regions. Investments in green technologies and a robust recycling infrastructure can mitigate the harmful influence of e-waste on the ecosystem and public well-being. Figure 4 depicts the risk of e-waste to the environment.

6 Environmental Impacts of E-Waste

The composition of e-waste depends on the type and volume of the disposed material, as well as its blend with other plastic and ceramic materials commonly used as coatings for electronic devices. During this process, the most frequently encountered heavy metals include hazardous ones like mercury, lead, cadmium, and nickel, as well as valuable materials such as copper, platinum, and gold. These metals are then collected by e-waste salvagers to be sold to manufacturers for use as resources (Das & Mishra, 2008; Das & Singh, 2011; Mishra & Das, 2008). However, although e-waste is collected for recycling, not all 100% is recycled. Nearly 79% of the e-waste is recyclable, and the remaining 21% of the e-waste is destined for landfills. This 21% portion contains materials like polyurethane, polyvinyl chloride (PVC), acrylonitrile butadiene styrene (ABS), along with other incombustible items like glass. These are major contributors to environmental pollution. They leech into the ecosystem, causing land, water, and atmospheric pollution. This creates a variety of issues, as discussed in the following section.

6.1 Toxic Substance Release

E-waste likely possesses hazardous substances like lead, mercury, cadmium, and chromium. When these substances are disposed of carelessly, they leach into the soil and reside in the water table, which causes water pollution and contaminates the ecosystem. This not only affects the water table but also nearby water bodies. Toxic substances reach and leach into bodies of water, which affects aquatic organisms. Similarly, the inflammable substances incorporated in the e-waste, like brominated flame-retardant plastics, release toxic substances like dioxins and furans during their incineration process, which pollute the air and contribute to global warming. The environment is polluted by the emission of greenhouse gases when e-waste is incinerated, which leads to enormous health issues in living organisms. Many plants and pollinators are also affected, which causes agricultural disruption. Exposure to toxic chemicals from e-waste can lead to health risks, including cancer, bronchitis, chronic respiratory illness, and polycystic ovary syndrome (PCOS) in women.

The environmental impact of e-waste is profound, encompassing pollution, resource depletion, and public health risks. E-waste, including discarded electronics like computers, cellphones, and televisions, contains dangerous materials such as lead, mercury, and cadmium. Illegitimate disposal of these items leads to contamination of soil, water, and air, resulting in ecological and health damage. Residue of heavy metals leaches into the water table, harming the environment and living organisms, especially human beings, by giving rise to a variety of dreadful diseases and organ failures (Ayeleru et al., 2023). Combustion of e-waste releases harmful chemicals into the air, causing air pollution and an unstable environment (Manna et al., 2024). Above all, dumping e-waste into landfills diminishes a ready supply of precious metals like gold, platinum, copper, and silver, whereas reclamation can retrieve these metals to relieve the resource strain (Kumar et al., 2024a, 2024b).

Reprocessing acts as a great remedy but remains underutilized due to cost-effectiveness and barriers to technological innovation. Research highlights the importance of environmentally conscious practices and modern machinery, which, when accompanied by either zero waste or a low cost, result in easy recycling production (Lee et al., 2024). Additionally, recent technologies like artificial intelligence (AI) and the Internet of Things (IoT) have enhanced e-waste management by boosting and categorizing waste according to resourcefulness recovery status (Sahoo, 2024). Public awareness, along with judicial laws and their framework, play a crucial role in this initiative to support and ensure that the collection and reclamation of e-waste is practiced universally. Significant barriers to progress include insufficient amenities, archaic waste management methods, and financial limitations. To bring about the necessary changes, collaboration is needed among governments, industries, and communities to effectively implement renewable e-waste management practices. Figure 5 depicts the impact of e-waste on the environment and ecosystem.

E-waste presents challenges to the ecosystem due to its toxic composition and improper disposal techniques. Low standards and poor recycling practices contribute to heavy metal pollution in aquatic systems, affecting both the environment and human health. Research in the Guiyu e-waste recycling area of China found declining though still elevated concentrations of heavy metals like lead and cadmium in rivers, connected to health risks for local children involving diminished growth from heavy metal exposure (Hu et al., 2024; Das & Mishra, 2010; Das et al., 2012). Apart from heavy metal contamination and pollution, e-waste also adversely affects air quality when noxious fumes are released during improper incineration of components, as noted in extensive discussions on renewable practices to control these effects (Castle-Green & Sailaja, 2024). Beyond the current strategies being employed to handle these problems, innovative methods, such as the manufacturing of nanoparticles from e-waste, are being crafted to lessen environmental consequences through the metamorphosis of waste into valuable materials, thus establishing environmentally friendly solutions for industrial applications (Anchani et al., 2024). However, universal e-waste commerce increases toxic environmental burdens, particularly in ASEAN+6 countries (ASEAN nations plus China, Japan,

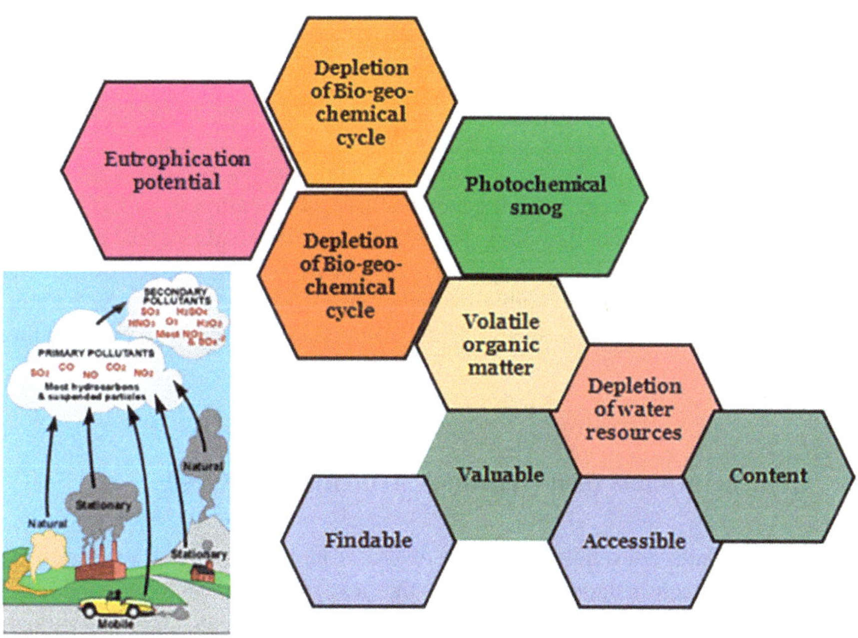

Fig. 5 Illustrates the environmental impact of E-waste

South Korea, India, Australia, and New Zealand), where the highest imports of waste lead to pollution hotspots, often due to insufficient regulation and lack of processing facilities (Lin et al., 2024). This intricate problem reinforces the need for firm international regulations, modern reclamation technologies, and localized renewable practices to lessen e-waste's environmental footprint.

7 E-Waste: A Contributor to Global Warming

Electronic waste (e-waste) contributes substantially to global warming due to its excessive.

carbon emissions and improper disposal techniques. Manufacturing and disposal electronic products provoke greenhouse gases (GHG) at various stages—mining raw materials, production, conveyance, and recycling. About 67% to 85% of the carbon footprint from information communication and technology (ICT) devices alone contributes to a lifetime of emissions of greenhouse gases. Many investigations and analyses assessed that the incorporation of greenhouse gases released from ICT e-waste increased to about 53% between 2014 and 2020. One study also projected that 580 million metric tons of carbon dioxide equivalent in 2020 could increase to approximately 852 million metric tons by 2030 without intervention (Singh & Ogunseitan, 2022). Irregular methods of disposal, like open burning and acid stripping, release hazardous pollutants, including polycyclic aromatic hydrocarbons (PAHs), dioxins, and furans, which intensify air pollution and global warming. These activities not only damage the local ecosystem but also contribute to the atmospheric accumulation of greenhouse gases (GHGs; Robinson, 2009). Renewable control approaches, including reclaiming the composite materials and increasing the longevity of the device, could yield as much as a 50% reduction of these gas releases. Renewable energy technologies also engage the recycling of electronic waste and provide a hopeful remedy to mitigate reliance on unsustainable resources (Seif et al., 2023). The regenerative economy approach, which includes ecodesign, mending, reconditioning, and reuse, is a crucial and essential part in the mitigation of e-waste's climate impact. The emphasis on universal policies, public awareness, and education is highly challenging, both in addressing the carbon-intensive nature of e-waste mishandling and in reducing its role in global warming (Fawole et al., 2023).

8 Health Impacts of E-Waste: A Multifaceted Crisis

The rise of health consequences from e-waste is basically because of hazardous components, which include heavy metals like lead, mercury, cadmium, and additional persistent organic pollutants (POPs). These substances can target various anatomical systems and may occasionally cause genetic mutations. Workers and residents who are forced to work and survive in an environment challenged by partial recycling and careless disposal of e-waste face the risk of exposure to hazardous substances. These conditions then generate diversified health consequences like respiratory disorders, cardiovascular diseases, improper secretion of the ductless glands, and other cutaneous diseases (Eckhardt & Kaifie, 2024). Reproductive health is particularly vulnerable, with studies linking exposure to e-waste with infertility, low birth weights, premature deliveries, and spontaneous abortions, as supported by research. Compounds like polycyclic aromatic hydrocarbons and brominated flame-retardants adversely affect reproductive health. Exposure of e-waste to pregnant women leads to infertility, premature deliveries, reduced birth weight, and occasionally, spontaneous abortions (Roy et al., 2024). Prolonged e-waste exposure in vulnerable populations such as children and pregnant women exacerbates toxic accumulation disorders, may cause miscarriages, impairs nutrient absorption, and perpetuates long-term health disparities (Jalca et al., 2024). Careless disposal methods always lead to open burning and acid percolation, releasing atmospheric noxious chemicals. Improper disposal methods may allow contaminants to trickle into the soil and a nearby aquatic ecosystem and cause bioaccumulation and contamination of the hierarchical nodes in the food chain. In China's Taizhou region, a center for e-waste reclamation discloses frightening information regarding the escalation of the malignancy rate and bronchitis in connection with environmental exposure (Wong Ming, 2023). Additionally, research shows that exposure to e-waste may affect male reproductive health issues like oligospermia, azoospermia, the lack or reduced secretion of male hormone testosterone, and DNA damage, especially at elevated exposure levels (Raja et al., 2023). To control these influences, international policy efforts are crucial for managing e-waste commerce, administering safe recycling strategies, and initiating a firm health surveillance system. A concerted effort between technology and public self-conscious crusades can further reduce human exposure to e-waste hazards. Exposure to electronic waste (e-waste) links to ductless gland imbalance secretion in women, primarily due to exposure to endocrine-disrupting chemicals (EDCs) such as polybrominated diphenyl ethers (PBDEs) and metal chemicals like lead (Pb) and cadmium (Cd) (Radović et al., 2024). These substances can interfere with the secretions of ductless glands, leading to fertility-dependent health issues. A research study analyzed the dust from e-waste dismantling sites and identified a mixture of decabromodiphenyl ether (DBDE), decabromodiphenyl ethane (DBDPE), tetrabromobisphenol A (TBBPA), lead, cadmium, and nickel as virulent contributors to infertility disorders in both males and females. The *in silico* analysis or docking study suggested that exposure to this combination of hazardous chemicals could escalate the danger of ductless gland secretion imbalances, highlighting the requirement for further wet laboratory analysis to investigate the specific genes and mechanisms involved in pathology. Additionally, PBDEs, commonly used as fire-resistant or incombustible materials in electronic devices, have been associated with alarming effects on thyroid hormones and reproductive fitness (Singh et al., 2022). A systematic review indicated that exposure to PBDEs might disrupt the secretion of hormones like thyroid hormones, affect the sperm count, and adversely affect neonatal health. The evidence for thyroid and reproductive hormone disruption was of intermediate quality, suggesting scant evidence of toxicity, while the proof for other hormonal effects was of low quality, indicating inadequate documentation of toxicity. These preliminary findings highlight the potential health risks for women in regions with significant e-waste recycling activities, where exposure to EDCs is more prevalent. Implementing safer recycling practices and conducting further research are crucial steps toward mitigating these potential health impacts.

E-waste directly and indirectly affects the food chain in the form of biomagnification. Aquatic plants from plankton to higher plants accumulate residues and hazardous toxins in them, organisms in the food chain consume them, and then it is passed along the hierarchy of the food matrix. Hence, insidious health consequences are a burgeoning concern of e-waste. These considerations must be addressed for the improved health of future generations.

9 Conclusion: Comprehensive Understanding of E-Waste Sources, Characteristics, and Environmental Impact

Electronic waste, born of rapid technological advancements and the needs of the consumer, originates from obsolete electronic items such as smartphones, computers, and household appliances. Its intricate configuration includes precious or costly redeemable metals like gold and rare earth elements, as well as hazardous components like lead, mercury, and brominated inflammables. Unsafe handling and disposal techniques, such as incineration and landfilling, release these toxic elements into the environment, contaminating the soil, water, and air. This results in rigorous environmental deterioration, affecting the biodiversity and sustainability of ecosystems. The environmental calamities of e-waste are multifaceted. Atmospheric pollution from burning e-waste increases greenhouse gas emissions, thus worsening climate change. The leeching of chemicals into the soil contaminates and clogs agricultural productivity, whereas aquatic contamination endangers the lives of aquatic organisms. Dangers to these organisms are not the only concern, as contamination maligns the food chain and web too. Unofficial recycling processes, which are highly prevalent in underdeveloped areas, often function without adequate safety protocols, intensifying human exposure to hazardous chemicals.

This results in severe health outcomes such as respiratory diseases, malignancy, fertility health issues, and nervous disorders. Handling the e-waste crisis must involve the government, nongovernmental organizations, manufacturing companies, and customers, employing a collaborative, multifaceted approach. Laws and regulations need to be strengthened through policies like extended producer responsibility (EPR) and banning transboundary e-waste dumping, which can totally minimize its proliferation. At the time of production, there should be a norm to recycle e-waste more efficiently and sustainably, such as through the use of modular electronics and biodegradable materials, offering pathways to reduce waste generation.

General public knowledge and education are equally essential to promoting responsible utilization and disposal practices.

The universal human community must collaborate to manage e-waste sustainably, ensuring environmental preservation and safeguarding human health and other living organisms in order to conserve biodiversity. This in-depth knowledge emphasizes the seriousness of developing and executing innovative, inclusive strategies to effectively reduce the rapid increase of e-waste problems. Collaboration of all contributing parties on technological, political, and social interventions can turn the e-waste challenge into an opportunity for a more sustainable and equitable future.

References

Anchani, A. H., Abishini, A. H., & Ashokkumar, T. (2024). A novel method for biosynthesis of copper oxide nanoparticles using banana blossom and E-waste: Characterization and application in malachite green dye removal through response surface methodology optimization. *Discover Biotechnology, 1*, 4. https://doi.org/10.1007/s44340-024-00004-9

Ayeleru, O. O., Modekwe, H. U., Adefiranye, O. O., Fafowora, B. L., Onu, M. A., Nyam, T. T., Adesiyan, I. A., & Olubambi, P. A. (2023). The environmental fate of E-waste. In *Electronic waste management: Policies, processes, technologies, and impact* (pp. 90–103). https://doi.org/10.1002/9781119891543.ch8

Bala, K., Anshul, Verma, S., Dogra, A., Das, A. P., Kumar, R., & Nandini. (2024). Sustainable management for urban plastic waste generation. In *Sustainable management of urban plastic waste through circular economic approaches* (pp. 211–236). https://doi.org/10.1201/9781003465447-13

Buchka, A. (2024). Classification of waste electrical and electronic equipment by ecological hazard. *Herald of Khmelnytskyi National University. Technical sciences, 337*(3(2)), 265–272. https://doi.org/10.31891/2307-5732-2024-337-3-40

Castle-Green, T., & Sailaja, N. (2024). Addressing E-waste. *Base Diseño e Innovación*. https://doi.org/10.52611/bdi.num10.2024.999

Chatterjee, A., & Abraham, J. (2017). Efficient management of e-wastes. *International journal of Environmental Science and Technology, 14*(1), 211–222. https://doi.org/10.1007/S13762-016-1072-6

Das, A. P., & Ghosh, S. (2018). Bioleaching of manganese from mining waste materials. *Materials Today Proceedings, 5*(1), 2381–2390. https://doi.org/10.1016/J.MATPR.2017.11.459

Das, A. P., & Mishra, S. (2008). Hexavalent chromium (VI): Health hazards and environmental pollutant. *Journal of Environmental Research and Development, 2*(3), 386–392.

Das, A., & Mishra, S. (2010). Biodegradation of the metallic carcinogen hexavalent chromium Cr(VI) by an indigenously isolated bacterial strain. *Journal of Carcinogenesis, 9*, 6. https://doi.org/10.4103/1477-3163.63584

Das, A. P., & Singh, S. (2011). Occupational health assessment of chromite toxicity among Indian miners. *Indian Journal of Occupational and Environmental Medicine, 15*(1), 6. https://doi.org/10.4103/0019-5278.82998

Das, A. P., Sukla, L. B., Pradhan, N., & Nayak, S. (2011). Manganese biomining: A review. *Bioresource Technology, 102*(16), 7381–7387. https://doi.org/10.1016/J.BIORTECH.2011.05.018

Das, A. P., Sukla, L. B., & Pradhan, N. (2012). Microbial recovery of manganese using *Staphylococcus epidermidis*. *International Journal of Nonferrous Metallurgy, 2012*(02), 9–12. https://doi.org/10.4236/IJNM.2012.12002

Das, A. P., Ghosh, S., Mohanty, S., & Sukla, L. B. (2015a). Advances in manganese pollution and its bioremediation. In L. B. Sukla, N. Pradhan, S. Panda, & B. K. Mishra (Eds.),

Environmental microbe biotechnology (1st ed., pp. 313–328). Springer. https://doi.org/10.1007/978-3-319-19018-1_16

Das, A. P., Ghosh, S., Mohanty, S., & Sukla, L. B. (2015b). Consequences of manganese compounds: A review. *Toxicological & Environmental Chemistry, 96*(7), 981–997. https://doi.org/10.1080/02772248.2015.1005428

Deng, W. J. (2019). E-waste management: Challenges and opportunities. In W. So, C. Chow, & J. Lee (Eds.), *Environmental sustainability and education for waste management. Education for sustainability*. Springer. https://doi.org/10.1007/978-981-13-9173-6_2

Dey, S., Dash, N., Ray, S. K., Priyadarsini, S., Chhotaray, C., & Das, A. P. (2024). Occurrence and quantification of Tire Wear plastics pollutants from Highway City dust of Bhubaneswar City of Odisha. In *Sustainable microbial technology for synthetic and cellulosic microfiber bioremediation* (pp. 77–89). https://doi.org/10.1007/978-3-031-62600-5_5

Eckhardt, B., & Kaifie, A. (2024). Bridging the knowledge gap! Health outcomes in informal e-waste workers. *Journal of Occupational Medicine and Toxicology, 19*, 11. https://doi.org/10.1186/s12995-024-00410-z

EWCRA. (2023). Retrieved from https://ewcra.org/2024/04/16/the-future-of-e-waste-management-innovation-and-solutions/

Fawole A A, Orikpete O F, Ehiobu N N, Ewim D R E. Climate change implications of electronic waste: strategies for sustainable management. Bulletin of the National Research Centre. 2023;47:147.

Fawole, A. A., Orikpete, O. F., Ehiobu, N. N., & Ewim, D. R. E. (2023). Climate change implications of electronic waste: Strategies for sustainable management. *Bulletin of the National Research Centre, 47*, 147. https://doi.org/10.1186/s42269-023-01124-8

Gundupalli, S. P., Hait, S., & Thakur, A. (2018). Classification of metallic and non-metallic fractions of e-waste using thermal imaging-based technique. *Process Safety and Environmental Protection, 118*, 32–39. https://doi.org/10.1016/J.PSEP.2018.06.022

Hu, H., Zeng, X., Dai, C., Xie, B., Zhang, J., Xu, X., & Huo, X. (2024). A decade-long comparison of heavy metal(loid)s in the river and children's health risk assessment of an E-waste recycling area. *Water, 16*(22), 3226. https://doi.org/10.3390/w16223226

Ilankoon, I. M. S. K., Ghorbani, Y., Chong, M. N., Herath, G., Moyo, T., & Petersen, J. (2018). E-waste in the international context—A review of trade flows, regulations, hazards, waste management strategies and technologies for value recovery. *Waste Management, 82*, 258–275. https://doi.org/10.1016/J.WASMAN.2018.10.018

IMPEL. (2023). *LIFE SWEAP—Shipments of Waste Enforcement Actions Project. European Union LIFE Programme*. https://www.sweap.eu

Jagatee, S., Priyadarshini, S., Rath, C. C., & Das, A. P. (2024). Synthetic microfiber: An enduring environmental problem linked to sustainable development. *Environmental Science and Engineering, Part F, 2333*, 93–112. https://doi.org/10.1007/978-3-031-51792-1_6

Jalca, J. E. C., Carreño, E. J. M., Suárez, K. V. M., & Moreira, M. G. (2024). E-waste: A global threat with an impact on nutrition. *Salud, Ciencia y Tecnología, 4*, 591. https://doi.org/10.56294/saludcyt2024.591

Kumar, A., Singh, P. P., & Khanna, M. K. (2024a). Sustainable E-waste management and its effect on environment and human health. In A. Gupta, R. Kumar, & V. Kumar (Eds.), *Integrated waste management*. Springer. https://doi.org/10.1007/978-981-97-0823-9_17

Kumar, R., et al. (2024b). Environmental impacts of E-waste: Pollution and resource depletion. In K. V. Rajesh Kumar et al. (Eds.), *Sustainable solutions for E-waste and development* (pp. 12–32). IGI Global. https://doi.org/10.4018/979-8-3693-1018-2.ch002

Lama, A., Vishnoi, V. K., Kumar, K., Kumar, B., & Saini, A. (2022). E-waste, a global challenge: Its management and impact on public health in India. In *2022 International Conference on Fourth Industrial Revolution Based Technology and Practices (ICFIRTP), Uttarakhand, India* (pp. 25–29). https://doi.org/10.1109/ICFIRTP56122.2022.10063211

Lee, J., Choi, H., & Kim, J. (2024). Environmental and economic impacts of e-waste recycling: A systematic review. *Chemical Engineering Journal, 494*, 152917. https://doi.org/10.1016/j.cej.2024.152917

Lin, C.-H., Yiting, C., & Yiwen, Y. (2024). Impact of electronic networks on E-waste trade: Evidence from ASEAN + 6 countries. *Scottish Journal of Political Economy, 72*, e12408. https://doi.org/10.1111/sjpe.12408

Manna, S., Mukherjee, S., Das, D., & Saha, A. (2024). Influence of E-waste Management in Green-Computing. *Journal of Information, Technology and Policy*. https://doi.org/10.62836/jitp.v1i1.183

Mishra, S., & Das, A. P. (2008). Hexavalent chromium reduction and 16S rDNA identification of bacteria isolated from a Cr (VI) contaminated site. *The Internet Journal of Microbiology, 7*(1).

Mishra, S., Singh, R. P., Rath, C. C., & Das, A. P. (2020). Synthetic microfibers: Source, transport and their remediation. *Journal of Water Process Engineering, 38*, 101612. https://doi.org/10.1016/J.JWPE.2020.101612

Mishra, S., Ghosh, S., van Hullebusch, E. D., Singh, S., & Das, A. P. (2023). A critical review on the recovery of base and critical elements from electronic waste-contaminated streams using microbial biotechnology. *Applied Biochemistry and Biotechnology, 2023*, 1–30. https://doi.org/10.1007/S12010-023-04440-X

Orisakwe, O. E., Frazzoli, C., Ilo, C. E., & Oritsemuelebi, B. (2019). Carga de salud pública de los electronicos en Africa. *Rev Salud y Contaminación, 9*(22), 190610. https://doi.org/10.5696/2156-9614-9.22.190610

Patil, Y., Bhat, V., & Rao, P. (2014). Management of electronic waste by employing combined technological strategies. *Global Journal of Finance and Management, 6*(6), 545–550.

Prabhakar, A., Mishra, S., & Das, A. P. (2019). Isolation and identification of Lead (Pb) solubilizing bacteria from automobile waste and its potential for recovery of Lead from end of life waste batteries. *Geomicrobiological Journal, 36*(10), 894–903. https://doi.org/10.1080/01490451.2019.1654044

Priyadarsini, S., & Das, A. P. (2024). Analytical and structural characterization of waste lithium-ion batteries for their effective recycling strategy. *Environmental Chemistry and Ecotoxicology, 7*, 182–191. https://doi.org/10.1016/j.enceco.2024.12.004

Priyadarsini, S., & Das, A. P. (2025). Lithium bioleaching: A review on microbial-assisted sustainable technology for lithium bio-circularity. *Journal of Water Process Engineering, 69*, 106744. https://doi.org/10.1016/j.jwpe.2024.106744

Radović, B., Baralić, K., Ćurčić, M., Marić, Đ., Živanović, J., Antonijević Miljaković, E., Buha Djordjevic, A., Ćosić, D. Đ., Bulat, Z., & Antonijević, B. (2024). Endocrine disruptors in e-waste dismantling dust: In silico prediction of mixture-induced reproductive toxicity mechanisms. *Science of the Total Environment, 917*, 170437. https://doi.org/10.1016/j.scitotenv.2024.170437

Raj, K., & Das, A. P. (2023). Lead pollution: Impact on environment and human health and approach for a sustainable solution. *Environmental Chemistry and Ecotoxicology, 5*, 79–85. https://doi.org/10.1016/J.ENCECO.2023.02.001

Raja, A., Costa, P., Blum, J. L., Doherty-Lyons, S., Kelechi Igbo, J., Meltzer, G., Orem, W., Mccawley, M., & Zelikoff, J. D. (2023). In vivo exposure to electronic waste (e-waste) leachate and hydraulic fracturing fluid adversely impacts the male reproductive system. *Reproductive Toxicology, 124*, 108533. https://doi.org/10.1016/j.reprotox.2023.108533

Robinson, B. (2009). E-waste: An assessment of global production and environmental impacts. *Science of the Total Environment, 408*(2), 183–191. https://doi.org/10.1016/J.SCITOTENV.2009.09.044

Roy, A., Mukherjee, S., Das, D., & Saha, A. (2024). *Impact of E-waste on reproduction* (pp. 293–299). https://doi.org/10.1002/9781394166923.ch15

Sahoo, P. K. (2024). Environmental aspects of e-waste management. In *Futuristic trends in chemical material sciences & nanotechnology* (pp. 306–318). https://doi.org/10.58532/v3bdcs1ch21

Sahoo, P. P., Singh, S., Rout, P. K., Mishra, S., & Das, A. P. (2022). Microbial remediation of plastic pollutants generated from discarded and abandoned marine fishing nets. *Biotechnology &*

Genetic Engineering Reviews, 39(2), 511–526. https://doi.org/10.1080/02648725.2022.2152629

Seif, R., Salem, F. Z., & Allam, N. K. (2023). E-waste recycled materials as efficient catalysts for renewable energy technologies and better environmental sustainability. *Environment, Development and Sustainability, 26*, 5473–5508. https://doi.org/10.1007/s10668-023-02925-7

Singh Jadaun, J., Bansal, S., Sonthalia, A., Rai, A. K., & Singh, S. P. (2022). Biodegradation of plastics for sustainable environment. *Bioresource Technology, 347*, 126697. https://doi.org/10.1016/J.BIORTECH.2022.126697

Singh, N., & Ogunseitan, O. A. (2022). Disentangling the worldwide web of e-waste and climate change co-benefits. *Circular Economy, 1*(2), 100011. https://doi.org/10.1016/j.cec.2022.100011

Singh, V., Cortes-Ramirez, J., Toms, L. M., Sooriyagoda, T., & Karatela, S. (2022). Effects of Polybrominated diphenyl ethers on hormonal and reproductive health in E-waste-exposed population: A systematic review. *International Journal of Environmental Research and Public Health, 19*(13), 7820. https://doi.org/10.3390/ijerph19137820

The Global E-Waste Monitor. (2024). Retrieved from https://unitar.org/about/news-stories/press/global-e-waste-monitor-2024-electronic-waste-rising-five-times-faster-documented-e-waste-recycling

Wirtu, Y. D., & Tucho, G. T. (2022). E-waste: Growing environmental and health problems and its management alternatives in developing countries. *Environmental Reviews, 30*(4), 524–536. https://doi.org/10.1139/er-2021-0120

Wong Ming, H. (2023). E-waste: A concise update on global movement, impacts, management, and site remediation. *Russian Journal of Earth Sciences*, 23, 1–19. https://doi.org/10.2205/2023es02si08

A Review on Environmental Electronic Waste Sources and Their Impact on Human Health

Prasamita Samantaray, Sailaja Priyadarsini, Kiran Bala, M. Janaki Ramaiah, and Alok Prasad Das

Abstract The electronics enterprise is the world's biggest and most hastily growing industrial sector, as humans rely heavily on contemporary technologies to live luxuriously. However, the production of an enormous quantity of e-waste due to its limited lifespan and development to more appealing features, designs, and compatibility concerns is concerning since it contains multiple hazardous substances that can harm the environment and, more importantly, human health. A record-breaking 62 million tonnes (Mt) of e-waste were created in 2022, up 82% from 2010, and are expected to climb another 32% to 82 million tonnes by 2030. Inappropriate disposal procedures, that involve open burning and acid leaching, release harmful compounds into the environment, including heavy metals, persistent organic pollutants, and brominated flame retardants. Toxic e-waste components have been related to several health difficulties, including neurological impairment, respiratory ailments, reproductive complications, and carcinogenic impacts. Children and workers are at great danger from extended exposure to poisonous fumes and direct contact with hazardous items. To address these issues, we need to use sustainable recycling systems, rigorous laws, and raise public awareness. Traditional technologies, such as pyrometallurgy and hydrometallurgy, are energy-intensive and emit hazardous byproducts. Bioleaching is a more environmentally friendly method of extracting rich metals like gold, copper, nickel, and lithium. This chapter discusses the sources of electronic waste and its impact on both the environment and human health. It also explores preventive measures and recycling technologies for effective e-waste management.

P. Samantaray · S. Priyadarsini · A. P. Das (✉)
Department of Life Sciences, Rama Devi Women's University, Bhubaneswar, Odisha, India

K. Bala
Department of Zoology Deshbandhu College, University of Delhi, New Delhi, Delhi, India

M. J. Ramaiah
Department of Biotechnology, Koneru Lakshmaiah Education Foundation (Deemed to be University), Guntur, Andhra Pradesh, India

A. P. Das, S. Priyadarsini (eds.), *Electronic Waste and Environmental Pollution*, Sustainable Environmental Waste Management Strategies,
https://doi.org/10.1007/978-3-031-90287-1_4

Keywords E-waste · Huaman health · Environment · Toxic · Recycling

1 Introduction

All demographic groups now utilize electrical and electronic equipment (EEE) more frequently than any other type of thing. End-of-life EEE can be extremely damaging to the environment, human health, and animal welfare, despite its shown usefulness for people and the advancement of economic activities (Priyadarsini & Das, 2025). It takes four centuries for the waste from EEE, alternatively referred to as used electrical and electronic equipment, or W3E, to disintegrate. Most of the time, they are discovered in landfills, whether they are controlled landfills or unofficial dumps. Occasionally, they are recovered and reprocessed (Ankit et al., 2021). For instance, barely 25% of the 25 million tons of WEEE that are collected annually in Europe are recycled. Sometimes WEEE is burned, and the greenhouse gasses that result have a major impact on the ozone layer's depletion. Because it poses one of the biggest risks to the globe and requires environmentally sound management, waste electronics, and waste have become a global concern. Nowadays, e-waste management is a substantial concern because of its global escalation. Because e-waste contains a many of dangerous resources, including heavy metals (Das & Mishra, 2008, 2009; Paul et al., 2023), radioactive chemicals, microplastics (Ghosh et al., 2023a; Mishra et al., 2021; Mohanty et al., 2024b; Sahu & Das, 2024), halogenated substances, micro- and nanoparticles, and more, it must be managed carefully (Mishra et al., 2023; Prabhakar et al., 2019; Raj & Das, 2023). Critical raw materials (CRM) include cobalt, palladium, manganese, and antimony (Das et al., 2015b; Das & Ghosh, 2023f; Ghosh et al., 2016b; Ghosh & Das, 2018). On the other side, there are several POP, PCBs, and brominated flame retardants, as well as heavy metals including lead, manganese, and mercury, (Das et al., 2011, 2015a; Das & Ghosh, 2023d; Das & Ghosh, 2023a; Mohanty et al., 2016). The estimated 20–50 million tons of WEEE produced worldwide each year are anticipated to be produced at a rate of 10% annually (Zhang et al., 2017). Additionally, it is among the trash categories with the fastest global growth. According to scientists, 5 billion people produce 3 billion tons of waste annually, which comes at a societal cost of $1.7 billion (Barney, 2013). These enormous waste volumes are becoming a global environmental concern. The most obvious effects include urban sprawl, garbage accumulation, biodiversity loss, depletion of resources, greater business and person movement, falling public budgets, and global warming (Das et al., 2024b; Dey et al., 2023a; Pradhan et al., 2023). Considering this circumstance, quick fixes are needed to lessen the effects of human activity. Additionally, e-waste contains certain persistent compounds such as heavy metals like lead, manganese, lithium, mercury, and others categorized as hazardous industrial waste (Das et al., 2012a; Das et al., 2012b; Das & Ghosh, 2023h; Das & Ghosh, 2023g; Ghosh et al., 2016a). If handled carelessly, these hazardous substances will seriously jeopardize human health as well as the health of the entire ecosystem (Toffel et al., 2004). When handling these hazardous materials in various stages, such as segregation, collection, storage, recycling, disposal, and treatment, caution must be used. Most countries manage e-waste through landfills (73%), followed by recycling (safe/unsafe; 87% of countries).

Many e-waste were recycled to recover valuable metal from the waste (Bal et al., 2018; Dey et al., 2023b; Ghosh et al., 2017; Mohanty et al., 2017; Sanket et al., 2016). Because wealthy countries face tight environmental regulations and costly disposal costs, e-waste from rich countries is sometimes illegally moved to emerging and undeveloped countries (like China, India, and Nigeria) for recycling. In these countries, the industrial infrastructure and recycling technologies are usually informal due to low income and economic crises. To profit from e-waste, workers utilize antiquated methods to remove assets from it in an unethical and illegal manner and to dispose of it illegally. Because e-waste comprises dangerous materials, exposure to it by humans may result in the development of many illnesses, such as cancers, thyroid disorders, mental and neurodevelopmental disorders, and deteriorations in function (DNA damage and altered gene expression). In developing and transitional countries, e-waste makes up 1–2 percent of total solid waste, and this percentage is predicted to increase significantly shortly.

The COVID-19 has increased in e-waste due to the need for work-from-home activities. Unlike conventional solid wastes such as household rubbish, e-waste is both hazardous and resourceful (Prabina et al., 2024; Priyadarshini et al., 2024; Tripathy et al., 2024). It is anticipated that waste generation will rise sharply shortly, reaching 121% between 2020 and 2030. Given the consequences and hazards associated with e-waste, it is crucial to create sustainable e-waste management and recycling technologies to provide a comprehensive e-waste solution (Ghosh & Das, 2017; Mishra et al., 2019; Mohanty et al., 2024a; Priyadarsini & Das, 2024). The bulk of electronic waste is made up of a complex mixture of combustible compounds, hazardous metals (Hg, Cd, and As), base metals (Cu, Al, Ni, and Mg), precious metals (Ag, Au, and Pt), and halogens (Das & Ghosh, 2023a; Das & Ghosh, 2023b; Das & Ghosh, 2023e; Ghosh et al., 2020). The recovery of valuable metals like nickel, copper, lead, zinc, manganese, and cobalt, as well as precious metals like lead, manganese, lithium, and nickel is the main economic reason for recycling e-waste (Das & Ghosh, 2018, 2023c; Ghosh et al., 2023b). It also includes nonmetallic elements with important secondary applications, such as glass fibers and technological polymers.

The creation and handling of e-waste are dangerous to the country's economic survival. As such, it is essential to characterize and implement suitable recycling techniques for e-waste. It is important to develop techniques to reduce pollution and promote resource preservation (Compd, 2002) (Das & Shreya, 2023). Pyrolysis is typically conducted at a temperature of about 430 °C and under pressure. The most efficient way to handle e-waste is believed to be recycling. Formal e-waste recycling involves three main steps: gathering waste, sorting it, and pre-processing (i.e., organizing, disassembling, and mechanically treating it), followed by treatment and disposal (Dutta et al., 2023). An "official take-back system" may serve as the foundation for the collection system. The primary category of e-waste is gathered by authorized collectors or governmental bodies such as municipalities, and it is gathered by national e-waste legislation. This is the most effective way to collect e-waste since it recovers valuable materials by using pre-treatment processes before sending the waste to advanced treatment amenities (Ha et al., 2009, Hou et al., 2020, Huo et al., 2019). The term “official take-back system” refers to the collection of e-waste by companies or licensed persons that handle garbage. Customers fall under the mixed residual waste collection category if they dispose of their e-waste in the same trash cans as other residential or business garbage. It is improbable that the e-waste collected via this technique will be segregated; instead, it will either be disposed of with other debris or burned in an incinerator. Studies show that there are several approaches for treating and disposing of e-waste, including burning, open dumping, landfilling, and mechanical-physical and chemical separation. Recycling procedures can also be classified as physical, chemical, or biological processes based on the processes involved (Li et al., 2012, Li et al., 2014).

Global solid waste management includes electronic waste as a significant component. E-waste, which makes up a sizable portion of solid waste, is full of toxic substances that harm humans, microorganisms, and plants (Bala et al., 2024; Das et al., 2024a; Mohanty et al., 2018). Mercury in soils can be converted by soil bacteria into methyl mercury, a substance that may be hazardous to humans and other animals. It gathers in their adipose tissues. It has been discovered that fetuses and young children are more vulnerable. Numerous other illnesses, including skin irritation, neurological disorders, memory problems, cognitive impairments, etc., are also brought on by it. As, Cu, Cr, Ni, and other heavy metals (HMs) can have a direct or indirect negative impact on humans and other living things(Coelho et al., 2013; Popoola et al., 2019).

2 Sources of E-Waste

Annually, up to 60 MT of e-waste are produced and an assessed of 74 MT of e-waste would be produced annually by 2030. Emerging nations, which are becoming more integrated into the world economy, are probably going to experience most of the growth. By contrast, the average for the 1.2 billion people living in Africa produced only 2.5 kg, while the EU’s figure is 16.2 kg. The EPA in the US estimates that the country produced around 4 MT of e-waste in 2015, with an expected 4–6% annual rise. The manufacture and usage of electronics, such as PCs, laptops, displays, cell phones, and tablets, and their production, will account for 14% of all emissions by 2040 (Sonawane et al., 2023). Table 1 describes the e-waste distribution in different nations.

The nation’s economic rate has improved because of the exponential growth of the IT sector’s economy, mostly driven by the need for electronic garbage, namely for laptops, computers, solar cells, and cell phones.

Table 1 Research on e-waste distribution in chosen nations

Principal region	Categories of e-waste examined	Study site	References
Behavior and disposal of e-waste	Cell phones, general	Malaysia, India, China	Wei and Liu (2012) and Wibowo and Deng (2013)
E-waste issues and difficulties	General	Malaysia	Ismail and Hanafiah (2019)
Generation of e-waste	TVs, air conditioners, microwaves, dryers, vacuum cleaners, washing machines, computer systems, and cell phones	India, Indonesia, Chaina, Australia	
Laws pertaining to e-waste	General, home appliances, printers, laptops, and mobile phones, household hazardous waste	China, Malaysia, India	Pathak et al. (2017)
Management of e-waste	Mobile phones, home appliances, solar panel and batteries for energy storage, TVs, laptops, printers, and IT accessories, as well as garbage from televisions and computers	Australia, India, China	Jaibee et al. (2015)
Health and environmental effects of e-waste	General	China, India	Medicine (2010)

2.1 E-Waste from Households

Considering that Families generate the least amount of electronic garbage. It is challenging to determine the precise capacity of e-waste generated. A review indicates that home utilizations such as refrigerators, generators, computers, and so on are not the main sources of output. Only about 20–21% of it comes from other sectors, and they are the ones who produce e-waste(Mundada et al., 2004).

2.2 Business Sector Electronic Waste

This industry comprises public and private sectors, government agencies, multinational corporations, etc. Considering that this industry is the primary driver of e-waste creation. According to the survey, they are the most affected of all the industries, accounting for almost 79% of all installed PCs. About 1.38 million old PCs are getting disposed of in homes and in this industry (Lebel, 2012; Mundada et al., 2004).

2.3 Manufacturers and Retailers

In this industry, a wide range of peripheral goods, motherboards, IC chips, PCs, and cathode ray tubes (CRTs) are created or manufactured. The e-waste produced by this industry is the second biggest donor on this list. It has been reported that the yearly disposal of these auxiliary devices exceeds 1050 million tonnes.

2.4 E-Waste Imports

It's widely accepted that the most lucrative industry is unquestionably rubbish imports. The major objective of importing old electronics is to recover precious metals and elements, such as steel, copper, mercury, aluminum, cadmium, and so on, that are mostly found in electronic waste. These goods offer valuable raw materials that can be used to create new goods. Various polymers and other materials are extracted from garbage products in numerous markets.

3 Causes of Electronics Waste

1. The world's waste problem is electronics waste.
2. This is not just a financial crisis; it's also a natural disaster caused by toxins that carry a danger to the environment and the associated physical conditions.
3. Rapid technological change in media (software, MP3, cassettes), falling prices, low initial cost, and increased rate of undesirability have caused a rapidly increasing global problem.
4. The rapid advancement of science and technology has led to a surge in electronic devices, which facilitates the disposal of outdated equipment, often known as e-waste.
5. Human attitudes also contribute to e-waste since people often buy new products to give as gifts.

4 Pollution of the Environment by Electronic Waste

Data from various nations has demonstrated the extensive extent of e-waste contamination of the environment. The degree of contamination raises questions about environmental health and necessitates immediate action and cleanup (Miyabara et al., 2005, Nishimura et al., 2017). The so-called "backyard activities," or unofficial recycling and dismantling of e-waste, have led to a rise in the environmental damage caused by e-waste. To recover materials from e-waste, these rudimentary recycling operations are being carried out across Asia and Africa. Among the main e-waste components that are sorted during this informal recycling are gold and silver.

Numerous investigations have demonstrated evidence of heavy metal contamination in soil resulting from a range of electronic operations, including manufacturing, selling, servicing, disposing of, and recycling. Comparing the levels of heavy metals in Guiyu, a significant recycling facility in China, to Bangalore, a similar study revealed higher levels. A workshop that facilitates informal e-waste recycling is located approximately 2 km from farmland that contains 191–9156 ng/g (dry weight) of PBDE. Pb concentrations in wastewater from Guiyu and India were found to be 17–247 times higher than in other locations where there were no known electronic activities. Heavy metals, particularly Pb, were found in higher amounts in sediments taken from cyanide leaching areas, the Lianjing River, and areas close to open burning sites in China than in other non-electronic activity sites. PBDE bioaccumulation has also been documented in the Nanyang River, which is close to Guiyu. The scientists concluded that the research areas' propensity for electronic activity was the cause of these increased concentrations. The Agbogboloshie market in Ghana has lagoon sediment next to disposal and burning areas. The sediment was found to have elevated levels of copper, lead, zinc, and copper. According to an Accra report, PCB levels in e-waste contaminated sediment ranged from 0.57 to 32.2 ng/g dw. These levels were either comparable to or greater than those reported in China's large e-waste sites but lower than those discovered in India, Senegal, Vietnam, and Japan.

5 Plant and Animal Poisoning Caused by E-Waste Pollution

There is proof that the components of e-waste have contaminated plants. It has been shown that plants and their surroundings contain high concentrations of PBDEs and PCBs (Liu et al., 2008). Research has demonstrated that PBDEs can go from the soil to plants. Even if the bioaccumulation coefficient of these pollutants is less than 0.01, plant absorption may provide a potential pathway for their entry into the food chain (Robinson, 2009). A study on the contamination of rice by e-waste revealed that the levels of Pb and Cd in food samples were 2–4 times higher than the maximum permitted limits of these metals (0.2 mg/kg) (Fu et al., 2008). Another study found that rice germination was slowed down in paddy soil next to a region that recycles electronic garbage. Furthermore, vegetables that are collected from e-waste sites have higher amounts of PAE (Ma et al., 2013). Edible amaranthus species with high concentrations of Cd, Cr, Ni, PAHs, PCBs, and PBDEs were cultivated in Nigeria on e-waste-contaminated soil (Alabi et al., 2012). *Eleusine indica* and *Ipomea batatas* from e-waste dumpsites had leaves, stems, and roots that contained considerably high quantities of Cr, Pb, Ni, and Cd. These concentrations showed a positive link with the metals in the soil where the plants were gathered. Studies on the toxicity of e-waste to plants are scarce. The cytogenotoxic effects of raw and simulated e-waste leachates from Alaba International Electronic Market and e-waste-contaminated subterranean water in *A. cepa* (Alabi et al., 2012).

Animals, both aquatic and terrestrial, have been observed to contain e-waste elements when they are located in or near e-waste processing facilities. Water snakes caught in an aquatic setting near an e-waste recycling factory, peregrine falcons, ducks, Virginia freshwater fish, fish taken from rivers, farmed rainbow trout, and wild whitefish have all been shown to bioaccumulate PBDEs(Hale et al., 2001; Zennegg et al., 2003). There have also been reports of fish, snails, and the surrounding environment containing elevated quantities of PCBs and PBDEs. When PBDEs were analyzed in chicken tissues, the concentration of the toxin was as high as 18 ng/g(Liang et al., 2008; Zhao et al., 2006). This high concentration of PBDEs raises concerns about potential threats to the ecosystem and public health. In addition to the chemical burden on animals found in e-waste, research has demonstrated the toxicity of e-waste exposure in a variety of animal models. It has been noted that mice treated with simulated and unprocessed leachates from unofficial e-waste disposal sites developed aberrant sperm morphology, MN, and chromosomal abnormalities (CA) (Alabi et al., 2014; Alabi & Bakare, 2011). The fish exposed to the e-waste elements in the water were found to have nuclear abnormalities and genetic damage, according to a study linking the two. Evaluation of oxidative biomarkers, including superoxide dismutase levels, lipoperoxidation, catalase, and reduced glutathione, suggested that oxidative stress could be a potential mechanism behind the genotoxicity that has been found (Bakare et al., 2013; Song et al., 2012).

6 Human Health Consequences Due to E-Waste

6.1 Routes of Human Exposure to E-Waste

The three major ways that people can be exposed to the dangerous materials that originate in e-waste are ingestion, skin contact, and inhalation. There are additional exposure hazards for fetuses, children, old individuals, persons with disabilities, and personnel in the e-waste recycling business. When parents bring e-waste home contaminated by it on their skin or clothing, it exposes kids to additional sources of exposure. This is made much worse when recycling is done within these houses because of the direct, high-level exposure.

7 The Amount of E-Waste Components in Human Bodies

Several nations have reported data on the presence of PBDEs in human biological samples. An increased level of this pollutant was found in the breast milk samples of women from Venice and Rome, Catalonia, Japan, Spain, the United States, and

Sweden, according to an analysis of the samples (Ingelido et al., 2007). Human adipose tissues, cord blood, and the liver tissues of stillborn fetuses have all been shown to contain elevated levels of PBDEs (Johnson-Restrepo et al., 2005; Schecter et al., 2007). Approximately 90% of the human burden is attributed to the five most common congeners identified in human tissues, namely BDE-47, 99, 100, 153, and 154 (Qu et al., 2007; Zhao et al., 2008). This concentration is approximately 50 times higher than that of participants exposed to contaminated ambient air in Ryugasaki and Tsukuba, and it is comparable to that discovered in hair samples taken from individuals in a substantially contaminated Chinese location. Since those who engage in open burning of e-waste typically have the highest blood Pb concentrations, open burning may present the greatest health danger. Studies revealed that employees at e-waste processing facilities had higher blood lead levels than non-e-waste employees. Fe, Cu, Pb, and Al concentrations in samples taken from e-waste factory workers exceeded the upper bounds established by the American Conference of Governmental Industrial Hygienists. In addition to heavy metals, researchers found that workers were directly exposed to PCDD, particularly PCDF combustion products. The length of time individuals worked at the e-waste site was correlated with their PCDD/F blood levels. Teenage scavengers at e-waste dumpsites had peripheral blood samples that contained noticeably elevated quantities of Pb, Ni, Cd, and Cr, according to a recent survey (Alabi et al., 2020; Kim et al., 2020). The length of time spent scavenging the e-waste was found to be related to the raised blood metal content. Figure 1 designates the e-waste sources, transport, and impact on the environment and human health.

8 A Case Study Demonstrating the Harmful Effects of E-Waste on People

An elevated level of placental Cd has been linked to the potential for e-waste to shorten placental telomeres. There was a correlation found between the telomere margarine of the fetus and newborn and a rise in leukemia cases among children in the study area. Workers at e-waste recycling facilities have reported experiencing symptoms such as blurred vision, rash/itching, numbness in the hands and feet, headaches, reduced gonadal hormones, symptoms related to the chest and respiratory tract, nausea, allergies, pains, migraines, and cancer. It has also been demonstrated that male humans may be susceptible to prostate cancer from exposure to e-waste. When compared to participants who were not exposed to e-waste, the exposed group had considerably greater levels of prostate-specific antigen and alpha-fetoprotein. Adults and children who are exposed to dust PBDEs have impaired thyroid and reproductive functions. Research has been done on the effects of e-waste exposure on pregnancy outcomes. Several investigations have shown that a persistent exposure impact is associated with a higher incidence of stillbirths, spontaneous abortions, early births, lower birth weights, and longer birth durations,

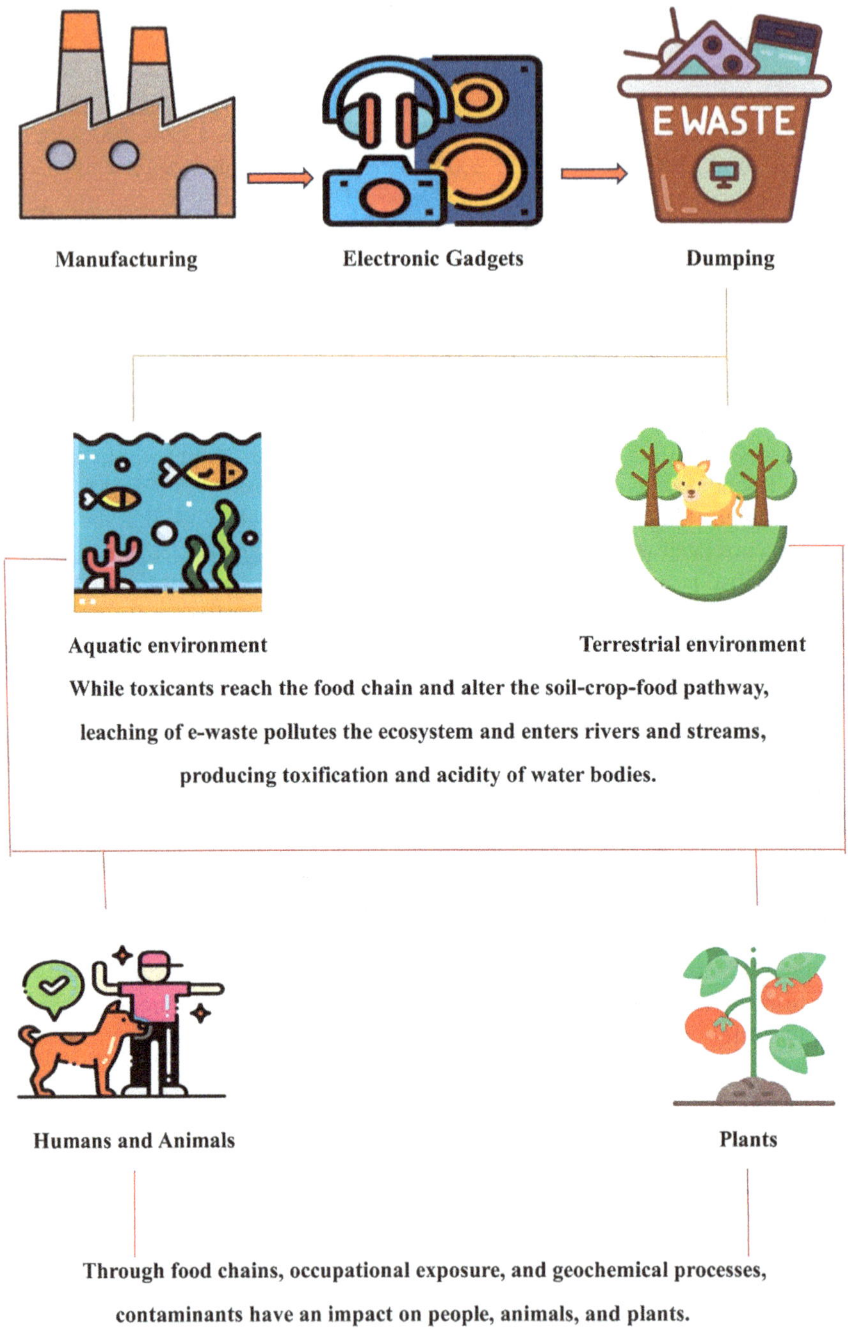

Fig. 1 Describes the e-waste sources, transport, and impact on the environment and human health

even in the face of variations in the toxicants examined and exposure circumstances. There is proof that higher levels of PAH and POPs, such as PCBs, perfluroalkyls, and PBDEs, are linked to unfavorable birth outcomes (Guo et al., 2012). Likewise, strong correlations have been observed globally between e-waste recycling initiatives and temperamental changes in kids. E-waste has been shown to have the capacity to harm DNA. It was recently discovered that the chromosomal abnormalities of the workers in the Junhai County e-waste recycling villages were 20 times more than those of the control villagers. Forty-nine e-waste workers showed noticeably greater amounts of micronucleated binucleated cells than the control group. When compared to participants without such a background, those who had previously worked with e-waste had a 28-fold higher chance of micronucleated binucleated cells. In addition, the workers' elevated levels of thyroid hormone were linked to the e-waste workers' elevated blood levels of PBDE. When compared to the control group, residents near the e-waste recycling plant experienced an increase in the MN rate due to e-waste. Additionally, it was shown that men had a higher frequency of genetic abnormalities than women, indicating that men might be more vulnerable to health issues linked to e-waste pollution. There is evidence of a rise in the frequency of e-waste exposed persons' peripheral blood micronucleated binucleated cells (He et al., 2015; Liu et al., 2009; Zhang et al., 2010).

It has also been demonstrated that exposure to e-waste can either up- or down-regulate the expression of specific genes.

8.1 Control Measures

One of the innovative methods that offers advantages for the environment, public health, and economy is the bioprospecting of e-waste. E-waste management and disposal is a significant undertaking in and of itself, especially because industrialization and population growth are increasing people's dependence on electronic devices and equipment. Nowadays, e-waste is a necessary component of solid trash. However, there are still relatively few approaches that deal with it, such as incineration and landfilling, each of which has pros and cons of its own.

The following suggestions are meant to protect the environment and public health from harmful impacts and pollution caused by e-waste:

1. The implementation of a well-coordinated management strategy by governments across many nations is necessary to prevent the careless disposal of electronic trash into the environment.
2. To protect workers and locals from additional exposure to e-waste toxins, it is necessary to clean up tainted soil and water in locations where e-waste is generated, among other comparable locations.
3. To keep residents safe from e-waste pollutants, residential areas, and recycling facilities for e-waste must be kept apart.

4. Governments in nations where e-waste recycling occurs are required to ensure that personnel are wearing safety gear and that the recycling process adheres to state-of-the-art standards to minimize or completely eradicate environmental contamination.
5. The general people must be made aware of the risks linked with handling e-waste. Both employees and inhabitants of e-waste activity zones must comply with this. Additionally, residents need to be taught not to store their electronic garbage in their attics or backyards or to dispose of it with regular trash.
6. Implement policies that will promote the use and import of EEE made using recyclable, non-hazardous chemicals in compliance with the EU RoHS Directive.
7. Extended Producer Responsibility must be put into effect, requiring importers, merchants, and producers to handle the recycling, disposal, and collection of end-of-life electronic equipment.
8. The duty of the citizen: The public's complete cooperation is necessary for EPR to succeed. To encourage recycling by the proper organizations, each citizen should be accountable for the kind of EEE they buy, including whether it contains harmful materials and how they release such products after using them.

9 Conclusion

Nearly every nation is struggling to contract with the rapidly rising amount of e-waste. Sustainable e-waste management has emerged as a pressing necessity. E-waste has several negative externalities associated with it. It extends beyond that since it has an influence on both environmental and human health. Many countries prioritize controlling municipal solid garbage over e-waste, so ignoring the compounding issue of e-waste. A significant quantity of e-waste is handled informally and without authorization in most emerging economies. The absence of official e-waste recycling facilities is a result of ineffective technology and a negligent attitude from the government. In addition, recyclers encounter difficulties obtaining industry certifications necessary for responsibly recycling e-waste. A turning point in this regard could come from the informal participants becoming small-time entrepreneurs. This can contribute to the attainment of the 2030 global sustainable development goals, which include fewer health issues and fewer fatalities from leachates, which are poisonous and dangerous materials. Campaigns to raise general awareness among suppliers and users may be advantageous. Massive amounts of garbage and leftover materials from repairs also wind up in roadsides, workshops, open fields, backyards, irrigation canals, ponds, and riverbanks. Respiratory and neurological issues are common among the local population. When processing e-waste, the presence of elements with low melting points and high vapor pressure, such as Sn, Pb, and Zn, should be kept to a minimum to reduce the emission of potentially harmful elements. To manage e-waste, it is imperative to replace antiquated, traditional methods with cutting-edge, environmentally friendly alternatives. It is also important to remember that no one problem can be solved with a single solution. Different HMs call for different removal procedures. Furthermore, there are two sides to the e-waste problem addressing process. To prevent the difficulties, both the government and the people must fulfill their respective roles.

References

Alabi, O. A., & Bakare, A. A. (2011). Genotoxicity and mutagenicity of electronic waste leachates using animal bioassays. *Toxicological and Environmental Chemistry, 93*(5), 1073–1088. https://doi.org/10.1080/02772248.2011.561949

Alabi, O., Bakare, A., Xu, X., Li, B., et al. (2012). *Comparative evaluation of environmental contamination and DNA damage induced by electronic-waste in Nigeria and China*. Retrieved August 22, 2024, from https://www.sciencedirect.com/science/article/pii/S0048969712001374

Alabi, O., et al. (2014). Cytogenotoxic effects and reproductive abnormalities induced by e-waste contaminated underground water in mice. *Cytologia*. https://doi.org/10.1508/cytologia.79.331.

Alabi, O. A., Adeoluwa, Y. M., & Bakare, A. A. (2020). Elevated serum Pb, Ni, cd, and Cr levels and DNA damage in exfoliated buccal cells of teenage scavengers at a major electronic waste dumpsite in Lagos, Nigeria. *Biological Trace Element Research, 194*(1), 24–33. https://doi.org/10.1007/S12011-019-01745-Z

Ankit, Saha, L., Kumar, V., Tiwari, J., Sweta, Rawat, S., Singh, J., & Bauddh, K. (2021). Electronic waste and their leachates impact on human health and environment: Global ecological threat and management. *Environmental Technology & Innovation, 24*, 102049. https://doi.org/10.1016/J.ETI.2021.102049

Bakare, A. A., Alabi, O. A., Gbadebo, A. M., Ogunsuyi, O. I., & Alimba, C. G. (2013). In vivo cytogenotoxicity and oxidative stress induced by electronic waste leachate and contaminated well water. *Challenges, 4*(2), 169–187. https://doi.org/10.3390/challe4020169

Bal, B., Ghosh, S., & Das, A. P. (2018). Microbial recovery and recycling of manganese waste and their future application: A review. *Geomicrobiological Journal, 36*(1), 85–96. https://doi.org/10.1080/01490451.2018.1497731

Bala, K., Anshul, Verma, S., Dogra, A., Das, A. P., Kumar, R., & Nandini. (2024). Sustainable management for urban plastic waste generation. In *Sustainable management of urban plastic waste through circular economic approaches* (pp. 211–236). https://doi.org/10.1201/9781003465447-13

Barney, G. O. (2013). Environment projections. In *The global 2000 report to the president of the U.S.* (pp. 227–449). https://doi.org/10.1016/B978-0-08-024618-5.50019-1

Coelho, M., Oliveira, T., & Fernandes, R. (2013). Biochemistry of adipose tissue: An endocrine organ. *Archives of Medical Science: AMS, 9*(2), 191. https://doi.org/10.5114/AOMS.2013.33181

Compd, W. L.-O. (2002). Polychlorinated dibenzodioxin and dibenzofuran (PCDD/PCDF) levels in environmental and human hair samples around electronic waste processing site. *Cir.Nii.Ac.Jp*. Retrieved August 22, 2024, from https://cir.nii.ac.jp/crid/1572543025258761856

Das, A. P., & Ghosh, S. (2018). Bioleaching of manganese from mining waste materials. *Materials Today: Proceedings, 5*(1), 2381–2390. https://doi.org/10.1016/J.MATPR.2017.11.459

Das, A. P., & Ghosh, S. (2023a). Cost-effectiveness of the biomining process. In *Manganese mining microorganisms* (pp. 143–151). https://doi.org/10.1016/B978-0-12-822147-1.00008-X

Das, A. P., & Ghosh, S. (2023b). Factors affecting bioleaching. In *Manganese mining microorganisms* (pp. 77–86). https://doi.org/10.1016/B978-0-12-822147-1.00004-2

Das, A. P., & Ghosh, S. (2023c). Manganese and its application. In *Manganese mining microorganisms* (pp. 29–40). https://doi.org/10.1016/B978-0-12-822147-1.00007-8

Das, A. P., & Ghosh, S. (2023d). Mechanism of manganese bioleaching. In *Manganese mining microorganisms* (pp. 87–98). https://doi.org/10.1016/B978-0-12-822147-1.00014-5

Das, A. P., & Ghosh, S. (2023e). Metagenomic insights into unculturable microbial diversity. In *Manganese mining microorganisms* (pp. 63–76). https://doi.org/10.1016/B978-0-12-822147-1.00002-9

Das, A. P., & Ghosh, S. (2023f). Microorganisms in manganese biomining processes. In *Manganese mining microorganisms* (pp. 55–61). https://doi.org/10.1016/B978-0-12-822147-1.00013-3

Das, A. P., & Ghosh, S. (2023g). Mixed culture bioleaching: An insight into manganese biomining process and efficacy. In *Manganese mining microorganisms* (pp. 129–141). https://doi.org/10.1016/B978-0-12-822147-1.00001-7

Das, A. P., & Ghosh, S. (2023h). Role of metabolites and metabolomics for understanding of manganese bioleaching. In *Manganese mining microorganisms* (pp. 117–127). https://doi.org/10.1016/B978-0-12-822147-1.00011-X

Das, A. P., & Mishra, S. (2008). Hexavalent chromium (VI): Health hazards and environmental pollutant. *J Environ Res Dev, 2*(3), 386–392.

Das, A. P., & Mishra, S. (2009). Hexavalent chromium [Cr (VI)]: Yellow water pollution and its remediation. *Sarovar Saurabh the ENVIS Newsletter on Wetland Ecosystems, 5*(2), 1.

Das, A. P., & Shreya, G. (2023). *Manganese mining microorganisms*. Elsevier.

Das, A. P., Sukla, L. B., Pradhan, N., & Nayak, S. (2011). Manganese biomining: A review. *Bioresource Technology, 102*(16), 7381–7387. https://doi.org/10.1016/J.BIORTECH.2011.05.018

Das, A. P., Sukla, L. B., & Pradhan, N. (2012a). Microbial recovery of manganese using *Staphylococcus epidermidis*. *International Journal of Nonferrous Metallurgy, 2012*(02), 9–12. https://doi.org/10.4236/IJNM.2012.12002

Das, A. P., Swain, S., Panda, S., Pradhan, N., & Sukla, L. B. (2012b). Reductive acid leaching of low grade manganese ores. *Geomaterials, 2012*(04), 70–72. https://doi.org/10.4236/GM.2012.24011

Das, A. P., Ghosh, S., Mohanty, S., & Sukla, L. B. (2015a). Advances in manganese pollution and its bioremediation. In L. B. Sukla, N. Pradhan, S. Panda, & B. K. Mishra (Eds.), *Environmental microbe biotechnology* (1st ed., pp. 313–328). Springer. https://doi.org/10.1007/978-3-319-19018-1_16

Das, A. P., Ghosh, S., Mohanty, S., & Sukla, L. B. (2015b). Consequences of manganese compounds: A review. *Toxicological & Environmental Chemistry, 96*(7), 981–997. https://doi.org/10.1080/02772248.2015.1005428

Das, A. P., Behera, I. D., & Das, N. P. (2024a). *Renewable energy generation and value addition from environmental microfiber pollution through advanced greener solution* (p. 278).

Das, A., Ray, S. K., Mohanty, M., Mohanty, J., Dey, S., & Das, A. P. (2024b). Ecotoxicity of microplastic wastes and their sustainable management: A review. *Environmental Chemistry and Ecotoxicology, 6*, 144–152. https://doi.org/10.1016/J.ENCECO.2024.05.003

Dey, S., Das, A., Mallick, K., Sahu, A., & Das, A. P. (2023a). *Environmental petroleum waste: Pollution, toxicity, sustainable remediation* (pp. 159–175). https://doi.org/10.1007/978-3-031-48220-5_7

Dey, S., Tripathy, B., Kumar, M. S., & Das, A. P. (2023b). Ecotoxicological consequences of manganese mining pollutants and their biological remediation. *Environmental Chemistry and Ecotoxicology., 5*, 55. https://doi.org/10.1016/J.ENCECO.2023.01.001

Dutta, D., Rautela, R., Gujjala, L. K. S., Kundu, D., Sharma, P., Tembhare, M., & Kumar, S. (2023). A review on recovery processes of metals from E-waste: A green perspective. *Science of the Total Environment, 859*, 160391. https://doi.org/10.1016/J.SCITOTENV.2022.160391

Fu, J., Zhou, Q., Liu, J., Liu, W., Wang, T., Zhang, Q., Chemosphere, G. J. (2008). *High levels of heavy metals in rice (Oryzasativa L.) from a typical E-waste recycling area in southeast China and its potential risk to human health*. Retrieved August 22, 2024, from https://www.sciencedirect.com/science/article/pii/S0045653507015317

Ghosh, S., & Das, A. P. (2017). Bioleaching of manganese from mining waste residues using Acinetobacter sp. *Geology, Ecology, and Landscapes, 1*(2), 77–83. https://doi.org/10.1080/24749508.2017.1332847

Ghosh, S., & Das, A. P. (2018). Metagenomic insights into the microbial diversity in manganese-contaminated mine tailings and their role in biogeochemical cycling of manganese. *Scientific Reports, 8*(1), 1–12. https://doi.org/10.1038/s41598-018-26311-w

Ghosh, S., Mohanty, S., Akcil, A., Sukla, L. B., & Das, A. P. (2016a). A greener approach for resource recycling: Manganese bioleaching. *Chemosphere, 154*, 628–639. https://doi.org/10.1016/J.CHEMOSPHERE.2016.04.028

Ghosh, S., Mohanty, S., Nayak, S., Sukla, L. B., & Das, A. P. (2016b). Molecular identification of indigenous manganese solubilising bacterial biodiversity from manganese mining deposits. *Journal of Basic Microbiology, 56*(3), 254–262. https://doi.org/10.1002/JOBM.201500477

Ghosh, S., Bal, B., & Das, A. P. (2017). Enhancing manganese recovery from low-grade ores by using mixed culture of indigenously isolated bacterial strains. *Geomicrobiology, 35*(3), 242–246. https://doi.org/10.1080/01490451.2017.1362080

Ghosh, S., Gandhi, M., van Hullebusch, E. D., & Das, A. P. (2020). Proteomic insights into Lysinibacillus sp.-mediated biosolubilization of manganese. *Environmental Science and Pollution Research, 28*(30), 40249–40263. https://doi.org/10.1007/S11356-020-10863-4

Ghosh, S., Sinha, J. K., Ghosh, S., Vashisth, K., Han, S., & Bhaskar, R. (2023a). Microplastics as an emerging threat to the global environment and human health. *Sustainability, 15*(14), 10821. https://doi.org/10.3390/SU151410821

Ghosh, S., Tripathy, B., Dey, S., & Das, A. P. (2023b). Comparative investigation of fungal and bacterial manganese biomining mechanisms. In *Manganese mining microorganisms* (pp. 99–115). https://doi.org/10.1016/b978-0-12-822147-1.00016-9

Guo, Y., Huo, X., Wu, K., Liu, J., & Zhang, Y. (2012). Carcinogenic polycyclic aromatic hydrocarbons in umbilical cord blood of human neonates from Guiyu, China. *Science of the Total Environment*. Retrieved August 23, 2024, from https://www.sciencedirect.com/science/article/pii/S0048969712005025

Ha, N. N., Agusa, T., Ramu, K., Tu, N. P. C., Murata, S., Bulbule, K. A., Parthasaraty, P., Takahashi, S., Subramanian, A., & Tanabe, S. (2009). Contamination by trace elements at e-waste recycling sites in Bangalore, India. *Chemosphere, 76*(1), 9–15. https://doi.org/10.1016/J.CHEMOSPHERE.2009.02.056

Hale, R. C., La Guardia, M. J., Harvey, E. P., Mainor, T. M., Duff, W. H., & Gaylor, M. O. (2001). Polybrominated diphenyl ether flame retardants in Virginia freshwater fishes (USA). *Environmental Science and Technology, 35*(23), 4585–4591. https://doi.org/10.1021/ES010845Q

He, X., Jing, Y., Wang, J., Li, K., Yang, Q., & Zhao, Y. (2015). Significant accumulation of persistent organic pollutants and dysregulation in multiple DNA damage repair pathways in the electronic-waste-exposed populations. *Environmental Research*. Retrieved August 23, 2024, from https://www.sciencedirect.com/science/article/pii/S0013935114004332

Hou, R., Huo, X., Zhang, S., Xu, C., & Huang, Y. (2020). Elevated levels of lead exposure and impact on the anti-inflammatory ability of oral sialic acids among preschool children in e-waste areas. *Science of the Total Environment*. Retrieved August 23, 2024, from https://www.sciencedirect.com/science/article/pii/S0048969719343712

Huo, X., Dai, Y., Yang, T., Zhang, Y., & Li, M. (2019). Decreased erythrocyte CD44 and CD58 expression link e-waste Pb toxicity to changes in erythrocyte immunity in preschool children. *Science of the Total Environment, 664*, 690–697. https://doi.org/10.1016/j.scitotenv.2019.02.040

Ingelido, A., Ballard, T., Dellatte, E., & Chemosphere, A. D. (2007). Polychlorinated biphenyls (PCBs) and polybrominated diphenyl ethers (PBDEs) in milk from Italian women living in Rome and Venice. *Chemosphere*. Retrieved August 23, 2024, from https://www.sciencedirect.com/science/article/pii/S0045653506016250

Ismail, H., & Hanafiah, M. M. (2019). Discovering opportunities to meet the challenges of an effective waste electrical and electronic equipment recycling system in Malaysia. *Journal of Cleaner Production, 238*, 117927. https://doi.org/10.1016/J.JCLEPRO.2019.117927

Jaibee, S., Khalil, A., Rahim, A., Mohamad, F., Jamian, S., Chee Kiong, S., Seiji, Y., Hisyamudin, N., & Nor, M. (2015). Review on current status of waste electric and electronic product in Malaysia. *Applied Mechanics and Materials, 773–774*, 898–907. https://doi.org/10.4028/www.scientific.net/AMM.773-774.898

Johnson-Restrepo, B., Kannan, K., Rapaport, D. P., & Rodan, B. D. (2005). Polybrominated diphenyl ethers and polychlorinated biphenyls in human adipose tissue from New York. *Environmental Science and Technology, 39*(14), 5177–5182. https://doi.org/10.1021/ES050399X

Kim, S., Xu, X., Zhang, Y., Zheng, X., et al. (2020). Birth outcomes associated with maternal exposure to metals from informal electronic waste recycling in Guiyu, China. *Environment International*. Retrieved August 23, 2024, from https://www.sciencedirect.com/science/article/pii/S0160412019337754

Lebel, S. (2012) Wasting the future: The technological sublime, communications technologies, and e-waste. *Scholarworks.Umass.Edu*. Retrieved August 22, 2024, from https://scholarworks.umass.edu/cgi/viewcontent.cgi?article=1003&context=cpo

Li, Y., Li, M., Liu, Y., Song, G., & Liu, N. (2012). A microarray for MicroRNA profiling in spermatozoa from adult men living in an environmentally polluted site. *Bulletin of Environmental Contamination and Toxicology, 89*(6), 1111–1114. https://doi.org/10.1007/S00128-012-0827-0

Li, K., Liu, S., Yang, Q., Zhao, Y., & Zuo, J. (2014). Genotoxic effects and serum abnormalities in residents of regions proximal to e-waste disposal facilities in Jinghai, China. *Ecotoxicology and Environmental Safety*. Retrieved August 23, 2024, from https://www.sciencedirect.com/science/article/pii/S0147651314001390

Liang, S. X., Zhao, Q., Qin, Z. F., Zhao, X. R., Yang, Z. Z., & Xu, X. B. (2008). Levels and distribution of polybrominated diphenyl ethers in various tissues of foraging hens from an electronic waste recycling area in South China. *Environmental Toxicology and Chemistry, 27*(6), 1279–1283. https://doi.org/10.1897/07-518.1

Liu, H., Zhou, Q., Wang, Y., & Zhang, Q. (2008). E-waste recycling induced polybrominated diphenyl ethers, polychlorinated biphenyls, polychlorinated dibenzo-p-dioxins and dibenzofurans pollution in the ambient. *Environment International*. Retrieved August 22, 2024, from https://www.sciencedirect.com/science/article/pii/S016041200700133X

Liu, Q., Cao, J., Li, K. Q., Miao, X. H., Li, G., Fan, F. Y., & Zhao, Y. C. (2009). Chromosomal aberrations and DNA damage in human populations exposed to the processing of electronics waste. *Environmental Science and Pollution Research, 16*(3), 329–338. https://doi.org/10.1007/S11356-008-0087-Z

Ma, T. T., Christie, P., Luo, Y. M., & Teng, Y. (2013). Phthalate esters contamination in soil and plants on agricultural land near an electronic waste recycling site. *Environmental Geochemistry and Health, 35*(4), 465–476. https://doi.org/10.1007/S10653-012-9508-5

Medicine, J. K. (2010). E-waste management: As a challenge to public health in India. Retrieved August 23, 2024, from https://journals.lww.com/ijcm/fulltext/2010/35030/e_waste_management__as_a_challenge_to_public.4.aspx

Mishra, S., Rout, P. K., & Das, A. P. (2019). Solar photovoltaic panels as next generation waste: A review. *Biointerface Research in Applied Chemistry, 9*(6), 4539.

Mishra, S., Swain, S., Sahoo, M., Mishra, S., & Das, A. P. (2021). Microbial colonization and degradation of microplastics in aquatic ecosystem: A review. *Geomicrobiology Journal, 39*, 259. https://doi.org/10.1080/01490451.2021.1983670

Mishra, S., Ghosh, S., van Hullebusch, E. D., Singh, S., & Das, A. P. (2023). A critical review on the recovery of base and critical elements from electronic waste-contaminated streams using microbial biotechnology. *Applied Biochemistry and Biotechnology, 2023*, 1–30. https://doi.org/10.1007/S12010-023-04440-X

Miyabara, Y., Nishimura, N., & Tohyama, C. (2005). Determination of dioxins in human hair: Estimation of external and internal exposure to dioxins. *Environmental Health and Preventive Medicine, 10*(2), 86–93. https://doi.org/10.1007/BF02897998

Mohanty, S., Ghosh, S., Nayak, S., & Das, A. P. (2016). Isolation, identification and screening of manganese solubilizing fungi from low-grade manganese ore deposits. *Geomicrobiological Journal, 34*(4), 309–316. https://doi.org/10.1080/01490451.2016.1189016

Mohanty, S., Ghosh, S., Nayak, S., & Das, A. P. (2017). Bioleaching of manganese by aspergillus sp. isolated from mining deposits. *Chemosphere, 172*, 302–309. https://doi.org/10.1016/J.CHEMOSPHERE.2016.12.136

Mohanty, S., Ghosh, S., Bal, B., & Das, A. P. (2018). A review of biotechnology processes applied for manganese recovery from wastes. *Reviews in Environmental Science and BioTechnology, 17*(4), 791–811. https://doi.org/10.1007/S11157-018-9482-1

Mohanty, J., Dey, S., Bala, K., & Das, A. P. (2024a). Recent advances in microplastic pollution for its sustainable management. In *Sustainable management of urban plastic waste through circular economic approaches* (pp. 27–48). https://doi.org/10.1201/9781003465447-3

Mohanty, M., Tripathy, B., Koteswara Reddy, G., & Das, A. P. (2024b). Greener technologies for establishing restraint over microplastic pollution. In *Sustainable microbial technology for synthetic and cellulosic microfiber bioremediation* (pp. 239–258). https://doi.org/10.1007/978-3-031-62600-5_13

Mundada, M. N., Kumar, S., & Shekdar, A. V. (2004). E-waste: A new challenge for waste management in India. *International Journal of Environmental Studies, 61*(3), 265–279. https://doi.org/10.1080/0020723042000176060

Nishimura, C., Horii, Y., Tanaka, S., Asante, K. A., Ballesteros, F., Viet, P. H., Itai, T., Takigami, H., Tanabe, S., & Fujimori, T. (2017). Occurrence, profiles, and toxic equivalents of chlorinated and brominated polycyclic aromatic hydrocarbons in E-waste open burning soils. *Environmental Pollution, 225*, 252–260. https://doi.org/10.1016/J.ENVPOL.2016.10.088

Pathak, P., Srivastava, R. R., & Ojasvi. (2017). Assessment of legislation and practices for the sustainable management of waste electrical and electronic equipment in India. *Renewable and Sustainable Energy Reviews, 78*, 220–232. https://doi.org/10.1016/J.RSER.2017.04.062

Paul, A., Dey, S., Ram, D. K., & Das, A. P. (2023). Hexavalent chromium pollution and its sustainable management through bioremediation. *Geomicrobiology Journal, 1*, 324. https://doi.org/10.1080/01490451.2023.2218377

Popoola, O., et al. (2019). Levels of awareness and concentrations of heavy metals in the blood of electronic waste scavengers in Nigeria. *Journal of Health and Pollution*. Retrieved August 23, 2024, from https://meridian.allenpress.com/jhp/article/9/21/190311/67558

Prabhakar, A., Mishra, S., & Das, A. P. (2019). Isolation and identification of lead (Pb) solubilizing bacteria from automobile waste and its potential for recovery of Lead from end of life waste batteries. *Geomicrobiological Journal, 36*(10), 894–903. https://doi.org/10.1080/01490451.2019.1654044

Prabina, K. E., Benya, A., Mohanty, S., Van Hullebusch, E. D., Das, A. P., & Singh, S. (2024). Recovery of multi-functional carbon fiber from COVID-19 generated waste for an effective management through circular economy. *Environmental Science and Engineering, Part F, 2827*, 83–97. https://doi.org/10.1007/978-3-031-50840-0_6

Pradhan, G., Tripathy, B., Ram, D. K., Digal, A. K., & Das, A. P. (2023). Bauxite mining waste pollution and its sustainable management through bioremediation. *Geomicrobiology Journal, 1–10*, 335. https://doi.org/10.1080/01490451.2023.2235353

Priyadarshini, B., Santhosh Kumar, M., & Das, A. P. (2024). Microbial degradation of synthetic plastics released from COVID-19 generated single use disposable face masks. *Environmental Science and Engineering, Part F, 2827*, 1–16. https://doi.org/10.1007/978-3-031-50840-0_1

Priyadarsini, S., & Das, A. P. (2024). Analytical and structural characterization of waste lithium-ion batteries for their effective recycling strategy. *Environmental Chemistry and Ecotoxicology., 7*, 182. https://doi.org/10.1016/j.enceco.2024.12.004

Priyadarsini, S., & Das, A. P. (2025). Lithium bioleaching: A review on microbial-assisted sustainable technology for lithium bio-circularity. *Journal of Water Process Engineering, 69*, 106744. https://doi.org/10.1016/j.jwpe.2024.106744

Qu, W., Bi, X., Sheng, G., Lu, S., Fu, J., & Yuan, J. (2007). Exposure to polybrominated diphenyl ethers among workers at an electronic waste dismantling region in Guangdong, China. *Environment International*. Retrieved August 23, 2024, from https://www.sciencedirect.com/science/article/pii/S0160412007001043

Raj, K., & Das, A. P. (2023). Lead pollution: Impact on environment and human health and approach for a sustainable solution. *Environmental Chemistry and Ecotoxicology, 5*, 79–85. https://doi.org/10.1016/J.ENCECO.2023.02.001

Robinson, B. H. (2009). E-waste: An assessment of global production and environmental impacts. *The Science of the Total Environment, 408*(2), 183–191. https://doi.org/10.1016/J.SCITOTENV.2009.09.044

Sahu, S., & Das, A. P. (2024). *Microplastics degradation mechanism by aquatic Plastiomes: A review* (pp. 213–226). https://doi.org/10.1007/978-3-031-50840-0_12

Sanket, A. S., Ghosh, S., Sahoo, R., Nayak, S., & Das, A. P. (2016). Molecular identification of acidophilic manganese (Mn)-solubilizing bacteria from mining effluents and their application in mineral beneficiation. *Geomicrobiological Journal, 34*(1), 71–80. https://doi.org/10.1080/01490451.2016.1141340

Schecter, A., Johnson-Welch, S., Tung, K. C., Harris, T. R., Päpke, O., & Rosen, R. (2007). Polybrominated diphenyl ether (PBDE) levels in livers of U.S. human fetuses and newborns. *Journal of Toxicology and Environmental Health. Part A, 70*(1), 1–6. https://doi.org/10.1080/15287390600748369

Sonawane, B., Kadry, A.-R. M., & Kancherla, J. (2023). Electronic waste: Human health and environmental risk. In *Patty's toxicology* (pp. 1–18). https://doi.org/10.1002/0471125474.TOX163

Song, Y., Wu, N., Tao, H., Tan, Y., Gao, M., Han, J., Shen, H., Liu, K., & Lou, J. (2012). Thyroid endocrine dysregulation and erythrocyte DNA damage associated with PBDE exposure in juvenile crucian carp collected from an e-waste dismantling site in Zhejiang Province, China. *Environmental Toxicology and Chemistry, 31*(9), 2047–2051. https://doi.org/10.1002/ETC.1915

Toffel, M., et al. (2004). Environmental implications of wireless technologies: News delivery and business meetings. *Environmental Science & Technology, 38*(11), 2961–2970. https://doi.org/10.1021/es035035o

Tripathy, B., Dey, S., Dubey, N. K., Lo, W.-C., & Das, A. P. (2024). *An omics-based approach for managing the risk of COVID-19-generated wastes* (pp. 51–66). https://doi.org/10.1007/978-3-031-50840-0_4

Wei, L., & Liu, Y. (2012). Present status of e-waste disposal and recycling in China. *Procedia Environmental Sciences, 16*, 506–514. https://doi.org/10.1016/J.PROENV.2012.10.070

Wibowo, S., & Deng, H. (2013). An intuitionistic fuzzy multicriteria group decision model for solid waste disposal site selection. *Pattern Recognition Image Process*. Retrieved August 23, 2024, from https://www.researchgate.net/profile/Santoso-Wibowo/publication/239731560_An_Intuitionistic_Fuzzy_Multicriteria_Group_Decision_Model_for_Solid_Waste_Disposal_Site_Selection/links/004635276cfd38c15a000000/An-Intuitionistic-Fuzzy-Multicriteria-Group-Decision-Model-for-Solid-Waste-Disposal-Site-Selection

Zennegg, M., Kohler, M., & Gerecke, A. (2003). Polybrominated diphenyl ethers in whitefish from Swiss lakes and farmed rainbow trout. *Chemosphere*. Retrieved August 23, 2024, from https://www.sciencedirect.com/science/article/pii/S004565350300047X

Zhang, J., Jiang, Y., Zhou, J., Wu, B., Liang, Y., Peng, Z., Fang, D., Liu, B., Huang, H., He, C., Wang, C., & Lu, F. (2010). Elevated body burdens of PBDEs, dioxins, and PCBs on thyroid hormone homeostasis at an electronic waste recycling site in China. *Environmental Science and Technology, 44*(10), 3956–3962. https://doi.org/10.1021/ES902883A

Zhang, S., Ding, Y., Liu, B., Chang, C., & chi. (2017). Supply and demand of some critical metals and present status of their recycling in WEEE. *Waste Management, 65*, 113–127. https://doi.org/10.1016/J.WASMAN.2017.04.003

Zhao, G., Xu, Y., Han, G., & Ling, B. (2006). Biotransfer of persistent organic pollutants from a large site in China used for the disassembly of electronic and electrical waste. *Environmental Geochemistry and Health, 28*(4), 341–351. https://doi.org/10.1007/S10653-005-9003-3

Zhao, G., Wang, Z., Dong, M. H., Rao, K., Luo, J., Wang, D., Zha, J., Huang, S., Xu, Y., & Ma, M. (2008). PBBs, PBDEs, and PCBs levels in hair of residents around e-waste disassembly sites in Zhejiang Province, China, and their potential sources. *Science of the Total Environment*. https://doi.org/10.1016/j.scitotenv.2008.03.010

E-Waste: An Invisible Threat to Global Health and Ecosystem

Pragya Yadav and Shailesh Kumar Srivastava

Abstract Electronic garbage (e-waste) is a rapidly expanding global concern as the demand for electronic devices rises and their lifespans reduce. The rapid growth of the electronics sector has worsened the worldwide problem, especially in growing nations where informal recycling processes worsen the situation. E-waste is a major environmental and health hazard, containing toxic components harmful to both humans and ecosystems. Improper disposal releases dangerous substances like polychlorinated biphenyls, heavy metals, and carcinogens, which can contaminate soil, water, and air. These pollutants may enter the food chain, causing serious health issues such as kidney damage, skin disorders, lung cancer, and brain damage. Managing e-waste effectively remains a major challenge, particularly in developing nations. Several obstacles are mentioned, including a lack of data and several elements in e-waste, and the absence of globally relevant standard procedures for measuring material concentrations in particular electronic equipment. Researchers are looking at eco-friendly metal recovery techniques and efficient recycling procedures as well as sustainable management strategies to address this problem.

Keywords Pollutants · E-waste · Heavy metals · Toxic · Sustainable management

1 Introduction

The creation of various electronic devices has raised people's standard of living because of growing industry and progress. However, it is concerning that a significant quantity of tech waste is produced due to their short lifespan and the development of more appealing features, designs, and compatibility problems (Priyadarsini & Das, 2025). The latest electronic devices, for instance, the latest smartphones, are

P. Yadav · S. K. Srivastava (✉)
Department of Zoology, Tripura University (A Central University), Agartala, Tripura, India
e-mail: shaileshsrivastava@tripurauniv.ac.in

A. P. Das, S. Priyadarsini (eds.), *Electronic Waste and Environmental Pollution*, Sustainable Environmental Waste Management Strategies,
https://doi.org/10.1007/978-3-031-90287-1_5

conceived for short usage periods and thus create e-waste. Beyond this continuous increase is anticipated to proliferate at a much higher level (Stevenin et al., 2024; Kumari & Samadder, 2022). The infrastructure for disposing of waste materials is under more stress because of the growing usage of electronical and electrical equipment (EEE) in modern society (Priyadarsini & Das, 2024). In the shape of computers, mobile phones, screens, and other domestic and business devices contribute significantly to solid trash (Saha et al., 2021).

Some upstream considerations, such as consumer volume and attitude, also influence purchasing and disposal decisions. Interestingly, a lot of gadgets are thrown away before they achieve their last stage of life and even fewer devices end up in a designated recycling facility (Kirwan, 1997). A significant portion of these electronic devices are either burned in a waste-to-energy operation or wind up in landfills. In riskier nations like China, India, and Africa, the majority of EEE is disposed of by developed nations in unsafe ways (e.g., inadequate safety measures, landfill disposal, and burning without considering the pollution of air and water). Both developed and developing countries suffer from this issue; however, the absence of government financing, legislation, and action to construct high-tech dumps is the reason for this in developing nations. Many developing nations, such as Nigeria, struggle with the absence of fundamental recycling infrastructure for e-waste management, which forces them to rely on unorganized sectors that use backyard recycling and rudimentary dismantling methods, open-pit burning of cables to recover copper and backyard disposing of and crude dismantling techniques (Sonawane et al., 2001). Humans are exposed to dangerous toxins through these actions. Furthermore, recycling and recovery are primary problems with managing e-waste disposal. Worldwide, about 20% to 30% of this garbage is recycled. Most e-waste is burnt at elevated temperatures or dumped in landfills. However, neither of these practices is environmentally safe, and too far, no comprehensive study or recording of e-waste disposal facilities has been conducted. A few typical pretreatment procedures. For example, component removal, physical disassembly, crushing, size reduction, etc., can result in serious local contamination (Elgarahy et al., 2024). Several recent research has demonstrated the impact of e-waste leachates on aquatic species, plants, humans, and soil microbes including bacteria and fungi. For example, mercury (Hg) is used in various electronic equipment like switches and display screens. Mercury can be changed by soil bacteria into methylmercury, a substance that may be hazardous to both individuals and other creatures. Studies have found that fetuses and young children are more vulnerable. Numerous other illnesses, including skin irritation, neurological disorders, memory problems, cognitive impairments, etc., are also brought on by it. Other toxic metals, like Cu, As, Cr, Ni, etc., can have a direct or secondary consequence on humans and other creatures (as shown in Fig. 1). These products' life cycle assessment (LCA) promotes better resource use and growth efficiency across industries. While numerous LCA frameworks were used, no consistent methodology has been established.

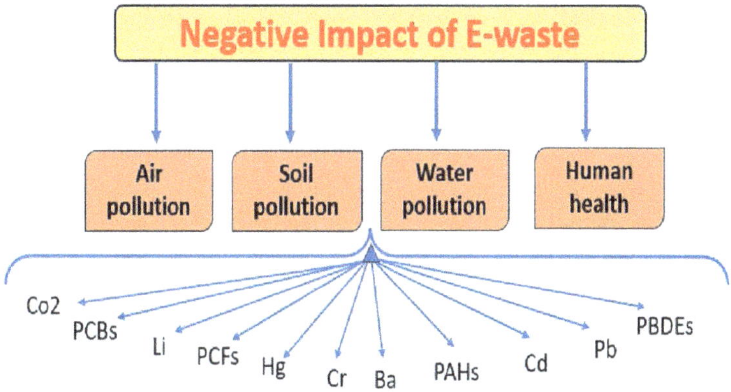

Fig. 1 Showing negative impact of e-waste

Economic input and output LCA, of toxic emission shows reliable comprehensive data (Birat, 2015). It must be recognized that the local community, nongovernmental organizations, commercial sector, and official institutions all share responsibility for safe electronic waste disposal. There are remedial methods effective and environmentally friendly for the e-waste problem. Presently, research on several aspects of electrical waste from birth to death is being carried out in several places around the world. In addition, there has been a steady increase in amounts of e-waste and other ongoing problems related to e-wastes. The formation of different physico-chemical–biological metal recovery techniques, problems of transboundary movement between countries, continuous changes of rules of law, as well as induction of hazardous metals within the environment cause major threats to human (Das & Mishra, 2008).

Comprehensive research has also been done on possible health hazards from e-waste to both biotic and abiotic environmental components. Additionally, efforts have been made to examine the effectiveness of existing methods for managing electronic trash and to investigate green technology of ecosystems which were contaminated by e-waste (Saha et al., 2021).

2 Epidemiology

Healthcare systems are fundamentally at the forefront of the fight against global warming. The advancements in public health over the last few decades and the capacity to deliver high-quality treatment are directly threatened by climate change, which poses a serious cost in terms of mortality and financial losses (Stevenin et al., 2024).

2.1 Global Status of E-Waste Produced

E-waste generation continues to evolve with technological advancements, both now and in the future (Debnath et al., 2023). A nation's GDP is often directly associated to its e-junk production and the total number of electronic devices it owns. The WHO estimates that between 2.9 and 12.9 million women, along with around 73 million children, engage in informal e-waste recycling globally (Liu et al., 2023). Africa's largest e-waste dump, Agbogbloshie in Ghana, discards about 250,000 tons of e-scrap annually (Kumar et al., 2013; Feldt et al., 2014).

2.2 E-Waste Generation in India

A recent study revealed that households in India have several heavy appliances like TVs, refrigerators, air conditioners, washing machines ovens, and others due to the growing economy as well as increasing standards of living among the people (Jana, 2023). Advancements in information technology, particularly through e-governance, have significantly increased the demand for computers and IT devices. Much of this e-waste comes from the informal sector, highlighting the need to integrate formal and informal recycling systems through incentive frameworks to prevent environmental degradation caused by improper recycling practices (Kumar et al., 2018). The e-waste generation pattern in India is shown in Table 1.

Damage is acquired due to migration because of nonofficial unrefined resource extraction of e-waste as well as those: collection emissions (including organic chemicals, metals). According to CSE 2015, heavy metals (HMs) have been studied in quantities that exceed the permissible limits as determined by a study conducted along Ramganga riverbank, in Moradabad, Uttar Pradesh, India. The open spaces and bins are used to dispose of the waste that is accepted for reuse and components post recycling (Arya & Kumar, 2020). Interestingly, only 8% are recyclers who are well informed about their profession and the type of e-waste they generate. Due to the prevailing ignorance of stakeholders toward these risks, they tend to overlook the dangers that come from poor e-waste disposal and face rudimentary methods that constantly expose them to hazardous fumes and physical toil, endangering their health and safety (Arya et al., 2021)

Table 1 Showing e-waste generation pattern

S. no	Region	E-waste generated (million metric tons)	E-waste generated (kg/person)
1	Asia	24.9	5.6
2	America	13.1	13.3
3	Europe	12	16.2
4	Africa	2.6	2.5
5	Oceania	0.7	16.1
6	India	1.6	2.4

3 Composition of E-Waste

Copper, in particular, is a commonly found precious metal (PMs) in e-waste (Torres et al., 2018). Larger devices, such as refrigerators and washing machines, tend to have lower metal content compared to smaller, lighter e-waste items (Berkhout and Herti, 2004). These heavy appliances predominantly contain steel, which is relatively less harmful to the environment. In contrast, PCs, notebooks, and mobile devices have higher concentrations of heavy metals (HMs) and flame retardants (Mishra et al., 2022).

Precious metals (PMs) are primarily utilized in electronic devices for their excellent electrical conductivity and resistance to oxidation, among other properties. Printed circuit boards (PCBs) are particularly valuable, containing metal concentrations up to ten times higher than commercially mined ores (Ilankoon et al., 2018).

E-waste also contains a unique mix of hazardous and complex pollutants not typically found in other types of waste. Products like silicon chips often include elements such as lithium (Li), lanthanum (La), beryllium (Be), cerium (Ce), antimony (Sb), indium (In), and gallium (Ga) (Sakhuja et al., 2022).

4 Causes of Electronic Waste Generation

Main causes of the ever-increasing garbage problem of the earth are electronic wastes, that damage both the environment and human health. Electronic goods change rapidly and become cheaper leading to high obsolescence, leading to rapid generation of electronic waste (Mishra et al., 2023b; Prabhakar et al., 2019; Raj & Das, 2023). This means that the rapid evolution of science and technology has, a substantial degree, provided the increased driving base for all electronic devices. Human desire to purchase newer products to replace older models also plays a huge role in electronic waste generation. Proliferation of technological devices at a rapid pace with the rising number of people in the world leads to high generation of e-scrap.

5 Toxicity of E-Waste

Heavy metals and organic elements or chemicals from e-waste can accumulate in the soil, affecting soil microbiomes, which in turn impacts plant and animal life (Das & Mishra, 2010; Das et al., 2012). These pollutants alter the genetic and functional traits of soil microorganisms, harming microbial diversity and reducing the efficiency of soil ecosystems (Ghosh & Das, 2020; Mishra & Das, 2023). Soil contamination can lead to changes in microbial communities and their activity, which is essential for maintaining soil fertility and plant growth.

Denaturing gradient gel electrophoresis assesses bacterial diversity in soil contaminated by e-waste. The findings revealed a significant decrease in populations of essential bacterial groups, such as Proteobacteria and Firmicutes, due to contamination. Furthermore, the study highlighted a decline in soil enzymatic activity and changes in the physical and chemical properties of the soil (Salam & Varma, 2019).

Leachates, a primary pathway for e-waste contaminants, easily pollute soil, water bodies, air, and the food chain, causing immediate harm to living organisms (Mohammadi, 2020). As many metals and chemicals existing in ionized forms are readily absorbed and assimilated by plants and animals, toxicant levels were greater in landfills containing e-waste compared to those without e-waste. Mobile phones are becoming the most rapidly increasing outmoded product within the flow of e-waste. Common harmful compounds which frequently blended with the plastics in EEE are BFRs (Brominated flame retardant) and toxic metals (Saha et al., 2021). The proportion of hazardous compounds has grown over the course of 15 years. Plastics from e-waste are expanding at a never-before-seen rate worldwide, which is quite concerning (Singh et al., 2020; Mishra et al., 2023a; Sahu & Das, 2024; Singh Jadaun et al., 2022). Different efforts of e-waste pollutants on human body are shown in Table 2.

Table 2 Showing adverse effect of pollutant present in e-waste

S. no	Toxicant	Existence in E-waste	Health effects
1	Halogenated compounds PCBs	Condenser, Transformer, adhesive in plastics	Cancer, immune system disorder, endocrine system damage
2	PVC	Cable Insulation	Cancer, birth defects, diabetes, learning and developmental delays, endometriosis, and immune system abnormalities
3	TBBA, PBB, PBDE	Fire retardants for plastics, insulation in cables	Impaired memory functions and learning in the fetus, endocrine system disorder
4	Arsenic (As)	Semiconductors, LED, IC, Solar cells	Damages the digestive system, skin problems, lung cancer
5	Barium (Ba)	Getters in spark plugs, CRT, fluorescent lamps	Brain swelling, muscle weakness
6	Cadmium (Cd)	Batteries, solders, CRTs, infrared detectors, Chips, Toner ink, photocopy machines, mobile phones	Fever, headache, sweating, and muscular pain, long exposure may lead to lung cancer, kidney damage, and lower cognitive skills in children
7	Copper (Cu)	Television, DVDs, cables, and wires	Hampers the liver function, nausea, diarrhea, chest pain
8	Lead (Pb)	Batteries, CRTs, cables, and wires	Asthma and decline in immune response, lower cognitive skill in children, and learning capabilities
9	Mercury (Hg)	Sensors, monitors, PCBs, Cathodes, Fluorescent lamps, LCDs, batteries in clock and pocket computers	Bioaccumulation causes brain and liver damage, respiratory, and skin disorders

5.1 Harmful Effects of E-Waste on Human Health

E-waste is deemed toxic in nature when amounts of heavy metals in EEE cross the permissible levels. E-waste can cause serious health hazards, including obstructed breathing, asphyxiation, coughing, tremors, pneumonia, neurological problems, comas, and death. Incineration or pyrolysis of halogenated polymers produces smoke containing dioxins, furans, and polychlorinated biphenyls. Such compounds may cause cancer of the lungs, liver, and other visceral organs.

Informal or crude methods for recovering resources from e-waste involve several hazardous practices. These include:

- *Collection*: improper handling of switches and light sources can release hazardous substances.
- *Manual disassembly*: open, manual processes expose workers to toxic emissions, particularly from CRT (Cathod Ray Tube) glass and LCDs (Liquid Crystal Displays).
- *Shredding*: this process releases carbon-based compounds and metals into the environment.
- *Burning and leaching*: these release pollutants that pose risks to human health and the ecosystem.
- *Uncontrolled dumping*: leads to further contamination and health hazards.

E-waste exposure typically occurs through three primary routes: consuming contaminated food, inhaling polluted air, ingesting soil or dust, or with direct skin contact. Unfortunately, inadequate awareness among stakeholders results in improper disposal practices and rudimentary handling methods that pose significant risks to worker's health and safety. Workers are exposed to harmful fumes, physical strain, and toxic dust, often through inhalation or ingestion of electronic debris.

Exposure to e-waste contaminants damages the immune and reproductive systems, disrupts metabolic processes, and increases the risk of cancer (Grant et al., 2013). Chlorofluorocarbons (CFCs), found in many devices, are a major contributor to ozone layer depletion, which can result in skin conditions and cataracts. Additionally, hazardous metals like mercury (Hg), cadmium (Cd), lead (Pb), barium (Ba), and hexavalent chromium [Cr (VI)] cause severe health issues, including liver and brain damage, dry eye syndrome, nausea, stomach discomfort, constipation, headaches and long-term conditions like blockages and other chronic ailments.

E-waste can impact human beings in various interests. Exposure to e-waste results in alteration of thyroid functioning, adverse effects on neonates, alteration in cellular expression and function, alteration in behavior and disposition psychologically, and reduction in lung function. Inhabitants of e-waste recycling towns have shown signs of having relatively more damage at the DNA level when compared to control towns. Studies have been done on bacterial and plant cells to characterize the genetic effects of leachate and has been further proven that leachate is toxic to mammalian cells and it also causes genetic damage. Leachates can enter cells and damage them as modifications are done both internally and externally to change the pH levels, which usually affects enzymes and changes DNA structures. Another postulation emerges wherein free radicals could very well react with lipids within the cell and therefore, have damage to the membrane. This then leads to the breaking of DNA. In addition, free radicals damage DNA, induce mutations, as well as some base substituents due to the charging of nucleic acid components such as purine and pyridine (Youcai, 2018). In Australia, dioxin and other toxic elements released from electronic junk absorbed in the human body through food, water, as well as air in huge amounts that could possible health risk for human. PCBs of E-waste were found in rice, through fish, eggs, meat and surface and ground waters to eventually arrive at the humans. In addition, it was also found that polybrominated diphenyl ethers (PBDEs), PCBs, and PBBs (Polybrominated biphenyl) contained in human hair were, respectively, 30, 182, and 58 ng/g from the areas where dismantling was done of e-waste. A huge proportion of the research on human biomonitoring indicates that the primary targets for all the hazardous chemicals emanating from recycling centers are pregnant women and children in developing countries. Harmful chemicals affect all age groups, but the effect varies between children and adults (Li et al., 2021). Before the start of breastfeeding and into infancy, contaminants such as lead and cadmium begin to affect babies once they start solid food. Higher levels of these contaminants in children were more strongly associated with a compromised ability to think than their association with sites of e-waste, cough being one of the pulmonary effects. Bio measurements such as IgE levels were found to be very high in children's blood. They also get asthma as well as other immune dysfunctions, including hepatitis (Cong et al., 2018).

The accurate estimation of the translocation of metals and semimetals into human food chain trophic levels can be employed in predicting the potential and significant danger from such soils contaminated by metals. A total of 131 serum samples were collected from Northern Vietnam's e-waste sites. The verification assessment ascertained that pollutants like thiocyanate (SCN−) and perchlorate (ClO − 4) coming from e-waste were capable of interfering with iodide (I−) uptake through the iodide–sodium symporter and affecting levels of thyroid hormones (THs) through their actions on the thyroid gland (Saha et al., 2021). The source of perchlorate (ClO − 4) is even more interesting, especially from the aspect of its significance. There are two sources of perchlorate formation: rocket propellants, fireworks, and explosives, and the production of polyvinyl chloride (PVC) as a dopant material. If "persistent" means that they do not decompose, metals exist in the environment for several years.

When different metals enter the human body through various entry routes, they cause harm to human health by entering the food chain. These heavy metals accumulated in plants are later transferred to animals and then to humans. These metals have different types of poisonous effects on the human body. For instance, lead (Pb) negatively impacts behavior and learning abilities, copper can lead to liver failure, and excessive exposure to cadmium may cause kidney failure and lung cancer (Ebrahimi et al., 2020; Li et al., 2021). Additionally, Grant et al. highlight that cadmium is not only a known carcinogen but also poses toxicity risks to the immune system, reproductive system, and various metabolic processes (Grant et al., 2013). It is one of the major ozone-depleting substances, eventually causing skin diseases

Table 3 Potential human health risks associated with various chemicals present in e-waste

Chemical component of e-waste suspected to cause human health effects	Health effects
Nickel, iron, cadmium	Liver damage
Cadmium	Bone damage
PBDEs, dioxins	Metabolic diseases
PBDEs, PCBs, dioxins, manganese, phthalates, bisphenols	Endocrine disruption
PCBs, dioxins, PAHs, PFOA, cadmium, arsenic, beryllium, chromium	Carcinogenic (cancer causing)
PBDEs, PCBs, dioxins, PFAS, PAHs, lead, cadmium, arsenic, chromium	Fetal growth and development
PBDEs, PCBs, PAHs, lead, mercury, cadmium, manganese	Neurodevelopment and cognitive function (IQ deficits)
Lead, PCBs, dioxins, PAHs.	Behavioral effects (shortened attention span, reduced ability to deal with frustration, hyperactivity, antisocial behavior, depression)
PBDEs, PCBs, dioxins, PFAS, lead, chromium, mercury, phthalates, bisphenols	Reproductive effects
PAHs, cadmium, arsenic, lithium	Lung damage
Lead, cadmium, mercury	Kidney damage
Dioxins, mercury, arsenic	Cardiovascular

and cataracts as they are emitted through various devices containing chlorofluorocarbon, resulting in liver and brain damage—dry eyes syndrome, vomiting, upset stomach, diarrhea, cerebral pains, blockage, and many chronic health issues (Table 3) (Arya et al., 2021).

5.2 *Health Risks Associated with the Disposal of E-Waste*

Consumption, inhalation, and dermal absorption of dust and dirt are the main exposure pathways for e-waste (Liu et al., 2020; Singh et al., 2020; Zhang et al., 2010). Health effects concerning exposure pathways of e-waste include cough, choking, pneumonia, tremors, neurological problems, and sometimes death for both respiratory-related and coma conditions. Afterburn or pyrolysis, halogenated polymers such as dioxins, furans, and polychlorinated biphenyl will produce smoke containing halogen species that are capable of cancer development. They are known to adversely affect the kidneys, circulatory system, and nerves in the human body. They also have developmental implications on the brain, especially for infants and young children.

Physical injury and chronic illnesses like asthma, skin conditions, neuritis, eye irritation, and stomach issues might result from such work-related dangers. It has been demonstrated that the particulate matter gathered from e-waste recycling facilities causes DNA damage, oxidative stress, and inflammation. Through conversion

Table 4 Sources and pathways of human exposure to hazardous substances found in waste electrical and electronic equipment

S. no	Hazardous substance	Ecological source of exposure	Component of electrical or electronic equipment
1	Cadmium (Cd)	Air, dust, soil, water, food (especially rice and vegetables)	Printer inks, toners, photocopying machines (printer drums), switches, springs, connectors, printed circuit boards, batteries, infrared detectors, semiconductor chips, cathode ray tubes, mobile phones
2	Lead (Pb)	Air, dust (ashes), water, soil	Printed circuit boards, cathode ray tubes, light bulbs, televisions, batteries
3	Chromium (Cr) or hexavalent chromium	Air, dust, water, soil	Anticorrosion coatings, data tapes, floppy disks
4	Mercury (Hg)	Air, vapor, water, soil, food (methylmercury bioaccumulates in fish and shellfish)	Switches, thermostats, sensors, monitors, cells, printed circuit boards, cold cathode fluorescent lamps, LCD screens
5	Nickel (Ni)	Air, soil, water, food (plants)	Batteries, cathode ray tubes
6	Lithium (Li)	Air, soil, water, food (plants)	Batteries
7	Barium (Ba)	Air, water, soil, food	Cathode ray tubes, fluorescent lamp.
8	Beryllium (Be) Power	Air, food, water	Power supply boxes, computers, X-ray machines, ceramic components of electronics

or transmission, pollutants may build up in nearby farms, where they may then accumulate in creatures through the food chain and spread (as shown in Table 4). The harmful impacts of materials associated with e-waste can persist for decades (Narayanasamy et al., 2021). Due to this, e-waste is now a significant worldwide ecological and health concern. Its effects considerably outweigh the risks of work exposure and put future generations and vulnerable groups at risk.

5.3 *Environmental Impact*

This chapter explores the various hazardous compounds present in e-junk, along with their various implications on both the ecological unit and animal human health. It highlights environmental contaminants and the consequences of e-waste processing at various stages, emphasizing the potential ecological challenges they pose.

5.3.1 Long-Term Environmental Hazards of E-Waste

The environmental and health impacts of heavy metals, persistent organic pollutants, flame retardants, and other hazardous chemicals in e-waste are substantial. Both formal and informal e-waste disposal methods pose significant risks to ecosystems, as they contribute to pollution. E-waste contaminants are released into various media, including air, dust, soil, and water, where they are absorbed by living organisms through geochemical cycles and biological pathways.

Dust, heavy metals, and persistent organic contaminants can be carried along in the air. Pollutants such as sulfur, nitrogen, dioxins, and heavy metals present in incineration flue gas get into the atmosphere. Pollutants will settle, transmigrate, and undergo transformation into a set of plants, animals, and humans through the water cycles: soils and groundwater that will eventually affect all living beings' health. An investigation done by a non-government organization for unregulated e-waste disposal sites in Vietnam revealed that children living in exposed towns consumed, on average, 3.90 times more heavy metals (Pb, Cd, Cr, Ni, and As) daily than children in a control village ($P < 0.01$) and cooked rice was identified as the primary contributor to heavy metal intake (Hollander, 2024).

5.3.2 Effect of E-Waste on Soil and Air Quality

Decomposing and shredding electronic waste release soil fragments or dioxin-like poisons into the environment, such as causing air pollution, and contributing to respiratory infections. Some cheap electronics are also burnt in the fire; on the contrary, that fire has another way of extracting copper and other precious metals from the devices. Burning electronic waste predisposes people to long-term health complications like cancers because it releases several small particles that can travel up thousands of kilometers and inflict various health complications on humans and plants. Precious metals like Ag and Au are usually extracted from complex devices with the use of acids, desolders, and other chemicals. However, smoke can result in areas where reuse is not properly managed. Reusing e-waste brings negative environmental consequences, posing threats to people who are into managing these wastes. Also, effluents tend to go beyond the recycling sites. Electronic trash pollution has different effects on various animal species, potentially threatening biodiversity in certain locations. Toxic environments can significantly affect the quality of water, soil, and plant species, resulting in irreversible impact. For example, the higher the Pb concentration in the air, the more Pb it brings in as it returns to water and soil. Dangerous toxins from electric garbage can cause unknown neurologic problems to animals, wildlife, and human beings and release toxic metals after combustion. Improper disposal of these toxic metals into landfills causes groundwater and soil toxicity. Toxic metals in the soil can create crop susceptibility to absorptive contamination, leading to problems that limit farming production. Very large particles from the process of burning, shredding, or dismantling e-waste redeposit into the environment very quickly due to their large size and range. Temperature, soil quality, pH levels, and soil composition are all factors that have to do with soil contamination levels (Okine, 2014). Poisonous compounds in the soil can kill microbes and flora after some time. Animals that rely on their habitat for existence may eat contaminated plants; hence, they will experience internal health problems. Different other poisonous metals following soil pollution could be observed as contained in electronic waste, such as Hg, Li, Pb, and Ba, that contaminate both soil and groundwater. Toxic metals easily migrate to groundwater and eventually enter lakes, rivers, and streams through such activities. Thus, the aquifer is acidified and poisoned, leading to health threats for plants, people, and animals, irrespective of their local relative to that of the reuse unit. Finding potable water may, therefore, be a difficult task. The ecological balance is disturbed by acidification and can obliterate species in both marine and freshwater environments. Further having the potential to majority of brain and reproductive health damage will severely contribute to diseases and birth defects.

There is a serious threat from inappropriate disposal of electronic waste to the global ecology. It is urgent to stress supplementing matters regarding this emerging challenge and the dire consequences that follow. E-cycling is increasingly becoming a lifeline for counteracting the evil effects of e-waste. This advocates the recovery and reuse attached to reselling or repurposing objects. Electronic waste is by its increasing volume an important reason for appropriate waste disposal. The pollutants are classified into three types: primary, secondary, and tertiary, according to the nature of the pollutants.

Primary pollutants: These primary pollutants are mostly found in hazardous e-waste that would be quite toxic. The toxic c-products usually manufacture electrical and electronic instruments of different types. The toxic constituents that pollute electronic waste include Pb, Sn, Hg, Ni, Cd, Cr, Ba, Zn, etc. Organic pollutants can also be present in e-waste and introduced into the environment or formed during disposal. Polybrominated biphenyls, among which can be named, are hexabromocyclododecane, polychlorinated biphenyls, flame retardants, polybrominated diphenyl ethers, tetra-bromophenol-A, and others.

Secondary pollutants: The processes generated secondary wastes, which consist of preprocessing byproducts like pyro-metallurgical byproducts (emission of gaseous byproduct), hydrometallurgical byproducts (acid waste and volatile product).

Tertiary pollutants: These are the elements that have been used in processing e-waste, either for restricting toxic metal emissions or for the separation of different materials. These products are harmful to human beings if not controlled properly (leaching agents) (Dey et al., 2023). Though the gases were detected in the gaseous compartment of samples collected for this study, furans and chlorinated dioxins were almost completely undetectable since they were not present at all.

5.3.3 Effect of E-Waste on Plant

E-waste's environmental effects on plants remain underexplored, but its toxic components, particularly heavy metals (Cd, Mn, Ni, Cr, Zn, Pd), are well-documented (refer to Table 5). These elements, often found in e-waste leachates, induce somatic mutations and cytotoxicity in *Allium cepa*. Testing with e-waste leachate concentrations (1%, 5%, 10%, 25%, and 50% v/v) over 48–72 h revealed maximum root growth at 1% and the least at 50% concentration (Bakare et al., 2009). Cytological abnormalities such as binucleate cells, anaphase bridges, and sticky chromosomes

Table 5 Showing toxic effects of E-waste on plants

S. no.	Element	Plants	Adverse effects
1.	Arsenic (As)	Clover (Trifolium pratense L. cv. Renova), Holcus lanatus L.	Increase in superoxide dismutase activity, peroxidase activity as well as decrease in chlorophyll and carotenoid concentrations, increase in lipid peroxidation, SOD activity and PC production correlated with increasing arsenate concentrations
2.	Cadmium (Cd)	Ricinus communis L., Brassica juncia L., Wheat (Triticum sp.)	Reduced the growth rate and biomass, decreased elongation of shoots and root Reduction in seed germination; decrease in plant nutrient content; reduced shoot and root length
3.	Chromium (Cr)	Wheat (Triticum sp.), Tomato (Lycopersicon esculentum)	Reduced the chlorophyll and carotenoid contents Decrease in plant growth and biomass
4.	Mercury (Hg)	Rice (Oryza sativa), Tomato (Lycopersicon esculentum)	Decrease in plant height; reduced tiller and panicle formation; yield reduction; bioaccumulation in shoot and root of seedlings Reduction in germination percentage; reduced plant height; reduction in flowering and fruit weight; chlorosis
5.	Lead (Pb)	Maize (Zea mays), Portia tree (Thespesia populnea), Oat (Avena sativa)	Reduction in germination percentage; suppressed growth; reduced plant biomass; decrease in plant protein content Reduction in number of leaves and leaf area; reduced plant height; decrease in plant biomass. Inhibition of enzyme activity which affected CO_2 fixation and decrease the plant biomass

were observed, highlighting e-waste leachates' potential to cause cytotoxicity and somatic mutations (Mallick & Moh, 2003; Nie et al., 2015).

Plants grown on 100 µg/L Cu media exhibit slower growth rates and lower biomass levels. *Allium* assays are commonly used to assess DNA damage caused by e-waste, including chromosomal abnormalities and mitotic cycle disruptions, due to their long cycle and ability to respond with various mutagenic chemicals. Assay results may indicate the existence of mutagenic, genotoxic, or cytotoxic chemicals in the ecosystem, posing both direct and indirect risks to living beings (Saha et al., 2021). Accumulated within the roots of plants, heavy metals may reach high concentrations and harm the plants; it may inhibit the intake of other nutrients necessary to grow. In contrast, metals like copper (Cu), iron (Fe), and aluminum (Al) can also have significant impacts on plant development. For example, Cu and Fe have been shown to cause notable reductions in tiller and panicle development in rice plants, as demonstrated in Table 5 (Saha et al., 2021). These metals disrupt essential metabolic pathways, particularly those involved in nutrient uptake and enzymatic activity, ultimately impairing plant growth.

The study found that the combined exposure to these heavy metals resulted in increased leaf death in *Lythrum salicaria*, a plant species commonly used for toxicity testing. This indicates that the interaction between different heavy metals can exacerbate the toxic effects on plants, leading to more severe physiological damage compared to individual metal exposure. The combined presence of these metals disrupts multiple cellular processes, including photosynthesis, respiration, and water absorption, further compromising plant health and development (Nicholls & Mal, 2003).

5.4 Impact of E-Waste on Microbial Community

Microorganisms play a significant role in soil quality as they are affected by pollutants. Microorganisms contribute to the preservation of productivity and primary yield through the decomposition of organic materials and the cycling of nutrients. Waste from electronics has toxins and chemicals that damage soil and bacteria. Toxic metals and chemicals denature soil microbial population and activities.

Accumulated heavy metals (HMs) are generally acknowledged to hinder the soil microbial biomass and enzyme activity. It results in a decreased functional variety of microorganisms in a soil ecosystem and alters the community structure of microorganisms (Saha et al., 2021). Continuous exposure studies on metals have also shown that they can induce the capacity of microbes to survive with those metals.

Microbiology can restore polluted soils. It is also demonstrated that cost-effective pollutants, copper, and decabromo diphenyl ether (BDE-209), accounted for about 70% of changes observed in microbial assembly (Liang, 2015).

E-waste pollution, when disposed of irresponsibly, can enormously damage microbial communities. Microbes act as drivers in biological cycles and ensure ecosystems work effectively. Their metabolism disturbances occur with small environmental changes. Most of the species, namely *Pseudomonas*, *Acinetobacter*, *Sphingomonas*, etc., have been reported to have resistance against heavy elements like cadmium, zinc, copper, etc. These bacteria may improve degradation capability as well as bioremediation efforts. Some laboratory investigations attempted collection to single and cumulative effects of some specific metals or chemicals on soil microorganisms.

5.5 Impact of E-Waste on the Aquatic Environment

Leachate from e-waste contaminates water bodies and this seriously affects the growth and reproduction of marine organisms (Saha et al., 2021). In contrast to the polluting e-waste, there is a strange relationship between the type and number of fungi, suggesting that in a very enlightening association, there is the case of personalized relationship between organisms and pollution (Saha et al., 2021). Many heavy metals have been exposed to fish and other aquatic species; it, therefore, get biomagnified into their tissues. Research by Luo revealed significant bioaccumulation of polybrominated diphenyl ethers (PBDEs) in aquatic life, particularly in carp fish from the Nanyang River in China, which had PBDE concentrations as high as 766 ng/g in fresh weight. Further investigations showed that in the same river, which had been polluted by electronic waste (e-waste), the PBDE concentrations in sediments reached even more alarming levels, peaking at 16,000 ng/g. (Saha et al., 2021). These findings highlight the widespread contamination of the aquatic ecosystem due to e-waste, with pollutants accumulating at various trophic levels.

6 Challenges Associated with E-Waste Management in India

In many regions, like Asia, most discarded electrical goods are repurposed through informal means, often by marginalized communities. Around 1000 families, typically from poorer backgrounds, rely on scavenging e-waste from dumpsites as their primary source of income. These families, often working in hazardous conditions, collect and sort through discarded electronics to extract valuable components and materials for resale. These segregated items are sold to unregulated "ragpickers," who then sort through and sell the materials to artisanal or commercial processors. This informal system facilitates the reuse and recycling of various materials but remains largely unregulated, with little oversight to ensure the safety or proper handling of hazardous substances.

Thousands of city-dwellers across these regions are involved in the e-waste recycling industry, performing tasks like collecting, sorting, repairing, refurbishing, and dismantling obsolete electronics. These individuals, many of whom work in informal recycling hubs or small workshops, often lack access to protective equipment or safety training, making them vulnerable to the harmful effects of toxic substances commonly found in e-junk, such as lead, Cd, and Hg.

In contrast to the informal sector, there are very few institutional electronics repurposing centers in countries like India. The absence of formal recycling infrastructure means that people have limited options for disposing of or repurposing their old electrical devices safely. More e-waste recycling applications may lead to the following major problems: monetary penalties for failing to comply with laws on the processing and handling of electronic waste to have little effect. It is because of inadequate training among low-wage workers that the public lacks an understanding of prices and safety issues in e-waste recycling. Though e-waste generation increases tremendously every year, so little investment is seen in industrial recycling and recovery facilities.

In India, the discarding and reprocessing of electronics are primarily done in the private sector. About a thousand poor families are relying on rag-picking for survival. Middle-class urban households generally take out their old newspapers and plastics, clothes, and metals and sell those to unregulated "ragpickers," who sort them and sell them to artisanal or commercial processors (Bala et al., 2024;

Priyadarshini et al., 2024; Sahu et al., 2024). Countless city dwellers participated in this informal e-waste recycling industry: gathering, sorting, repairing, refurbishing, and dismantling unwanted electronics. In industrialized economies such as India, people have nowhere to go to surrender old electrical products because there are no institutions regarding electronics repurposing.

Significant issues for better e-waste recycling applications would pose are: the monetary sanctions for noncompliance with electronic waste processing and handling laws have not acted sufficiently. The public grasp of e-recycling pricing and safety is also deficient owing to the inadequately trained staff of low-wage workers. Despite the greatly rising quantity of e-waste produced annually, there is very little investment in industrial recycling and recovery facilities.

6.1 *Inadequate Facilities for Repurposing Electronics*

An inadequate framework within India manages e-waste on a wide scale. Less than one-fifth of the total electronic waste produced per annum by the USA is recycled at incredibly few government-authorized facilities within the country. From the facility perspective, the formal sector of electronic waste recycling offers mechanical disassembling and human sorting. No directors of commercial electronic waste exist with the right specifications to monitor the recovery of base and precious metals. Poor processing capabilities characterize a small active entrepreneur who recovers gold from e-scrap (Dey et al., 2023). E-waste is most formally managed in the formal sector and comes from countries having important infrastructures for metal extraction at a large scale. On the other hand, uncontrolled business practices and hazardous methods like acid extraction and combustion in the open air reduce environmental damage and health hazards. The flame suppressant and other persistent organic pollutants present make recycling plastic e-waste extremely difficult (Kumar et al., 2023).

6.2 *Limited Data on E-Waste Generation Rates*

State Pollution Control Boards (SPCBs) in India are collecting information regarding the responsibilities of E-waste and addressing its insufficient management across country. Very tough state-level inventories for managing e-waste often fall under federal jurisdiction. E-waste is not generated only in the country but also illegally from developed nations. The people do not know what kind and how much of their electronic waste is being imported but the data on production and composition should be reasonably precise for the garbage collection, transportation, and treatment.

6.3 Unsustainable Practices in the Informal Sector That Harm the Environment

Unsustainable practices are some of the constraints that private companies will have to deal with to prevent e-scrap handling systems from developing in the formal sector. For example, to administer e-junks, countries such as India will require capital in large amounts for implementing effective recycling technologies (Turaga et al., 2019; Bigum et al., 2017). Therefore, private enterprises cannot justify such investments due to doubts about their access to substantial amounts of e-waste. Similarly, they face informational barriers in such markets. While there has been a healthy growth in the recycling and dismantling industry within formal sector, it addresses only a minuscule portion of waste. Most significantly, by providing home collection and financial incentives, the private sector encourages customers to return their waste. On the other hand, the formal sector cannot rival the informal sector; investments have not yet been made in collection and preprocessing facilities that can be said to be effective. A sustained success in any such initiative concerning the management of electronic waste in India will depend upon resolving the ethical dilemma posed for public policy (Dey et al., 2023).

6.4 Insufficient Regulatory Framework and Enforcement

E-waste management efforts have hardly advanced because manufacturers have no incentive to take responsibility for the waste recovery initiative goals in the first place. There were several amendments and if passed, they could have ensured stakeholder control over reliability by prescribing stringent collection targets. However, such a controlling framework is weighty on organizations not geared simultaneously (Turaga et al., 2019). Furthermore, the legislation has to provide comprehensive standard protocols for other players like dismantlers, collectors, recyclers, and wholesale buyers.

6.5 Lack of Technical Expertise in the E-Waste Sector

Having no professional knowledge is one of most important issues which confront India's e-waste disposal system. E-Junk from households is thrown along with other household rubbish in the same disposal bins and most of the unofficial groups miss proper education and professionalism which fails to handle this waste in a responsible manner. Therefore, e-waste ends up being treated interminably mixed in ordinary refuse (Arya & Kumar, 2020). It is mostly at the next level transferred to Municipal Solid Waste (MSW) incinerators or trash dumps, where separation is likely done before reaching its destination.

7 Management of E-Waste

The collection of e-junk through a network of smart equipment's, while an innovative solution, proves to be an effective method for managing electronic waste. This approach directly contributes to addressing several United Nations Sustainable Development Goals (SDGs). Specifically, it aligns with Goal 12, which focuses on responsible production and consumption, Goal 11 on creating sustainable cities and communities, Goal 14 concerning life below water, and Goal 6 advocates for clean water and sanitation by 2030 (Saha et al., 2021). Among the various methods for handling e-waste, bioprospecting has emerged as a promising strategy. This technique not only provides environmental benefits but also promotes public health and offers economic advantages. By utilizing bio-based approaches, such as the extraction of valuable materials from e-waste, we can minimize environmental harm while creating opportunities for sustainable economic development (Awasthi et al., 2019).

7.1 Landfilling

Using this technique, e-waste is buried in trenches dug out on land and covered with a thick layer of dirt to keep it from being exposed to the air. This approach, however, is not appropriate for leachate, or the liquid that results from waste drainage which contaminates nearby water and soil. This issue is avoided by enhancing landfill technology, such as a controlled landfill. Leachate disposal can be partially addressed by collecting leachates in the basin and transporting them to a wastewater treatment plant (Agamuthu et al., 2015). Since horizontal barriers serve as underlying liners, they may be useful in preventing HM pollutants from moving below. It is possible to use vertical barriers to halt the flow of tainted water via deep aquifers. To stop the leachate from moving, the liner needs to be flexible, strong, impermeable, and have a high tensile strength. However, it has some shortcomings because the landfilling technique does not address many issues due to uncontrolled fires, chemical reactions, vaporization, leaking landfills, etc.

7.2 Incineration

Incineration is a highly regulated procedure that involves boiling or heating hot waste at extremely high temperatures (900–1000 °C) in specialized incinerators. One of its benefits is the ability to harness the heat from flammable substances and reduce the overall volume of e-waste (Saha et al., 2021). E-waste incineration facilities are a significant source of mercury and cadmium emissions annually (Khan & Saadat, 2019).

7.3 Reuse

Another environmentally friendly way to keep e-waste from quickly becoming outdated is to reuse it. It can be either used second-hand or with small adjustments to the original. Many electronic devices, including laptops, smartphones, and other gadgets, can be repurposed by those who cannot afford new ones. It not only makes it easier for consumers to utilize electronic devices but also promotes the reuse of e-waste, reducing environmental impact and conserving valuable resources (Forti et al., 2020).

7.4 E-Waste Recycling

The creation, assembly, purchase, or promotion of new electronic items made from waste materials is known as e-waste recycling. As markets expand in developing countries and devices are frequently replaced, fully functional electronic devices quickly turn into e-waste. Besides this, a considerable quantity of e-trash is produced by quick technological advancements and improvement, shifting fashion and status, attractive deals on electronic goods, and a limited lifespan of electrical equipment. Recycling e-waste has several advantages, including: recovering valuable materials from end-of-life (EOL) gadgets; utilizing recycled materials to create new equipment; assisting in the preservation of finite natural resources; preventing pollution, and boosting the creation of jobs. Recycling e-waste is critical because it reduces a significant threat to essential natural resources. It is executed as a multi-step procedure.

Implementation of all end-of-service life problems with electronics or e-waste should be carried out in a problem-solving manner. It is therefore vital to establish viable competitive recycling and reuse industries for this purpose. There are good reasons for developing new ways and systems for collecting, reusing, and reprocessing electronics (Lenkiewicz & Webster, 2017). Conclusively, future frameworks must include collection agents, recycling experts, and reconditioners, who play a crucial role in managing the materials throughout the lifecycle of electrical items. A lot of innovations and advancements in queries have been happening in India at the government level or corporate level or societal level mainly owing to the concern of availability and access of critical resources. Synergy between resource competency and product output attempts gain power-packed manufacturing in terms of EEE. Given such a backdrop, India needs to create propitious and constructive plans for the entire e-waste value chain in the nation. Primarily, a solid and efficient top-level design system is required for the repurposing of e-trash. It is recommended that governments create a fair recycling system that takes into account their specific geographic, urban and economic circumstances. To create an environmentally responsible system for recycling electronics, items across the board and governments must enhance the pertinent laws and regulations, with the expanded manufacturer accountability serving as a key motivator (Borthakur & Singh, 2021; Liu et al., 2023).

It is necessary for the recycling framework to include extended producer responsibility systems, risk prevention strategies and possible environmental effects into the existing regulatory framework. The prolonged manufacturer liability system has emerged as a potent and successful instrument for managing transboundary e-waste movements (Liu et al., 2023). The primary barrier to recycling e-waste is still the collection process. The actual rate of recycling technical rate can be quickly raised by the government by providing financial incentives to recycling businesses and encouraging the involvement of outside funding. This will assist in addressing the core cause of the e-waste recycling issue. Including producers of electronics, marketing departments, and recycling businesses might result in a thriving sector and significant new projects.

7.5 Advances in Collection

The variety of electronic items is increasing in tandem with the evolution of societal demands. Technical difficulties for recycling electronic products are also brought on by shifting product kinds and ongoing product upgrades. In general, the three essential components of recycling electronic products are collection, disassembly, and fine segregation. Constructing an e-scrap repurposing network using the Internet as the primary knowledge transfer is the key to tackling the issue, given the Internet's rapid development (Razip et al., 2022). We suggest developing an online/offline integrated e-trash gathering system as this technology allows e-scrap intelligence to be coded for visual management recording and observing. Electronic devices must be codified by manufacturers and merchants must register the products they sell with their personal data. On a third-party platform offered by the company, buyers provide details on trash electrical and electronic devices in exchange for cash and credit points. Disposers provide services through Internet information acceptance, offline transfers, and recycling revenue. The "closed-loop" administration of trashed electronic items is implemented by the government through online oversight and offline cash assistance.

7.6 Plastics from E-Waste for Application in Cell Culture

In spite of their complicated composition and the harmful chemicals that have a deleterious effect on both human and environmental health, plastics found in e-waste are rarely recyclable, despite the economic and environmental importance of recovering them. As a result, researchers have discovered a novel application for the plastics contained in e-waste as a substitute for the polymers used in laboratory petri dishes and other cell culture containers. A group from Singapore's Nanyang

Technology University recycled plastics from e-waste by merely sterilizing them before using them in experiments. Besides it will decrease the massive volume of plastic trash produced by biomedical research. In a different investigation, e-plastics were particularly recovered from computer components that were destroyed, like prismatic sheets and keyboards which were utilized to direct the discrimination and proliferation of human adipose-derived mesenchymal stem cells after being thoroughly cleansed, disinfected, and described. According to the study, the e-plastics that were used demonstrated a capacity similar to that of commercial tissue-culture treatment plates in supporting cell proliferation and stem cell phenotype for a duration of 2 weeks (Fathi et al., 2022). Along with a decrease in waste production, other comparable studies have also shown notable reductions in single-use plastics.

7.7 *Eco-Friendly Product Design and Management: An Alternative Solution*

Due to enormous resource consumption and trash production, the rapidly rising desire for electronic devices is expensive for the atmosphere. Furthermore, the development of nongreen manufacturing technology poses a risk to both the environment and people needs in-depth research. Thus, environmentally friendly goods and procedures are now essential for sustainable growth. The goal of sustainable innovation is to meet customer demands while having the least negative environmental impact. This necessitates that all production businesses take into account all environmental factors pertaining to energy consumption, waste creation, and the consumption of raw materials in their goods. This might necessitate some adjustments to manufacturing, services, and design. By definition, there are two ways to accomplish this: (1) enhancement of the structure and (2) innovation or portion enhancement. Therefore, engineers and designers ought to create electrical devices that benefit the environment, economy, and society. This can be accomplished by implementing environment-friendly, sustainable solutions and creating goods that use sustainable materials and less energy during the manufacturing and use phases, as well as generating goods that are basic to discard or reuse when they become obsolete. The volume of e-waste and the use of nonbiodegradable crude materials that are harmful to both human and environmental health are the main problems associated with it. Based on previous studies, over 80% of the environmental implications of e-scrap may be predicted early in the electronic item design process. Product developers can therefore affect the product's whole lifecycle and the possibility of managing it after it turns into e-waste. The ultimate product, the production process, and the choice of materials are all included in the early design stage. Thus, the usage of less power and less raw and renewable materials should be taken into account by the future green electronics sector. Eco-design aims to use materials with less of an impact on the environment, fewer resources, and less material consumption during the manufacturing process. The ecology is seriously harmed by the adaptable plastic substrates that are frequently employed in printed electronics.

Biodegradable materials are a potential material class for eco-friendly electronics. Such materials should easily break down into smaller, harmless components that are broken down by bacteria or enzymes at rates similar to those of garbage (Fathi et al., 2022). Such a deterioration process ought to be initiated by a trigger during an electronic device's end-of-life phase and proceed at a regulated pace under practical environmental circumstances. Because these natural materials have several advantages, including availability, low immunoreactivity and low cost, environmentally friendly electronics are therefore required as a green and sustainable approach for managing e-waste. More significantly, these devices decrease the negative environmental effects by producing no pollution because they decompose readily.

7.8 *Sustainable Strategies for the Proper Handling of E-Waste*

Sustainable e-waste handling is the global interest for recent researchers and it focuses on eco-friendly and unpolluted procedures for regaining chemicals from raw and processed material comprising e-scrap, to address current and future environmental concerns (Fathi et al., 2022). E-waste's diverse composition and rapid growth necessitate specialized waste management strategies, including pollution control and resource recovery. The circular economy (CE) strategy offers a viable alternative for achieving sustainability. CE encourages reusing valuable materials to reduce pollution and handle secondary materials effectively. Extending producer responsibility can help employ secondary raw materials from e-waste streams to mitigate future supply chain hazards. So, CE is a kind of economic development-movement activity generally done through other activities such as material reuse (recycle, recovery) and prolongation of e-waste life (reuse, repair, renovation) instead of using the standard line of activity. Smart using of EEE and e-junks, namely reduce, rethink and refuse the activity plays a vital role in minimizing waste generation and fostering sustainable practices. The practice of the 3Rs indicates reduction of e-devices and e-waste, reuse of items and recycling the nonrepairable materials (Fathi et al., 2022). 3Rs approaches to e-waste management lacks the needed interdisciplinary collaboration among government, nongovernment, and industry stakeholders, as well as enterprise and the social sector.

7.9 *Managing Electronic Waste in India in Relation to the Sustainable Development Goals*

Extraction of valuable metals is a widely practiced activity in India (Arya & Kumar, 2020). Such practices seem to be harmful for both human beings and the environment as depicted in Fig. 1. Moreover, these harmful activities result in losses of precious metals from ineffective extraction processes. Sustainability goals have taken a great toll in the cases of Good Health and Well-being (SDG-3) and Decent

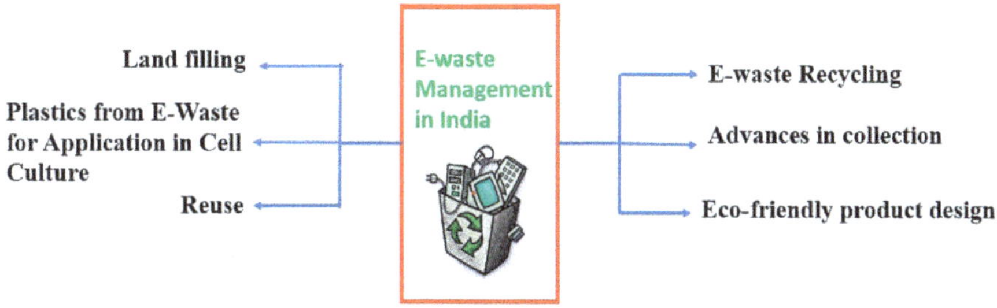

Fig. 2 Methods of managing electronic waste in India

Work and Economic Growth (SDG-8). Two segments address e-waste management in India from the perspective of a domestic generation and e-junk imported into India (legal and illegal) (Arya & Kumar, 2020). However, the Extended Producer Responsibility (EPR) has not fully complied with. Private players who violate these restrictions engage in unlicensed or informal recycling of e-waste to maximize revenues, ignoring the dangerous effect on human health and the ecosystem. However, producers may not find it profitable to execute an EPR system. Illegal/informal e-waste management does not fully exploit disassembly and recycling capabilities. E-waste is often considered as regular waste due to lack of information, which can lead to negative consequences. Unawareness about e-waste causes it to be regarded as regular waste, damaging water bodies and hindering progress towards SDGs 6 and 14 about clean water and sanitation (Kumar et al., 2022). Megacity dumps containing e-waste pose a threat to SDG 11 on Sustainable Cities and Communities, in addition to water bodies. Those who collect, deconstruct, and recycle rubbish (including e-waste) face significant risks to their health and safety. Fig. 2 describes the different methods for the management of e-waste in India.

References

Agamuthu, P., Kasapo, P., & Nordin, N. A. M. (2015). E-waste flow among selected institutions of higher learning using material flow analysis model. *Resources, Conservation and Recycling, 105*, 177–185.

Arya, S., & Kumar, S. (2020). E-waste in India at a glance: Current trends, regulations, challenges and management strategies. *Journal of Cleaner Production, 271*, 122707.

Arya, S., Rautela, R., Chava, D., & Kumar, S. (2021). Evaluation of soil contamination due to crude E-waste recycling activities in the capital city of India. *Process Safety and Environmental Protection, 152*, 641–653.

Awasthi, A. K., Hasan, M., Mishra, Y. K., Pandey, A. K., Tiwary, B. N., Kuhad, R. C., & Thakur, V. K. (2019). Environmentally sound system for E-waste: Biotechnological perspectives. *Current Research in Biotechnology, 1*, 58–64.

Bakare, A. A., Okunola, A. A., Adetunji, O. A., & Jenmi, H. B. (2009). Genotoxicity assessment of a pharmaceutical effluent using four bioassays. *Genetics and Molecular Biology, 32*, 373–381.

Bala, K., Anshul., Verma, S., Dogra, A., Das, A. P., Kumar, R., & Nandini. (2024). Sustainable management for urban plastic waste generation. In *Sustainable management of urban plas-*

tic waste through circular economic approaches (pp. 211–236). https://doi.org/10.1201/9781003465447-13

Berkhout, F., & Herti, J. (2004). De-materialising and re-materialising: digital technologies and the environment. *Futures, 36*(8), 903–920.

Bigum, M., Damgaard, A., Scheutz, C., & Christensen, T. H. (2017). Environmental impacts and resource losses of incinerating misplaced household special wastes (WEEE, batteries, ink cartridges and cables). *Resources, Conservation and Recycling, 122*, 251–260.

Birat, J. P. (2015). Life-cycle assessment, resource efficiency and recycling. *Metallurgical Research & Technology, 112*(2), 206.

Borthakur, A., & Singh, P. (2021). The journey from products to waste: a pilot study on perception and discarding of electronic waste in contemporary urban India. *Environmental Science and Pollution Research, 28*, 24511–24520.

Cong, X., Xu, X., Xu, L., Li, M., Xu, C., Qin, Q., & Huo, X. (2018). Elevated biomarkers of sympatho-adrenomedullary activity linked to e-waste air pollutant exposure in preschool children. *Environment International, 115*, 117–126.

Das, A. P., & Mishra, S. (2008). Hexavalent Chromium (VI): Health hazards and environmental pollutant. *Journal of Environmental Research and Development, 2*(3), 386–392.

Das, A., & Mishra, S. (2010). Biodegradation of the metallic carcinogen hexavalent chromium Cr(VI) by an indigenously isolated bacterial strain. *Journal of Carcinogenesis, 9*, 6. https://doi.org/10.4103/1477-3163.63584

Das, A. P., Sukla, L. B., & Pradhan, N. (2012). Microbial recovery of manganese using *Staphylococcus Epidermidis*. *International Journal of Nonferrous Metallurgy, 2012*(02), 9–12. https://doi.org/10.4236/IJNM.2012.12002

Debnath, B., Das, A., Chowdary, P. A., & Bhattacharyya, S. (2023). *Technological Advancement in E-waste Management*. CRC Press.

Dey, S., Veerendra, G. T. N., Padavala, S. A. B., & Manoj, A. P. (2023). Recycling of e-waste materials for controlling the environmental and human heath degradation in India. *Green Analytical Chemistry, 7*, 100085.

Ebrahimi, M., Khalili, N., Razi, S., Keshavarz-Fathi, M., Khalili, N., & Rezaei, N. (2020). Effects of lead and cadmium on the immune system and cancer progression. *Journal of Environmental Health Science and Engineering, 18*, 335–343.

Elgarahy, A. M., Eloffy, M. G., Priya, A. K., Hammad, A., Zahran, M., Maged, A., & Elwakeel, K. Z. (2024). Revitalizing the circular economy: An exploration of E-waste recycling approaches in a technological Epoch. *Sustainable Chemistry for the Environment, 7*, 100124.

Fathi, B. M., Ansari, A., & Ansari, A. (2022). Threats of internet-of-thing on environmental sustainability by E-waste. *Sustainability, 14*(16), 10161.

Feldt, T., Fobil, J. N., Wittsiepe, J., Wilhelm, M., Till, H., Zoufaly, A., & Göen, T. (2014). High levels of PAH-metabolites in urine of e-waste recycling workers from Agbogbloshie, Ghana. *Science of the Total Environment, 466*, 369–376.

Forti, V., Baldé, C. P., Kuehr, R., & Bel, G. (2020). *The global e-waste monitor 2020* (Vol. 120). ISWA.

Ghosh, S., & Das, A. P. (2020). Microbial metagenomics: current advances in investigating microbial ecology and population dynamics. In S. K. Nayak & B. B. Mishra (Eds.), *Frontiers in Soil and Environmental Microbiology* (1st ed., pp. 247–254). CRC Press.

Grant, K., Goldizen, F. C., Sly, P. D., Brune, M. N., Neir, M., van den Berg, M., & Norman, R. E. (2013). Health consequences of exposure to e-waste: a systematic review. *The Lancet Global Health, 1*(6), e350–e361.

Hollander, H. (2024). The decarbonisation of the maritime transport sector: assessing the transition and systemic barriers (Master's thesis).

Ilankoon, I. M. S. K., Ghorbani, Y., Chong, M. N., Herath, G., Moyo, T., & Petersen, J. (2018). E-waste in the international context–A review of trade flows, regulations, hazards, waste management strategies and technologies for value recovery. *Waste Management, 82*, 258–275.

Jana, T. K. (2023). E–waste recycling technology patents filed in India-An analysis. *Journal of Intellectual Property Rights (JIPR), 19*(5), 315–324.

Khan, M. A. R., & Saadat, A. H. M. (2019). Status of electronic waste generation in Bangladesh: a review. *International Journal of Innovative Science and Research Technology, 4*(8), 918–926.
Kirwan, M. (1997). Open university community education in the west of Scotland. *Open Learning: The Journal of Open, Distance and e-Learning, 12*(3), 49–53.
Kumar, S., Singh, R., Singh, D., Prasad, R., & Yadav, T. (2013). Electronics-waste management. *International Journal of Environmental Engineering and Management, 4*(4), 389–396.
Kumar, A., Saini, H. S., & Kumar, S. (2018). Bioleaching of gold and silver from waste printed circuit boards by Pseudomonas balearica SAE1 isolated from an e-waste recycling facility. *Current Microbiology, 75*, 194–201.
Kumar, S., Agarwal, N., Anand, S. K., & Rajak, B. K. (2022). E-waste management in India: A strategy for the attainment of SDGs 2030. *Materials Today Proceedings, 60*, 811–814.
Kumar, D., Kumar, A., Kumar, N., Sharma, A., Choudhury, R., Faisal, N., & Mane, B. K. (2023). Current trends, regulations, challenges, and management strategies of e-waste in India. In *Sustainable approaches and strategies for e-waste management and utilization* (pp. 1–25).
Kumari, R., & Samadder, S. R. (2022). A critical review of the pre-processing and metals recovery methods from e-wastes. *Journal of Environmental Management, 320*, 115887.
Lenkiewicz, Z., & Webster, M. (2017). Making waste work: A toolkit. *International Journal of Science and Engineering Research, 5*(4), 322–330.
Li, J., Wei, Y., Zhao, L., Zhang, J., Shangguan, Y., Li, F., & Ho, H. (2021). Bioaccessibility of antimony and arsenic in highly polluted soils of the mine area and health risk assessment associated with oral ingestion exposure. *iEcotoxicology and Environmental Safety*, 110/i, 308–315.
Liang, L. (2015). *The electronic waste management systems–a comparison of the e-waste management policies, processes and practices among selected countries in the greater mekong subregion*. Doctoral dissertation,. Thammasat University.
Liu, Y., Luo, X., Zeng, Y., Tu, W., Deng, M., Wu, Y., & Mai, B. (2020). Species-specific biomagnification and habitat-dependent trophic transfer of halogenated organic pollutants in insect-dominated food webs from an e-waste recycling site. *Environment International, 138*, 105674.
Liu, K., Tan, Q., Yu, J., & Wang, M. (2023). A global perspective on e-waste recycling. *Circular Economy, 2*(1), 100028.
Mallick, N., & Moh, F. H. (2003). Use of chlorophyll fluorescence in metal-stress research: a case study with the green microalga Scenedesmus. *Ecotoxicology and Environmental Safety, 55*(1), 64–69.
Mishra, S., & Das, A. P. (2023). Microbial remediation of synthetic microfiber contaminated wastewater. In M. P. Shah (Ed.), *Microbial technologies in industrial wastewater treatment* (pp. 337–350). Springer. https://doi.org/10.1007/978-981-99-2435-6_17
Mishra, S., Dash, D., Al-Tawaha, A. R. M. S., & Das, A. P. (2022). A review on heavy metal ion adsorption on synthetic microfiber surface in aquatic environments. *Applied Biochemistry and Biotechnology, 194*(10), 4639–4654. https://doi.org/10.1007/S12010-022-04029-W/METRICS
Mishra, S., Dash, D., & Das, A. P. (2023a). Aquatic microbial diversity on plastisphere: Colonization and potential role in microplastic biodegradation. *Geomicrobiology Journal, 1–12*, 312. https://doi.org/10.1080/01490451.2023.2209750
Mishra, S., Ghosh, S., van Hullebusch, E. D., Singh, S., & Das, A. P. (2023b). A critical review on the recovery of base and critical elements from electronic waste-contaminated streams using microbial biotechnology. *Applied Biochemistry and Biotechnology, 2023*, 1–30. https://doi.org/10.1007/S12010-023-04440-X
Mohammadi, E. (2020). *E-waste in the Caribbean: Is there a potential for a circular economy?*. Master's thesis,. University of Waterloo.
Narayanasamy, M., Dhanasekaran, D., & Thajuddin, N. (2021). Bioremediation of noxious metals from e-waste printed circuit boards by Frankia. *Microbiological Research, 245*, 126707.
Nicholls, A. M., & Mal, T. K. (2003). Effects of lead and copper exposure on growth of an invasive weed, Lythrum salicaria L (Purple Loosestrife) 1. *The Ohio Journal of Science, 103*(5), 129–133.

Nie, X., Fan, C., Wang, Z., Su, T., Liu, X., & An, T. (2015). Toxic assessment of the leachates of paddy soils and river sediments from e-waste dismantling sites to microalga, *Pseudokirchneriella subcapitata*. *Ecotoxicology and Environmental Safety, 111*, 168–176.

Okine, H. A. (2014). *E-waste imports and management practices in Ghana: A case study of Accra-Tema metropolitan area.* Doctoral dissertation, M. Phil thesis,. University of Ghana.

Prabhakar, A., Mishra, S., & Das, A. P. (2019). Isolation and identification of lead (Pb) solubilizing bacteria from automobile waste and its potential for recovery of lead from end of life waste batteries. *Geomicrobiology Journal, 36*(10), 894–903. https://doi.org/10.1080/01490451.2019.1654044

Priyadarshini, B., Santhosh Kumar, M., & Das, A. P. (2024). Microbial degradation of synthetic plastics released from COVID-19 generated single use disposable face masks. Impact of COVID-19 waste on environmental pollution and its sustainable management. *Environmental Science and Engineering, Part F, 2827*, 1–16. https://doi.org/10.1007/978-3-031-50840-0_1

Priyadarsini, S., & Das, A. P. (2024). Analytical and structural characterization of waste lithium-ion batteries for their effective recycling strategy. *Environmental Chemistry and Ecotoxicology, 7*, 182. https://doi.org/10.1016/j.enceco.2024.12.004

Priyadarsini, S., & Das, A. P. (2025). Lithium bioleaching: A review on microbial-assisted sustainable technology for lithium bio-circularity. *Journal of Water Process Engineering, 69*, 106744. https://doi.org/10.1016/j.jwpe.2024.106744

Raj, K., & Das, A. P. (2023). Lead pollution: Impact on environment and human health and approach for a sustainable solution. *Environmental Chemistry and Ecotoxicology, 5*, 79–85. https://doi.org/10.1016/J.ENCECO.2023.02.001

Razip, M. M., Savita, K. S., Kalid, K. S., Ahmad, M. N., Zaffar, M., Rahim, E. E. A., & Ahmadian, A. (2022). The development of sustainable IoT E-waste management guideline for households. *Chemosphere, 303*, 134767.

Saha, L., Kumar, V., Tiwari, J., Rawat, S., Singh, J., & Bauddh, K. (2021). Electronic waste and their leachates impact on human health and environment: Global ecological threat and management. *Environmental Technology & Innovation, 24*, 102049.

Sahu, S., & Das, A. P. (2024). Microplastics degradation mechanism by aquatic plastiomes: A review. Impact of COVID-19 waste on environmental pollution and its sustainable management. *Environmental Science and Engineering, Part F, 2827*, 213–226. https://doi.org/10.1007/978-3-031-50840-0_12

Sahu, A., Dey, S., Mallick, K., Priyadarsini, S., Behera, I. D., & Das, A. P. (2024). A review on advanced technology for sustainable management of synthetic microplastic waste. In *Sustainable management of urban plastic waste through circular economic approaches* (pp. 262–279). https://doi.org/10.1201/9781003465447-15

Sakhuja, D., Ghai, H., Bhatia, R. K., & Bhatt, A. K. (2022). Management of e-Waste: technological challenges and opportunities. In *Handbook of solid waste management: Sustainability through circular economy* (pp. 1523–1557). Springer.

Salam, M. D., & Varma, A. (2019). A review on impact of e-waste on soil microbial community and ecosystem function. *Pollution, 5*(4), 761–774.

Singh Jadaun, J., Bansal, S., Sonthalia, A., Rai, A. K., & Singh, S. P. (2022). Biodegradation of plastics for sustainable environment. *Bioresource Technology, 347*, 126697. https://doi.org/10.1016/J.BIORTECH.2022.126697

Singh, N., Duan, H., & Tang, Y. (2020). Toxicity evaluation of E-waste plastics and potential repercussions for human health. *Environment International, 137*, 105559.

Sonawane, B., Kadry, A. R. M., & Kancherla, J. (2001). Electronic waste: Human health and environmental risk. In *Patty's Toxicology* (pp. 1–18).

Stevenin, G., Canonge, J., Gervais, M., Fiore, A., Lareyre, F., Touma, J., & Sénémaud, J. (2024). e-Health and environmental sustainability in vascular surgery. *Seminars in Vascular Surgery, 37*(3), 333–341.

Torres, R., Segura-Bailón, B., & Lapidus, G. T. (2018). Effect of temperature on copper, iron and lead leaching from e-waste using citrate solutions. *Waste Management, 71*, 420–425.

Turaga, R. M. R., Bhaskar, K., Sinha, S., Hinchliffe, D., Hemkhaus, M., Arora, R., & Sharma, H. (2019). E-waste management in India: Issues and strategies. *Vikalpa, 44*(3), 127–162.

Youcai, Z. (2018). Leachate generation and characteristics. In *Pollution control technology for leachate from municipal solid waste* (pp. 1–30).

Zhang, W., Wang, H., Zhang, R., Yu, X. Z., Qian, P. Y., & Wong, M. H. (2010). Bacterial communities in PAH contaminated soils at an electronic-waste processing center in China. *Ecotoxicology, 19*, 96–104.

Impact of E-Waste on Human Health and Environment

Kiran Bala, Arushi Dogra, Rajkumar, Nandini, Alok Prasad Das, and Sarfaraz Ali

Abstract The environment and human health are seriously threatened by the exponential growth of electronic garbage, or "e-waste." By 2050, it is anticipated that the amount of e-waste produced by outdated electronic gadgets will exceed 110 million metric tons per year. Toxic substances including lead, mercury, and cadmium are released into the environment through improper disposal, which contaminates soil, water, and air, especially in developing countries. Particularly among susceptible groups, this results in serious health problems such as neurological abnormalities, respiratory conditions, and developmental problems. Because of insufficient safety measures, informal recycling makes these risks worse. Additionally, e-waste pollutes biodiversity, upsets ecosystems, and depletes resources. Strong international waste management regulations, cutting-edge recycling technologies, and public education campaigns are needed to address this challenge. Governments, businesses, and communities working together can lessen the ecological impact of e-waste, promoting sustainable development and protecting the environment and public health. This report emphasizes how urgent it is to implement systemic e-waste solutions to lessen its growing worldwide effects.

K. Bala (✉)
Department of Zoology Deshbandhu College, University of Delhi, New Delhi, Delhi, India
e-mail: kbala@db.du.ac.in

A. Dogra
Faculty of Life Sciences and Biotechnology, South Asian University, New Delhi, Delhi, India

Rajkumar
S.N. Sinha College Tekari, Gaya Magadh University, Bodh Gaya, Bihar, India

Nandini
Department of Zoology, University of Delhi, New Delhi, Delhi, India

A. P. Das
Department of Life Sciences, Rama Devi Women's University, Bhubaneswar, Odisha, India

S. Ali
P. G. Department of Biotechnology, Magadh University, Bodh Gaya, Bihar, India

A. P. Das, S. Priyadarsini (eds.), *Electronic Waste and Environmental Pollution*, Sustainable Environmental Waste Management Strategies,
https://doi.org/10.1007/978-3-031-90287-1_6

Keywords E-waste · Health · Environment · Management

1 Introduction

In an increasingly digital age, the quick development of electronic gadgets has also led to a sharp rise in electronic waste, or "e-waste," endangering both the environment and human health. E-waste is already a major issue; as the use of electronics increases, its production may reach 110 million metric tons or more by 2050. Developing nations bear the brunt of this issue, as many of them dispose of it in a very inefficient manner. Electronic waste breaks under the terms of obsolete electrical and electronic devices which do not work and are no longer useful (Saha et al., 2021). This category includes a wide range of products, such as computers, cellphones, televisions, and household appliances, all of which have several parts that, if illegally disposed of, could be harmful to both human health and the environment (Mishra et al., 2023). Due to significant technological advancements and changes in end-user behavior, e-waste—the second-fastest growing waste category in the world—generated around 53.6 megatonnes of rubbish in 2019 and is predicted to continue its upward trend in the years to come. When improperly marketed, these gadgets can also release harmful metals like lead, mercury, and cadmium into the environment, which can have detrimental effects on health, including neurotoxicity and severe after-effects (Hodge et al., 2012).

The status of e-waste across the globe is extremely alarming, with around 53.6 mega tonnes of e-waste having been generated in the year 2019, which is expected to reach around 110 mega tonnes by the year 2050. This increase mirrors shifts in the tech landscape, as well as a dynamic consumer culture that nudges consumers toward upgrading and disposing of devices often. In many developing countries, where informal recycling is common, a substantial amount of this waste is unmanaged (Barbiani, 2024). That frequently results in acute pollution and health dangers. Heavy metals, including lead and mercury, can leach into soil and water, which is a vital risk for ecosystems and human health. This signifies the need for various actions in waste management globally as the correlation between electronic waste generation and its effects is a complicated issue and needs immediate thought regarding e-waste management.

1.1 Importance of the Study of the Effects of E-Waste

Since e-waste is present in both developing and developed nations, it is crucial to understand its impact. A major environmental hazard, electronic waste, contains toxic materials including but not limited to lead, mercury, and cadmium (Prabhakar et al., 2019; Raj & Das, 2023). They can seep into soil and water, which presents grave health dangers to humans and wildlife (Das & Singh, 2011; Dey et al., 2023a). Considering e-waste production is expected to rise to 110 million metric tons by 2050, knowing its life cycle is important to create successful management plans (Dey et al., 2023b). In addition, how e-waste gets disposed of and the detrimental processes employed through their disposal (e.g., landfilling e-waste or burning e-waste) contribute to pollution and the ability for rapid public health intervention to be performed to reduce exposure to harmful materials. Examining these effects provides insight into the social and economic challenges of e-waste disposal— but it also raises awareness of more sustainable ways to dispose of electronics that can inform health- and environmentally oriented policies in the wake of shifting technology (Kumar et al., 2022).

2 Human Impact of E-Waste

E-waste refers to the ways we exhaust electronics— the destruction they cause to the ecology and to human health— the overall effects being grave and complicated, particularly in lower-income countries, where recycling is done in the informal sector (Fevrier, 2020). Simple recycling, which often lacks any safety protocols, exposes workers to toxic materials including lead, mercury, and cadmium, endangering their lives and those of nearby communities. Thus, health problems appear not only as immediate difficulty breathing and cognition but become chronic conditions like stunting in children and chronic illness (Hassan & Saleh, 2022). That e-waste recycling, in general, happens for financial motives rather than environmental ones has only compounded these hazards. Human beings are not the only ones who are being harmed by them, but the environment is too, and therefore, these need to be acted on in the form of a policy change so that their effect is minimized (Côté et al., 2021). Sources of e-waste is summarized in Fig. 1.

2.1 Contaminants Found in E-Waste

They are even major threats due to the harmful contents of electronic devices that can negatively influence human health and ruin the environment while generally considered e-waste (Kumar & Singh, 2014). Upon analyzing the actual components of electronic devices, these components often include toxic heavy metals such as lead, mercury, lithium, manganese, or cadmium, which naturally leach out into soil and water if not without sufficient and proper disposal and are potentially injurious to the environment (Stuart et al., 2012; Priyadarsini & Das, 2024, 2025). This problem is exacerbated at informal recycling sites, particularly in low-income countries, where workers are exposed to hazardous waste without adequate protective measures (Staines et al., 2004; Das & Mishra, 2008, 2009; Paul et al., 2023). There are

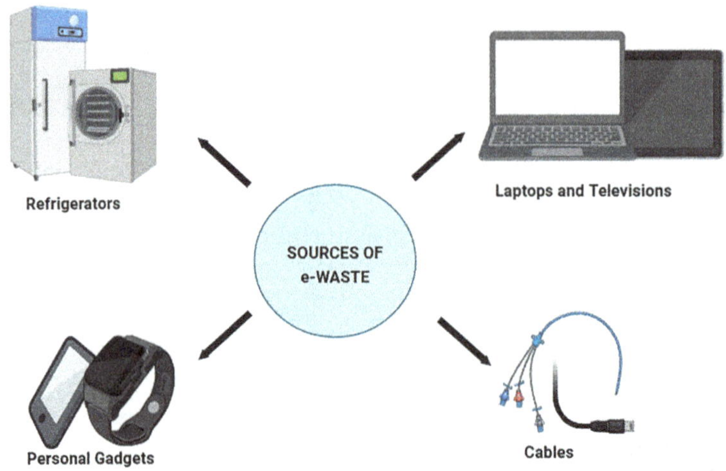

Fig. 1 Sources of e-waste

at least a few ways people can be exposed, through inhalation of dust and fumes during e-waste work, or ingestion of unsafe water and food. All of these toxic effects add up to serious illness, including brain development and cancer (Huang et al., 2016; Mishra & Das, 2008). This highlights the urgency to manage and recycle e-waste to minimize adverse impacts on health and ecology.

2.2 Health Effects Associated with E- Waste Exposure

The unsafe disposal of e-waste leads to serious health problems among sensitive population groups that perform informal recycling in underdeveloped countries. E-waste contains hazardous substances, including lead, mercury, and brominated flame retardants, which can leach into the environment and are detrimental to human health. People living near e-waste, either through direct contact with the waste itself or through toxic soil and air runoff, experience increased rates of respiratory illnesses (Ling et al., 2022), neurological disorders (Swanston, 1993), and reproductive-system diseases (Xu et al., 2014). Less resilient populations, such as women and children, are also hit harder due to their higher susceptibility to toxicity. Middle-class socioeconomic circumstances that restrict access to health care and legal support compound this issue (Swanston, 1993). While these communities contend with economic needs associated with e-waste recycling, strong regulatory frameworks set in place can establish health/ safety (including reduced risk from exposure to e-waste) and increase better recycling practices (Widmer et al., 2005).

2.3 Vulnerable Populations in the E-Waste Context

The dangerous consequences of e-waste fall heavily on the most vulnerable, particularly among low-income and marginalized communities that reside near informal recycling operations. These populations are more exposed to toxic chemistries, including heavy metals and persistent organic pollutants that accumulate in the environment and result from poor waste management, thus putting them at increased risk (Tsydenova & Bengtsson, 2011). Many studies show that the workers are taking on high body burdens of dangerous pollutants, and so are residents in those areas, leading to severe health problems including a range of neurological, endocrine, and developmental illnesses, especially in children (Liu, 2023). The socioeconomic challenges faced by these groups with poor education and health literacy increase their vulnerability to the negative health effects of e-waste exposure (Swanston, 1993). Thus, it highlights the need for proper governance across the spectrum to save these communities from the devastating effects of e-waste.

2.4 Health Effects of E-Waste: Evidence from Case Studies

Health effects from mismanaged e-waste have cast a bright light on the issues of disposal. When it comes to informal recycling jobs, harmful substances such as lead, mercury, and cadmium the worker is exposed to the potential to cause brain damage and chronic disease in local communities in many developing countries (Leung et al., 2006). For instance, Guiyu, China, has been responsible for the processing of a substantial amount of e-waste and given the available literature, studies have shown high blood lead levels in local children, which have been associated with developmental deficiencies (Hodge et al., 2012) The information found in this paper highlights the pressing necessity of regulations for mandatory HIA when considering e-waste management within areas of poor disposal systems. Based on these case studies, e-waste has been causing detrimental impacts on the environment and human health and is a serious problem to be researched, and governments have taken measures to alert them (Tetteh & Lengel, 2017).

3 E- Waste and Its Efficacy on the Environment

Given that we are consuming more technology, that is quickly becoming obsolete, electronic waste (or e-waste) is a significant environmental issue. E-waste typically contains harmful materials including metals like lead, manganese, and cadmium as well as toxic substances that can leach into soil and water, contaminating ecosystems (Das et al., 2011; Ghosh et al., 2016; Mohanty et al., 2016). This pollution alters the habitat and can impact human health as well, particularly where improper disposal approaches contaminate the environment (Louis, 2019). Furthermore, developing countries are often e-waste dumping for developed nations which adds to their problems by contaminating air and fresh water. Therefore, a systemic cycle refers to the need for eco-friendly environments while environmental impact can be controlled with recycling and the right environment disposal measures with fossil-free information technology approaches (Mwaguni et al., 2017). This step helps cater to the sustainable growth of technology that is eco-friendly and human-friendly as well as eliminating the growth of e-waste.

3.1 E-Waste Disposal and Soils Contamination

The soil-bearing pollution due to e-waste disposal can pose massive health drawbacks to the environment and agricultural productivity. Badly managed e-waste also contains hazardous substances, including lead cadmium and brominated flame retardants that can seep toxic chemicals into the soil (Yang et al., 2019). Biologically, this translates into dire ecological problems and health risks. More serious is that these metals can leach out from e-waste and disturb the chemical properties of soil, a situation that would hinder plants′ root development and impair soil-associated microbes (Staines et al., 2004). And that can damage human health when crops absorb these pollutants. People who consume contaminated food may suffer from chronic diseases like neurotoxicity and carcinogenicity (Chen et al., 2011). Cities can tackle soil pollution from e-waste and protect environmental and public health, through stricter regulations and innovative recycling technology. Overcoming this challenge is critical for sustainable land use and for food security (Dórea, 2021).

3.2 E- Waste and Water Pollution

The growing world production of e-waste threatens the quality of water, and an evident link exists between the expansion of technologies and environmental degradation. Electronics have poisonous components such as heavy metals and toxic organic chemicals, and if disposed of inappropriately, it adds to the pollution of the water as these materials seep into the ground or are burned (Gadi & Gadi, 2015). Metals (lead, mercury, cadmium, etc.) accumulate in these environments and leach easily into the water systems and are thus very harmful to the quality of water (Gulati et al., 2022). If, say, water is unsafe because it contains e-waste, fewer fish will hatch or they will have reproductive issues which in turn means there are food safety and public health implications. So, tackling e-waste helps protect our environment but also protects people from the invisible danger of water pollution linked to electronic waste (Abalansa et al., 2021).

3.3 Atmospheric Pollution Due to E-Waste Processing

Electronic waste (e-waste) is recycled, which however is known to highly reduce air quality (air quality is highly degraded), as the recycling of e-waste produces a large amount of harmful pollutants (Gangwar et al., 2019). Because people often burn e-waste and mine metals where they must, toxic chemicals like dioxins and heavy metals end up in the air, contributing to local and global air pollution. Beyond that, air pollution can exacerbate respiratory ailments and other health problems for those who live near such facilities, especially children and seniors (Akram et al., 2019). Audible emanations from these unregulated electronics recycling operations have been connected to compromised air quality as well, making it difficult, if not dangerous, to live in the neighborhood. Addressing the issues associated with e-waste is thus important not just for the environment, but also for the health of people living in surrounding communities (Chen et al., 2009).

3.4 The Increasing Amounts of Technological Waste and the Loss of Biodiversity

The amount of e-waste generated is bringing with it a newer form of toxicity into the ecosystem, which is beginning to pose a threat to biodiversity. Toxic compounds typically found in e-waste heavy metals, flame retardants, and so on seep into the dirt and water, poisoning ecosystems and depleting natural resources necessary to billions of organisms (Wang et al., 2022). Along with human behaviors that accompany e-waste, this may cause further damage to ecosystems by impairing species migrations and genetic diversity. These adverse effects are particularly exacerbated by improper e-waste disposal methods (burning, dumping, etc.) to the extent that it is viable for biodiversity conservancy concerns in these regions (Rawat et al., 2020). Additionally, e-waste and shrinking habitats are a source of contention for the hoped-for exchange of technological advancements and the nurturing of environmental ingredients under less effective regulation in informal sectors, where a greater number of people and environmental fitness will presumably be tight due to the much poorer adaptiveness (Robinson, 2009). The impact of improper e-waste on humans and the environment is summarized in Fig. 2.

4 E-Waste Management and Mitigation Strategies

Ensuring responsible e-waste management and reduction is crucial not only for human health but also for the environment. Responsible recycling and refurbishment of electronic materials greatly reduce the number of toxic substances being released into ecosystems and thus into the pollution of air, soil, and water. Finally,

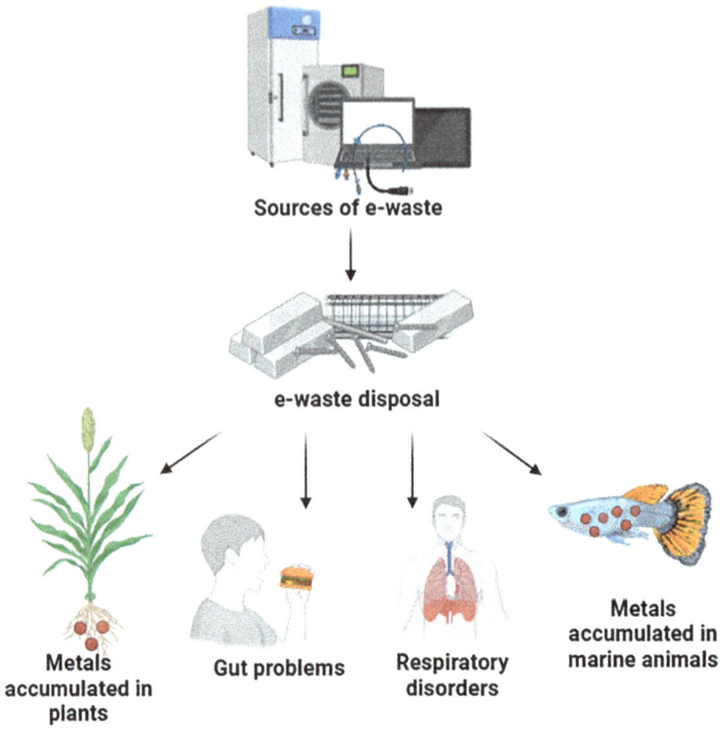

Fig. 2 Impact of e-waste on the environment and humans

e-waste disposal for general recycling should be more closely regulated, especially in low-income countries where e-waste is often recycled in such an occurrence that it is in a way such to aggravate the hazardous impact of the toxic substances (Agarwal et al., 2012). Conclusion certifies e-waste treatment plants; awareness campaign promoting sustainable consumer behavior motivates consumers, to visit certified e-waste treatment plants. In addition, collaboration between government, nonprofit, and commercial spheres in this area will help to restore the resource recovery methods optimized, resulting in high-value components and materials into the manufacturing chain again to save resources and fight the environmental problem of e-waste, by adopting a circular economy (Kahhat et al., 2008).

4.1 Step-for-Step Treatment Templates Regulating the E-Waste

While several governments are now formulating policies to tackle the global challenge posed by e-waste and mitigate its adverse effects in environmental and public health terms, there is weak implementation and poor coordination among stakeholders. These could all be developed as standard tools (basically alternatives with similar purposes of sustainable behavior), but these will be highly dependent on each country having the ability to put them into place adapted to the country (Huber et al., 1998). In the first place, existing laws in many instances have insufficient coverage considering the complexity of e-waste streams, where improper disposal may result in hazardous problems in terms of contamination of soil or water. Therefore, such monumental reforms to ensure accountability with limitations to minimize the frequency of replacements, and a misconception of sustainability and robust recycling models should be tailored to achieve better e-waste management and safeguard public and environmental health (Mor et al., 2021).

4.2 Technologies for Recycling and Recovery

Recycling and recovery technologies are critical to tackling the major environmental and health threats posed by electronic waste (e-waste). Hydrometallurgical processes and bioleaching processes are new techniques; they not only successfully obtain valuable components from the waste but also are eco-friendly and financially profitable as they decrease the liberation of fossils that make the cycle of an economy. For example, the process of extracting precious metals from e-waste reduces the need for new materials and reduces the damaging impacts of quarrying, which is recognized as a risk to the environment and the health of people (Wysokińska, 2010). As an example, phosphorus is necessary to sustain farming; therefore, the recovery of phosphorus from the e-waste is needed (Mayer et al., 2016), which shows how waste management strategies could also address issues of resource scarcity and environmental degradation. However, these technologies can only function based on legitimate policies and awareness among the public regarding sustainable electronic waste disposal and recuperation methods, which in the long run will not only uplift environmental standards but also public health (Li & Achal, 2020).

4.3 Public Outreach and Engagement Initiatives

Awareness and education are key to countering the threats that e-waste (electronic waste) poses to human health and the environment. It discusses the hazards of disposing of e-waste, and what proper recycling practices and disposal encompass. Studies indicate that the health risks associated with the leaching of toxic metals such as Pb and Hg in the illegal disposal of e-waste could be mitigated by implementing a public awareness campaign emphasizing the health hazards of the former metals involved (Wahga et al., 2015). Integrating sustainability and e-waste in the curriculum can also promote responsibility among students (Boone et al., 2011). Furthermore, if indeed community initiatives play a role in promoting local governance and organizations, by keeping the call for sustainable solutions more prevalent among the citizens themselves, such awareness campaigns can become all the more potent, helping not only to make at-risk populations aware of the issues that e-waste disposal brings, but even incentivize populations at risk to embrace sustainable e-waste practices to mitigate the adverse effects of e-waste should it be done in an organized manner and at scale.

4.4 *Role of Industries in Management E-Waste*

It made it that much more important to find sensible ways to deal with e-waste not just to extract its negative impact on human and planetary health, but to close the recycling loop and tackle those nasty global supply chains through industry engagement. Through interfirm best practices, recycling useful materials and reducing waste around the pedal are more cyclical economy compatible (Pickren, 2015). Instead, the substances hazardous in e-waste such as lead and mercury which have human health and environmental degradation effects are discharged into industries and encouraged to develop new treatment methods. They are also leaders in advancing sustainable design guidelines, reducing toxic materials in electronics, and extending product life. Certainly, adoption of sustainable e-waste management practices in industry systems can significantly reduce the threat to public health and keep the environment viable for future (Fu et al., 2020).

5 Conclusion

Technology contributes to the growth of society, but everything has a limit. The adverse impacts of improper waste management methods such as landfilling and incineration, which according to extensive research cause pollution and contamination of soils and water bodies with hazardous heavy metals and toxic substances, permeate natural systems. E-waste toxic materials, including lead, mercury, and cadmium, are health risks from exposure to such items, adding to the problem of waste disposal and showing how important better solutions to waste are. Ignoring the e-waste problems, this chapter argues, not only threatens the environment but is a significant risk to public health, especially for vulnerable populations living in poorer regions. So new methods of disposal and recycling need to at least be introduced in parallel with regulation to minimize the health risks and environmental harm from the growing scourge of e-waste.

References

Abalansa, S., El Mahrad, B., Icely, J., & Newton, A. (2021). Electronic waste, an environmental problem exported to developing countries: The GOOD, the BAD and the UGLY. *Sustainability, 13*(9), 5302.

Agarwal, B., Archer, C., Castañeda, I., Sonia, D., Dankelman, I., Elson, D., et al. (2012). *Powerful-synergies: Gender equality*. Sustainability.

Akram, R., Natasha, Fahad, S., Hashmi, M. Z., Wahid, A., Adnan, M., et al. (2019). Trends of electronic waste pollution and its impact on the global environment and ecosystem. *Environmental Science and Pollution Research, 26*, 16923–16938.

Barbiani, C. (2024). *The right to Bank: A proposal for introducing a new human right under international law*. Ethics International Press.

Boone, J. W., Carson, A. H., Sarah, F., Hughes, E. J., Leech, C. A., McGrath, T. J., Moody, N. L., Neumann, C. I., Ranii, M. K., Ryder, J. D., Sherman, M. Tower, S. W., & Weinstein, S. (2011). *Gettysburg College Sustainability Proposal*.

Chen, D., Bi, X., Zhao, J., Chen, L., Tan, J., Mai, B., et al. (2009). Pollution characterization and diurnal variation of PBDEs in the atmosphere of an E-waste dismantling region. *Environmental Pollution, 157*(3), 1051–1057.

Chen, A., Dietrich, K. N., Huo, X., & Ho, S. M. (2011). Developmental neurotoxicants in e-waste: An emerging health concern. *Environmental Health Perspectives, 119*(4), 431–438.

Côté, D., Gravel, S., Gladu, S., Bakhiyi, B., & Gravel, S. (2021). Worker health in formal electronic waste recycling plants. *International Journal of Workplace Health Management, 14*(3), 292–309.

Das, A. P., & Mishra, S. (2008). Hexavalent chromium (VI): Health hazards and environmental pollutant. *J Environ Res Dev, 2*(3), 386–392.

Das, A. P., & Mishra, S. (2009). Hexavalent chromium [Cr (VI)]: Yellow water pollution and its remediation. *Sarovar Saurabh ENVIS Newsl Wetl Ecosyst, 5*(2), 1.

Das, A. P., & Singh, S. (2011). Occupational health assessment of chromite toxicity among Indian miners. *Indian Journal of Occupational and Environmental Medicine, 15*(1), 6. https://doi.org/10.4103/0019-5278.82998

Das, A. P., Sukla, L. B., Pradhan, N., & Nayak, S. (2011). Manganese biomining: A review. *Bioresource Technology, 102*(16), 7381–7387. https://doi.org/10.1016/J.BIORTECH.2011.05.018

Dey, S., Das, A., Mallick, K., Sahu, A., & Das, A. P. (2023a). *Environmental petroleum waste: Pollution, toxicity, sustainable remediation* (pp. 159–175). https://doi.org/10.1007/978-3-031-48220-5_7

Dey, S., Veerendra, G. T. N., Padavala, S. S. A. B., & Manoj, A. P. (2023b). Recycling of e-waste materials for controlling the environmental and human heath degradation in India. *Green Analytical Chemistry, 7*, 100085.

Dórea, J. G. (2021). Neurodevelopment and exposure to neurotoxic metal (loid) s in environments polluted by mining, metal scrapping and smelters, and e-waste recycling in low and middle-income countries. *Environmental Research, 197*, 111124.

Fevrier, K. M. (2020). *Race and waste: The politics of electronic waste recycling & scrap metal recovery in Agbogbloshie*. Accra.

Fu, J., Zhong, J., Chen, D., & Liu, Q. (2020). Urban environmental governance, government intervention, and optimal strategies: A perspective on electronic waste management in China. *Resources, Conservation and Recycling, 154*, 104547.

Gadi, R. N., & Gadi, N. A. N. (2015). E-Waste: An Emerging Problem of Innovative Society. *International Journal of Students Research in Technology & Management, 2*, 163–167.

Gangwar, C., Choudhari, R., Chauhan, A., Kumar, A., Singh, A., & Tripathi, A. (2019). Assessment of air pollution caused by illegal e-waste burning to evaluate the human health risk. *Environment International, 125*, 191–199.

Ghosh, S., Mohanty, S., Nayak, S., Sukla, L. B., & Das, A. P. (2016). Molecular identification of indigenous manganese solubilising bacterial biodiversity from manganese mining deposits. *Journal of Basic Microbiology, 56*(3), 254–262. https://doi.org/10.1002/JOBM.201500477

Gulati, R., Kour, A., & Sharma, P. (2022). Ecological impact of heavy metals on aquatic environment with reference to fish and human health. *Journal of Applied and Natural Science, 14*(4), 1471–1484.

Hassan, A. I., & Saleh, H. M. (2022). Toxicity and hazardous waste regulations. In *Hazardous waste management* (pp. 165–182). Elsevier.

Hodge, J., Fuse Brown, E., Scanlon, M., & Corbett, A. (2012). Legal review concerning the use of health impact assessments in non-health sectors. *Journal of Allied Health, 41*(2), 52.

Huang, C. L., Bao, L. J., Luo, P., Wang, Z. Y., Li, S. M., & Zeng, E. Y. (2016). Potential health risk for residents around a typical e-waste recycling zone via inhalation of size-fractionated particle-bound heavy metals. *Journal of Hazardous Materials, 317*, 449–456.

Huber, R. M., Ruitenbeek, H. J., & Da Motta, R. S. (1998). *Market-based instruments for environmental policymaking in Latin America and the Caribbean: Lessons from eleven countries* (Vol. 381). World Bank Publications.

Kahhat, R., Kim, J., Xu, M., Allenby, B., Williams, E., & Zhang, P. (2008). Exploring e-waste management systems in the United States. *Resources, Conservation and Recycling, 52*(7), 955–964.

Kumar, U., & Singh, D. N. (2014). Electronic waste: Concerns & hazardous threats. *International Journal of Current Engineering and Technology, 4*(2), 802.

Kumar, A., Gaur, D., Liu, Y., & Sharma, D. (2022). Sustainable waste electrical and electronic equipment management guide in emerging economies context: A structural model approach. *Journal of Cleaner Production, 336*, 130391.

Leung, A., Cai, Z. W., & Wong, M. H. (2006). Environmental contamination from electronic waste recycling at Guiyu, Southeast China. *Journal of Material Cycles and Waste Management, 8*, 21–33.

Li, W., & Achal, V. (2020). Environmental and health impacts due to e-waste disposal in China–A review. *Science of the Total Environment, 737*, 139745.

Ling, S., Lu, C., Fu, M., Zhou, S., Qiao, Z., Peng, C., et al. (2022). Distribution characteristics and risks assessment of brominated flame retardants in surface soil from both a legacy and a new e-waste dismantling site. *Journal of Cleaner Production, 373*, 133970.

Liu, Z. (2023). Informal electrical and electronics waste recycling and its health impacts: Evidence from developing countries.

Louis, Y. S. (2019). *IT Dumping in Africa: How Green Technology Can Be an Antidote.*

Mayer, B. K., Baker, L. A., Boyer, T. H., Drechsel, P., Gifford, M., Hanjra, M. A., et al. (2016). Total value of phosphorus recovery. *Environmental Science & Technology, 50*(13), 6606–6620.

Mishra, S., & Das, A. P. (2008). Hexavalent chromium reduction and 16S rDNA identification of bacteria isolated from a Cr (VI) contaminated site. *The Internet Journal of Microbiology, 7*(1).

Mishra, S., Ghosh, S., van Hullebusch, E. D., Singh, S., & Das, A. P. (2023). A critical review on the recovery of base and critical elements from electronic waste-contaminated streams using microbial biotechnology. *Applied Biochemistry and Biotechnology, 2023*, 1–30. https://doi.org/10.1007/S12010-023-04440-X

Mohanty, S., Ghosh, S., Nayak, S., & Das, A. P. (2016). Isolation, identification and screening of manganese solubilizing fungi from low-grade manganese ore deposits. *Geomicrobiological Journal, 34*(4), 309–316. https://doi.org/10.1080/01490451.2016.1189016

Mor, R. S., Sangwan, K. S., Singh, S., Singh, A., & Kharub, M. (2021). E-waste management for environmental sustainability: An exploratory study. *Procedia CIRP, 98*, 193–198.

Mwaguni, S., Ayiemba, E., & Onyari, J. (2017). On-site sanitation and its effects on the groundwater resources of Nyali-Bamburi-Shanzu and Diani-Chale, Kenya. *Journal of Environment and Earth Science, 7*(8), 47–57.

Paul, A., Dey, S., Ram, D. K., & Das, A. P. (2023). Hexavalent chromium pollution and its sustainable management through bioremediation. *Geomicrobiology Journal, 41*, 1. https://doi.org/10.1080/01490451.2023.2218377

Pickren, G. (2015). Making connections between global production networks for used goods and the realm of production: A case study on e-waste governance. *Global Networks, 15*(4), 403–423.

Prabhakar, A., Mishra, S., & Das, A. P. (2019). Isolation and identification of Lead (Pb) solubilizing bacteria from automobile waste and its potential for recovery of Lead from end of life waste batteries. *Geomicrobiological Journal, 36*(10), 894–903. https://doi.org/10.1080/01490451.2019.1654044

Priyadarsini, S., & Das, A. P. (2024). Analytical and structural characterization of waste lithium-ion batteries for their effective recycling strategy. *Environmental Chemistry and Ecotoxicology., 7*, 182. https://doi.org/10.1016/j.enceco.2024.12.004

Priyadarsini, S., & Das, A. P. (2025). Lithium bioleaching: A review on microbial-assisted sustainable technology for lithium bio-circularity. *Journal of Water Process Engineering, 69*, 106744. https://doi.org/10.1016/j.jwpe.2024.106744

Raj, K., & Das, A. P. (2023). Lead pollution: Impact on environment and human health and approach for a sustainable solution. *Environmental Chemistry and Ecotoxicology, 5*, 79–85. https://doi.org/10.1016/J.ENCECO.2023.02.001

Rawat, S., Verma, L., & Singh, J. (2020). Environmental hazards and management of E-waste. In *Environmental concerns and sustainable development: Volume 2: Biodiversity, soil and waste management* (pp. 381–398).

Robinson, B. H. (2009). E-waste: An assessment of global production and environmental impacts. *Science of the Total Environment, 408*(2), 183–191.

Saha, L., Kumar, V., Tiwari, J., Rawat, S., Singh, J., & Bauddh, K. (2021). Electronic waste and their leachates impact on human health and environment: Global ecological threat and management. *Environmental Technology & Innovation, 24*, 102049.

Staines, A., Crowley, D., Bruen, M., & O'Connor, P. (2004). *Public health and landfill sites*. Department of. Public Health Eastern Regional Health.

Stuart, M., Lapworth, D., Crane, E., & Hart, A. (2012). Review of risk from potential emerging contaminants in UK groundwater. *Science of the Total Environment, 416*, 1–21.

Swanston, S. F. (1993). Race, gender, age, and disproportionate impact: What can we do about the failure to protect the most vulnerable. *Fordham Urb LJ, 21*, 577.

Tetteh, D., & Lengel, L. (2017). The urgent need for health impact assessment: Proposing a transdisciplinary approach to the e-waste crisis in sub-Saharan Africa. *Global Health Promotion, 24*(2), 35–42.

Tsydenova, O., & Bengtsson, M. (2011). Chemical hazards associated with treatment of waste electrical and electronic equipment. *Waste Management, 31*(1), 45–58.

Wahga, A., Blundel, R., & Schaefer, A. (2015). Human capital and environmental engagement of SMEs in Pakistan: A comparative analysis of the leather industry.

Wang, X., Teng, Y., Wang, X., Xu, Y., Li, R., Sun, Y., et al. (2022). Effects of combined pollution of organic pollutants and heavy metals on biodiversity and soil multifunctionality in e-waste contaminated soil. *Journal of Hazardous Materials, 440*, 129727.

Widmer, R., Oswald-Krapf, H., Sinha-Khetriwal, D., Schnellmann, M., & Böni, H. (2005). Global perspectives on e-waste. *Environmental Impact Assessment Review, 25*(5), 436–458.

Wysokińska, Z. (2010). Completion of the common internal market of recycling in the EU-position of new member states. *Comparative economic research. Central and Eastern Europe, 13*(4), 47–59.

Xu, X., Zhang, Y., Yekeen, T. A., Li, Y., Zhuang, B., & Huo, X. (2014). Increase male genital diseases morbidity linked to informal electronic waste recycling in Guiyu, China. *Environmental Science and Pollution Research, 21*, 3540–3545.

Yang, S., He, M., Zhi, Y., Chang, S. X., Gu, B., Liu, X., & Xu, J. (2019). An integrated analysis on source-exposure risk of heavy metals in agricultural soils near intense electronic waste recycling activities. *Environment International, 133*, 105239.

Environmental Impacts of E-Waste and the Challenges and Opportunities for Its Effective Management

Ankita Mohanty, Aditya Kishore Dash, and Alok Prasad Das

Abstract The present-day exponential increase in technological development to fulfil the needs of the expeditious information era resulted in the discarding of electronic and electrical equipment at a much faster rate than in the past. E-waste which is classified under hazardous wastes, if not collected and disposed of properly, can pose a threat to the environment and the health of living beings. The world produced 53.6 Mt of e-waste during the year 2019, that is, 21% more in the last 5 years and it is predicted that, it will reach 74 Mt globally by the year 2030. It is reported that, with 24.9 Mt, Asia topped the list with respect to generation of the e-waste followed by America (13.1 Mt), Europe (12 Mt), Africa (2.9 Mt) and Oceania (0.7 Mt). The common e-waste components include cathode ray tubes, dismantled PCB processing, wires, gold-plated substances, chips, printed circuit board, plastic from computers and peripherals. The potential environmental impacts of e-waste include the direct release of hydrocarbons leaching of heavy metals like Pb, Ba, Cd, and Hg into both the surface and groundwater, causing water pollution. Also, heavy metals, hydrocarbons, brominated dioxins, ashes, and PAHs are released into the air and soil, causing environmental pollution. Silicosis, skin contamination, and inhalation of Pb, Cd, Hg, and brominated dioxin among the employees and nearby residents are common health hazards. Acid mists and fumes, SO_2, and Cl_2 gas released during the process can cause pulmonary oedema and circulatory failure, causing death of the organism. Proper disposal and management including the old and discarded electronic gadgets, by implementing the 5Rs steps, that is, refuse, reuse, reduce, recycle and repurpose, must be followed. In the present review, an attempt has been made for an in-depth review of environmental impacts and the challenges and opportunities associated with the effective management of e-waste in an eco-friendly way.

A. Mohanty · A. K. Dash (✉)
Department of Chemistry, ITER, Siksha 'O' Anusandhan (Deemed to be University), Bhubaneswar, Odisha, India
e-mail: adityadash@soa.ac.in

A. P. Das
Department of Life Sciences, Rama Devi Women's University, Bhubaneswar, Odisha, India

A. P. Das, S. Priyadarsini (eds.), *Electronic Waste and Environmental Pollution*, Sustainable Environmental Waste Management Strategies,
https://doi.org/10.1007/978-3-031-90287-1_7

Keywords E-waste management · Health impact · Sustainability · Environment · Public health

1 Introduction

The rapid development of information technology and electronics has led to a substantial production increase and eventual disposal of electronic items. E-waste is a term to describe old, end of life, electronic items that are no longer in use. Discarded laptops, TVs, computer, radios, refrigerator, etc., are called as e-waste (Mishra et al., 2023). E-waste is a growing source of pollution that can also harm the environment. It can be reused, recycle, and refurbished. The manufacture of electrical and electronic equipment relies on various raw materials, such as metals, plastics, glass, semiconductors, oil and wood. Many of these materials can be both reused and recycled (Minoja & Romano, 2021). As resource consumption and demand continue to grow, the relevance of addressing e-waste is amplified due to its environmental and health impacts. Additionally, e-waste serves as a secondary source of valuable metal items (Gunarathne et al., 2020). Recent United Nations reports in 2021 indicated that, each individual worldwide generates an average value of 7.6 kg of e-waste/day, totalling approximately 57.4 million metric tons globally (Forum, 2021). As modern technology plays an essential role in people's lives, the demand for electronic devices has surged globally (Arya & Kumar, 2020). Approximately 7 billion people inhabit the earth, with around 5 billion able to meet their basic needs and over 6 billion owning mobile phones, now a necessity for many (Krishnamoorthy et al., 2018).

Technological progress, lifestyle changes, and world economic growth have led to continual rise in e-waste generation, creating major health and environmental challenges (Matsakas et al., 2017; Bhuiya et al., 2020). During 2019, approximately 83% of e-waste was untracked and likely disposed of through open burning or illegal dumping, presenting severe risks to both the environment and human health issues (Baldé et al., 2016) and only 17% of e-waste was properly collected and recycled. It is reported that, with 24.9 Mt, Asia topped the list with respect to the generation of e-waste, followed by America (13.1 Mt), Europe (12 Mt), Africa (2.9 Mt) and Oceania (0.7 Mt). Although Asia produces the most total e-wastes, the per person generation is 5.6 kg where, as in Oceania (16.1 kg), Europe (16.2 kg), and America (13.3 kg), due to its larger population (4.4 billion people). The per capita generation is lowest in Africa, that is, 2.5 kg (Forti et al., 2020).

In some developing countries, particularly those with low to middle incomes, a significant quantity of e-waste is disposed of improperly, creating public health risks and landfills. Kuang et al., 2023 documented health issues affecting children due to the recycling of e-waste in China. Informal recycling is common in such regions, where wires are often burned openly to recover copper, releasing harmful chemicals into the air. Similarly, the extract of valuable metals like Au, Ag, Pd, and Pt is in practice in countries like India, Vietnam and China (Forti et al., 2020).

E-waste contains about 1000 substances, most of which are toxic materials (Dhir et al., 2021). According to the 'Associated Chambers of Commerce of India', India's e-waste is increasing at a compounded annual growth of around 25% (Garlapati, 2016). Within India, Mumbai is the top contributor, producing 1.2 MT of e-waste yearly, followed by Delhi with 0.98 MT (Srivastava & Pathak, 2020). United Nations reported that annual e-waste production had surpassed 44 million metric tons by the end of 2020, equivalent to approximately 4500 Eiffel Towers (Bocken & Short, 2021). E-waste is growing faster than any other type of municipal waste, with projections indicating it could reach 74.7 million MT by 2030 (Onac et al., 2020). By 2050, the cumulative global quantity of e-waste is estimated to hit 78 million metric tons (Asefi et al., 2019). This issue has garnered significant attention from scientists, policymakers, and practitioners due to the hazardous nature of e-waste and its substantial risks to the surroundings. On the basis of their impact on health and the environment, e-waste is classified into different types. These are hazardous, non-hazardous, toxic, recyclable and mixed e-waste.

2 E-Waste Sources of Generation

Nowadays, rapid changes in the digital environment, technological innovations and the rising demand for new devices create the problem of e-waste. It is for societal important to confess the sources of e-waste and take immediate action to reduce its effect on the environment. Different types of sources of generation include:

Discarded Electronics: The discarded consumer electronics are significantly increased. These items are smartphones, laptops, tablets, and other devices. These substances are quickly replaced by newer models, adding to the growing volume of e-waste. To reduce this, consumers must be aware of and utilize available recycling programmes and responsible disposal options.

Medical and Laboratory Equipment: The health sector and research industries depend on electronic devices for diagnostics, treatment, and experimentation. But rapid advancements in medical and laboratory technologies result in the rapid disposal of advanced equipment.

Outdated Appliances: Outdated appliances are like refrigerators, washing machines, microwave, and AC. In outdated appliances, to alleviate these matters, it concentrates on designing eco-friendly appliances, enhancing reparability, and fulfilling take-back programmes.

E-Waste from Manufacturing: The production of e-waste like leftover materials, defective items, and obsolete parts takes place because of the production of electronic devices and components. To decrease the environmental impact of manufacturing processes, companies can embrace sustainable practices such as optimizing resource use and executing effective recycling programmes.

Entertainment and Leisure Devices: For entertainment, few industries offer wide varieties of electronic items such as televisions, gaming consoles and audio systems. Due to the advanced technology these devices become outdated. Raising consumer awareness about recycling the e-waste, the industry adopted sustainable design to control the environmental pollution.

Network Infrastructure Upgrades: With the increase of global connectivity and communication networks, regular upgrades to network infrastructure are needed. Network infrastructure can contribute to e-waste, but these are ways to reduce the environmental impact. It should be done by replacing equipment and disposing of old equipment.

Battery Disposal: Batteries are the most important part of electronic devices which present significant challenges in e-waste management. Inappropriate disposal can result in harmful materials contaminating the environment. Battery disposal is important for controlling e-waste pollution because batteries can contain harmful substances like lead, lithium, mercury, nickel, and cadmium. Recycling of batteries can reduce environmental pollution (Priyadarsini & Das, 2024, 2025).

Solar Power Systems: The components of solar power systems are panels and inverters; power systems are panels composed of 80% glass and metals, which are recyclable. The other components such as charge controllers, cables, and metres also produce e-waste, which can be recycled.

Lighting: Nearly 150 years ago, light bulbs were involved. Fluorescent lamps contain mercury, which can contaminate the environment and cause neurotoxicity. Many modern household lights contain LEDs made from hazardous elements such as indium and gallium. Improper disposal of LED lights can lead to environmental harm, as can other types of lighting waste. It is crucial to recycle these items responsibly.

3 E-Waste Production Scenario

Worldwide, several legal frameworks have come forward to manage and regulate e-waste. Europe has taken a leading role in safeguarding the environment from the hazards of e-waste by introducing two significant directives, that is, "Waste Electrical and Electronic Equipment (WEEE) Directive" and the "Restriction of Hazardous Substances Regulation". These directives aim to minimize the environmental effect of e-waste which is adopted by many nations.

In **China**, e-waste generated has three primary destinations: second-hand markets, donations and recycling by private companies.

In the United Kingdom, many consumers, particularly in areas like Sharston and Manchester, are open to purchasing products with similar functionality to replace faulty electronics. The UK's e-waste management is governed by the "WEEE Directive", an EU directive incorporated into UK law in 2007.

In **the United States**, domestic e-waste is legally classified as non-toxic waste, with its management primarily handled by municipalities and state authorities. The most common method of dealing with end-of-life electronics is storage, followed by disposal, with only a small portion being recycled. The United States adopts an ECR approach, placing financial responsibility for e-waste management to the consumers rather than producers. A survey by the EPA found that around 80% of users agreed to pay a recycling fee of up to $5 for outdated electronics.

Switzerland was the first country to establish an e-waste management system which is now a global leader in e-waste collection. Its pioneering system is based on the EPR principle, holding manufacturers and exporters financially and operationally accountable for the environmentally friendly recycling, treatment, as well as disposal of e-waste. Switzerland's success is attributed to its environmentally conscious and law-abiding consumers, who regularly return discarded electronics to retailer outlets, collection centres, or recyclers.

In **Germany**, households are legally required to take obsolete electronics to designated collection points. Alternatively, consumers can pay for their e-waste collection. Despite mandates for separate collection sorting, studies reveal that few electronics still constitute between 0.4% and 1.5% of the residual household waste stream. This indicates that the current system has not fully succeeded in focusing on the e-waste crisis, raising concerns about the effectiveness of existing collection schemes (Awasthi et al., 2016a).

Japan and the European Economic Community's standards are closely aligned for e-waste management. Since 2001, Japanese policymakers have set up laws and regulations to transform e-waste into valuable resources. In 2001, Japan set up different laws to promote recycling. The Home Appliances Recycling law was introduced in June 1998 and was enforced in April 2001. The aim of this law is to reduce the environmental impact of mass production, consumption, and waste production. The Home Appliances Recycling law encourages people to value their items and to achieve a recycling-oriented society. This has resulted in the loss of valuable materials from Japan's domestic recycling system, hindering the country's e-waste management initiatives (Awasthi et al., 2016b).

3.1 India's E-Waste Production Scenario

The issue of e -waste in India started to surface following the country's initial phase of economic liberalization. The growth of the electronics and consumer goods industries, driven by fierce competition among both Indian and foreign companies, contributed to an increase in e-waste. The post-liberalization era, marked by lower prices and improved purchasing power, saw a boom in demand for electronic products such as domestic appliances (refrigerators, air conditioners, washing machines, TVs, etc.), telecommunication devices, IT equipment, and computers. The rapid expansion of the IT sector, coupled with infrastructure reforms and e-governance, further accelerated the generation of e-waste. The Indian telecom market, for instance, saw significant growth, with 113.26 million new cellular users added in 2008, bringing the total to 261.97 million by 2007–2008. In addition, sales of household appliances like AC and microwave ovens grew by approximately 25% in 2006. The transactions of white goods reached 4 million in 2006–2007, with production increasing by 17% compared to the previous year. The market for coloured televisions saw a threefold increase by 2007. The states that contribute the most to discarded electrical and electronic items generation in India are Maharashtra, AP, Tamil Nadu, UP, WB, Delhi, Gujarat, and Punjab. The major cities generating the highest amount of e-waste include Kolkata, Delhi, Bangalore, Mumbai, Pune, Hyderabad, Surat, Nagpur, Chennai and Ahmedabad (Awasthi et al., 2016b).

4 E-Waste Collection Scenario

The unhygienic handling of e-waste contributes to global pollution, human health risks, and loss of valuable resources. Addressing this requires coordinated international efforts, innovative policies, and advanced technologies. Global challenges include unregulated disposal and resource loss. Many countries, particularly in the Global South, lack robust systems to manage e-waste. Only 17.4% of e-waste is formally collected and then recycled, leading to a significant loss of rare and precious materials such as gold, palladium, and cobalt. Few global solutions are like international treaties, collaboration initiatives and technological innovations. Frameworks like Basel Convention aim to regulate transboundary movements of hazardous waste, including e-waste. The United Nations' Step Initiative promotes sustainable e-waste management processes globally. Developing advanced recycling technologies for extracting valuable materials efficiently and safely can solve the issue to some extent.

India is one of the world's top e-waste generators, producing over 3.2 million metric tons annually. Challenges for e-waste collection in India include informal sector dominance, low public awareness, and inadequate infrastructure. Nearly 95% of e-waste recycling is handled by the informal sector, often employing unsafe methods. Many consumers are unaware of proper e-waste disposal channels. Despite regulations, there is a shortage of authorized collection and recycling centres. Government of India has taken many interventions in this regard like by framing E-Waste (Management) Rules, 2016, organizing awareness campaigns, establishing authorized centres. As a result of successful initiatives, cities like Bengaluru and Pune have initiated e-waste collection programmes through partnership with NGOs and private companies such as Dell and Samsung who have set up take-back programmes to recycle old electronics.

5 E-Waste Pollution a Global Challenge

Improper disposal of electronic waste significantly contributes to pollution of environment, affecting water, air, soil, and living organisms. This contamination largely arises from inadequate management practices at disposal and recycling sites, which often release heavy metals as well as particulate matter into the environment (Chatterjee, 2008). Unscientific recycling practices result in the release of heavy metals and persistent environment pollutants (Pradhan & Kumar, 2014). Impact of e-waste on various segments of environment is as follows.

Impact on Biodiversity: Once the hazardous substances are introduced into the environment, they become involved in processes such as bioaccumulation, contamination of the food chain, and disruption of ecological systems. This widespread contamination causes serious risks to ecology and human health also (Chen et al., 2011). E-waste pollution is a critical global issue, significantly affecting the natural environment, including plant and animal life. When improperly disposed of, the toxic elements seep into the soil, altering its composition and reducing its ability to support plant growth. Plants absorb these toxins from the soil, leading to uptake and damage to their physiological processes such as photosynthesis. In chemical runoff, the e-waste stored or discarded in landfills often leads to harmful chemicals leaching into nearby water bodies. This contamination reduces water quality and threatens aquatic plants and animals. Mercury is a toxic substance which affects fish and amphibians. The toxic effect of mercury causes behavioural changes, reproductive changes, and mortality. When e-waste burns, it releases hazardous gases and fine particles. These pollutants affect the plants, creating stunted growth and entering the respiratory system of animals, causing health problems. E-waste toxins contaminated by plants are consumed by herbivores, and it is transferred through the food chain. These toxins accumulate in predators, leading to bioaccumulation and biomagnification, which can cause organ damage and decrease fertility. Most of the animals are affected by e-waste pollution (Das & Mishra, 2008, 2009; Paul et al., 2023). The contaminated plants and water are ingested by land animals, leading to bioaccumulation of toxic substances in their body. It causes health problems such as organ damage and reproductive issues and finally leads to death. The chemical food of the e-waste flows into rivers, lakes, and oceans to affect aquatic ecosystems, which are particularly vulnerable. These pollutants also harm marine organisms, interrupt the food chain and threaten species survival e-waste pollution affects plants and animals.

Impact on Air Quality: E-waste pollution reduces the air quality. It releases harmful substances into the atmosphere causing air pollution. Air pollution due to unscientific disposal of e-waste or from any other sources causes health hazards. Therefore, to know the air quality, determination of the air quality index and dispersion of pollutants of an area, it is important to monitor the ambient air and to find suitable control measures (Parida et al., 2015; Sahu et al., 2015; Dash & Dash, 2015a, b; Dash & Dash, 2017, 2018a, b; Kumar et al., 2018; Sahoo et al., 2017; Dash et al., 2019, 2023). E-waste contaminants often spread into the air as dust or fumes, posing health risks through skin absorption, ingestion, and inhalation (Mielke & Reagan, 1998). Both children and infants are particularly vulnerable to e-waste exposure due to their higher breathing rates, frequent hand-to-mouth behaviour, high intake of food, lower efficiency in eliminating toxins, and breastfeeding (Guo et al., 2012). Hazardous substances like PCBs, Hg, Pb, Cd and secondary toxicants such as PAHs, dioxins, and furans are frequently released during e-waste recycling and processing indifferent regions of the world (Sthiannopkao & Wong, 2013). The extent of contamination in the air varies based on nearby activities, such as recycling and dumpling of electrical components, which release significant amounts of Cu and Pb into the atmosphere (Brigden, 2005). These contaminants, carried as particulate matter, can travel long distances and persist in the air which can inhaled in to the lungs. Eventually, they settle on soil and water surfaces, entering ecosystems and bioaccumulation in soil organisms, aquatic life, and plants. Prolonged exposure to fine particles can lead to respiratory diseases, including asthma, bronchitis, and lung cancer. This is especially concerning in areas where informal e-waste recycling is common, as the workers and surrounding populations are exposed to dangerous levels of airborne pollutants.

Impact on Water Quality: E-waste contains different types of toxic substances, heavy metals such as Pb, Hg, and Cd causing water pollution. If not managed properly can go down into the soil and water system, contaminating groundwater, river, and ocean. It has both short-term and long-term impacts on the ecosystem and on human life. Therefore, the treatment of water and wastewater plays an important role. Hazardous substances released during these operations pollute local water sources, rendering them unfit for drinking, farming, and aquatic ecosystems (Kumar et al., 2023; Mishra et al., 2022; Mishra & Das, 2022). For example, heavy metals like Pb and Hg can disrupt aquatic environments, harm marine and freshwater species, and accumulate in the food chain, causing human health risks (Prabhakar et al., 2019; Raj & Das, 2023). Additionally, e-waste contributes to water pollution through microplastics. Besides the major hazardous elements, the plastic component of e-waste is also a warning to the environment. A lot of literature is available on the occurrence, fate, environmental impacts and management of plastic waste worldwide (Dash et al., 2024; Naik et al., 2024a, b; Parida & Dash, 2024; Panigrahi & Dash, 2024). Degrading components like cables and plastic casings break down into tiny particles that persist in water bodies and harm aquatic life (Mishra et al., 2024a, b; Mishra & Das, 2021). The release of certain chemicals also causes water acidification, further endangering biodiversity and complicating water treatment efforts. Physio-chemical method of treatment of contaminants from water/wastewater is cost-intensive which attracts the scientist to analyse the water/wastewater samples and to go for low-cost biological treatments (Mishra et al., 2012; Dash, 2012; Pradhan et al., 2013; 2015a; 2017; Mahapatra et al., 2019; Jena et al., 2022; Parida et al., 2023; Amreen et al., 2024; Shreya et al., 2022). The increasing volume of e-waste worldwide highlights the urgent need for improved waste-handling solutions. Initiatives such as enhanced recycling systems, stricter regulations, and public awareness campaigns are critical. Without these measures, the negative impacts on water quality will continue to escalate, threatening ecosystems and public health in the long term.

Impact on Soil Quality: Electronic waste contains toxic chemicals such as lead, mercury, and cadmium etc., which can harm plants, animals, and human beings. E-waste contaminates soil during landfill and when burned, large particles can contaminate the soil. Heavy metals from e-waste are persistent and non-biodegradable, which creates long-term soil contamination. They disturb microbial activity, which is important for maintaining soil health and nutrient cycle. In soil, accumulation of these pollutants harms plants growth and reduces agricultural productivity. These practices release pollutants into the air, which eventually settle in the soil, further degrading its quality. However, wastewater and fly ash containing contaminants can support the growth of plants as they contain few essential nutrients required for plant growth (Pradhan et al. 2015b; Dash et al., 2015; Pradhan et al., 2016). Soil fertility study (Nayak et al., 2018) and evaluation of enzyme activities near solid waste dump sites (Panigrahi et al., 2024) play an important role in this context.

Impact on Human Health: E-waste has a good impact on human health. Defective disposal and recycling methods release heavy metals like Pb, Hg and Cd and toxic chemicals such as dioxins and brominated flame retardants into the surroundings. These pollutants enter the human body through inhalation, ingestion, and dermal contact, leading to health problems. Heavy metal damages the organ system. For example, lead affects the kidneys of adult mercury affects the nervous system and leads to neurological and developmental disorders, and cadmium affects the kidneys and bones. It also degrades the bone. Toxic chemicals create respiratory problems, skin irritation, and cancer. Informal e-waste recycling workers and local communities are particularly vulnerable, as they are often exposed to high concentrations of these substances without adequate protective measures. Moreover, the release of hazardous substances into the soil as well as water further amplifies the risks, as they can contaminate food and drinking water sources. The detection (Das et al. 2014a), biodegradation (Das & Mishra, 2010), health assessment (Das & Mishra, 2008; Das & Singh, 2011), environmental effect (Das et al. 2014b; Das et al., 2011) of various elements in the environment has been studied by different scientists. Addressing the health impacts of e-waste pollution requires international collaboration to establish safer recycling practices, enforce regulations, and raise awareness about the dangers of improper e-waste handling. Table 1 shows the sources of common hazardous chemicals like Hg, Pb, As, Li, Ba, Sb, dioxins, PAH, asbestos, PVC and polychlorinated hydrocarbons with their corresponding health hazards. The related environmental issues like air, water and soil contamination, food contamination and biomagnifications are also reflected (Jain et al., 2023; Srivastav et al., 2023). Figure 1 shows the impact of environmental contaminants on various human body parts. The common body parts that are affected are the brain, body system, nervous system, lungs, heart, liver, kidney, endocrine system, and bones.

6 Present Practice for Management of E-Waste

Inadequate disposal of electronic waste significantly contributes to environmental pollution, affecting water, air, soil, and living organisms. This contamination largely arises from inadequate management practices at disposal and recycling sites, which often release heavy metals and also particulate matter into the surrounding (Chatterjee, 2008). Some of the sustainable e-waste management practices are;

Metal Recovery: Fiberglass and epoxy resin, though non-recyclable, must be managed effectively to improve the economic feasibility of recycling of e-waste.

Table 1 Sources of common hazardous chemicals with their corresponding health hazards. The related environmental issues are air, water, soil contamination, food contamination and biomagnifications

Sl. No.	Hazardous chemicals	Sources	Common health hazards	Environmental issues	References
1	Hg	Thermometer, batteries, copper machines, circuit boards, monitors	Brain disorder, retardation in child growth, impact on kidney and central nervous system	Water and soil pollution, biomagnification	Shamim et al. (2015), Kumar and Singh (2014), Gangwar et al. (2019)
2	Pb	Electronic devices that utilize cathode ray technology, such as certain types of monitors, as well as components like batteries and lamps, fall into the category of cathode ray-based appliances. Similarly, thermal components designed to transform heat into electrical energy including devices like thermocouples, thermistors, and thermoelectric elements	Damage to kidneys, reproductive system, and nervous system may occur, potentially leading to decreases of the blood and brain. In children, this might result in delayed mental development, as well as problems affecting the nerves	Pollution of the air, water, and soil	Arya and Kumar, (2020), Yang and Zhang (2018), Grant et al. (2013), Zeng et al. (2020), Ilankoon et al. (2018), Wu et al. (2019)
3	As	Electronic components like light-emitting diodes (LEDs), diodes, microwave technologies, photovoltaic cells, and semiconductors	Prolonged asbestos exposure can damage the nervous system and skin, and increase the risk of developing lung cancer	Air and water contamination	El-Ghiaty & El-Kadi (2021), Yang et al. (2020), Kumarathilaka et al. (2018)
4	Li	Lithium ion rechargeable batteries like camera batteries, mobile phone batteries	Extended exposure to lithium vapours may result in symptoms like confusion, muscle weakness, vomiting, nausea, and digestive issues	Soil degradation	Saha et al. (2021), Zhang et al. (2012), Li and Achal (2020), Closer look (2022)
5	Ba	Substances used to reinforce plastics and rubbers, along with electronic tubes	This can impact the heart muscle	Environmental pollution	Pinto (2008), Ari (2016)
6	Sb	Industrial processes, smelters	It can harm the liver, kidneys, cause eye irritation, lead to hair loss, lungs, and heart	Pollution of the atmosphere, water sources, and land	Kumar and Singh (2014)

7	Dioxins	Printed wiring boards (PWB), various types of electric motors, ceiling fans, cables, metal smelting, products, generators, transformers, capacitors, plastics from computers, and their peripherals	Issues affecting reproduction, neurobehavioral functions, immune system, along with carcinogenic, mutagenic, teratogenic effect, and an elevated risk of developing cancer	Pollution of environment, water, soil, dust, and food	Tue et al. (2019), Pavithra et al. (2020)
8	Asbestos	Heat-resistant materials used in heating equipment	Severe side effects include difficulty breathing, lung damage, coughing, even cancer	Soil and water contamination	Disposal of Asbestos (2022), Debnath et al. (2021)
9	PAH	Electrical connections and printed circuit assemblies	Possible side effects of this medication include confusion, vomiting, diarrhoea, nausea, and eye irritation	Environmental pollution	Sánchez-Quiles and Tovar-Sánchez (2015), Wang et al. (2012)
10	Polychlorinated hydrocarbons	Microchips, synthetic materials, and electrical cables	Genetic mutations, lung cancer, and developmental defects.	Pollution of the air and soil	Pavithra et al. (2020)
11	PVC	Wiring and protective insulation	It can lead to harm to the respiratory and immune system	Environmental contamination	Kurup and Senthil Kumar (2017), Stapleton et al. (2008)

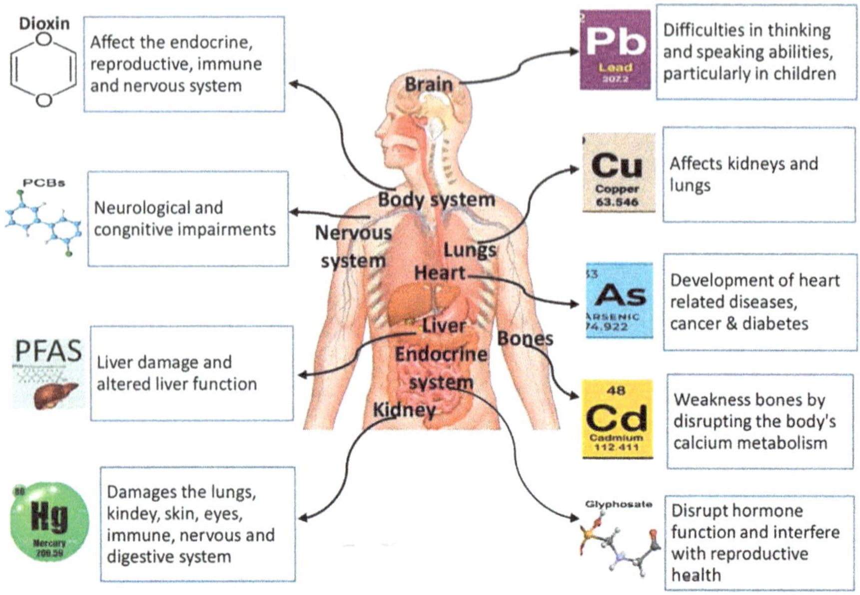

Fig. 1 Impact of environmental contaminants on various human body parts. The common body parts are the brain, body system, nervous system, lungs, heart, liver, kidney, endocrine system and bones

Their removal is necessary for reducing the bulk of e-waste, thereby concentrating on the metal content. This concentration increases the efficiency and value of the recycling process, as a smaller volume of waste with a higher density of recoverable metals can be transported for further extraction and processing. This approach not only enhances resource recovery but also makes the overall recycling process more cost-effective (Yang et al., 2021). Ferrous materials are separated from e-waste using either a permanent magnet or an electromagnet, which attract and isolates these components. For non-ferrous metals like copper and aluminium, an eddy current separator is utilized. This technology provides electromagnetic forces to differentiate metals from non-metallic materials. The following equations describe electrochemical reactions that are relevant for processing involved in e-waste recycling:

$$4Fe^{2+} + O_2 + 4H^+ \rightarrow 4Fe^{3+} + 2H_2O \tag{1}$$

$$2Fe^{2+} + Cu^{2+} \rightarrow 2Fe^{3+} + Cu \tag{2}$$

$$2Cu^{+}_{(l)} + Fe_{(s)} \rightarrow Fe^{2+}_{(l)} + 2Cu_{(s)} \tag{3}$$

Acid bath method for e-waste management is used to remove heavy metals from the circuits. The metals can then be recycled and used to make new products. It is widely used to recover metals like silver, copper, lead and gold from circuit boards. This is done by immersing circuit boards in sulphuric acid for approximately 12 h to dissolve the metal. The valuable metals are recovered through this approach (Ma, 2014).

Recycle and Reuse: Recycling is the process of treating and reusing discarded materials to recover either their original quality or a lower-grade form. It involves extracting reusable components, known as secondary materials, from waste products. These materials can either undergo upcycling, where they are transformed into items of equal or greater value and functionality, or down cycling, where they are repurposed into products of lesser quality. While upcycling is generally more desirable, it is not always feasible in practice (Morseletto, 2020). E-waste management typically involves five key stages: collection, toxic materials removal, pre-processing, final processing, and the throwing away of non-recyclable waste. Effectiveness of recycling electronic waste largely depends on the methods employed, particularly during the dismantling and material separation phases (Lucier & Gareau, 2019). Economically, e-waste recycling is often seen as highly valuable because it yields various precious materials, like gold, silver, platinum, palladium, copper, and iron. Many of these materials in electronic waste can be effectively recovered and reused through proper recycling processes. As per the United Nations University, the global e-waste stream contains raw materials valued at over USD 62 billion (Aboelmaged, 2021). Recycling e-waste can also address employment challenges by creating job opportunities. For instance, in China, a single e-waste recycling facility employs approximately 100,000 workers. Additionally, e-waste recycling helps to safeguard the environment by preventing the release of harmful materials and toxic fumes.

Dismantling: Dismantling is a process of separating hazardous materials such as polychlorinated biphenyls, chlorofluorocarbons, and mercury-containing switches from e-waste. These valuable materials include copper-containing cables and iron. Dismantling may be carried out either manually or using mechanical methods to separate dangerous materials and valuable components (Ghimire & Ariya, 2020). It is a manual procedure which includes tasks such as segregating, sorting, emptying, cleaning, and separating components to eliminate pollutants. In this process, hammering and shredding are involved (Vats & Singh, 2014b). According to Huang et al. (2014), e-waste recycling in Ghana is mainly conducted by the informal sector. They utilize rudimentary methods like open burning and dismantling. This is the process of extracting copper and other metals by processing electronic devices into individual parts. Materials deemed worthless are disposed of at dump sites, where they are stacked in large heaps and occasionally burned to minimize their volume. Manual dismantling of e-waste facilitates the recovery of recyclable plastics, metals, and various components. It also allows for the separation of hazardous substances, like lead and mercury, found in items such as LCDs, CRT glass, and batteries, which require specialized treatment.

Processing: During processing, e-waste is either gradually broken down using a smasher or cut into smaller pieces with a cross-flow shredder. Mechanical processing of e-waste is favoured in Europe due to the high cost of labour (Ghimire &

Ariya, 2020). Following mechanical processing, the next step involves separating valuable materials and hazardous substances. Additional separation and crushing are carried out using equipment such as sieves, eddy current separators, hammer mills, separators, and magnetic. This process yields metal concentrates for recovered plastics and metal refineries (Salhofer et al., 2016). Guo et al. (2009) noted that e-waste processing takes place indoors and involves tasks such as handling circuit boards, shredding plastics, and melting materials.

End Processing: During end processing, the copper/lead, ferrous, and aluminium functions are forwarded to smelters. Ferrous materials are processed at steel plants for aluminium, iron recovery is directed to circuit boards, copper/lead fractions, aluminium smelters, and other materials with precious metals are sent to integrated metal smelters. Non-recyclable and hazardous waste is disposed of accordingly (Ghimire & Ariya, 2020). End processing of e-waste consists of mechanically concentrating fractions through methods like grinding, segregation, milling, and shredding, along with specific treatments. Recovery primarily relies on mechanical processes like magnetic and eddy current separation, shredding, and grinding (Vats & Singh, 2014a).

Landfill: Landfills are commonly used for disposing of e-waste due to their simplicity and convenience (Ghimire & Ariya, 2020). However, Sivaramanan (2013) described landfills containing e-waste as 'toxic time bombs' due to their long-term environmental risks. E-waste in landfills is significantly more toxic than non-e-waste landfill materials. Decomposition in these sites produces leachate, a harmful liquid that renders the land unusable for future purposes. Toxic substances such as heavy metals like arsenic, iron, lead, nickel, aluminium and polyhalogenated organics are released during this process, causing environmental impacts. The additional components like batteries and electronic circuits release acid and toxic metals. They cause serious threats to both humans and ecosystems (Ghimire & Ariya, 2020). E-waste pollutants mix through soil and groundwater spreading contamination beyond the landfill site (Sivaramanan, 2013).

7 E-Waste Management: Its Opportunities and Challenges

The challenges and opportunities of e-waste management are due to the lack of proper e-waste legislation. Many developing countries don't have e-waste legislation. In Asia, the most polluted country is China. Other countries such as Pakistan and Bangladesh have no laws at this time. Due to the lack of regulation on e-waste management, that is necessary to collect and recycle, it is still challenging to handle e-waste because of its complicated mixture of hazardous, precious, base, and other toxic materials. E-waste consists of 40% metal, 30% plastic polymers, and 30% oxides of various materials (Sahajwalla & Hossain, 2020). E-waste has precious materials like Au, Al, Ni, Sn, Fe, Bi, Sb, and hazardous materials like Hg, Be, Pb,

Table 2 E-waste management challenges and possible solutions. Lack of infrastructure, recycling facilities, public awareness, policies and regulations, and health issues are the common challenges

Sl. No.	Challenges	Possible solutions
1	Lack of infrastructure	Allocate resources for establishment of waste collection points and recycling opportunities.
2	Recycling facilities	Build alliances with global organizations to obtain financial support and technical expertise.
3	Informal recycling sector	Structure and govern the informal sector by delivering training and resources for safe recycling practices.
4	Limited public awareness	Organize awareness campaigns to educate the public on proper e-waste disposal and recycling techniques.
5	Inadequate policies and regulations	Formulate comprehensive guidelines and regulations that cover all stages of electronic products' lifecycle.
6	Health and environmental hazards	Enforce tougher health and safety guidelines for e-waste management and disposal to reduce potential risks.
7	Lack of financial resources	Establish global collaborations and secure funding for infrastructure development.
8	Limited research and development	Promote research and innovation in advanced technologies for effective e-waste resource recovery, and recycling.
9	Illegal export of e-waste	Enhance the implementation of regulations and improve border monitoring to stop the illegal export of e-waste.
10	Data security concerns	Disposing of electronic devices may result in sensitive data being leaked if not properly wiped from devices.

Cd, and As; halogen materials like Br, F, Cl, and plastic, glass, and ceramics (Sahle-Demessie et al., 2018). Table 2 shows the detailed of e-waste management challenges and their corresponding possible solutions.

8 Circular Economy in E-Waste Management

During the year people across the world have been facing severe consequences of various environmental challenges such as climatic change, deforestation, environmental pollution, degradation of soil and depletion of natural resources. Such environmental concerns not only harm ecosystems but also hinder sustainable social and economic development. To meet the requirements of increasing population and improving living standards, the demand for natural resources has surged. The traditional linear economic model adopted by most industries has significantly accelerated the extraction and processing of natural resources, resulting in their rapid depletion and raising concerns about their eventual exhaustion. The linear economic model operates on the principle of technique and disposal is sustainable and describes the natural resources organized by different organizations. Different companies follow this model and extract raw materials from the environment. They use energy and labour to produce goods and finally sell them to consumers. The consumers dispose of the product after a single use, which leads to environmental waste (Goyal et al., 2018). According to Ravindra and Mor in 2019the model can be sustained over time if it leads to a significant depletion of natural resources and disregards the immense value contained within materials (Tiwari et al., 2019) that the circular economy has emerged as a widely adopted and effective model. When electronics and electrical items are raised to produce, there is a corresponding increase in waste generated during manufacturing and consumer use. Experts forecast that e-waste will grow by approximately 80 million tonnes/year by 2030 (Ahirwar & Tripathi, 2021). Integration of circular economy principles in e-waste management offers several advantages, like reduced dependence on raw materials, job creation, lower resource consumption, and a decrease in the harmful effects linked to resource extraction and processing. Notably, for every 1000 tonnes of e-waste processed, approximately 12.6 direct employment and 20 indirect employments are generated. With a 20% global e-waste recycling rate, in comparison with an ideal 100% recycling rate, job creation could vary between 0.30 and 1.50 million employments. Moreover, Safdar et al. (2020) emphasize that various stages of e-waste management such as refurbishment, assessment, recycling, collection, and transportation centres can create employment opportunities at the facility level. The number of jobs generated depends on factors like the processing capacity/worker and the amount of e-waste handled at each centre. Figure 2 shows the circular economy for e-waste handling. The important steps involved are resource conservation, recycling, refuse, reduce and product life increase (Srivastav et al., 2023).

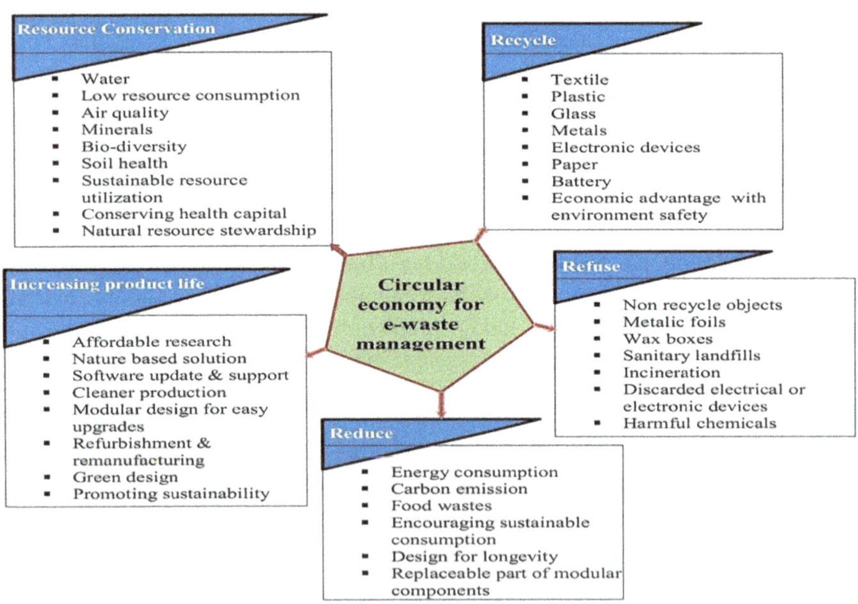

Fig. 2 Circular economics for e-waste management. It involves the processes like resource conservation, recycling, refuse, reduction, product life increase

9 Legal Aspects for Management E-Waste

In India, laws related to handling e-waste have been placed since the year 2011. According to this, only the e-waste can be collected by the authorized recyclers and dismantlers. Further, the E-waste (Management) Rule, 2016 was enacted in India during the year 2017. The e-waste rules are applicable to all the consumers, procedures, recyclers, collection centres and dismantlers of e-waste. The rule clearly defines the responsibility of the manufacturer, consumer, dealer, producer, collection centres, refurbisher, recycler and dismantler. It also explains the responsibility of the state government for environmental good management of e-waste. Schedule I of these Rules says that, it is applicable to those who are involved in the sale, processing, purchase and manufacturing of either electrical or electronic components. Extended producer responsibility is an important feature of these rules. Here, for collection and exchange of e-waste, the producers are made responsible. In order to formalize the recycling sectors, the government of India has taken a number of steps. Central Pollution Control Board (CPCB), under the government of India, has framed various guidelines and standard operating procedures to process e-waste in our country. Both CPCBs and SPCBs are responsible for monitoring the units. To mainstream and modernize e-waste recycling industries in India, both CPCB and SPCBs were taking steps with the help of the Electronics and Information Technology Ministry in India. The E-Waste (Management) Rule has been amended from time to time, and the central government, MoEF & CC, has issued the E-waste (Management) Amended Rules, 2024, on 8 March 2024.

10 Conclusion

The challenge for management of e-waste is linked with broader environmental sustainability and climate change issues that our planet is facing today. It has been studied that informal sectors handle most of the e-waste in developing countries. Collection, shorting, waste washing, and treatment are the important steps for e-waste recycling and recovery. The e-waste contains a mixture of toxic organic and inorganic materials like heavy metals, brominated flame retardants, PAHs, and polychlorinated biphenyls. The unscientific management of e-waste causes severe negative effects on the economy, human health and environment. To protect the society and environment from the threat of e-waste, its management must be given priority. Since there is an alarming rate of generation of e-wastes worldwide, it is very crucial to look into the growing global concern. In addition to having different recycling and valuable materials, e-waste also contains some toxic elements. E-waste containing radioactive materials can harm plant tissues and also inhibit plant growth. The plant nutrient uptake capacity is also affected due to the presence of radioactive materials in the soil. There is a scope for economic opportunities in recovering various valuable metals through sustainable management of different e-wastes. E-waste should not merely be recognized as a waste having management issues, but rather an opportunity for mitigation of climate change, resource conservation and sustainable economic development. It can be minimized by utilizing in various possible ways following the concept of 3Rs. Approaches like physical, biotechnological and electrochemical processes can be adopted for e-waste management for metal recovery. Regulatory and legislative measures, social motivation, eco-design, adequate funds, enhanced recycling technologies, scientific management facilities, behavioural change like good practices and robust international cooperation can minimize the challenges related to e-waste. Application of advanced computational tools such as artificial intelligence and machine learning can play important roles in better management of e-waste. This will help in achieving sustainable goals and accomplishing circular economic concepts.

References

Aboelmaged, M. (2021). E-waste recycling behaviour: An integration of recycling habits into the theory of planned behaviour. *Journal of Cleaner Production, 278*, 124182. https://doi.org/10.1016/j.jclepro.2020.124182

Ahirwar, R., & Tripathi, A. K. (2021). E-waste management: A review of recycling process, environmental and occupational health hazards, and potential solutions. *Environmental Nanotechnology, Monitoring and Management, 15*, 100409. https://doi.org/10.1016/j.enmm.2020.100409

Amreen, S., Dash, S. K., Naik, B., Dash, A. K., Verma, A. K., & Pradhan, A. (2024). Marine macro algae-derived activated carbon an efficient bio-sorbent for Malachite green removal from wastewater. *Environmental Quality Management*. https://doi.org/10.1002/tqem.22213

Ari, V. (2016). A review of technology of metal recovery from electronic waste. In *E-waste in transition—From pollution to resource*. https://doi.org/10.5772/61569

Arya, S., & Kumar, S. (2020). E-waste in India at a glance: Current trends, regulations, challenges, and management strategies. *Journal of Cleaner Production, 271*, 122707. https://doi.org/10.1016/j.jclepro.2020.122707

Asefi, H., Shahparvari, S., & Chhetri, P. (2019). Integrated municipal solid waste management under uncertainty: A tri-echelon city logistics and transportation context. *Sustainable Cities and Society, 50*, 101606. https://doi.org/10.1016/j.scs.2019.101606

Awasthi, A. K., Zeng, X., & Li, J. (2016a). Relationship between e-waste recycling and human health risk in India: A critical review. *Environmental Science and Pollution Research, 23*, 11509–11532.

Awasthi, A. K., Zeng, X., & Li, J. (2016b). Environmental pollution of electronic waste recycling in India: A critical review. *Environmental Pollution, 211*, 259–270. https://doi.org/10.1016/j.envpol.2015.11.027

Baldé, C. P., Wang, F., Kuehr, R. (2016). Transboundary movements of used and waste electronic and electrical equipment: Estimates from the European Union using trade statistics.

Bhuiya, M. M. K., Rasul, M., Khan, M., Ashwath, N., & Mofijur, M. (2020). Comparison of oil extraction between screw press and solvent (n-hexane) extraction technique from beauty leaf (*Calophyllum inophyllum* L.) feedstock. *Industrial Crops and Products, 144*, 112024.

Bocken, N. M. P., & Short, S. W. (2021). Unsustainable business models: Recognising and resolving institutionalised social and environmental harm. *Journal of Cleaner Production, 312*, 127828. https://doi.org/10.1016/j.jclepro.2021.127828

Brigden, K. (2005). Recycling of electronic wastes in China and India: Workplace and environmental contamination. Greenpeace. http://www.greenpeace.org/raw/content/china/en/press/reports/recycling-of-electronic-wastes.pdf. Accessed 14 July 2022.

Chatterjee, P. (2008). Health costs of recycling. *BMJ, 337*, a296. https://doi.org/10.1136/bmj.a296

Chen, A., Dietrich, K. N., Huo, X., & Ho, S. M. (2011). Developmental neurotoxicants in e-waste: An emerging health concern. *Environmental Health Perspectives, 119*(4), 431–438.

Closer Look: Lithium-Ion Batteries in E-waste. (2022). https://www.simslifecycle.com/2019/05/21/a-closer-look-lithium-ion-batteries-in-E-waste/

Das, A. P., & Mishra, S. (2008). Hexavalent Chromium (VI): Health hazards and environmental pollutant. *Journal of Environmental Research and Development, 2*(3), 386–392.

Das, A. P., & Mishra, S. (2009). Hexavalent chromium [Cr (VI)]: Yellow water pollution and its remediation. *Sarovar Saurabh ENVIS Newsletter Wetland Ecosystem, 5*(2).

Das, A. P., & Mishra, S. (2010). Biodegradation of the metallic carcinogen hexavalent chromium Cr(VI) by an indigenously isolated bacterial strain. *Journal of Carcinogenesis, 9*, 6.

Das, A. P., & Singh, S. (2011). Occupational health assessment of chromite toxicity among Indian miners. *Indian Journal Of Occupational And Environmental Medicine, 15*(1), 6–17.

Das, A. P., Sukla, L. B., Pradhan, N., & Nayak, S. (2011). Manganese biomining: A review. *Bioresource Technology, 102*(16), 7381–7387.

Das, A. P., Ghosh, S., Mohanty, S., & Sukla, L. B. (2014a). Consequences of manganese compounds: A review. *Toxicological & Environmental Chemistry, 96*(7), 981–997.

Das, A. P., Kumar, P. S., & Swain, S. (2014b). Recent advances in biosensor-based endotoxin detection. *Biosensors and Bioelectronics, 51*, 62–75.

Dash, A. K. (2012). Impact of domestic wastewater on seed germination and physiological parameters of wheat and rice. *International Journal for Research and Review in Applied Sciences, 12*(2), 280–286.

Dash, S. K., & Dash, A. K. (2015a). Assessment of ambient air quality with reference to particulate matter (PM10 and PM2.5) and gaseous (SO2 and NO2) pollutants near Bileipada, Joda area of Keonjhar, Odisha, India. *Pollution Research, 34*(4), 817–824.

Dash, S. K., & Dash, A. K. (2015b). Determination of Air Quality Index status near Bileipada, Joda Area of Keonjhar, Odisha, India. *Indian Journal of Science and Technology, 8*(35), 1–7. https://doi.org/10.17485/ijst/2015/v8i35/81468

Dash, A. K., & Dash, S. K. (2017). Atmospheric Pollution Load Assessment through Air Quality Index: A case study. *Indian Journal of Environmental Protection, 37*(9), 736–741.

Dash, S. K., & Dash, A. K. (2018a). Air Pollution Tolerance Index to assess the pollution tolerance level of plant species in industrial areas. *Asian Journal of Chemistry, 30*(1), 219–222. https://doi.org/10.14233/ajchem.2018.20991

Dash, S. K., & Dash, A. K. (2018b). Seasonal variation of air quality index and assessment. *Global Journal of Environmental Science and Management, 4*(4), 483–492. https://doi.org/10.22034/gjesm.2018.04.008

Dash, A. K., Pradhan, A., Das, S., & Mohanty, S. S. (2015). Fly ash as a potential source of soil amendment in agriculture and a component of an integrated plant nutrient supply system. *Journal of Industrial Pollution Control, 31*(2), 251–259.

Dash, S. K., Sahu, H. B., & Dash, A. K. (2019). Atmospheric dispersion modeling using AERMOD to predict the impact of PM10 near Bileipada, Odisha. *Indian Journal of Environmental Protection, 39*(4), 299–306.

Dash, S. K., Pradhan, A., Panda, B. P., & Dash, A. K. (2023). Ambient air quality monitoring and assessment of an industrial area as a tool for sustainable environmental management. *Materials Today: Proceedings*. (In press). https://doi.org/10.1016/j.matpr.2023.10.024

Dash, A. K., Pradhan, A., & Sukla, L. B. (2024). Environmental occurrence and contemporary health issues of microplastics. In A. P. Das, I. D. Behera, & N. P. Das (Eds.), *Renewable energy generation and value addition from environmental microfiber pollution through advanced greener solution. Environmental science and engineering* (pp. 113–136). Springer, . ISBN: 978-3-031-51792-1. https://doi.org/10.1007/978-3-031-51792-1_7

Debnath, B., Saha, I., Mukherjee, T., Mitra, S., Das, A., & Das, A. (2021). Sorbents from waste materials: A circular economic approach. In *Sorbents materials for controlling environmental pollution* (pp. 285–322). Elsevier.

Dhir, A., Koshta, N., Goyal, R. K., Sakashita, M., & Almotairi, M. (2021). Behavioral reasoning theory (BRT) perspectives on e-waste recycling and management. *Journal of Cleaner Production, 280*, 124269. https://doi.org/10.1016/j.jclepro.2020.124269

Disposal of Asbestos, Chemicals, Oil, Trade and e-Waste - Parkes Shire Council. https://www.parkes.nsw.gov.au/environment/waste-recycling/waste-facilities/disposal-of-chemicals-oil-trade-and-E-waste/. Accessed 14 Feb 2022.

El-Ghiaty, M. A., & El-Kadi, A. O. (2021). Arsenic: Various species with different effects on cytochrome P450 regulation in humans. *EXCLI Journal, 20*, 1184.

Forti, V., Baldé, C. P., Kuehr, R., & Bel, G. (2020). The Global E-waste Monitor 2020: Quantities, flows, and the circular economy potential. United Nations University (UNU)/United Nations Institute for Training and Research (UNITAR)—Co-hosted SCYCLE Programme, International Telecommunication Union (ITU) & International Solid Waste Association (ISWA), Bonn/Geneva/Rotterdam. Available at: https://www.itu.int

Forum WEEE. (2021). Report on International E-Waste Day 2021. Available online: https://weee-forum.org/iewd-about/

Gangwar, C., Choudhari, R., Chauhan, A., Kumar, A., Singh, A., & Tripathi, A. (2019). Assessment of air pollution caused by illegal e-waste burning to evaluate the human health risk. *Environment International, 125*, 191–199.

Garlapati, V. K. (2016). E-waste in India and developed countries: Management, recycling, business, and biotechnological initiatives. *Renewable and Sustainable Energy Reviews, 54*, 874–881.

Ghimire, H., & Ariya, A. A. (2020). E-wastes: Bridging the knowledge gaps in global production budgets, composition, recycling and sustainability implications. *Sustainable Chemistry, 2*, 154–182. https://doi.org/10.3390/suschem1020012

Goyal, S., Esposito, M., & Kapoor, A. (2018). Circular economy business models in developing economies: Lessons from India on reduce, recycle, and reuse paradigms. *Thunderbird International Business Review, 60*(5), 729–740. https://doi.org/10.1002/tie.21883

Grant, K., Goldizen, F. C., Sly, P. D., Brune, M. N., Neira, M., van den Berg, M., & Norman, R. E. (2013). Health consequences of exposure to e-waste: A systematic review. *The Lancet Global Health, 1*(6), e350–e361.

Gunarathne, N., de Alwis, A., & Alahakoon, Y. (2020). Challenges facing sustainable urban mining in the e-waste recycling industry in Sri Lanka. *Journal of Cleaner Production, 251*, 119641. https://doi.org/10.1016/j.jclepro.2019.119641

Guo, Y., Changjiang, H., Hong, Z., & Qiaoxiang, D. (2009). Heavy metal contamination from electronic waste recycling at Guiyu, Southeastern China. *Journal of Environmental Quality, 38*, 1617–1626.

Guo, Y., Huo, X., Wu, K., Liu, J., Zhang, Y., & Xu, X. (2012). Carcinogenic polycyclic aromatic hydrocarbons in umbilical cord blood of human neonates from Guiyu, China. *Science of the Total Environment, 427*, 35–40. https://doi.org/10.1016/j.scitotenv.2012.04.007

Huang, J., Philip, N. N., Desmond, O. A., & Ebenezer, M. (2014). E-waste disposal effects on the aquatic environment, Accra, Ghana. In D. M. Whitacre (Ed.), *Reviews of environmental contamination and toxicology* (p. 229). Springer International Publishing.

Ilankoon, I. M., Ghorbani, Y., Chong, M. N., Herath, G., Moyo, T., & Petersen, J. (2018). E-waste in the international context–A review of trade flows, regulations, hazards, waste management strategies and technologies for value recovery. *Waste Management, 82*, 258–275.

Jain, M., Kumar, D., Chaudhary, J., Kumar, S., Sharma, S., & Verma, A. S. (2023). Review on e-waste management and its impact on the environment and society. *Waste Management Bulletin, 1*, 34–44. https://doi.org/10.1016/j.wmb.2023.06.004

Jena, P. S., Pradhan, A., Nanda, S. P., Dash, A. K., & Naik, B. (2022). Biosorption of heavy metals from wastewater using *Saccharomyces cerevisiae* as a biosorbent: A mini review. *Materials Today Proceedings, 67*(8), 1140–1146. https://doi.org/10.1016/j.matpr.2022.07.306

Krishnamoorthy, Y., Vijayageetha, M., Sakthivel, M., & Sarveswaran, G. (2018). Emerging public health threat of E-waste management: Global and Indian perspective. *Review of Environmental Health, 33*(4), 321–329.

Kuang, H. X., Li, M. Y., Li, L. Z., Li, Z. C., Wang, C. H., Xiang, M. D., & Yu, Y. J. (2023). Co-exposure levels of volatile organic compounds and metals/metalloids in children: Implications for e-waste recycling activity prediction. *Science of the Total Environment, 863*, 160911.

Kumar, U., & Singh, D. (2014). Electronic waste: Concerns & hazardous threats. *International Journal of Current Engineering and Technology, 4*, 802–811.

Kumar, D. S., Bhushan, S. H., & Kishore, D. A. (2018). Atmospheric dispersion model to predict the impact of gaseous pollutants in an industrial and mining cluster. *Global Journal of Environmental Science and Management, 4*(3), 351–358. https://doi.org/10.22034/gjesm.2018.03.008

Kumar, M., Pandit, S., Patel, V., Khanna, N., Nag, M., Lahiri, D., Ray, R. R., Das, A., & Das, D. (2023). Maximization of energy recovery from starch processing wastewater by thermophilic dark fermentation coupled with microbial fuel cell technology. *Geomicrobiology Journal*, 1–12. https://doi.org/10.1080/01490451.2023.2209555

Kumarathilaka, P., Seneweera, S., Meharg, A., & Bundschuh, J. (2018). Arsenic speciation dynamics in paddy rice soil-water environment: Sources, physico-chemical, and biological factors-A review. *Water Research, 140*, 403–414.

Kurup, A. R., & Senthil Kumar, K. (2017). Effect of recycled PVC fibres from electronic waste and silica powder on shear strength of concrete. *Journal of Hazardous, Toxic, and Radioactive Waste, 21*(3), 06017001.

Li, W., & Achal, V. (2020). Environmental and health impacts due to e-waste disposal in China–a review. *Science of Total Environment, 737*, 139745.

Lucier, C. A., & Gareau, B. J. (2019). Electronic waste recycling and disposal: An overview. In *Assessment and management of radioactive and electronic wastes* (pp. 1–12). Books on Demand.

Ma, H. K. (2014). Eco-recycling of precious metal from e-waste. In Conf. on electronics Recycling Asia, organized by World Recycling Forum on 11th to 14th Nov 2014, 33.

Mahapatra, B., Dhal, N. K., Dash, A. K., Panda, B. P., Panigrahi, K. C. S., & Pradhan, A. (2019). Perspective of mitigating atmospheric heavy metal pollution: Using mosses as biomonitoring and indicator organisms. *Environmental Science and Pollution Research, 26*(29), 29620–29638. https://doi.org/10.1007/s11356-019-06270-z

Matsakas, L., et al. (2017). Green conversion of municipal solid waste into fuels and chemicals. *Electronic Journal of Biotechnology, 26*, 69–83.

Mielke, H. W., & Reagan, P. L. (1998). Soil is an important pathway of human lead exposure. *Environmental Health Perspectives, 106*(1), 217–229. https://doi.org/10.4236/ss.2011.24042

Minoja, M., & Romano, G. (2021). Managing intellectual capital for sustainability: Evidence from a re-municipalized, publicly owned waste management firm. *Journal of Cleaner Production, 279*, 123213. https://doi.org/10.1016/j.jclepro.2020.123213

Mishra, S., & Das, A. P. (2021). Current treatment technologies for removal of microplastic and microfiber pollutants from wastewater. In *Wastewater treatment: Cutting-edge molecular tools, techniques and applied aspects* (pp. 237–251). https://doi.org/10.1016/B978-0-12-821881-5.00011-8

Mishra, S., & Das, A. P. (2022). Treatment of the wastewater polluted with synthetic microfiber released from washing machine. In B. B. Das, H. Hettiarachchi, P. K. Sahu, & S. Nanda (Eds.), *Lecture notes in civil engineering* (Vol. 207, pp. 109–117). Springer Science and Business Media Deutschland GmbH. https://doi.org/10.1007/978-981-16-7509-6_9/COVER

Mishra, P. C., Dash, A. K., & Pradhan, K. (2012). Metals in environmental segments at Hirakud of Odisha, India. *ISCA Journal of Biological Sciences, 1*(1), 7–27.

Mishra, S., Singh, R. P., Rout, P. K., & Das, A. P. (2022). Membrane bioreactor (MBR) as an advanced wastewater treatment technology for the removal of synthetic microplastics. In

Development in wastewater treatment research and processes (pp. 45–60). https://doi.org/10.1016/B978-0-323-85583-9.00022-3

Mishra, S., Ghosh, S., van Hullebusch, E. D., Singh, S., & Das, A. P. (2023). A critical review on the recovery of base and critical elements from electronic waste-contaminated streams using microbial biotechnology. *Applied Biochemistry and Biotechnology, 2023*, 1–30. https://doi.org/10.1007/S12010-023-04440-X

Mishra, S., Dash, D., & Das, A. P. (2024a). Isolation and molecular characterization of polyester-degrading bacterial strains from municipal wastewater and their application in microplastic degradation in contaminated wastewater. *Geomicrobiology Journal*. https://doi.org/10.1080/01490451.2024.2376101

Mishra, S., Dash, D., & Das, A. P. (2024b). Microbial degradation of polyester microfibers using indigenously isolated bacterial strain Exiguobacterium Sp. *CLEAN—Soil, Air, Water, 52*(12), e202300343. https://doi.org/10.1002/CLEN.202300343

Morseletto, P. (2020). Targets for a circular economy. *Resources, Conservation and Recycling, 153*, 104553. https://doi.org/10.1016/j.resconrec.2019.104553

Naik, B., Sukla, L. B., & Dash, A. K. (2024a). Sources of synthetic microfiber pollutants in the environment and their fate and toxicity. In A. P. Das, I. D. Behera, & D. Bhanja (Eds.), *Sustainable microbial technology for synthetic and cellulosic microfiber bioremediation*. Springer. https://doi.org/10.1007/978-3-031-62600-5_1

Naik, B., Sukla, L. B., & Dash, A. K. (2024b). Innovative methods and greener technology for remediation of microfiber pollutants. In *Prospects of science, technology and applications* (pp. 295–315). https://doi.org/10.1201/9781003489443-32

Nayak, K. P., Mishra, A., Pradhan, A., Dash, A. K., & Saren, S. (2018). Soil fertility study as a tool for sustainable agriculture. *NPI International Research Journal of Industrial and Environmental Biotechnology (IRJIEB), 1*(1 & 2), 57–74.

Onac, C., Kaya, A., Atar, N., Sener, I., & Alpoguz, H. K. (2020). Superiority of modified polymeric membrane with nanomaterial on temperature and mechanical stability and application in industrial wastewater. *ECS Journal of Solid State Science and Technology, 9*(6), 061019. https://doi.org/10.1149/2162-8777/aba725

Panigrahi, B., & Dash, A. K. (2024). Sustainable management for reduction of plastic waste through a circular economy. In A. P. Das & A. John (Eds.), *Sustainable management of urban plastic waste through circular economic approaches*. CRC Press. https://doi.org/10.1201/9781003465447

Panigrahi, B. R., Pradhan, S. C., Dash, A. K., & Mishra, P. C. (2024). Evaluation of protease activity in the soil of municipal solid waste dumping sites in Balasore district of Odisha, India. *Pollution Research, 43*(3–4), 267–269.

Parida, B., & Dash, A. K. (2024). Sources, characteristics, environmental impact, and sustainable remediation of synthetic microfiber. In A. P. Das, I. D. Behera, & D. Bhanja (Eds.), *Sustainable microbial technology for synthetic and cellulosic microfiber bioremediation*. Springer. https://doi.org/10.1007/978-3-031-62600-5_2

Parida, S., Das, M., & Dash, A. K. (2015). Ambient air study near the industrial complex 'Nisha' at Angul, Odisha, India. In *Advances in environmental sciences and engineering* (pp. 514–537). Daya Publishing House, a division of Astral Int. Pvt., Ltd. ISBN: 978-93-5124-336-6.

Parida, B., Dash, A. K., & Verma, A. (2023). Technologies for treatment of landfill leachate: A brief review. *Recent Innovations in Chemical Engineering, 16*(4), 260–284. https://doi.org/10.2174/0124055204276847231013040437

Paul, A., Dey, S., Ram, D. K., & Das, A. P. (2023). Hexavalent chromium pollution and its sustainable management through bioremediation. *Geomicrobiology Journal, 1–11*. https://doi.org/10.1080/01490451.2023.2218377

Pavithra, K. G., Rajan, P. S. S., Balaji, D., & Gopinath, K. P. (2020). Sustainable electronic-waste management: Implications on environmental and human health. In *E-waste recycling and management* (pp. 201–218). Springer.

Pinto, V. N. (2008). E-waste hazard: The impending challenge. *Indian Journal Of Occupational And Environmental Medicine, 12*(2), 65.

Prabhakar, A., Mishra, S., & Das, A. P. (2019). Isolation and identification of Lead (Pb) solubilizing bacteria from automobile waste and its potential for recovery of Lead from end of life waste batteries. *Geomicrobiological Journal, 36*(10), 894–903. https://doi.org/10.1080/01490451.2019.1654044

Pradhan, J. K., & Kumar, S. (2014). Informal e-waste recycling: Environmental risk assessment of heavy metal contamination in Mandoli industrial area, Delhi, India. *Environmental Science and Pollution Research International, 21*(13), 7913–7928. https://doi.org/10.1007/s11356-014-2713-2

Pradhan, A., Sahu, S. K., & Dash, A. K. (2013). Changes in pigment content (chlorophyll and carotenoid), enzyme activities (catalase and peroxidase), biomass, and yield of rice plant (*Oriza sativa* L.) following irrigation of rice mill wastewater under pot culture conditions. *International Journal of Scientific & Engineering Research, 4*(6), 2706–2717.

Pradhan, A., Dash, A. K., Mohanty, S. S., & Das, S. (2015a). Potential use of fly ash in floriculture: A case study on the photosynthetic pigments content and vegetative growth of *Tagetes erecta* (MARIGOLD). *Ecology, Environment and Conservation, 21*(August Suppl), 369–376.

Pradhan, A., Sahu, S. K., & Dash, A. K. (2015b). Changes in physico-chemical characteristics of soil following application of rice mill wastewater under field irrigation conditions. In *Advances in environmental sciences and engineering* (pp. 480–492). Daya Publishing House, a division of Astral Int. Pvt., Ltd. ISBN: 978-93-5124-336-6.

Pradhan, A., Sahu, S. K., & Dash, A. K. (2016). Changes in physico-chemical characteristics of soil following application of rice mill wastewater under pot irrigation conditions. *Advanced Science Letters, 22*(2), 504–510. https://doi.org/10.1166/asl.2016.6838

Pradhan, A., Kumari, S., Dash, S., Biswal, D. P., Dash, A. K., & Panigrahi, K. C. S. (2017). Heavy metal absorption efficiency of two species of mosses (*Physcomitrella patens* and *Funaria hygrometrica*) studied in mercury-treated culture under laboratory conditions. *IOP Publishing, IOP Conf. Series: Materials Science and Engineering, 225*, 012225. https://doi.org/10.1088/1757-899X/225/1/012225

Priyadarsini, S., & Das, A. P. (2024). Analytical and structural characterization of waste lithium-ion batteries for their effective recycling strategy. *Environmental Chemistry and Ecotoxicology*. https://doi.org/10.1016/j.enceco.2024.12.004

Priyadarsini, S., & Das, A. P. (2025). Lithium bioleaching: A review on microbial-assisted sustainable technology for lithium bio-circularity. *Journal of Water Process Engineering, 69*, 106744. https://doi.org/10.1016/j.jwpe.2024.106744

Raj, K., & Das, A. P. (2023). Lead pollution: Impact on environment and human health and approach for a sustainable solution. *Environmental Chemistry and Ecotoxicology, 5*, 79–85. https://doi.org/10.1016/J.ENCECO.2023.02.001

Ravindra, K., & Mor, S. (2019). E-waste generation and management practices in Chandigarh, India and economic evaluation for sustainable recycling. *Journal of Cleaner Production, 221*, 286–294. https://doi.org/10.1016/j.jclepro.2019.02.158

Safdar, N., Khalid, R., Ahmed, W., & Imran, M. (2020). Reverse logistics network design of e-waste management under the triple bottom line approach. https://doi.org/10.1016/j.jclepro.2020.122662

Saha, L., Kumar, V., Tiwari, J., Rawat, S., Singh, J., & Bauddh, K. (2021). Electronic waste and their leachates impact on human health and environment: Global ecological threat and management. *Environment Technology Innovation, 24*, 102049.

Sahajwalla, V., & Hossain, R. (2020). The science of microrecycling: A review of selective synthesis of materials from electronic waste. *Master Today Sustainability, 9*, 100040. https://doi.org/10.1016/j.mtsust.2020.100040

Sahle-Demessie, E., Glaser, J., & Richardson, T. (2018). *Electronics waste management challenges and opportunities*. American Chemical Society, National Meeting.

Sahoo, D., Dash, A. K., & Sahu, S. K. (2017). Ambient air quality monitoring and health impact study of air pollution near Joda of Keonjhar, Odisha, India. *International Journal of Engineering Sciences & Research Technology, 6*(1), 429–434. https://doi.org/10.5281/zenodo.260069

Sahu, S. K., Mishra, A., & Dash, A. K. (2015). Ambient air quality around OCL India Limited at Rajgangpur, Odisha, India. In *Advances in environmental sciences and engineering* (pp. 459–479). Daya Publishing House, a division of Astral Int. Pvt., Ltd. ISBN: 978-93-5130-300-8.

Salhofer, S., Steuer, B., Ramusch, R., & Beigl, P. (2016). WEEE management in Europe and China: A comparison. *Waste Management, 57*, 27–35.

Sánchez-Quiles, D., & Tovar-Sánchez, A. (2015). Are sunscreens a new environmental risk associated with coastal tourism? *Environment International, 83*, 158–170.

Shamim, A., Mursheda, A. K., & Rafiq, I. (2015). E-waste trading impact on public health and ecosystem services in developing countries. *Journal of Waste Resource, 5*(4), 1–18.

Shreya, Verma, A. K., Gilron, J., Oren, Y., Ronen, Z., Dash, A. K., Bhunia, P., & Dash, R. R. (2022). Effect of biological treatment on perfluoroalkyl and polyfluoroalkyl (PFASs) substances degradation. In *Current developments in biotechnology and bioengineering: Sustainable treatment Technologies for pre- and polyfluoroalkyl substances*. https://doi.org/10.1016/B978-0-323-99906-9.00004-8

Sivaramanan, S. (2013). E-waste management, disposal and its impact on the environment. *Universal Journal of Environmental Research and Technology, 3*, 531–537.

Srivastav, A. L., Markandeya, Patel, N., Pandey, M., Pandey, A. K., Dubey, A. K., Kumar, A., Bhardwaj, A. K., & Chaudhary, V. K. (2023). Concepts of circular economy for sustainable management of electronic wastes: Challenges and management options. *Environmental Science and Pollution Research, 30*, 48654–48675. https://doi.org/10.1007/s11356-023-26052-y

Srivastava, R. R., & Pathak, P. (2020). Policy issues for efficient management of e-waste in developing countries. In *Handbook of electronic waste management* (pp. 81–99). Butterworth-Heinemann.

Stapleton, H. M., Allen, J. G., Kelly, S. M., Konstantinov, A., Klosterhaus, S., Watkins, D., McClean, M. D., & Webster, T. F. (2008). Alternate and new brominated flame retardants detected in US house dust. *Environmental Science and Technology, 42*(18), 6910–6916.

Sthiannopkao, S., & Wong, M. H. (2013). Handling e-waste in developed and developing countries: Initiatives, practices, and consequences. *Science of the Total Environment, 463*, 1147–1153. https://doi.org/10.1016/j.scitotenv.2012.06.088

Tiwari, D., Raghupathy, L., Khan, A. S., & Dhawan, N. G. (2019). A study on the e-waste collection systems in some Asian countries with special reference to India. *Nature Environment and Pollution Technology, 18*(1), 149–156. https://doi.org/10.12944/cwe.18.3.28

Tue, N. M., Matsushita, T., Goto, A., Itai, T., Asante, K. A., Obiri, S., Mohammed, S., Tanabe, S., & Kunisue, T. (2019). Complex mixtures of brominated/chlorinated diphenyl ethers and dibenzofurans in soils from the Agbogbloshie E-waste site (Ghana): Occurrence, formation, and exposure implications. *Environmental Science and Technology, 53*(6), 3010–3017.

Vats, M. C., & Singh, S. K. (2014a). E-waste characteristics and its disposal. *International Journal of Ecology, Science and Environmental Engineering, 1*(2), 49–61.

Vats, M. C., & Singh, S. K. (2014b). Status of e-waste in India—A review. *International Journal of Innovative Research in Science, Engineering and Technology, 3*(10), 16917–16931.

Wang, Y., Tian, Z., Zhu, H., Cheng, Z., Kang, M., Luo, C., Li, J., & Zhang, G. (2012). Polycyclic aromatic hydrocarbons (PAHs) in soils and vegetation near an E-waste recycling site in South China: Concentration, distribution, source, and risk assessment. *Science of the Total Environment, 439*, 187–193.

Wu, Q., Leung, J. Y., Du, Y., Kong, D., Shi, Y., Wang, Y., & Xiao, T. (2019). Trace metals in e-waste lead to serious health risk through consumption of rice growing near an abandoned e-waste recycling site: Comparisons with PBDEs and AHFRs. *Environmental Pollution, 247*, 46–54.

Yang, J., & Zhang, B. (2018). Air pollution and healthcare expenditure: Implication for the benefit of control in China. *Environment International, 120*, 443–455.

Yang, J., Bertram, J., Schettgen, T., Heitland, P., Fischer, D., Seidu, F., Felten, M., Kraus, T., Fobil, J. N., & Kaifie, A. (2020). Arsenic burden in E-waste recycling workers–A cross-sectional study at the Agbogbloshie E-waste recycling site. Ghana. *Chemosphere, 261*, 127712.

Yang, W.-D., Sun, Q., & Ni, H.-G. (2021). Cost-benefit analysis of metal recovery from e-waste, implications for international policy. *Waste Management, 123*, 42–47. https://doi.org/10.1016/j.wasman.2021.01.023

Zeng, X., Huo, X., Xu, X., Liu, D., & Wu, W. (2020). E-waste lead exposure and children's health in China. *Science of Total Environment, 734*, 139286.

Zhang, K., Schnoor, J. L., & Zeng, E. Y. (2012). E-waste recycling: Where does it go from here? *Environment Science and Technology, 46*(20), 10861–10867.

Review on E-Waste Plastics in Construction: A Circular Economy Solution

Sweta Sinha

Abstract Electronic waste (e-waste) has become one of the fastest-growing waste streams globally, creating critical environmental and resource management challenges. Plastics, which make up approximately 20% of e-waste, present a significant opportunity for integration into construction materials, aligning with circular economy principles. This review explores the potential of e-waste plastics to serve as sustainable construction materials, highlighting current research, technological advancements, and practical applications in the field. E-waste plastics have been effectively utilized in various construction applications, such as concrete composites, polymer-modified materials, and insulation products, where they act as partial substitutes for conventional materials or introduce novel functionalities. Studies reveal that incorporating e-waste plastics at replacement rates of 10–20% by volume in concrete can enhance critical properties, including thermal insulation, reduced density, and improved ductility. Beyond concrete, applications extend to polymer-modified bitumen, plastic lumber and composite panels. These innovations contribute to environmental benefits, such as decreased landfill waste, reduced carbon emissions, and conservation of finite natural resources, while also offering economic advantages through lower material costs and new revenue opportunities in waste management. There are several challenges persist, including technical issues related to material heterogeneity, contamination, and quality control, as well as regulatory and market barriers. Addressing these challenges requires advancements in sorting and processing technologies, the development of durable materials, and supportive policy frameworks, including standardized guidelines and incentives for adoption.

This review emphasizes the need for collaborative efforts across research, policy, and industry to realize the full potential of e-waste plastics in construction. By overcoming existing barriers, e-waste plastics can serve as a sustainable alternative to traditional materials, fostering a more resilient and environmentally friendly built environment while addressing global waste management challenges.

S. Sinha (✉)
Department of Chemistry, Amity University Jharkhand, Ranchi, Jharkhand, India

A. P. Das, S. Priyadarsini (eds.), *Electronic Waste and Environmental Pollution*, Sustainable Environmental Waste Management Strategies,
https://doi.org/10.1007/978-3-031-90287-1_8

Keywords E-waste plastics · Sustainable management · Circular economy · Ecosystem

1 Introduction

The growing global plastic waste problem demands innovative approaches to mitigate its environmental consequences (Priyadarshini et al., 2024; Singh Jadaun et al., 2022). However, the collective effort to draw public attention toward this issue dates back to the 1970s, with numerous discussions by environmentalists centering on ecological balance and sustainability (Sinha 2022). While the rapid development across the globe has undoubtedly improved living standards led to a significant increase in waste generation (Akhtar & Sarmah, 2018). Human activities now produce waste at exponentially higher rates compared to earlier periods (Das et al., 2019). The growing volume of waste has brought about a significant increase in its complexity and variety, making recycling and disposal more challenging (Abdel-Shafy & Mansour, 2018). As a result, the environment faces mounting difficulty in assimilating waste without undergoing severe degradation. Waste can be broadly categorized based on criteria such as its physical form, origin, degradability, and potential environmental impact. Among the diverse waste types, non-biodegradable materials like plastics (Bala et al., 2024; Sinha 2025; John & Das, 2024; Sahoo et al., 2022) and metals (Sinha 2024, Das et al., 2011; Mishra et al., 2022), along with hazardous waste such as biomedical (Khatun et al., 2023; Sadasivuni et al., 2019) and electronic waste (Mishra et al., 2023; Priyadarsini & Das, 2025), are particularly concerning due to their long-term and often harmful effects on ecosystems. These types of waste persist in the environment for prolonged periods, making their reuse and disposal particularly challenging. Therefore, adopting effective waste management strategies has become a pressing need to ensure sustainable development.

Countries worldwide are facing the intertwined challenges of addressing rising energy needs while working to combat climate change. With 18% of the global population, India contributes 7% to global emissions, reflecting both its developmental aspirations and the associated environmental pressures (Shahbaz et al., 2021) At the COP26 Summit in Glasgow, nations, including India, committed to ambitious climate targets for 2030. These include reducing projected carbon emissions, transitioning to renewable energy sources, and significantly cutting the carbon intensity of economies. Achieving these goals demands innovative and collaborative approaches worldwide to reduce carbon dependency while sustaining economic growth. The construction industry, which accounts for approximately 39% of global carbon emissions, stands out as a sector with significant potential for transformation (Hassan et al., 2024). A combination of population-specific solutions, government policies, and efforts to mobilize individuals and communities can foster global capacity for reducing carbon footprints. Practices such as waste segregation, reduced reliance on non-biodegradable materials, and promoting circular economies are crucial steps toward a sustainable future. However, reaching such ambitious goals will take time and concerted efforts across all sectors and geographies.

This chapter explores a globally relevant solution to the mounting challenge of non-biodegradable waste, particularly plastic. By repurposing such waste into economically and environmentally valuable applications, we can address two pressing issues simultaneously—managing waste sustainably and reducing carbon emissions. This integrated approach can serve as a model for nations worldwide striving to balance development and environmental responsibility.

2 Challenges

2.1 Population Growth

As of 2024, the global population is approximately 8.05 billion and is projected to rise to 9.8 billion by 2050, according to United Nations data. Asia remains home to around 59% of the global population, while Africa, experiencing the highest growth rate, is expected to contribute significantly to future population increases. India and China continue to lead as the most populous nations. Recent estimates suggest that India, with a population of 1.43 billion, has overtaken China's 1.42 billion due to China's declining growth rate, which has been influenced by policies such as the one-child policy. The dramatic global population increase is attributed to factors such as improved economic conditions, advancements in healthcare leading to higher life expectancy, rising urbanization, and significant migration patterns. These changes bring both opportunities and challenges. On one hand, population-driven economic growth supports industrial development and market expansion. On the other, rapid population growth strains natural resources and infrastructure, leading to increased waste generation, pollution, and ecological instability.

United States, China, and India, are also major centers of manufacturing, production, and consumption (Cresswell et al., 2021). These nations rank among the largest contributors to global waste streams, including plastic, electronic waste, radioactive materials, and biomedical waste. Additionally, China and India are significant consumers of fossil fuels, such as coal and oil, making them key contributors to greenhouse gas emissions and global warming. While population growth can stimulate economic progress, it poses significant challenges to achieving sustainable development. Balancing population-driven economic gains with environmental sustainability will require innovative policies and collaborative efforts to reduce pollution, manage waste effectively, and minimize ecological disruptions (Davis, 2015).

2.2 Role of the Construction Industry in Waste Generation

The global population continues to grow at an unprecedented rate, presenting challenges to sustainable development, particularly in developing nations. While controlling population growth raises ethical concerns, it also complicates efforts to ensure equitable development for all. While some developed countries have managed to stabilize their populations, nations like India, which houses a large proportion of the global population, face significant obstacles due to financial limitations and infrastructure gaps. According to the United Nations' World Population Report, India is witnessing the fastest rate of urbanization globally, driven primarily by migration from rural areas in search of better opportunities (Davis, 2015). This shift places immense pressure on urban infrastructure, which often lacks adequate planning, creating resource strains and escalating environmental challenges.

India's construction industry plays a critical role in supporting this urban growth, ranking as the third-largest construction sector in the world and contributing over 10% to the nation's GDP. However, the sector is also a major contributor to carbon emissions, which contradicts international commitments like the COP26 agenda aimed at reducing carbon intensity globally. The demand for raw materials such as sand, rocks, and cement has surged, contributing to the overexploitation of natural resources and leading to their alarming depletion (Mondal & Palit, 2022). Additionally, the scale of construction has resulted in a significant increase in waste generation, which further harms the environment. A key contributor to the environmental impact of construction is cement. India consumed around 396.7 million tons in 2023–24. Projections indicate that by 2025, cement consumption could rise to 475 million tons, driven by ongoing demand in infrastructure and housing sectors. Cement industries are responsible for approximately 8% of global emissions (Ahmed et al., 2021), which contributes as a major cause of global climate change.

Although the pandemic (COVID-19) temporarily (from 2020 to 2021) slowed construction activities, easing pressure on overstressed ecosystems. The construction and cement industries must explore alternative practices to reduce their carbon footprints, conserve resources, and minimize ecological damage. Prioritizing environmental protection will be key as India works toward balancing growth with sustainable development goals.

2.3 E-Plastic: A Digital Waste

The global digital revolution has transformed technological landscapes, with rapid internet penetration and device proliferation reshaping multiple sectors. Emerging economies like India are experiencing significant technological transitions, characterized by widespread adoption of electronic devices such as smartphones,

Table 1 Types and contribution of Digital Waste

Electronic gadgets	68%
Medical equipment	7%
Telecommunications devices	14%
Electrical equipment	8%
Other categories	5%

Table 2 E-waste management strategies

Addressing electronic waste requires comprehensive strategies	Promoting public awareness
	Implementing stringent regulatory frameworks
	Developing robust recycling infrastructures
	Investing in advanced waste processing technologies
	Encouraging sustainable technological design

smartwatches, and smart televisions. Electronic waste (e-waste) represents a critical environmental challenge arising from technological advancement (Shahabuddin et al., 2023). Devices with typical service lifecycles of 5–7 years generate substantial waste containing complex material compositions (Table 1).

These discarded electronics contain hazardous substances like mercury, arsenic, cadmium, and lead, alongside valuable materials such as gold, platinum, silver, and copper (Cui & Zhang, 2008). The intricate waste composition presents both environmental risks and potential resource recovery opportunities.

Plastic waste, a significant e-waste component, poses unique disposal challenges. Recent data suggests India generated approximately 3.5 million tons of plastic waste, with global recycling rates around 25%. The low economic value of recycled plastics further complicates waste management efforts.

Effective e-waste management demands collaborative efforts across governments, industries, and consumers, emphasizing the critical 3R principles: Reduce, Reuse, Recycle (Table 2) (Ilankoon et al., 2018).

3 E-Plastic Blended Concrete

3.1 Global Electronic Device Landscape

The contemporary technological ecosystem is characterized by unprecedented electronic device proliferation. By 2024, global household penetration of digital technologies has reached extraordinary levels, with approximately 80% of households worldwide owning multiple electronic devices (Carmody, 2024). This digital transformation encompasses smartphones, computers, televisions, printers, and interconnected smart devices, reflecting a fundamental shift in technological consumption patterns. Developed and emerging economies alike are experiencing rapid technological integration, driven by decreasing device costs, enhanced connectivity, and increasing digital literacy.

3.2 E-Plastic Composition and Challenges

Electronic waste plastics represent a complex and heterogeneous waste stream with multifaceted environmental implications. Thermoplastic polymers dominate this landscape, comprising approximately 78% of global plastic production. The primary polymers encountered in electronic waste include polyethylene terephthalate (PET), polypropylene (PP), polystyrene (PS), and acrylonitrile butadiene styrene (ABS). Each polymer type presents unique chemical properties, recycling challenges, and potential applications.

The intrinsic complexity of e-plastic waste stems from several critical factors:

- Technological manufacturing variations.
- Potential chemical contamination.
- Diverse material compositions.
- Degradation during previous device usage.
- Limited standardized recycling protocols.

3.3 Sustainable Integration Strategies and Innovation

Emerging sustainable waste management approaches focus on transforming electronic waste from an environmental liability into a valuable resource. Interdisciplinary research is exploring innovative strategies that transcend traditional waste disposal methodologies. These strategies encompass circular economy principles, emphasizing material recovery, reprocessing, and value creation.

Pioneering research demonstrates promising pathways for integrating e-plastic waste into construction materials, particularly concrete composites. By systematically replacing traditional aggregates with processed electronic waste plastics, researchers aim to develop innovative, sustainable construction materials. This approach addresses two interconnected challenges: plastic waste management and sustainable infrastructure development.

The integration of e-plastic waste into construction materials offers multifaceted benefits:

- Enhanced material circularity.
- Potential cost efficiencies in construction.
- Decreased reliance on virgin aggregate extraction.
- Significant reduction in plastic environmental pollution.
- Reduced carbon footprint associated with waste management.

3.4 Technological Considerations

Technological advancement plays a pivotal role in successfully repurposing electronic waste plastics. Thermoplastics, comprising approximately 78% of global plastic production, offer unique recycling opportunities. Their molecular structure allows repeated processing with minimal property degradation, making them ideal candidates for sustainable material integration.

Critical considerations include:

- Development of standardized recycling methodologies.
- Advanced material characterization techniques.
- Comprehensive understanding of polymer chemistry.
- Sophisticated sorting and processing technologies.
- Ensuring structural integrity of composite materials.

3.5 Recycling Complexity

The recycling of electronic waste plastics presents significant technological challenges. The complex mixture of polymers, potential chemical contaminants, and varied material states require sophisticated separation and processing techniques. Existing recycling infrastructures often struggle with the intricate nature of e-plastic waste, necessitating continuous technological.

Therefore, integrating e-plastic waste into construction materials represents a promising avenue for sustainable waste management. Continued interdisciplinary research and technological innovation will be crucial in transforming electronic waste from an environmental challenge to a valuable resource. Common types of plastics used in the various electronic devices are described in the Table. 3.

In 2024, plastic waste generation remains a critical environmental challenge, with India producing approximately 15,342.6 tons of plastic waste daily. This substantial volume reflects broader global trends in plastic consumption and waste management. Low-density polyethylene (LDPE) continues to pose significant recycling challenges, while other plastics used in electronic devices demonstrate promising characteristics for material repurposing (Mishra et al., 2024a, 2024b). Recent experimental studies have shown that integrating plastic waste into concrete mixes can be an effective solution (Table 4). Many studies globally have explored the use of plastics such as HDPE, PVC, PP, and PET as substitutes for conventional concrete components, yielding positive results for concrete durability and performance (Ahmed et al., 2021; Liu et al., 2024; Sinha & Hamdan, 2023). For example, replacing fine aggregates with PET plastic bottles has been shown to improve shear strength, particularly when superplasticizers are added to the mix.

E-plastic waste, like polyurethane (PE), ABS, PP, and PS, makes up about 70% of the plastic found in electronic devices globally. Using e-plastic in concrete is a relatively new idea, but early studies show that while it may slightly lower the

Table 3 Common types of plastics used in electronics and their applications (Stevens & Goosey, 2008; Liu et al., 2024)

Plastic type	Key properties	Industries	Percentage of global production	Common applications
Polyethylene terephthalate (PET)	Lightweight, good chemical resistance, recyclability	Packaging, textiles, automotive	30%	Bottles, packaging, some electronic components
Polyvinyl chloride (PVC)	Durable, good chemical resistance, low cost	Construction, automotive, electronics	25%	Cables, insulation, some housings
Polypropylene (PP)	High melting point, good chemical resistance, flexibility	Automotive, packaging, medical	25%	Battery cases, connectors, switches
Polycarbonate (PC)	High impact strength, transparency, heat resistance	Optics, automotive, medical	20%	CD/DVD cases, lenses, connectors
Polystyrene (PS)	Low cost, easy to mold, good insulation	Packaging, construction, electronics	15%	Packaging, insulation, some electronic components
Acrylonitrile butadiene styrene (ABS)	High impact strength, good chemical resistance, rigidity	Consumer electronics, automotive	8%	Enclosures, housings, connectors, buttons

strength of concrete, the impact is less compared to using regular plastic waste. Adding e-plastic to concrete can help reduce plastic pollution and support eco-friendly construction. Although more research is needed to improve concrete mixes and set industry standards, this approach has great potential to cut waste, save natural resources, and promote sustainable building practices.

4 Strength Characteristics of E-Plastic Concrete

The integration of e-plastic waste into concrete offers a promising solution to tackle environmental issues related to plastic waste. However, for it to be a viable alternative to conventional concrete, the resulting material must demonstrate sufficient strength to bear heavy loads, especially in stress-intensive structural applications. Therefore, evaluating the strength properties of e-plastic concrete, including compressive strength, tensile strength, and flexural strength, is critical to its potential adoption in the global construction market.

Table 4 Properties of concrete incorporating e-plastic

E-plastic form and type	Particle size (mm)	Optimal replacement (%)	Water-cement ratio	Concrete grade	Additional notes
Crushed, PHB	100	20%	0.46	M30	OPC 53 grade
Crushed, non-metallic	4.75–20	5–6% ideal, 10% allowable	0.50	M20, M25	OPC 43 grade cement, 60 specimens
Crushed	5–20	10%	0.50	M20	OPC 53 grade cement
Shredded (gravel size)	20	15% E-plastic, 10% Fly ash	0.50	M20	10% Fly ash by cement weight
Chips and flakes, HIPS	6–12	20%	0.53, 0.49, 0.45	–	486 concrete specimens
Fibers, cable outer casing, PVC	4 dia, 35 length, 0.8 thick	0.6%	0.4	M30	10% silica powder by cement volume
Crushed, ground	4–24 (by weight of coarse aggregate)	12% E-plastic, 10% Fly ash	1.18–2.36	M20	10% Fly ash by cement weight
Shredded wire insulation, PVC	4 dia (30, 40, 50) length	–	0.5	Mix ratio 1:1.31:3.10	OPC 53 grade
Angular	–	10% (fine and coarse both)	0.5	M20	50% GGBS replacement by cement weight, OPC 53 grade
Grinded, ABS	9.5–19	–	0.53	–	–
Pellets, ABS & PP	3–6	5%	0.45	M40	OPC 53 grade

4.1 Compressive Strength

Compressive strength is a key property in determining the performance of concrete when integrated with e-plastic waste. This parameter measures the material's ability to withstand axial loads without failure, making it crucial for structural applications. Experimental studies provide detailed insights into how e-plastic waste affects this characteristic.

The optimal amount of e-plastic waste to be incorporated into concrete typically ranges from 0% to 2% by the total weight of the concrete. Within this range, the concrete retains approximately 85% to 95% of the compressive strength of conventional concrete. This retention is significant, as it demonstrates that minor adjustments in mix composition can enable the use of e-plastic without substantial compromise to structural integrity.

Several factors influence the compressive strength of e-plastic concrete. The size of the plastic particles plays a critical role, as smaller particles tend to blend more uniformly, resulting in better load distribution. The type of polymer used in the e-plastic also affects the strength, with certain polymers contributing more positively to the overall mix. The method used to process the e-plastic, such as shredding or granulating, determines how well it integrates into the concrete matrix. Lastly, the overall concrete mix design, including the ratios of cement, aggregates, and water, impacts the performance of the final product.

These findings underscore the potential of e-plastic concrete as a sustainable material, provided the mix proportions and processing conditions are carefully optimized.

Table presents the comparative strength characteristics of different plastic types in concrete.

Plastic type	Compressive strength impact	Optimal replacement percentage
HDPE	Moderate reduction (5–8%)	0.5–1%
PP	Slight strength enhancement	1–2%
PET	Minimal reduction (2–5%)	1–1.5%
ABS	Marginal strength variation	1–1.5%

4.2 *Splitting Tensile Strength*

Splitting tensile strength is a vital property that evaluates a concrete's ability to resist tensile forces, particularly those that may cause cracking. When e-plastic waste is incorporated into concrete, it has a noticeable impact on this property, providing insights into its suitability for applications requiring tensile load-bearing capacity.

Studies have shown that the splitting tensile strength of concrete with e-plastic waste generally decreases compared to conventional concrete. The extent of this reduction depends on several factors, such as the type and size of e-plastic particles, their distribution within the concrete mix, and the overall mix design. Smaller, well-distributed plastic particles tend to perform better, as they create fewer weak points in the concrete matrix. Additionally, the polymer type significantly influences tensile strength, with some polymers providing better bonding with the cement matrix than others.

Despite the reduction in tensile strength, concrete with e-plastic waste still demonstrates adequate performance for non-structural and some structural applications where tensile stresses are not the primary concern. This makes it a feasible option for projects emphasizing sustainability and resource conservation. Proper processing of e-plastic waste and careful optimization of the mix design can help mitigate strength loss and ensure a more balanced performance.

These findings suggest that while incorporating e-plastic into concrete may present challenges in tensile strength, its potential for reducing waste and contributing to eco-friendly construction remains significant. Further research and innovation can help refine the process and improve the material's performance under tensile stress.

4.3 Flexural Strength Test

Flexural strength is an important property that measures a concrete's ability to resist bending or flexural loads, which is crucial for structural elements like beams, slabs, and pavements. The incorporation of e-plastic waste into concrete can influence its flexural strength, offering insights into its potential for load-bearing applications.

Research indicates that the flexural strength of e-plastic concrete generally decreases compared to conventional concrete, though the extent of reduction depends on several factors. The type and size of e-plastic particles play a significant role, with smaller, uniformly distributed particles providing better results due to improved bonding with the cement matrix. The type of polymer used in the e-plastic also affects flexural performance, as certain polymers enhance adhesion within the mix, while others may act as stress concentrators, leading to earlier failure under bending loads.

Processing methods of e-plastic, such as shredding or granulation, also influence the final performance. Properly processed e-plastic waste ensures a more uniform distribution, which helps maintain structural integrity under flexural stress. Additionally, the overall mix design, including the water-to-cement ratio and the proportion of plastic waste, is critical for achieving a balance between strength and sustainability.

While the flexural strength of e-plastic concrete may be slightly lower, it remains suitable for certain applications where high bending resistance is not a primary requirement. This property, combined with the environmental benefits of reducing plastic waste and conserving natural resources, makes e-plastic concrete a promising material for sustainable construction practices. Further advancements in mix optimization and processing techniques can help improve flexural performance, expanding its applicability in construction.

5 Research Findings

The integration of e-plastic waste into concrete has gained significant attention in the global research community, with emerging research hubs located in Europe, North America, and East Asia. These regions are at the forefront of exploring innovative solutions to address plastic waste challenges while advancing sustainable construction practices. The primary focus areas of research revolve around several critical aspects of e-plastic concrete. One key area is material compatibility, where researchers investigate how different types of e-plastics interact with traditional concrete components such as cement, aggregates, and water. Understanding this compatibility is crucial for achieving uniformity and ensuring structural performance. Long-term durability is another central theme, as studies aim to assess the performance of e-plastic concrete over extended periods. Factors such as resistance to environmental conditions, load-bearing capacity over time, and potential degradation of plastic components are examined to ensure the material's reliability in practical applications.

Sustainability metrics are also a significant area of exploration. Researchers are analyzing the environmental benefits of using e-plastic waste in concrete, including reductions in plastic pollution, conservation of natural resources, and a lower carbon footprint during production. These metrics are critical for promoting the adoption of e-plastic concrete in industries striving for eco-friendly solutions. Finally, performance optimization remains a key research objective. Efforts are focused on refining mix designs, processing methods, and plastic treatment techniques to enhance the mechanical properties of e-plastic concrete, such as compressive, tensile, and flexural strengths. This includes exploring innovative additives, alternative curing methods, and advanced polymer processing to maximize performance without compromising sustainability goals. The global research landscape demonstrates a collective effort to address the dual challenges of plastic waste management and sustainable construction. With continued innovation and collaboration, e-plastic concrete holds great potential as a transformative material in the building industry.

6 Results and Discussion

6.1 Comprehensive Performance Evaluation

The evaluation of e-plastic concrete's performance provides a comprehensive understanding of its mechanical properties, environmental benefits, and economic viability. Research findings across various studies consistently highlight its potential as a sustainable alternative in construction.

6.2 Mechanical Performance

The mechanical performance of e-plastic concrete has shown encouraging results in experimental investigations. Strength characteristics, including compressive, tensile, and flexural strengths, demonstrate consistent trends across multiple studies, ensuring a predictable performance envelope. This consistency makes it a reliable material for various applications, especially in non-structural and semi-structural contexts. Controlled experimental conditions reveal minimal variability in performance metrics, underscoring the material's stability when factors such as mix design, particle size, and polymer type are optimized. While there is a slight reduction in mechanical strength compared to conventional concrete, it remains within *acceptable limits for practical use.*

6.3 Environmental Impact Assessment

The environmental benefits of integrating e-plastic waste into concrete are significant. One of the primary advantages is the substantial reduction in plastic waste accumulation, addressing the growing global challenge of electronic waste management. Additionally, using e-plastic in construction materials has the potential to lower CO_2 emissions, particularly by replacing traditional aggregates and reducing dependency on energy-intensive raw materials. Resource conservation metrics further highlight the material's contribution to sustainable development by extending the lifecycle of plastic waste and minimizing the extraction of natural resources like sand and gravel. These environmental benefits align with global sustainability goals, making e-plastic concrete an attractive choice for green building practices.

6.4 Economic Considerations

From an economic perspective, e-plastic concrete offers multiple advantages. The use of plastic waste as a partial replacement for traditional materials can lead to cost savings in construction projects, particularly in regions where the cost of conventional aggregates is high. Additionally, reduced expenses in waste management, such as landfill fees and transportation costs, further enhance its economic appeal. The incorporation of e-plastic also promotes material circularity, enabling the repurposing of waste into valuable construction components. This circular approach not only reduces costs but also supports the development of a more sustainable and resource-efficient construction industry.

7 Conclusion and Future Directions

The integration of e-plastic waste into concrete presents a promising solution to address the dual challenges of plastic waste management and sustainable construction. Comprehensive studies demonstrate that e-plastic concrete is a viable material, offering considerable mechanical, environmental, and economic benefits. Its predictable mechanical performance, marked by consistent compressive, tensile, and flexural strength trends, ensures reliability for various applications. Environmentally, the material contributes to significant reductions in plastic waste accumulation, CO_2

emissions, and resource extraction, aligning with global sustainability goals. Economically, it offers potential cost savings through reduced reliance on traditional aggregates and lower waste management expenses while promoting material circularity.

Despite these promising attributes, further research and technological innovation are required to fully realize the potential of e-plastic concrete. Future work should focus on optimizing mix designs to improve mechanical properties, particularly tensile and flexural strengths, to broaden its application scope. Advanced processing techniques for e-plastic waste, such as pre-treatment and particle size control, can enhance its compatibility with cementitious materials.

Long-term durability studies are essential to assess the performance of e-plastic concrete under various environmental conditions, including extreme temperatures, moisture, and chemical exposure. Standardization efforts to establish guidelines for the safe and efficient use of e-plastic in construction materials will also be critical for industry-wide adoption.

Additionally, interdisciplinary research should explore innovative additives and hybrid materials to further enhance the material's properties. Life cycle assessments (LCAs) should be conducted to comprehensively evaluate the environmental impact of e-plastic concrete from production to disposal, ensuring its sustainability.

In conclusion, e-plastic concrete holds substantial potential to transform the construction industry into a more sustainable and eco-friendly sector. Continued collaboration among researchers, policymakers, and industry stakeholders will be vital in overcoming existing challenges and driving the adoption of e-plastic concrete on a global scale. With sustained effort, this innovative material can play a pivotal role in building a more sustainable future.

References

Abdel-Shafy, H. I., & Mansour, M. S. (2018). Solid waste issue: Sources, composition, disposal, recycling, and valorization. *Egyptian Journal of Petroleum, 27*(4), 1275–1290.

Ahmed, M., Bashar, I., Alam, S. T., Wasi, A. I., Jerin, I., Khatun, S., & Rahman, M. (2021). An overview of Asian cement industry: Environmental impacts, research methodologies and mitigation measures. *Sustainable Production and Consumption, 28*, 1018–1039.

Akhtar, A., & Sarmah, A. K. (2018). Construction and demolition waste generation and properties of recycled aggregate concrete: A global perspective. *Journal of Cleaner Production, 186*, 262–281.

Bala, K., Anshul, Verma, S., Dogra, A., Das, A. P., Kumar, R., & Nandini. (2024). Sustainable management for urban plastic waste generation. In *Sustainable management of urban plastic waste through circular economic approaches* (pp. 211–236). https://doi.org/10.1201/9781003465447-13

Carmody, P. (2024). 'Digital provide' or product consumption diffusion cycle? The diffusion impact and potential of digital technology in the global south. *Information Technology for Development*, 1–14.

Cresswell, I. D., Janke, T., & Johnston, E. L. (2021). *Australia state of the environment 2021: Overview, independent report to the Australian government minister for the environment* (Vol. 10). Commonwealth of Australia.

Cui, J., & Zhang, L. (2008). Metallurgical recovery of metals from electronic waste: A review. *Journal of Hazardous Materials, 158*(2–3), 228–256.

Das, A. P., Sukla, L. B., Pradhan, N., & Nayak, S. (2011). Manganese biomining: A review. *Bioresource Technology, 102*(16), 7381–7387. https://doi.org/10.1016/J.BIORTECH.2011.05.018

Das, S., Lee, S. H., Kumar, P., Kim, K. H., Lee, S. S., & Bhattacharya, S. S. (2019). Solid waste management: Scope and the challenge of sustainability. *Journal of Cleaner Production, 228*, 658–678.

Davis, K. (2015). The urbanization of the human population. In *The City reader* (pp. 43–53). Routledge.

Hassan, Q., Viktor, P., Al-Musawi, T. J., Ali, B. M., Algburi, S., Alzoubi, H. M., et al. (2024). The renewable energy role in the global energy transformations. *Renewable Energy Focus, 48*, 100545.

Ilankoon, I. M. S. K., Ghorbani, Y., Chong, M. N., Herath, G., Moyo, T., & Petersen, J. (2018). E-waste in the international context–A review of trade flows, regulations, hazards, waste management strategies and technologies for value recovery. *Waste Management, 82*, 258–275.

John, A., & Das, A. P. (2024). Government strategies for achieving circular economy for plastic waste Management in Southeast Asia. In *Sustainable Management of Urban Plastic Waste Through Circular Economic Approaches* (pp. 72–82). https://doi.org/10.1201/9781003465447-6/GOVERNMENT-STRATEGIES-ACHIEVING-CIRCULAR-ECONOMY-PLASTIC-WASTE-MANAGEMENT-SOUTHEAST-ASIA-AKBAR-JOHN-ALOK-PRASAD-DAS

Khatun, R., Singh, S., Dubey, N. K., & Prasad Das, A. (2023). A review on marine-based phytochemicals and their application in biomedical research. In *Recent frontiers of phytochemicals: Applications in food, pharmacy, cosmetics, and biotechnology* (pp. 383–395). https://doi.org/10.1016/B978-0-443-19143-5.00013-X

Liu, Q., Martinez-Villarreal, S., Wang, S., Tien, N. N. T., Kammoun, M., De Roover, Q., & Richel, A. (2024). The role of plastic chemical recycling processes in a circular economy context. *Chemical Engineering Journal, 155227*, 155227.

Mishra, S., Dash, D., Al-Tawaha, A. R. M. S., & Das, A. P. (2022). A review on heavy metal ion adsorption on synthetic microfiber surface in aquatic environments. *Applied Biochemistry and Biotechnology, 194*(10), 4639–4654. https://doi.org/10.1007/S12010-022-04029-W/METRICS

Mishra, S., Ghosh, S., van Hullebusch, E. D., Singh, S., & Das, A. P. (2023). A critical review on the recovery of base and critical elements from electronic waste-contaminated streams using microbial biotechnology. *Applied Biochemistry and Biotechnology, 2023*, 1–30. https://doi.org/10.1007/S12010-023-04440-X

Mishra, S., Dash, D., & Das, A. P. (2024a). Isolation and molecular characterization of polyester-degrading bacterial strains from municipal wastewater and their application in microplastic degradation in contaminated wastewater. *Geomicrobiology Journal, 41*, 732. https://doi.org/10.1080/01490451.2024.2376101

Mishra, S., Dash, D., & Das, A. P. (2024b). Microbial degradation of polyester microfibers using indigenously isolated bacterial strain Exiguobacterium sp. *CLEAN—Soil, Air, Water, 52*(12), e202300343. https://doi.org/10.1002/CLEN.202300343

Mondal, S., & Palit, D. (2022). Challenges in natural resource management for ecological sustainability. In *Natural resources conservation and advances for sustainability* (pp. 29–59). Elsevier.

Priyadarshini, B., Santhosh Kumar, M., & Das, A. P. (2024). Microbial degradation of synthetic plastics released from COVID-19 generated single use disposable face masks. *Environmental Science and Engineering, Part F, 2827*, 1–16. https://doi.org/10.1007/978-3-031-50840-0_1

Priyadarsini, S., & Das, A. P. (2025). Lithium bioleaching: A review on microbial-assisted sustainable technology for lithium bio-circularity. *Journal of Water Process Engineering, 69*, 106744. https://doi.org/10.1016/j.jwpe.2024.106744

Sadasivuni, K. K., Rattan, S., Waseem, S., Brahme, S. K., Kondawar, S. B., Ghosh, S., Das, A. P., Chakraborty, P. K., Adhikari, J., Saha, P., & Mazumdar, P. (2019). Silver nanoparticles and its polymer nanocomposites—Synthesis, optimization, biomedical usage, and its various applications. In *Lecture Notes in Bioengineering* (pp. 331–373). https://doi.org/10.1007/978-3-030-04741-2_11

Sahoo, P. P., Singh, S., Rout, P. K., Mishra, S., & Das, A. P. (2022). Microbial remediation of plastic pollutants generated from discarded and abandoned marine fishing nets. *Biotechnology and Genetic Engineering Reviews, 39*, 511. https://doi.org/10.1080/02648725.2022.2152629

Shahabuddin, M., Uddin, M. N., Chowdhury, J. I., Ahmed, S. F., Uddin, M. N., Mofijur, M., & Uddin, M. A. (2023). A review of the recent development, challenges, and opportunities of electronic waste (e-waste). *International Journal of Environmental Science and Technology, 20*(4), 4513–4520.

Shahbaz, M., Sharma, R., Sinha, A., & Jiao, Z. (2021). Analyzing nonlinear impact of economic growth drivers on CO2 emissions: Designing an SDG framework for India. *Energy Policy, 148*, 111965.

Singh Jadaun, J., Bansal, S., Sonthalia, A., Rai, A. K., & Singh, S. P. (2022). Biodegradation of plastics for sustainable environment. *Bioresource Technology, 347*, 126697. https://doi.org/10.1016/J.BIORTECH.2022.126697

Sinha, S. (2022). Biopolymer derived superabsorbent for environmental sustainability: A review. *Environmental Quality Management, 32*(2), 177–185. https://doi.org/10.1002/tqem.21936

Sinha, S., & Hamdan, A. (2023). E-plastic waste—A sustainable construction material: An Indian perspective. In *Handbook of sustainable materials: Modelling, characterization, and optimization* (pp. 213–232). CRC Press.

Sinha, S. (2024). Biopolymer-Derived Superabsorbent for Environmental Sustainability: A Review. *Bio-Based Polymers and Composites*, 1–19

Sinha, S. (2025). An Overview of Phytoremediation for the Treatment of Wastewater. *Recent Advances in Bioremediation and Phytoremediation*, 273–284.

Stevens, G. C., & Goosey, M. (2008). Materials used in manufacturing electrical and electronic products. In *Electronic waste management* (pp. 33–65).

Bioprospecting of Indigenous and Engineered Microbiome for Heavy Metals Extraction from Electronic Wastes

Suganthi Rajendran, Priyadharsini Sengottaiyan, Poornima Ramesh, and Gowtham Sornam

Abstract Electric and electronic items in day-to-day life are indispensable which favors the current livelihood resulted in huge quantum of wastes without proper disposal. E-waste is heterogeneous which entails various subunits depending upon composition, type and their applications threatens the environment and the survival of living flora and fauna. Physicochemical approaches are not quite competent in recovering heavy metals from wastes enlisting certain constraints such as high cost, pollution, and loss of heavy metals during the process. In order to overcome these limitations, biohydrometallurgy or bioleaching that involves bacteria, fungi and their metabolites in mobilizing and recovery of heavy metals is considered as eco-friendly and sustainable technology. Type of microorganism, environmental factor and the composition of e-waste together determine the leaching efficiency. Microbes have various mechanisms such as oxidation, acidolysis, redoxolysis, cyanogenesis, organic acids generation, chelation, and complexation. These bioleaching techniques have a quite impact on the microbes involved. Hence, the process has to be modified in such a way that it has higher leaching efficiency. Synthetic biology opens up avenues for improving catalytic capacity of the enzymes, changes in substrate specificity, toxicity tolerance, enhanced compatibility, etc.

Keywords E-waste · Heavy metals · Microbes · Bioleaching · Bacteria · Biorecovery · Biotransformation

S. Rajendran (✉) · P. Sengottaiyan · G. Sornam
Department of Environmental Sciences, Tamil Nadu Agricultural University, Coimbatore, India

P. Ramesh
Department of Environmental Sciences, Vanavarayar Institute of Agriculture, Coimbatore, India

A. P. Das, S. Priyadarsini (eds.), *Electronic Waste and Environmental Pollution*, Sustainable Environmental Waste Management Strategies,
https://doi.org/10.1007/978-3-031-90287-1_9

1 Introduction

Improving lifestyle with sophistication led to the arrival for numerous electronic gadgets in day-to-day life to ease our burden is recognized as rapidly mounting global solid waste sector. Digitalization arising a revolution in the global information technology (IT) field accounts for the generation of electrical as well as electronic goods which results in a large quantum of waste (Saha et al., 2021; Forti et al., 2020). Global e-waste accounts for 62 million metric tonnes in the year 2022 increasing from 53.6 million metric tonnes in 2019 and only 22.3% is properly identified, collected and recycled (Balde et al., 2024; Forti et al., 2020). The life expectancy, usage of the products, development of better-featured devices with subsequent versions etc. results in influencing the consumers to upgrade their products leading to vast quantity of waste. In case of developed and developing nations, Most of the dumped electronics ends up in landfills rather than recycling units. It is quite a misfortune in case of underdeveloped countries who lacks basic infrastructural layout for the treatment of waste materials. Burning of e-wastes results in the emission of various toxic gases, which will expose the humans toward those gases and degrade their health. E-wastes are not handled properly mainly due to lack of knowledge and negligent implementation of laws (Odılı et al., 2018; Althaf et al., 2021). Life cycle assessment of electronic waste has to be studied to manage the waste generation, the toxicity of the components, and its management. The most widely recognized concepts of tackling waste is recycling, recovery and disposal. Waste of meager quantity alone is being recycled and precious metals are being recovered under unsafe practices, which put pressure on human health and work efficiency. Management methods such as landfills, incineration at higher temperatures are not environmentally safe which could be overcome by introducing various pre-treatment methods which includes manual dismantling, separation of components, crushing, size reduction, etc. (Khanna et al., 2020).

Electronic waste encompasses hazardous materials such as lead, mercury, beryllium, cadmium, and Polychlorinated Biphenyls, which present substantial risks of ecotoxicity. The leaching of these hazardous substances leads to environmental pollution and adversely impacts ecosystems, as illustrated in Fig. 1. The improper management and disposal of electronic waste, particularly in developing nations such as India, exacerbate these challenges, leading to enduring negative consequences for environmental quality and public health. In addition to electronic waste, it contains phthalates, esters, benzene, dioxins, and polycyclic aromatic hydrocarbons, which exhibit significant toxicity upon ingestion. The carcinogenic characteristics of these compounds underscore the pressing necessity for effective electronic waste management (Khan et al., 2021; Rautela et al., 2021; Ahirwar & Tripathi, 2021). The accumulation of these toxic agents in ecosystems can disrupt biological functions and inflict harm upon various organisms. Consequently, the prudent management of electronic waste is imperative to alleviate these environmental hazards and safeguard both human health and ecological integrity. Awareness of the potential toxic

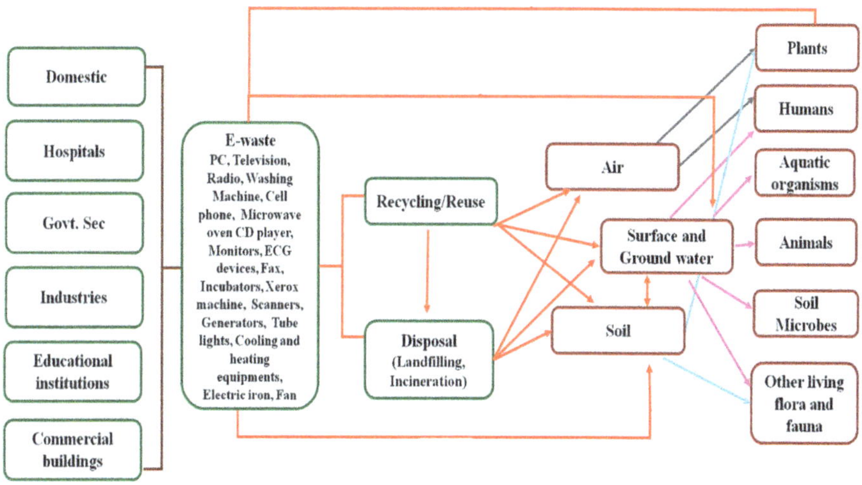

Fig. 1 Fate and transport of e-wastes in the ecosystem

ramifications associated with electronic components should be disseminated among manufacturers, consumers, recyclers, and disposers to mitigate the environmental impact of such products (Woo et al., 2015).

Numerous studies have been conducted to explore the prospects of sustainable waste management, as well as the impact of electronic waste and its leachates on soil biodiversity (Jiang et al., 2019; Bao et al., 2020; Purchase et al., 2020; Saha et al., 2021). For example, lead is incorporated in batteries, whereas mercury finds applications in lamps, laptops, displays, and batteries, among others. The methylation of mercury, mediated by microbial action, produces methylmercury, a potent neurotoxin detrimental to both humans and other organisms. Mercury exhibits a high affinity for sulfur groups, resulting in its accumulation in adipose tissues. Human exposure to mercury can lead to a variety of disorders, including cognitive impairment, neurological dysfunction, memory disturbances, irritation, and anxiety, with pregnant women and children being particularly susceptible (Wang et al., 2020; Sharma et al., 2020; Suganthi et al., 2021; Rajendran et al., 2024).

The life cycle assessment (LCA) of electronic waste elucidates the sources and limitations inherent in waste management practices, thereby facilitating the development of enhanced methodologies aimed at achieving greater efficiency and improved resource recovery. A range of LCA methodologies has been investigated, among which the Economic Input-Output LCA is deemed reliable due to its incorporation of comprehensive big data analyses regarding the release of toxic compounds (Kaya, 2019; Saha et al., 2021). Despite the Basel Convention's efforts to limit the transboundary movement and disposal of hazardous wastes, there is a critical need for stringent monitoring of illegal exports and imports, as well as ocean dumping of such wastes. The safe disposal of electronic waste constitutes a collective responsibility that must be shared among local communities, non-governmental organizations, the private sector, and public institutions (Forti et al., 2020).

In the context of metal recovery from electronic waste, bioleaching or bio-hydrometallurgy emerges as a straightforward process that utilizes microorganisms, including bacteria and fungi, to extract metals from minerals and waste materials (Das & Ghosh, 2018; Ghosh et al., 2016a; Priyadarsini & Das, 2025). This method offers several advantages, including cost-effectiveness, enhanced efficiency, and safety, as well as ease of handling, minimal maintenance requirements, operation at atmospheric pressure and room temperature, environmental friendliness, and independence from skilled labor that does not compromise process efficiency (Vakilchap et al., 2016; Zeng et al., 2016; Das & Ghosh, 2023b; Ghosh & Das, 2017). It involves microbial production of organic and inorganic acids, cyanide which are responsible for enzymatic oxidation, reduction. Proton transfer, ligand formation/chelation resulting in metal extraction. Bioleaching microorganisms can be classified into three categories: chemolithoautotrophic bacteria, which operate through mechanisms such as acidolysis and redoxolysis; organic acid-producing fungi, which employ acidolysis and complexolysis; and cyanogenic bacteria, which utilize complexolysis (Vakilchap et al., 2016; Abhilash et al., 2015; Xiang et al., 2010; Priyadarsini & Das, 2024). Commercial extraction of metals from minerals are catalyzed by well recognized chemolithotrophs such as *Acidothiobacillus ferroxidans* and *Acidothiobacillus thiooxidans* (Johnson et al., 2021; Johnson, 2018). Besides bacillus being the most effective species, other organisms such as fungi—*Aspergillus niger*, *Penicillium simplicissimum*, and cyanogenic bacteria such as *Chromobacterium violaceum*, *Bacillus megatherium* can also be utilized in the metal extraction process (Wang et al., 2015; Vakilchap et al., 2016; Baniasadi et al., 2019).

Different chemolithotrophs has been tested against metal extraction from e-wastes such as printed circuit boards, lithium-ion batteries, Zinc batteries, Cathode ray tubes etc. (Arshadi & Mousavi, 2014; Arshadi & Mousavi, 2015; Chen et al., 2015; Fu et al., 2016). Printed circuit boards contains a mixture of metals (base metals, heavy metals, alkali metals, precious metals, Transition elements) and non-metals. Bioleaching studies focuses on maximum recover of metals without degrading the environmental quality (Das & Ghosh, 2023a). In order to increase the metal yield, various indigenous and engineered microorganisms are being employed in recovering metal from electronic wastes. Engineered bacteria such as *C. violaceum* with pBAD and pTAC strains were employed in gold recovery from electronic scraps which exhibited 68% and 51% more cyanide production than non-engineered bacteria which ultimately resulted in improved gold recovery (Tay et al., 2013; Baniasadi et al., 2019). This chapter delves into ecotoxicity of e-waste, indigenous and engineered microbes involved in bioleaching of e-wastes, factors governing bioleaching, challenges in recovery of metals from e-wastes that will maximize the efficiency as well as gives a better green technology for large-scale operations.

2 Ecotoxicity of E-Waste

An array of governmental and non-governmental sectors such as industries, research organizations, educational institution, hospitals, households, commercial buildings, theme parks etc. raises environmental and health concern due to the presence of hazardous materials such as heavy metals (Mishra et al., 2022), persistent organic pollutants, plastics (Priyadarshini et al., 2024; Singh Jadaun et al., 2022), glass fiberes, metal oxides, ceramics in discarded electronic devices contaminating different environmental matrices and its flora and fauna. Apart from metals, other hazardous chemicals are also being released during e-waste recycling such as brominated flame retardants and polyaromatic hydrocarbons (Tembhare & Bhanvase, 2023; Sonawane, 2023). The non-biodegradable nature of these e-waste components results in long-term environmental pollution, necessitating urgent management solutions (Najar et al., 2024). Informal e-waste recycling mainly occurs in household and small-scale industrial waste disposal that exacerbate these issues. It was reported that majority of the electronic wastes about 30% are covered by washing machine, dryer, vacuum cleaner, air conditioner, coffee machines, irons etc. followed by refrigerators with 20%, 30% by computers, mobile phone, printers, compact disc, fax etc., 10% by television and 10% exclusively by monitors (Kaya, 2016; Saha et al., 2021). While comparing toxic pollutants from electrical and electronic items, lighter electronic items such as computers, laptop, mobile phones contains more contaminants. Smart phone usage has been increasing over time as their disposal. These discarded e-wastes contains huge prospects of precious metals recovery such as gold, palladium, platinum, silver and copper (Ilankoon et al., 2018). E-waste and its components ought to affect physicochemical properties of soil, soil microbial diversity and dynamics, phylogenetics and functions (Tang et al., 2013), thereby entering the ecosystem through food chain. Various unsound practices at waste sites such as scavenging, open burning, acid leaching, stripping/shredding of plastic coatings and landfilling with regular waste results in release of contaminants into the environment. In addition to discarded e-wastes, leachates from landfills, incineration plants, e-waste processing units, etc., are seeking attention since they contain a cocktail of persistent organic pollutants, nitrogen compounds, ammonia, heavy metals, aromatic compounds such as polychlorinated biphenyl (Youcai, 2018; Saha et al., 2021).

Humans exposed to toxic metals will exhibit thyroid impairment, respiratory disorders, psychological disorders, mutation, etc., which is illustrated in Fig. 2. People in and around e-waste recycling unit are often exposed and their DNA gets damaged rather than people from normal regions (Grant et al., 2013). Leachates also tend to affect cellular function and DNA structure (Youcai, 2018). Long-term exposure may result in the accumulation of these toxic substances within the body and exhibit Parkinson's, Alzheimer's, multiple sclerosis, etc. Perchlorate and thiocyanate compounds from e-waste were identified in the serum samples collected in North Vietnam, and it affects the iodine transport and uptake, resulting in thyroid impairment (Eguchi et al., 2014). Several epidemiological studies state that children

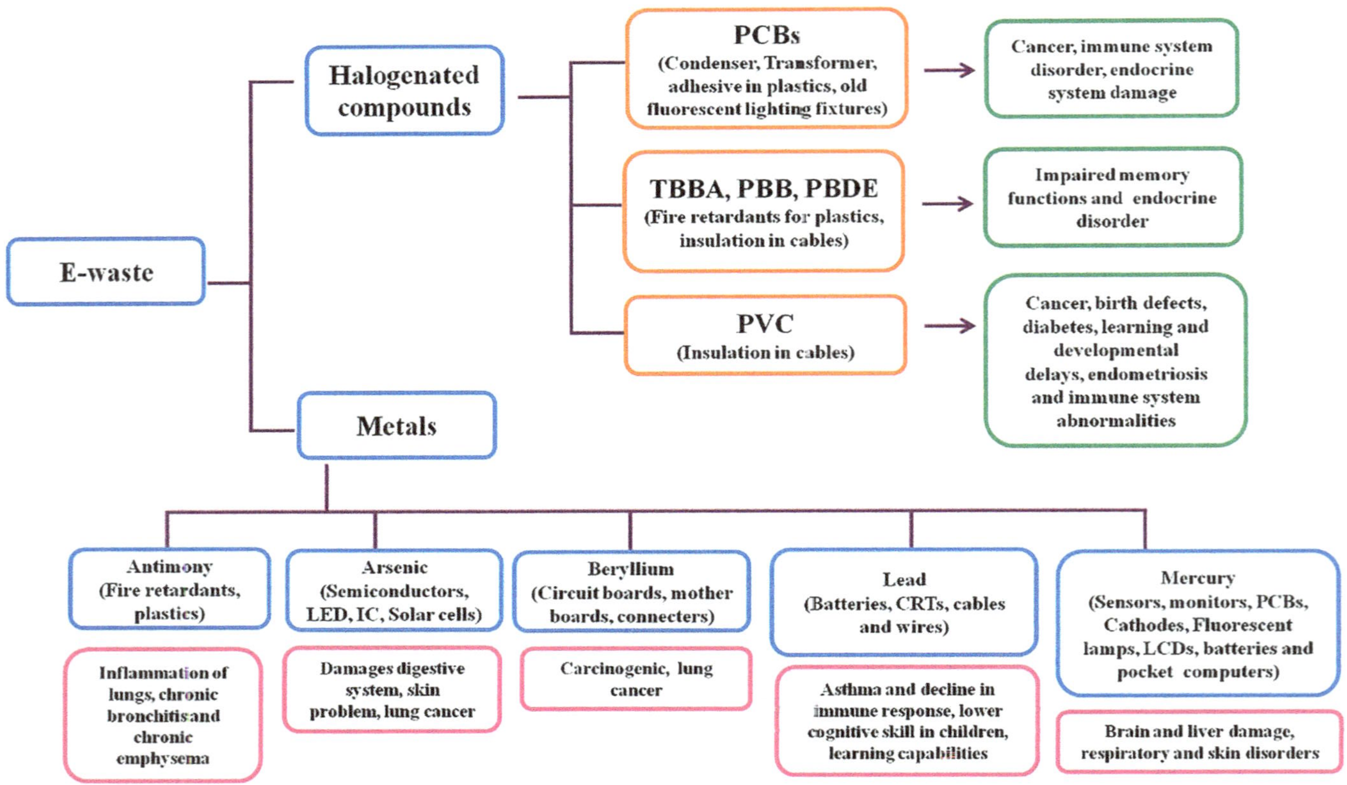

Fig. 2 Sources and effects of e-waste components on human health

and pregnant women are very vulnerable to the health hazards linked with improper e-waste recycling. The International Labour Organization (ILO) estimated that 16.5 million children are found to be working in the industrial sector alone, which includes waste processing, in 2020 (ILO, 2021). On e-waste and hazardous waste exposure, effects observed are adverse neonatal outcomes such as stillbirth and premature birth, neurodevelopment, learning and behavior outcomes, reduced lung and respiratory function such as asthma incidence (WHO, 2021). Young children and infants are highly sensitive to various pollutants which directly affects immune, respiratory and central nervous system.

Several researches have studied the accumulation, translocation and transformation of contaminants within the plant system grown in e-waste-contaminated soil. Bioaccumulation of contaminants such as PCBs and heavy metals such as lead, nickel, copper and cadmium were evident in Corn (Luo et al., 2015), *Echinocloa*, *Phragmites*, *Sesbania*, *Cucurbita*, *Glycine*, *Sesamum* and *Solidago* (Li et al., 2021). In addition, biodegradable ligands and plant growth-promoting bacteria were added to the soil for improved extraction. These compounds on accumulation tend to induce cell damage and mutation in onion (Bakare et al., 2012). E-waste leachate affects chlorophyll synthesis thereby reducing primary production and activity of Photosystem II. These chemicals induce oxidative stress and hinder the antioxidant enzyme production within *Pseudokirchneriella subcapitata* (Nie et al., 2015). Heavy metals from e-wastes hamper the essential nutrient uptake by the plants, carbon dioxide fixation, seed germination, flowering and fruiting, cytoplasmic enzymes, plant growth and biomass production (Karthik et al., 2021; Dradrach et al., 2020; Yeboah et al., 2021; Okereafor et al., 2020).

E-wastes and their leachates alters the physicochemical properties of water as well as soil. These wastes, on accumulation, modify the soil structure and their dynamics. Metals such as copper, cadmium, nickel and chromium were reported in water samples (Das & Bissoyi, 2011; Das & Mishra, 2008, 2009). Water sources adjacent to the recycling units and industries are prone to contamination with various contaminants, confirming the higher concentration of pollutant in the areas adjacent to landfill sites (Gupta & Nath, 2020). Thus water is becoming unsafe for potable and non-potable purposes because of the contamination. For instance, e-waste handling is banned in New Delhi but informal sector is involved in disposal of e-wastes. Monitoring studies were studied at different points in Krishna vihar near mandolin which recorded heavy metals such as lead, cadmium and copper to an extent (Panwar & Ahmed, 2018). These wastes affect aquatic organisms growth, survival and reproduction. Bioaccumulation of metals in fish was studied by various researchers asserting the e-waste contamination in water. Even PCBs and PBDEs were reported in aquatic birds such as water fowl in China (Luo et al., 2009). While the ecotoxicity of e-waste presents a daunting challenge, advancements in green computing and alternative testing methodologies offer promising avenues for mitigating these risks. These approaches aim to reduce the environmental footprint of electronic devices and improve the safety and efficacy of e-waste management practices (Najar et al., 2024; Eze & Vinken, 2024).

3 E-Waste Recycling Methods

The increased production of electrical and electronic equipment, which turns into e-waste at the end of its useful life, has resulted in a sizable amount of e-waste in the modern period of rapid technological advancement. Both throughout the manufacturing phase and after use, efficient methods are required for ecologically sound e-waste treatment. Recycling is a good way to cut down on the increasing quantity of e-waste, but improper recycling practices are increasing the levels of dangerous chemicals in the environment. The present trend in e-waste management is unsustainable and necessitates a broad plan to give a holistic solution, as 82.6% of the world's e-waste was treated informally in 2019Understanding the e-waste collection and recycling process better is a crucial first step in this direction (Ahirwar and Tripathi, 2021). Meskers et al. (2009) state that recycling e-waste comprises (i) collecting the e-waste, (ii) handling it, and (iii) retrieving valuable resources (Fig. 3). In order to efficiently collect residential or institutional e-waste, it is necessary to develop regulatory agreements, government regulations, and easily accessible collection facilities. Furthermore, raising public awareness is an essential first step in gathering and compiling outdated EEE. Producers and/or government-approved organizations collect e-waste through stores, pick-up services, and municipal collection sites under official collection. Some nations have put in place a take-back scheme called the Extended

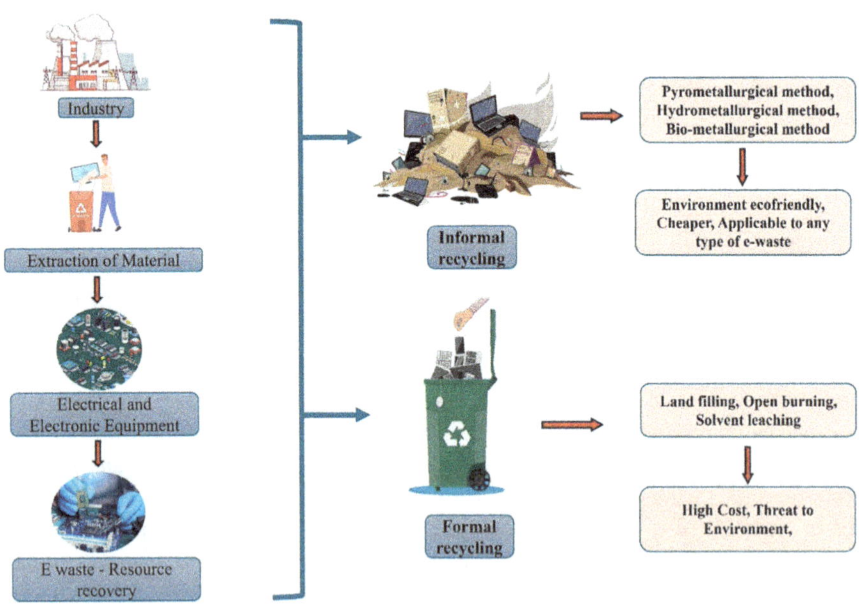

Fig. 3 Process of e-waste management

Producer Responsibility (EPR) standard, which mandates that original equipment manufacturers recover a certain portion of their sales.

In the next step of recycling e-waste, non-essential and non-recyclable components are removed in order to enrich precious metals. The EEE must be disassembled in order to collect its constituent pieces. An e-waste may contain varying percentages of ferrous components (like steel and iron), non-ferrous metals (including copper, aluminum, platinum, and silver), polymers, glass, wood, rubber, and other materials, depending on the kind, manufacturer, and age of the equipment. Before being further shredded into smaller pieces (e.g., 0.017 to 5 cm), these components are separated via mechanical processes into different batches of metals, plastics, ceramics, paper, wood, and categories including capacitors, batteries, LCDs, and PCBs (Cui & Forssberg, 2003). Industrial-scale granulation machines are used for these purposes, and the e-waste is enriched for recovery by separating it into metallic and non-metallic fractions using density, magnetic, or current-based separation techniques (Li et al., 2007). The vast majority of plastic found in e-waste gets recycled into energy-dense plastic bottles.

4 Pyrometallurgical Methodology

Among the pyrometallurgical methods of metal reclamation used to recover valuable metals from e-waste are conflagration, smelting in furnaces or alkali chemicals, drossing, sintering, plasma melting, and other high-temperature solid-liquid-gas reactions (Ding et al., 2019; Li et al., 2019). Pyrometallurgical processes use the same techniques used in mechanical or hydrometallurgical processes, including separation, crushing, upgrading, and purifying. Because of flame retardants and different plastic mixes, it is challenging to recycle plastic components of e-waste (Dey et al., 2024). However, smelting processes consume less energy since plastics and other combustible components in e-waste are burned. Antrekowitsch et al. (2006) state that most industrial-scale e-waste recycling, especially for lead, copper and precious metals, is done at Cu and Pb smelters using pyrometallurgical processes (Prabhakar et al., 2019; Raj & Das, 2023). These methods employ hydrometallurgical and electrochemical technologies to partially or fully recover metals by separating them into a solvent metal phase (molten bath) based on their chemical and metallurgical proportions.

Pyrometallurgy techniques are often used for the initial segregation and upgrading of precious metals into base metals, followed by hydrometallurgical and electrometallurgical processes to recover pure base and precious metals (Das et al., 2011; Ghosh et al., 2016b). For example, platinum group metals (such as palladium (Pd), rhodium (Rh), platinum (Pt), and ruthenium (Ru)) are recovered from e-waste using a combination of hydrometallurgical, pyrometallurgical, and electrometallurgical processes. Khaliq et al. (2014) state that copper is the most common metal in the composition of most e-waste streams, and that it is reduced at a temperature of approximately 1250 °C. Wasted lithium-ion batteries (LIB) are often handled separately because cathode materials are highly stable at high temperatures and a good reduction can only be achieved by smelting between 1400 and 1700 °C (Assefi et al., 2020; Liu et al., 2019). Moreover, LIBs' lithium content is easily oxidized and transferred to the slag. This metal cannot be recycled if it is in a complex slag with several other impurities (Dang et al., 2018).

For most smelters that deal with e-waste, the fundamental idea is to decrease the oxides and keep them in a single large collector (either copper or lead). Reducing metal loss and protecting important metals are the main objectives of recycling e-waste (Ghodrat et al., 2016). Because e-waste (from screens, boards, casings, etc.) has higher concentrations of ceramics, glass, and oxides, a considerable volume of slag is foreseen. This can result in a considerable loss of metals and make the process more difficult overall (Dang et al., 2018). This is prevented by reducing the amount of e-waste in the feed to no more than 50% by diluting it with ore concentrates (or copper scraps). The "black copper route" refers to the process of incorporating rich copper-bearing materials (scraps or concentrates) into the feed of large-scale e-waste recycling plants (Ghodrat et al., 2018). A combination of CaO, SiO_2, Al_2O_3, and sporadically MgO is used as flux in smelters for the recycling of wasted LIBs (Anindya et al., 2013). Because Mn oxidizes and migrates to slag, smelters are unable to recycle rechargeable batteries of the Mn type (Ren et al., 2017). Additionally, from any source of used LIBs, Li builds up in the slag as a difficult-to-treat phase of LiAl (SiO_3) (Dang et al., 2020). To address this issue, researchers have proposed using alternative fluxes (like pyrolusite) and altering the environment (like vacuum) to produce MnO and $Li_2O/LiCO_3$ in the slag as some treatable phases for further operations. In order to convert $LiAl(SiO_3)_2$ into a water-soluble material that can subsequently be refined using hydrometallurgical procedures, it is also recommended to roast the slag by chlorination and sulfation (Dang et al., 2020).

Pyrometallurgical processes are believed to be economical and make full use of precious metal recovery. They do, however, have certain drawbacks, such as the release of hazardous dioxins from flame retardants and the melted polyvinyl chloride (PVC) present in e-waste, which necessitates particular emission controls to safeguard the environment (Hageluken, 2006; Cui & Zhang, 2008). Additionally, additional hydrometallurgical and electrochemical procedures are needed to extract pure metals from base metals. Finally, process control for the smelting processes is difficult due to the material complexity of e-waste.

5 Hydrometallurgical Methods

Solvent extraction, leaching, adsorption, and ion exchange processes are used in hydrometallurgical methods to recover metal from e-waste. This is followed by chemical reduction or electro-refining to selectively dissolve valuable metals

(Sadegh Safarzadeh et al., 2007). Pure base and valuable metals have been recovered from e-waste using a variety of mineral acids and oxidants, such as aqua regia ($3HCl:HNO_3$), HCl, H_2SO_4, HNO_3/H_2O_2, $HClO_4$, and NaClO, as well as lixiviants, such as cyanide, thiosulfate, thiourea, and halide (Ding et al., 2019; Islam et al., 2020). According to Priya and Hait (2017) and Widmer et al. (2005), e-waste streams are made up of 50% iron steel, 13% non-ferrous metals, 21% plastics, and 16% ceramic, glass, and rubber. They need to come into touch with the reagents in order to recover the metal. In order to somewhat remove the plastics and ceramics that are encasing the metals, a pretreatment step must be conducted (Gramatyka et al., 2007; Tuncuk et al., 2012). Despite their complexity, the majority of metallic elements can be recycled using hydrometallurgical techniques because they are present in either native form or alloy. Waste printed circuit board (WPCB) are regarded as one of the most important forms of e-waste because of the high concentrations of valuable components they contain (Nekouei et al., 2019). WPCBs can be chemically leached by cyanide, halide, thiourea, thiosulfate, aqua regia, ionic liquid acid leaching, and other leaching processes with the use of oxidizing reagents and HNO_3, H_2SO_4, and HCl (Li et al., 2018; Wu et al., 2017).

The metallic components of Waste Electrical and Electronic Equipment (WEEE), including gold, silver, copper, lead and zinc, can be extracted using this method. This method for recovering valuable metals from e-waste has been investigated by a number of researchers (Gonçalves et al., 2015). Here, a leaching solution, such as a powerful acid or alkali, is used to dissolve the metallic portion of e-waste. Following leaching, the pregnant solution is gathered, and contaminants are eliminated using solvent extraction, adsorption, or ion exchange enrichment procedures. Lastly, chemical reduction procedures or electro-refining (electrometallurgy) are used to recover metals. In the chemical leaching process, several well-known leaching agents were used (Ghasem & Khoramnejadian, 2015; Guo et al., 2015). Silveira et al. (2015) extracted 613 mg of indium kg^{-1} of liquid crystal display (LCD) screen powder using 1.0 M sulfuric acid at 90 °C. In an experiment with a leaching method employing a 2.0 M HCl solution, Kim and Paulos (2011) recovered up to 98% zinc, 96% lead, 96% tin and 71% copper from waste printed circuit board (PCB) in 4 h at 50 °C and 400 rpm.

Hydrometallurgical operations generate a lot of solid waste and effluents and use combustible, hazardous, or extremely acidic/alkaline reagents. Hydrometallurgy requires off-gas treatment because the halogenated flame retardants used in printed circuit boards produce harmful volatile metals, dioxin, and furans (Hageluken, 2006). However, no single process can meet the requirements of high extraction efficiency, recyclability, low emissions of harmful components into the environment, and low extracting costs at the same time. To reduce the amount of hazardous waste that is disposed of in the environment, methods like landfilling and incineration are used to treat effluent and other garbage.

6 Biometallurgical Processes

Microbes may be used to dissolve valuable components from e-waste, making it a cost-effective and contemporary recycling method (Faraji et al., 2018). Through certain methods, living organisms like fungus and bacteria are employed in this procedure to solubilize metals. There are several bioleaching paths recommended for each type of e-waste stream due to the vast range of devices and metals that are contained in e-waste (Baniasadi et al., 2019). Pre-waste is frequently recycled using a variety of biological species, including cyanogenic microbes, fungi, and acidophilic bacteria (Srichandan et al., 2019). Biometallurgy is divided into two subfields: bioleaching and biosorption. Both living and dead organisms can be used thanks to the physical and chemical interactions between the bacteria and ions in solution during biosorption. Numerous microorganisms, including bacteria (de Vargas et al., 2004), yeasts (Wang & Chen, 2006), fungus (Sag & Kutsal, 2001), and algae (Romera et al., 2006), are known to actively acquire heavy and valuable metals.

Due to their ability to continuously produce sulfuric acid and convert Fe^{2+} to Fe^{3+}, sulfur- and iron-oxidizing bacteria (such as *Acidithiobacillus thiooxidans* and *Acidobacillus ferrooxidans*, respectively) are two well-known species of bacteria that have been widely used in mining and e-waste recycling (Priya & Hait, 2017). These bacteria can provide an excellent dissolving environment for the solubilization of a variety of metals found in e-waste (Mostafavi et al., 2018). Fungal bioleaching is a more practical method for recycling waste because it can metabolize over a larger pH range with a shorter lag phase, produce organic acids that are suitable chelating agents, and withstand higher concentrations of metallic contamination (Faraji et al., 2018). The well-known fungus *Aspergillus niger* has been used in a variety of bioleaching procedures and e-waste recycling because it can metabolize over a larger pH range with a shorter lag phase, produce organic acids that are suitable chelating agents, and withstand higher concentrations of metallic contamination. (Islam et al., 2020). Since fungal bioleaching involves a mixture of acids, the leaching recovery is expected to be larger than the application of individual acids. The organic acids leach metals through the three mechanisms of acidolysis, complexolysis, and redoxolysis (Seh-Bardan et al., 2012). Almost any type of e-waste, including cathode materials from wasted LIBs that need a reducing agent, can be handled by fungus bioleaching. A portion of the organic acids in this process function as reducing agents by breaking down into CO_2 and H_2O. The reduced form of metals would then be mobilized by the acidolysis and complexolysis processes of the residual organic acids (Bahaloo-Horeh et al., 2018). Fungi such as *Aspergillus niger* and *Penicillium* are examples of eukaryotic microorganisms used in bioleaching during metal recovery from industrial wastes (Cui & Zhang, 2008). Metals from electronic trash may often be bioleached with the help of the microorganism. This approach is less damaging to the environment, safer, more economical, and more sustainable than old processing methods. However, it takes longer, recovery rates aren't always great, and the microbes need to meet specific requirements (food, oxygen, mineral salts, etc.) (Roshanfar et al., 2019).

7 Biological Intervention in Biohydrometallurgy

Since electronic waste (e-waste) contains both hazardous materials and valuable metals like gold, silver, copper, and rare-earth elements, it is becoming a major global problem. One promising method for recycling e-waste is bioleaching, which uses microorganisms to extract metals from solid waste in an environmentally friendly and sustainable manner. The mechanism by microbes can mobilize metals, including acidolysis, redoxolysis, complexation and bioaccumulation (Fig. 4).

Acidolysis involves the secretion of organic acids by microbes to protonate metal oxides. Redoxolysis uses microbial enzymes to reduce highly oxidized metal compounds. Complexation forms soluble metal-ligand complexes, while bioaccumulation involves the uptake of metals into microbial cells (Adetunji et al., 2023). The two main strategies for bioleaching metals from e-waste: one step bioleaching, two-step bioleaching and spent medium bioleaching (Fig. 5). In one-step bioleaching, a freshly grown microbial culture is inoculated directly into a medium containing the e-waste. In two-step bioleaching, the microbes are cultivated to maximize growth and metabolite production, before the e-waste is added and helps mitigate the toxic effect of e-waste on the microbes. A third approach, spent medium bioleaching, involves using the cell-free culture supernatant containing the microbial metabolites to leach the metals without the presence of the live cell (Baniasadi et al., 2019). This eliminates interaction between the microbes and e-waste, and allows for higher pulp densities.

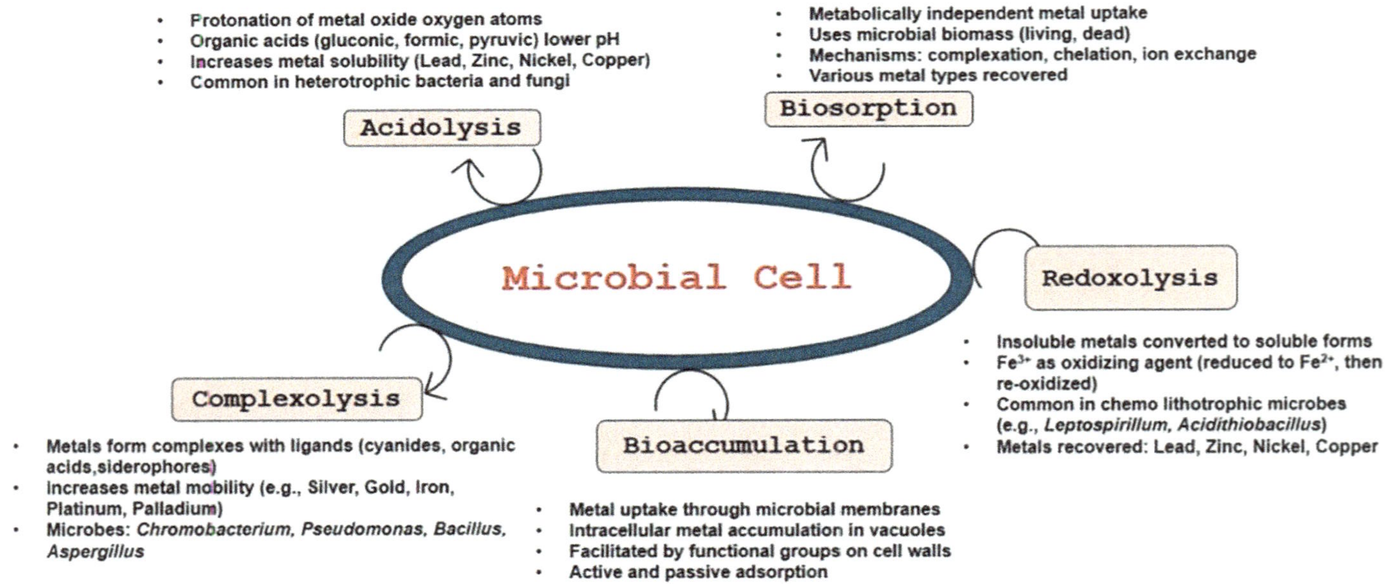

Fig. 4 Mechanism of bioleaching

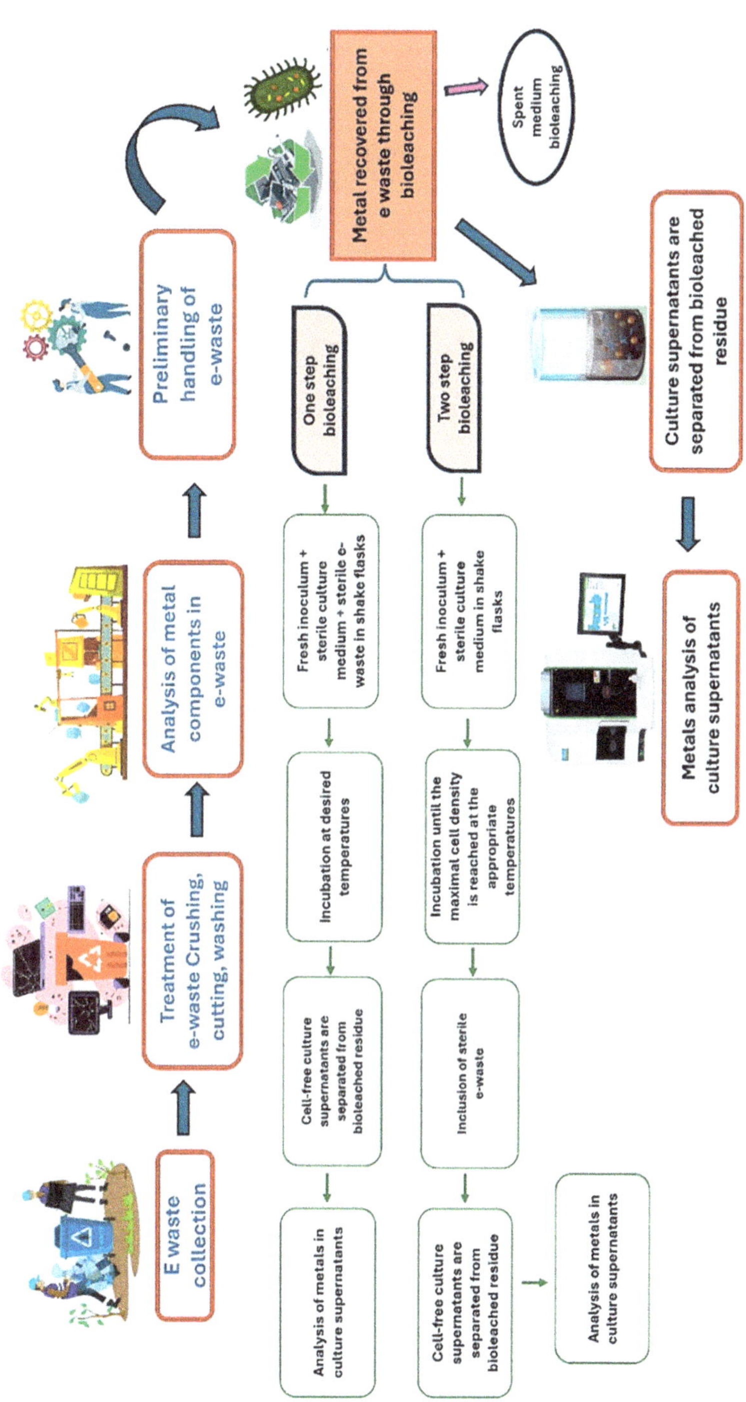

Fig. 5 Strategies for bioleaching metals from e-waste

8 Native Microbes in Bioleaching

A sustainable approach of bioleaching, utilizes native microorganisms like bacteria and fungi to extract valuable metals like copper, nickel, gold and rare-earth elements (REEs) from e-waste (Balaram, 2024). These microbes possess metabolic pathway that enable them to tolerate toxic environments, secrete lixiviants, and solubilize metals, making them a cost-effective and ecofriendly alternative to chemical-based methods (Si, 2024). Mechanisms such as acidolysis, redoxolysis, and complexolysis facilitate metal recovery under ambient conditions, reducing energy consumption and environmental impact (Srichandan et al., 2019). The mechanism employed main groups of microbes involved are chemolithotrophic bacteria, cyanogenic bacteria, and heterotrophic fungi. Chemolithotrophic bacteria such as *Acidithiobacillus ferrooxidans* and *Acidithiobacillus thiooxidans*, use inorganic compounds like iron and sulfur as energy sources. Those microbes can achieve high metal recovery efficiencies (96%) for heavy metals like copper, nickel, zinc and silver from e-waste through mechanisms like acidolysis and redoxolysis (Priya & Hait, 2020). Cyanogenic bacteria like *Chromobacterium violaceum* and *Pseudomonas* sp. can produce cyanide as a metabolite, which can be from soluble metal-cyanide complexes to extract precious metals like gold and silver from e-waste (Merli et al., 2022). Heterotrophic bacteria, including cyanogenic species like *Pseudomonas fluorescens* and *Chromobacterium violaceum*, produce metabolites like cyanide to recover metals, with *C. violaceum* ATCC 12471 achieving 70.6% gold recovery from PCBs (Li et al., 2015). Heterotrophic fungi like *Aspergillus niger* and *Penicillium simplicissimum* secrete organic acids that can solubilize metals through acidolysis. They can also accumulate metals intracellulary or from metal-chelating complexes. Fungal bioleaching has shown recoveries of 80–100% for metals like copper, nickel, zinc and indium (Cui et al., 2021; Arshadi et al., 2022). Using mixed cultures or consortia of microorganisms and synergistic interactions between different species can lead to improved leaching efficiencies (Netpae & Suckley, 2020). *Phanerochaete chrysosporium*, a ligninolytic fungus, recovered 60.96% copper through enzymatic activity (Liu et al., 2020). Adapted to local conditions, native microbes are resilient to e-waste toxicity, cost-effective, and environmentally friendly, reducing secondary pollution and supporting sustainable metal recovery (Gopikrishnan et al., 2020). Key factors influencing the bioleaching process include pH, temperature, pH, pulp density and aeration. Optimal pH and temperature vary depending on the microorganisms used, with acidophilic bacteria preferring lower pH and mesophilic or thermophilic conditions (Adetunji et al., 2023). Adequate aeration is crucial for aerobic microbes, while the culture medium composition provides the necessary nutrients for microbial growth and metabolite production (Khayum et al., 2022). Their role in bioleaching underscores their potential to drive circular economy initiatives, making them pivotal for sustainable urban mining with further research and optimization. Different types of microorganisms used in biohydrometallurgy or bioleaching are described in Table 1.

Table 1 Microorganisms in biohydrometallurgy

S.no	Microorganism	Types of microorganism	Metals	Source of waste	Removal efficiency	Mechanism	Reference
Bacteria in bioleaching							
1.	*Acidithiobacillus thiooxidans*	Chemolithotrophic microorganisms (iron and sulfur-oxidizing bacteria or acidophilic bacteria)	Gold	Telecommunication printed circuit boards	(65 ± 0.78%).	Bio-Thio leaching method	Pourhossein and Mousavi (2023)
2.	*Acidithiobacillus ferrooxidans* and *Acidiphilium acidophilum*	Acidophilic bacteria	Heavy metals (Cu, Zn, Ni, and Pb), rare-earth elements (Ag, Au, Sc, Ce, La and Nd)	Waste-printed circuit boards of computer	Cu-96%, Zn-94.5%, Ni-75%, and Pb-74.5% Ag-46%, Au-38%, Sc-31%, Ce -27% 24% -La and Nd	Bioleaching by the mixed culture	Priya and Hait (2020)
3.	*Acidithiobacillus ferrooxidans*,	Acidophilic bacteria	Copper and nickel	Electronic waste	Cu 90% and Ni 88%	Bioleaching	Arshadi and Yaghmaei (2020)
4.	*Leptospirillum ferriphilum* and *Sulfobacillus thermosulfidooxidans*	Acidophilic bacteria	Copper	Printed circuit board (PCB)	–	Bioleaching	Wu et al. (2018)
5.	*Consortium* (*Ferroplasma* sp., *Sulfobacillus* sp., *Leptospirillum ferriphilum* and *Acidithiobacilluscaldus*)	Thermophilic microorganisms	Cobalt, nickel and lithium	Lithium-ion batteries	99.9% of cobalt, 99.7% of nickel and 84% of lithium	Bioleaching	Ghassa et al. (2020)
6.	*Sulfobacillus thermosulfidooxidans*	–	Copper, nickel and zinc	Waste printed circuit boards (WPCBs)	Cu-93%, Ni-91%, and Zn-94%	Twofold bioleaching	Ilyas et al. (2022)

(continued)

Table 1 (continued)

S.no	Microorganism	Types of microorganism	Metals	Source of waste	Removal efficiency	Mechanism	Reference
7.	*Roseovarius tolerans* and *Roseovarius mucosus*	Iodide oxidizing bacteria	Gold	Printed circuit boards	100% and 34%	Two-step leaching	Kudpeng et al. (2020)
8.	*Chromobacterium violaceum*	Cyanogenic bacteria	Silver	Mobile-PCBs waste	62.40%	Bioleaching using 2 L of batch-stirred tank reactor (BSTR)	AbdolJani et al. (2023)
9.	*Bacillus megaterium*	Heterotrophic microorganisms (metabolic engineering bacteria)	Gold	Gold and copper from telecommunication printed circuit boards	85%	Bioleaching	Aminian-Dehkordi et al. (2020), Golzar-Ahmadi et al. (2024)
10.	*Pseudomonas balearica* SAE1		Copper silver	Low-grade shredded dust e-waste	Cu-95% and Ag-36%	Bioleaching	Thakur and Kumar (2021)
Fungi in bioleaching							
11.	*Aspergillus niger*	Heterotrophic fungi	Copper, nickel and zinc	Printed circuit board (PCB)	71% Cu, 32% Ni, and 79% Zn	Bioleaching of metals by organic acid	Trivedi and Hait (2020)
12.	*Aspergillus tubingensis*	Heterotrophic fungi	Copper, nickel and zinc	Printed circuit board (PCB)	54% Cu, 41% Zn, and 14% Ni.		
13.	*Aspergillus niger*	Heterotrophic fungi	Indium	LCD (liquid crystal display) panels	100%.	Fermentation method optimization	Cui et al. (2021)
14.	*Penicillium simplicissimum*	Heterotrophic fungi	Cu, Al, Ni, Zn, Mn, Mg, Ba, Cr, and Fe	Computer printed circuit boards waste	94%, 99/8%, 100%, 100%, 61%, 81%, 46%, 26%, and 100%	Bioleaching	Arshadi et al. (2022)

15.	*Aspergillus niger* and *Penicillium simplicissimum*	Heterotrophic fungi	Copper, nickel and zinc	Mobile phone printed circuit boards	94%, 100%, and 100% using *A. niger* and 100%, 95%, and 87% using *P. simplicissimum*	Bioleaching	Arshadi et al. (2022)
16.	*Phanerochaete chrysosporium*	Heterotrophic fungi	Copper	Printed circuit boards waste	60.96%	Leaching by bio-enzyme	Liu et al. (2020)
17.	*Rhizopus oligosporus* and *Aspergillu sniger*	Heterotrophic fungi	Ni and Cd	Battery	578.13 ± 7.02 mg Ni g^{-1} and 128.35 ± 11.42 mg Cd g^{-1}.	One step and two-step bioleaching	Netpae and Suckley (2020)
18.	*Penicillium expansum*	Heterotrophic fungi	Rare-earth elements	Printed circuit boards waste	Pr,Tb and Nd -70%, Gd-50% and La-40%	Bioleaching	Baez et al. (2024)
19.	*Saccharomyces cerevisiae*	Heterotrophic fungi	(Au)	PCB scrap	98%	Bioleaching	Díaz-Martínez et al. (2019)

9 Genetic Engineering of Microbes in Bioleaching

Genetic engineering has revolutionized bioleaching by enabling the development of microbes with enhanced abilities to tolerate toxic e-waste environments, produce higher quantities of lixiviants, and specifically target metals for extraction (Arslan, 2021). Techniques like plasmid transfection, conjugation, and mutation can increase metal tolerance, enhance metabolite production, and expand the pH range of the organisms (Liu et al., 2022). Key microbial species like *Acidithiobacillus ferrooxidans* TFBk, *Chromobacterium violaceum* ATCC 12472, *Pseudomonas balearica* SAE1, and *Aspergillus niger* facilitate metal recovery through mechanisms such as acidolysis, redoxolysis, and complexolysis (Thakur & Kumar, 2021; Yachkula et al., 2022; bashina et al., 2023). Engineered strains, such as modified *Chromobacterium violaceum*, exhibit increased hydrogen cyanide production, crucial for solubilizing 62% of precious metals silver (AbdolJani et al., 2023). For instance, genetically enhanced *Chromobacterium* sp. Dyh27s2016 achieved an80, 94, 52 and 56% improvement in Cu, Fe, Zn and Ag recovery from e-waste (Sandrasaigaran et al., 2022). Similarly, engineered *Acidithiobacillus ferrooxidans* has leached copper with over 90% efficiency by improving tolerance to toxic metals (Arshadi & Yaghmaei, 2020). Moreover, consortium-based bioleaching using genetically engineered *Leptospirillum ferriphilum* and *Sulfobacillus thermosulfidooxidans* has recovered more than 94% of copper from printed circuit boards by optimizing ferric ion production (Rouchalová et al., 2024). Additionally, modified fungal strains like *Aspergillus niger* have effectively extracted rare-earth elements (REEs) and base metals, owing to their enhanced organic acid secretion, further showcasing the potential of genetic engineering in sustainable e-waste management (Cui et al., 2021) strain mobilizes respectively from e-waste.

10 Bioleaching Mechanism

Acidophiles and archaea are the key microbes involved in the leaching of sulfide minerals. Among them, *A. ferrooxidans* was the first sulfide-oxidizing bacterium to be isolated (Trudinger et al., 1980; Sand et al., 1992). There are three main bioleaching mechanisms:

1. **Indirect mechanism:** Bacteria oxidize ferrous ions in solution to ferric ions.
2. **Indirect contact mechanism:** Bacterial cells attached to the surface oxidize ferrous ions within the extracellular polymeric substances (EPS), and the ferric ions in the bacterial layer leach the sulfide.
3. **Direct contact mechanism:** Bacteria directly oxidize the sulfide without using ferric ions (Silverman, 1967).

10.1 Direct Contact Mechanism

In contact bioleaching, cells adhere to the surface of solid particles, forming biofilms within the EPS produced by sessile cells, while planktonic cells float in the fluid. Microbial communities can produce metabolites that benefit other organisms and protect cells from environmental stressors, such as antimicrobials. This process is largely driven by microbial enzymes. In the presence of oxygen, *A. ferrooxidans* oxidizes FeS_2,using O_2 as the terminal electron acceptor (Kappler et al., 2015). However, there are limited studies detailing how contact leaching occurs, where microorganisms break the metal-sulfide bonds:

$$2FeS^2 + 7O^2 + 2H^2O \rightarrow 2FeSO_4 + H_2SO_4$$

The adsorption of *A. ferrooxidans* depends on factors such as pH, interaction time, surface area, cell density, and mineral composition. Adsorption decreases beyond neutral pH in the following order: pyrite > chalcopyrite > arsenopyrite (Chandraprabha et al., 2007).

10.2 Indirect Mechanism

In indirect bioleaching, microorganisms do not require direct contact with solid particles. Metal solubilization occurs through three methods:

- Secretion of organic and inorganic acids
- Redox reactions
- Emission of complex agents that form ligands

Sulfuric acid, produced by *Acidithiobacillus* species, is a common inorganic acid used in chemical leaching, while ferric ions serve as a strong oxidizing agent. Microorganisms produce ligands such as oxalate, citrate, succinate, and malonate. Biofilm formation on solid particles can improve bioleaching by providing a barrier that concentrates metabolites like organic acids on the particle surface (Li et al., 2016):

$$4FeSO_4 + O_2 + 2H_2SO_4 \rightarrow 2Fe_2\left[SO_4\right]_3 + 2H_2O$$

$$FeS_2 + Fe_2\left[SO_4\right]_3 \rightarrow 3FeSO_4 + 2S$$

This sulfide oxidation mechanism, termed "galvanic interaction" (Natarajan, 1998), involves electrostatic, hydrophobic, and chemical forces in bacterial attachment to sulfide mineral surfaces. Electrostatic and hydrophobic interactions are significant during initial cell adhesion and are reversible, while chemical interactions, involving polymeric bridges, are irreversible (Barr et al., 1992):

Anodic reaction

$$MS \rightarrow M^{+} + S^{0} + 2e^{-} \left(M \text{ is the bivalent metal}\right)$$

Cathodic reduction of oxygen

$$O_2 + 4H^{+} + 4e^{+} \rightarrow 2H_2O$$

10.3 Polysulfate and Thiosulfate Mechanism

The dissolution of metal sulfides can be classified based on their solubility in acids:

1. **Polysulfide mechanism:** Acid-soluble metal sulfides (e.g., ZnS, $CuFeS_2$, FeAsS, PbS and MnS_2) are dissolved by Fe^{3+} and H^+. Proton attack breaks metal-sulfur bonds, forming $H_2S_2^{+}$ radicals that oxidize to form elemental sulfur.
2. **Thiosulfate mechanism:** Acid-insoluble metal sulfides (e.g., FeS_2 and MoS_2) are oxidized solely by Fe^{3+} (Mishra et al., 2005).

   The primary leaching mechanism involves ferric ions secreted by the EPS, which facilitate metal dissolution by concentrating ferric ions on the bacterial cell wall. The specific pathway depends on the mineral structure and metal solubility:

- **Thiosulfate pathway:** Ferric ions attack the surface of FeS_2, MoS_2 and WS_2 (Fig. 6)
- **Polysulfide pathway:** Ferric ions or protons attack the surface of ZnS, $CuFeS_2$, PbS and MnS_2 (Vera et al., 2013; Schippers & Sand, 1999; Natarajan, 2018) (Fig. 6)

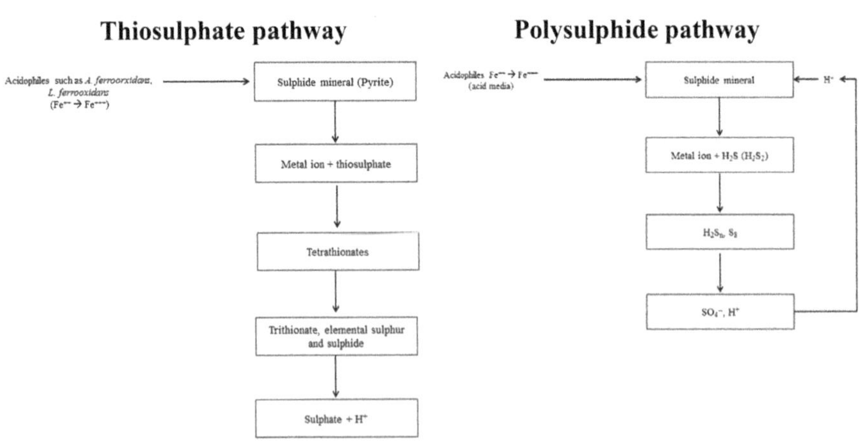

Fig. 6 Thiosulpfate pathway and polysulfide pathway

11 Factors Influencing Bioprocess Parameters

Bioleaching efficiency is often limited and can be time-consuming. Several factors such as interruptions in bacterial growth, formation of precipitates, metal toxicity, and pH fluctuations can affect the process. The key factors influencing bioprocessing are outlined below.

12 pH

When bioleaching with acidophiles, pH is crucial. By oxidizing sulfur, sulfur-oxidizing bacteria (SOB) in the leaching medium contribute to the system's acidity, which helps to maintain the ideal pH range of 2.0 to 3.5 for *A. ferrooxidans* growth. Erust et al. (2020) found that Cu and Ni leaching was most effective at an initial pH of 1.8, while Al and Zn had similar efficiencies at pH 1.8 and 2.0. Higher pH levels (>2 or 2.25), however, caused jarosite formation, which impeded leaching. Gu et al. (2017) found that pH rose quickly during the first 5 days of bioleaching and then stabilized because H^+ was consumed by both *A. ferrooxidans* metabolism and the metal ion formation process.

13 Inoculum Volume

The inoculum volume influences the metal dissolution rate. Erust et al. (2020) found that increasing the inoculum volume from 5% to 20% (v/v) did not significantly improve metal recovery compared to using 10% inoculum. High inoculum levels led to rapid Fe^{2+} oxidation, maintaining a strong oxidizing environment. Fu et al. (2021) showed that a higher inoculum volume increased metal leaching in the early stages, but in the middle-late stage, the growth of bacteria and metal dissolution caused jarosite precipitation and nutrient depletion, reducing efficiency. A 5% inoculum was deemed optimal for industrial applications.

14 Fe^{2+} Concentration

Iron-oxidizing bacteria produce Fe^{3+}, a strong oxidizing agent for metal leaching, by using the energy from Fe^{2+} oxidation. Erust et al. (2020) showed that Cu leaching was increased by 20% and other metal leaching by 10% when the Fe2+ content was raised from 7 to 9 g/L. Higher starting Fe2+ concentrations (8.09 g/L) resulted in a larger concentration drop during bioleaching, which improved metal recovery, according to Gu et al. (2017).

15 Pulp Density

Pulp density has an impact on metal recovery because of the higher concentration of toxic substances and pH changes. Garg et al. (2019) found that Ni recovery was 47.9% at a pulp density of 15%, while Erust et al. (2020) found that Cu leaching was highest at a pulp density of 2.5% (w/v) and that Al, Zn and Ni leaching was optimal at 10% (w/v). A pulp density of 10% was thought to be ideal for maximum recovery of multiple metals because it allowed for effective Fe^{2+} oxidation and metal dissolution.

16 Particle Size

The efficacy of bioleaching is increased by physicochemical changes brought on by mechanical activation during crushing. Microcracks created during crushing promoted bacterial growth and improved Cu bioleaching, according to Gu et al. (2019). In order to decrease particle size and enhance bioleaching activity, Yan et al. (2020) suggested a two-step crushing procedure with heat pretreatment.

17 Catalyst

Graphite has been shown to improve bioleaching rates: Tong et al. (2020) showed that adding 2.5 g/L of graphite increased Cu leaching from 90% to 100%; Gu et al. (2017) reported that adding graphene increased the Cu leaching rate from 74% to 84%; and nitrogen-doped carbon nanotube-modified electrodes improved Cu leaching by 20%. Mehrabani et al. (2017) discovered that graphite influenced microbial populations during sphalerite bioleaching.

18 Bioreactor

In bench-scale bioreactors, Erust et al. (2020) tested ideal bioleaching conditions and obtained recovery rates of 98.1% Cu, 55.9% Al, 79.5% Ni, and 66.9% Zn in 8 days. These results were in line with laboratory-scale experiments and showed that scaling up bioleaching processes is feasible.

19 Hybrid Bioleaching

Bioleaching is still slower than hydrometallurgy and pyrometallurgy, even under ideal conditions. Hybrid bioleaching research has shown encouraging findings. Sheel and Pant (2018) used *Lactobacillus acidophilus* and ammonium thiosulfate to recover 90% of the Au from e-waste. In order to achieve 92.7% Cu recovery, Sinha et al. (2018) created a biorecovery approach that was followed by electrochemical treatment. Chemical-biological hybrid systems were used by Dolker and Pant (2019) to improve cobalt biosorption by 98% and Li leaching by 25%. Compared to employing individual cultures, hybrid approaches—such as electrodialytic remediation in conjunction with microbial metabolism—have demonstrated higher bioleaching capacity.

20 Challenges in E-Waste Remediation

Particle size, pH, temperature, inoculum strength, nutrient availability, metal concentration, and pulp density are some of the factors that affect the bioleaching of e-waste. For example, Fe^{3+}, pH, and inoculum strength have a significant impact on the bioleaching of copper from used printed circuit boards by *Acidithiobacillus ferroxidans*. 100 % inoculum with 6.6 g L^{-1} Fe^{3+} and pH 1.5 resulted in high leaching rate. Several models and approaches were employed to identify and exploit suitable parameters which will project a clear understanding on the process and widen the scope for optimization techniques (Naseri et al., 2023). Extremophiles were examined for their survivability in hostile conditions. The introduction of extremophiles, which have a greater affinity for metals, is anticipated to be beneficial in recycling e-waste in a favorable environment. However, genetic engineering could assist in enhancing tolerance of the microbes for positive output. Leachates from e-waste dumping, processing and recycling can raise concerns against human health. These leachates are heterogeneous and infiltrates the soil contaminating ground water with various contaminants and enters food chain. For instance, a significant amount of barium, beryllium, aluminum, cadmium, copper, cobalt, vanadium, nickel and lead were reported in the leachates from broken e-wastes of a pilot study (Kiddee et al., 2013, 2014). Even though we have a number of laws and regulations regarding the use and disposal of e-waste, there is still a serious problem with the general public's ignorance of its components, usage, and toxicity in both rural and urban areas. Monitoring of e-waste collection, processing, recycling, and disposal should be done by professionals.

21 Conclusion

In the recovery of heavy metals from electronic waste materials, rare-earth elements, precious metals, and greenhouse gas reduction, bioleaching is showing promise as a green method. Scientific community is revolving around microbes to increase the bioleaching process by metabolic and genetical modification. These novel omics approaches which involves genome sequencing facilitates the study of entire consortium Microbiome. This chapter summarized the challenges in e-waste management and omics approaches, with an emphasis on future research that will improve our knowledge of bioleaching kinetics, microbial adaptability, and appropriate genetic manipulation for the creation of new, effective microbes that are more adept at bioleaching metals solely from e-waste. Although the use of genetically modified microorganisms to recover metals from e-waste is highly promising, a great deal of research, from lab to pilot to field investigation, must be done before the technique is commercially used. Cell tolerance as well as higher leaching efficiency for multiple metals at the same time, has to be a vital criterion in designing microorganisms for remediation, facilitating commercial deployment in the foreseeable future.

References

AbdolJani, W. N. F., Suja', F., SayedJamaludin, S. I., Mohamad, N. F., & Abdul Rani, N. H. (2023). Optimization of precious metals recovery from electronic waste by Chromobacteriumviolaceum using response surface methodology (RSM). *Bioinorganic Chemistry and Applications, 2023*(1), 4011670.

Abhilash, P., Pandey, B. D., & Natarajan, A. K. (2015). *Microbiology for minerals, metals, materials and the environment*. CRC Press.

Adetunji, A. I., Oberholster, P. J., & Erasmus, M. (2023). Bioleaching of metals from e-waste using microorganisms: A review. *Minerals, 13*(6), 828.

Ahirwar, R., & Tripathi, A. K. (2021). E-waste management: A review of recycling process, environmental and occupational health hazards, and potential solutions. *Environmental Nanotechnology, Monitoring & Management, 15*, 100409.

Althaf, S., Babbitt, C. W., & Chen, R. (2021). The evolution of consumer electronic waste in the United States. *Journal of Industrial Ecology, 25*(3), 693–706.

Aminian-Dehkordi, J., Mousavi, S. M., Marashi, S. A., Jafari, A., & Mijakovic, I. (2020). A systems-based approach for cyanide overproduction by Bacillus megaterium for gold bioleaching enhancement. *Frontiers in Bioengineering and Biotechnology, 8*, 528.

Anindya, A., Swinboume, D. R., Reuter, M. A., & Matusewicz, R. W. (2013). Distribution of elements between copper and FeOx-CaO-SiO2 slags during pyrometallurgical processing of WEEE: Part 1-Tin. *Mineral Processing and Extractive Metallurgy, 122*, 165–173.

Antrekowitsch, H., Potesser, M., Spruzina, W., & Prior, F. (2006). Metallurgical recycling of electronic scrap. In *EPD congress* (pp. 899–908).

Arshadi, M., & Mousavi, S. M. (2014). Simultaneous recovery of Ni and Cu from computer-printed circuit boards using bioleaching: Statistical evaluation and optimization. *Bioresource Technology, 174*, 233–242.

Arshadi, M., & Mousavi, S. M. (2015). Multi-objective optimization of heavy metals bioleaching from discarded mobile phone PCBs: Simultaneous Cu and Ni recovery using Acidithiobacillus ferrooxidans. *Separation and Purification Technology, 147*, 210–219.

Arshadi, M., & Yaghmaei, S. (2020). Advances in bioleaching of copper and nickel from electronic waste using *Acidithiobacillus ferrooxidans*: Evaluating daily pH adjustment. *Chemical Papers, 74*, 2211–2227.

Arshadi, M., Esmaeili, A., & Yaghmaei, S. (2022). Bioleaching of basic metals from computer printed circuit boards waste using *penicillium simplicissimum. Iranian Chemical Engineering Journal, 21*(121), 55–65.

Arslan, V. (2021). Bacterial leaching of copper, zinc, nickel and aluminum from discarded printed circuit boards using acidophilic bacteria. *Journal of Material Cycles and Waste Management, 23*(5), 2005–2015.

Assefi, M., Maroufi, S., Yamauchi, Y., & Sahajwalla, V. (2020). Pyrometallurgical recycling of Li-ion, Ni-Cd and Ni-MH batteries: A mini-review. *Current Opinion in Green and Sustainable Chemistry, 24*, 26–31.

Baez, A. G., Muñoz, L. P., Timmermans, M. J., Garelick, H., & Purchase, D. (2024). Molding the future: Optimization of bioleaching of rare earth elements from electronic waste by Penicillium expansum and insights into its mechanism. *Bioresource Technology, 402*, 130750.

Bahaloo-Horeh, N., Mousavi, S. M., & Baniasadi, M. (2018). Use of adapted metal tolerant Aspergillus niger to enhance bioleaching efficiency of valuable metals from spent lithium-ion mobile phone batteries. *Journal of Cleaner Production, 197*, 1546–1557.

Bakare, A. A., Adeyemi, A. O., Adeyemi, A., Alabi, O. A., & Osibanjo, O. (2012). Cytogenotoxic effects of electronic waste leachate in Allium cepa. *Caryologia, 65*(2), 94–100.

Balaram, V. (2024). Sustainable recovery of rare earth elements by recycling of E-waste for a circular economy: Perspectives and recent advances. *Environmental Materials and Waste*, 499–544.

Balde, C. P., Kuehr, R., Yamamoto, T., McDonald, R., D'Angelo, E., Althaf, S., et al. (2024). *The global E-waste monitor 2024*. International Telecommunication Union, United Nations Institute for Training and Resources. https://ewastemonitor.info/

Baniasadi, M., Vakilchap, F., Bahaloo-Horeh, N., Mousavi, S. M., & Farnaud, S. (2019). Advances in bioleaching as a sustainable method for metal recovery from e-waste: A review. *Journal of Industrial and Engineering Chemistry, 76*, 75–90.

Bao, S., Pan, B., Wang, L., Cheng, Z., Liu, X., Zhou, Z., & Nie, X. (2020). Adverse effects in Daphnia magna exposed to e-waste leachate: Assessment based on life trait changes and responses of detoxification-related genes. *Environmental Research, 188*, 109821.

Barr, D. W., Jordan, M. A., Norris, P. R., & Phillips, C. V. (1992). An investigation into bacterial cell, ferrous iron, pH and Eh interactions during thermophilic leaching of copper concentrates. *Minerals Engineering, 5*(3–5), 557–567.

Chandraprabha, M. N., Modak, J. M., & Natarajan, K. A. (2007). Effect of LPS removal on electrophoretic softness of Acidithiobacillus ferrooxidans cells. *Advanced Materials Research, 20*, 341–344.

Chen, S., Yang, Y., Liu, C., Dong, F., & Liu, B. (2015). Column bioleaching copper and its kinetics of waste printed circuit boards (WPCBs) by Acidithiobacillus ferrooxidans. *Chemosphere, 141*, 162–168.

Cui, J., & Forssberg, E. (2003). Mechanical recycling of waste electric and electronic equipment: A review. *Journal of Hazardous Materials, 99*, 243–263.

Cui, J., & Zhang, L. (2008). Metallurgical recovery of metals from electronic waste: A review. *Journal of Hazardous Materials, 158*, 228–256.

Cui, J., Zhu, N., Mao, F., Wu, P., & Dang, Z. (2021). Bioleaching of indium from waste LCD panels by Aspergillus niger: Method optimization and mechanism analysis. *Science of the Total Environment, 790*, 148151.

Dang, H., Wang, B., Chang, Z., Wu, X., Feng, J., Zhou, H., Li, W., & Sun, C. (2018). Recycled lithium from simulated pyrometallurgical slag by chlorination roasting. *ACS Sustainable Chemistry & Engineering, 6*, 13160–13167.

Dang, H., Li, N., Chang, Z., Wang, B., Zhan, Y., Wu, X., Liu, W., Ali, S., Li, H., & Guo, J. (2020). Lithium leaching via calcium chloride roasting from simulated pyrometallurgical slag of spent lithium ion battery. *Separation and Purification Technology, 233*, 116025.

Das, A. P., & Bissoyi, A. (2011). Computational approach for comparative phylogenetic analysis of isolated chromium resistant strain Brevibacterium casei. *Journal of Engineering and Technology Research, 3*(3), 82–87. https://doi.org/10.5897/JETR.9000020

Das, A. P., & Ghosh, S. (2018). Bioleaching of manganese from mining waste materials. *Materials Today: Proceedings, 5*(1), 2381–2390. https://doi.org/10.1016/J.MATPR.2017.11.459

Das, A. P., & Ghosh, S. (2023a). Factors affecting bioleaching. In *Manganese mining microorganisms* (pp. 77–86). https://doi.org/10.1016/B978-0-12-822147-1.00004-2

Das, A. P., & Ghosh, S. (2023b). Role of metabolites and metabolomics for understanding of manganese bioleaching. In *Manganese mining microorganisms* (pp. 117–127). https://doi.org/10.1016/B978-0-12-822147-1.00011-X

Das, A. P., & Mishra, S. (2008). Hexavalent Chromium (VI): Health hazards and environmental pollutant. *The Journal of Environment & Development, 2*(3), 386–392.

Das, A. P., & Mishra, S. (2009). Hexavalent chromium [Cr (VI)]: Yellow water pollution and its remediation. *Sarovar Saurabh ENVIS Newsletter Wetland Ecosystem, 5*(2).

Das, A. P., Sukla, L. B., Pradhan, N., & Nayak, S. (2011). Manganese biomining: A review. *Bioresource Technology, 102*(16), 7381–7387. https://doi.org/10.1016/J.BIORTECH.2011.05.018

De Vargas, I., Macaskie, L. E., & Guibal, E. (2004). Biosorption of palladium and platinum by sulfate-reducing bacteria. *Journal of Chemical Technology and Biotechnology, 79*, 49–56.

Dey, S., Dash, N., Ray, S. K., Priyadarsini, S., Chhotaray, C., & Das, A. P. (2024). Occurrence and quantification of tire wear plastics pollutants from highway city dust of Bhubaneswar city of Odisha. *Sustainable Microbial Technology for Synthetic and Cellulosic Microfiber Bioremediation, 77–89*. https://doi.org/10.1007/978-3-031-62600-5_5

Díaz-Martínez, M. E., Argumedo-Delira, R., Sánchez-Viveros, G., Alarcón, A., & Mendoza-López, M. R. (2019). Microbial bioleaching of Ag, Au and Cu from printed circuit boards of mobile phones. *Current Microbiology, 76*, 536–544.

Ding, Y., Zhang, S., Liu, B., Zheng, H., Chang, C., & Ekberg, C. (2019). Recovery of precious metals from electronic waste and spent catalysts: A review. *Resources, Conservation and Recycling, 141*, 284–298.

Dolker, T., & Pant, D. (2019). Chemical-biological hybrid systems for the metal recovery from waste lithium ion battery. *Journal of Environmental Management, 248*, 109270.

Dradrach, A., Karczewska, A., & Szopka, K. (2020). Arsenic uptake by two tolerant grass species: Holcus lanatus and Agrostis capillaris growing in soils contaminated by historical mining. *Plants, 9*(8), 980.

Eguchi, A., Kunisue, T., Wu, Q., Trang, P. T. K., Viet, P. H., Kannan, K., & Tanabe, S. (2014). Occurrence of perchlorate and thiocyanate in human serum from e-waste recycling and reference sites in Vietnam: Association with thyroid hormone and iodide levels. *Archives of Environmental Contamination and Toxicology, 67*, 29–41.

Erust, C., Akcil, A., Tuncuk, A., & Panda, S. (2020). Intensified acidophilic bioleaching of multi-metals from waste printed circuit boards (WPCBs) of spent mobile phones. *Journal of Chemical Technology & Biotechnology, 95*(8), 2272–2285.

Eze, C., & Vinken, M. (2024). E-waste: Mechanisms of toxicity and safety testing. *FEBS Open Bio, 14*(9), 1420–1440.

Faraji, F., Golmohammadzadeh, R., Rashchi, F., & Alimardani, N. (2018). Fungal bioleaching of WPCBs using Aspergillus niger: Observation, optimization and kinetics. *Journal of Environmental Management, 217*, 775–787.

Forti, V., Balde, C. P., Kuehr, R., & Bel, G. (2020). The global E-waste monitor 2020: Quantities, flows and the circular economy potential.

Fu, K., Wang, B., Chen, H., Chen, M., & Chen, S. (2016). Bioleaching of Al from coarse-grained waste printed circuit boards in a stirred tank reactor. *Procedia Environmental Sciences, 31*, 897–902.

Fu, K., Tian, L., Hou, P., Long, M., Chen, S., & Lin, H. (2021). Stirred-tank leaching of coarse-grained waste, printed circuit boards with Acidithiobacillus ferrooxidans. *Physicochemical Problems of Mineral Processing, 57*(5), 153–163.

Garg, H., Nagar, N., Ellamparuthy, G., Angadi, S. I., & Gahan, C. S. (2019). Bench scale microbial catalysed leaching of mobile phone PCBs with an increasing pulp density. *Heliyon, 5*(12).

Ghasem, A., & Khoramnejadian, S. (2015). The extraction of gold from e-waste by hydrometallurgy. *Oriental Journal of Chemistry, 31*(1), 113–120.

Ghassa, S., Farzanegan, A., Gharabaghi, M., & Abdollahi, H. (2020). Novel bioleaching of waste lithium ion batteries by mixed moderate thermophilic microorganisms, using iron scrap as energy source and reducing agent. *Hydrometallurgy, 197*, 105465.

Ghodrat, M., Rhamdhani, M. A., Brooks, G., Masood, S., & Corder, G. (2016). Techno economic analysis of electronic waste processing through black copper smelting route. *Journal of Cleaner Production, 126*, 178–190.

Ghodrat, M., Rhamdhani, M. A., Khaliq, A., Brooks, G., & Samali, B. (2018). Thermodynamic analysis of metals recycling out of waste printed circuit board through secondary copper smelting. *Journal of Material Cycles and Waste Management, 20*, 386–401.

Ghosh, S., & Das, A. P. (2017). Bioleaching of manganese from mining waste residues using Acinetobacter sp. *Geology, Ecology, and Landscapes, 1*(2), 77–83. https://doi.org/10.1080/24749508.2017.1332847

Ghosh, S., Mohanty, S., Akcil, A., Sukla, L. B., & Das, A. P. (2016a). A greener approach for resource recycling: Manganese bioleaching. *Chemosphere, 154*, 628–639. https://doi.org/10.1016/J.CHEMOSPHERE.2016.04.028

Ghosh, S., Mohanty, S., Nayak, S., Sukla, L. B., & Das, A. P. (2016b). Molecular identification of indigenous manganese solubilising bacterial biodiversity from manganese mining deposits. *Journal of Basic Microbiology, 56*(3), 254–262. https://doi.org/10.1002/JOBM.201500477

Golzar-Ahmadi, M., Bahaloo-Horeh, N., Pourhossein, F., Norouzi, F., Schoenberger, N., Hintersatz, C., et al. (2024). Pathway to industrial application of heterotrophic organisms in critical metals recycling from e-waste. *Biotechnology Advances*, 108438.

Gonçalves, M. C. A., Garcia, E. M., Taroco, H. A., Gorgulho, H. F., Melo, J. O. F., Silva, R. R. A., & Souza, A. G. (2015). Chemical recycling of cell phone Li-ion batteries: Application in environmental remediation. *Waste Management, 40*, 144–150.

Gopikrishnan, V., Vignesh, A., Radhakrishnan, M., Joseph, J., Shanmugasundaram, T., Doble, M., & Balagurunathan, R. (2020). Microbial leaching of heavy metals from e-waste: Opportunities and challenges. In *Biovalorisation of wastes to renewable chemicals and biofuels* (pp. 189–216).

Gramatyka, P., Nowosielski, R., & Sakiewicz, P. (2007). Recycling of waste electrical and electronic equipment. *Journal of Achievements in Materials and Manufacturing Engineering, 20*, 535–538.

Grant, K., Goldizen, F. C., Sly, P. D., Brune, M. N., Neira, M., van den Berg, M., & Norman, R. E. (2013). Health consequences of exposure to e-waste: A systematic review. *The Lancet Global Health, 1*(6), e350–e361.

Gu, W., Bai, J., Dong, B., Zhuang, X., Zhao, J., Zhang, C., et al. (2017). Enhanced bioleaching efficiency of copper from waste printed circuit board driven by nitrogen-doped carbon nanotubes modified electrode. *Chemical Engineering Journal, 324*, 122–129.

Gu, W., Bai, J., Lu, L., Zhuang, X., Zhao, J., Yuan, W., et al. (2019). Improved bioleaching efficiency of metals from waste printed circuit boards by mechanical activation. *Waste Management, 98*, 21–28.

Guo, X., Liu, J., Qin, H., Liu, Y., Tian, Q., & Li, D. (2015). Recovery of metal values from waste printed circuit boards using an alkali fusion-leaching-separation process. *Hydrometallurgy, 156*, 199–205.

Gupta, N., & Nath, M. (2020). Groundwater contamination by E-waste and its remedial measure-A literature review. *Journal of Physics: Conference Series, 1531*(1), 012023. IOP Publishing.

Hageluken, C. (2006). Improving metal returns and eco-efficiency in electronics recycling—A holistic approach for interface optimisation between pre-processing and integrated metals smelting and refining. In: *Proceedings of the IEEE international symposium on electronics and the environment*, Scottsdale, AZ, USA.

Ilankoon, I. M. S. K., Ghorbani, Y., Chong, M. N., Herath, G., Moyo, T., & Petersen, J. (2018). E-waste in the international context–A review of trade flows, regulations, hazards, waste management strategies and technologies for value recovery. *Waste Management, 82*, 258–275.

Ilyas, S., Srivastava, R. R., Kim, H., & Ilyas, N. (2022). Biotechnological recycling of hazardous waste PCBs using Sulfobacillus thermosulfidooxidans through pretreatment of toxicant metals: Process optimization and kinetic studies. *Chemosphere, 286*, 131978.

International Labour Office and United Nations Children's Fund. (2021). Child labour: Global estimates 2020, trends and the road forward. https://www.ilo.org/ipec/ChildlabourstatisticsSIMPOC/lang%2D%2Den/index.htm

Islam, A., Ahmed, T., Awual, M. R., Rahman, A., Sultana, M., Aziz, A. A., Monir, M. U., Teo, S. H., & Hasan, M. (2020). Advances in sustainable approaches to recover metals from e-waste-A review. *Journal of Cleaner Production, 244*, 118815.

Jiang, B., Adebayo, A., Jia, J., Xing, Y., Deng, S., Guo, L., et al. (2019). Impacts of heavy metals and soil properties at a Nigerian e-waste site on soil microbial community. *Journal of Hazardous Materials, 362*, 187–195.

Johnson, D. B. (2018). The evolution, current status, and future prospects of using biotechnologies in the mineral extraction and metal recovery sectors. *Minerals, 8*(8), 343.

Johnson, D. B., Smith, S. L., & Santos, A. L. (2021). Bioleaching of transition metals from limonitic laterite deposits and reassessment of the multiple roles of sulfur-oxidizing acidophiles in the process. *Frontiers in Microbiology, 12*, 703177.

Kappler, A., Emerson, D., Gralnick, J. A., Roden, E. E., & Muehe, E. M. (2015). Geomicrobiology of iron. In *Ehrlich's geomicrobiology* (6th ed., p. 635).

Karthik, C., Kadirvelu, K., Bruno, B., Maharajan, K., Rajkumar, M., Manoj, S. R., & Arulselvi, P. I. (2021). Cellulosimicrobium funkei strain AR6 alleviate Cr (VI) toxicity in Lycopersicon esculentum by regulating the expression of growth responsible, stress tolerant and metal transporter genes. *Rhizosphere, 18*, 100351.

Kaya, M. (2016). Recovery of metals and nonmetals from electronic waste by physical and chemical recycling processes. *Waste Management, 57*, 64–90.

Kaya, M. (2019). *Electronic waste and printed circuit board recycling technologies*. Springer.

Khaliq, A., Rhamdhani, M. A., Brooks, G., & Masood, S. (2014). Metal extraction processes for electronic waste and existing industrial routes: A review and Australian perspective. *Resources, 3*, 152–179.

Khan, J., Kumar, A., Giri, A., Pal, D. B., Tripathi, A., & Giri, D. D. (2021). Impact of electronic waste pollutants on underground water. In *Groundwater geochemistry: Pollution and remediation methods* (pp. 265–281).

Khanna, R., Saini, R., Park, M., Ellamparuthy, G., Biswal, S. K., & Mukherjee, P. S. (2020). Factors influencing the release of potentially toxic elements (PTEs) during thermal processing of electronic waste. *Waste Management, 105*, 414–424.

Khayum, A., Kolar, A. B., Banu, D. S., & Kensa, M. (2022). A summary of the use of microbial cell culture. *IJFANS International Journal of Food and Nutritional Sciences, 11*, 604–612.

Kiddee, P., Naidu, R., & Wong, M. H. (2013). Metals and polybrominated diphenyl ethers leaching from electronic waste in simulated landfills. *Journal of Hazardous Materials, 252*, 243–249.

Kiddee, P., Naidu, R., Wong, M. H., Hearn, L., & Muller, J. F. (2014). Field investigation of the quality of fresh and aged leachates from selected landfills receiving e-waste in an arid climate. *Waste Management, 34*(11), 2292–2304.

Kim, S., & Paulos, E. (2011). Practices in the creative reuse of e-waste. In Proceedings of the SIGCHI Conference on Human Factors in Computing Systems (pp. 2395–2404).

Kudpeng, K., Bohu, T., Morris, C., Thiravetyan, P., & Kaksonen, A. H. (2020). Bioleaching of gold from sulfidic gold ore concentrate and electronic waste by Roseovarius tolerans and Roseovarius mucosus. *Microorganisms, 8*(11), 1783.

Li, J., Lu, H., Guo, J., Xu, Z., & Zhou, Y. (2007). Recycle technology for recovering resources and products from waste printed circuit boards. *Environmental Science & Technology, 41*, 1995–2000.

Li, J., Liang, C., & Ma, C. (2015). Bioleaching of gold from waste printed circuit boards by Chromobacteriumviolaceum. *Journal of Material Cycles and Waste Management, 17*, 529–539.

Li, Y., Jia, R., Al-Mahamedh, H. H., Xu, D., & Gu, T. (2016). Enhanced biocide mitigation of field biofilm consortia by a mixture of D-amino acids. *Frontiers in Microbiology, 7*, 896.

Li, H., Eksteen, J., & Oraby, E. (2018). Hydrometallurgical recovery of metals from waste printed circuit boards (WPCBs): Current status and perspectives-A review. *Resources, Conservation and Recycling, 139*, 122–139.

Li, H., La Guardia, M. J., Liu, H., Hale, R. C., Mainor, T. M., Harvey, E., & Peng, P. A. (2019). Brominated and organophosphate flame retardants along a sediment transect encompassing the Guiyu, China e-waste recycling zone. *Science of the total environment, 646*, 58–67.

Li, J. J., Zhao, X. Q., Wang, J. L., & Shen, R. F. (2021). Strategies of cadmium and copper uptake and translocation in different plant species growing near an E-waste dismantling site at Wenling, China. *Environmental Science and Pollution Research, 28*, 62562–62571.

Liu, C., Lin, J., Cao, H., Zhang, Y., & Sun, Z. (2019). Recycling of spent lithium-ion batteries in view of lithium recovery: A critical review. *Journal of Cleaner Production, 228*, 801–813.

Liu, Q., Bai, J. F., Gu, W. H., Peng, S. J., Wang, L. C., Wang, J. W., & Li, H. X. (2020). Leaching of copper from waste printed circuit boards using Phanerochaete chrysosporium fungi. *Hydrometallurgy, 196*, 105427.

Liu, Y., Feng, J., Pan, H., Zhang, X., & Zhang, Y. (2022). Genetically engineered bacterium: Principles, practices, and prospects. *Frontiers in Microbiology, 13*, 997587.

Luo, X. J., Liu, J., Luo, Y., Zhang, X. L., Wu, J. P., Lin, Z., et al. (2009). Polybrominated diphenyl ethers (PBDEs) in free-range domestic fowl from an e-waste recycling site in South China: Levels, profile and human dietary exposure. *Environment International, 35*(2), 253–258.

Luo, C., Wang, S., Wang, Y., Yang, R., Zhang, G., & Shen, Z. (2015). Effects of EDDS and plant-growth-promoting bacteria on plant uptake of trace metals and PCBs from e-waste–contaminated soil. *Journal of Hazardous Materials, 286*, 379–385.

Mehrabani, J. V., Shafaei, S. Z., Noaparast, M., & Mousavi, S. M. (2017). Bioleaching of different pyrites and sphalerite in the presence of graphite. *Geomicrobiology Journal, 34*(2), 97–108.

Merli, G., Becci, A., & Amato, A. (2022). Recovery of precious metals from printed circuit boards by cyanogenic bacteria: Optimization of cyanide production by statistical analysis. *Journal of Environmental Chemical Engineering, 10*(3), 107495.

Meskers, C., Hagelüken, C., Salhofer, S., Spitzbart, M., 2009. Impact of pre-processing routes on precious metal recovery from PCs.

Mishra, D., Kim, D. J., Ahn, J. G., & Rhee, Y. H. (2005). Bioleaching: A microbial process of metal recovery; a review. *Metals and Materials International, 11*, 249–256.

Mishra, S., Dash, D., Al-Tawaha, A. R. M. S., & Das, A. P. (2022). A review on heavy metal ion adsorption on synthetic microfiber surface in aquatic environments. *Applied Biochemistry and Biotechnology, 194*(10), 4639–4654. https://doi.org/10.1007/S12010-022-04029-W/METRICS

Mostafavi, M., Mirazimi, S., Rashchi, F., Faraji, F., & Mostoufi, N. (2018). Bioleaching and kinetic investigation of WPCBs by A. ferrooxidans, A. thiooxidans and their mixtures. *Journal of Chemical and Petroleum Engineering, 52*, 81–91.

Najar, I. N., Sharma, P., Das, R., Mondal, K., Singh, A. K., Tamang, S., et al. (2024). In search of Poly-3-Hydroxybutyrate (PHB): A comprehensive review unveiling applications and progress in fostering a sustainable bio-circular economy. *Food and Bioproducts Processing, 148*, 11–30.

Naseri, T., Beiki, V., Mousavi, S. M., & Farnaud, S. (2023). A comprehensive review of bioleaching optimization by statistical approaches: Recycling mechanisms, factors affecting, challenges, and sustainability. *RSC Advances, 13*(34), 23570–23589.

Natarajan, K. A. (1998). An integrated biotechnological approach to gold processing-An Indian experience. *Mineral Procesing and Extractive Metallurgy Review, 19*(1), 235–251.

Natarajan, K. A. (2018). *Biotechnology of metals: Principles, recovery methods and environmental concerns*. Elsevier.

Nekouei, R. K., Pahlevani, F., Golmohammadzadeh, R., Assefi, M., Rajarao, R., Chen, Y.-H., & Sahajwalla, V. (2019). Recovery of heavy metals from waste printed circuit boards: Statistical

optimization of leaching and residue characterization. *Environmental Science and Pollution Research, 26*, 24417–24429.
Netpae, T., & Suckley, S. (2020). Comparison of three culture media for one-step and two-step bioleaching of nickel and cadmium from spent Ni-Cd batteries by Aspergillus niger. *Advances in Environmental Technology, 6*(3), 167–172.
Nie, X., Fan, C., Wang, Z., Su, T., Liu, X., & An, T. (2015). Toxic assessment of the leachates of paddy soils and river sediments from e-waste dismantling sites to microalga, Pseudokirchneriella subcapitata. *Ecotoxicology and Environmental Safety, 111*, 168–176.
Odılı, J. B., Norazıah, A., Ambar, R., Abd Wahab, M. H., & Maı, F. (2018). E-waste generation, awareness and management in third world countries: Prospects and challenges. *The Eurasia Proceedings of Educational and Social Sciences, 9*, 349–353.
Okereafor, U., Makhatha, M., Mekuto, L., Uche-Okereafor, N., Sebola, T., & Mavumengwana, V. (2020). Toxic metal implications on agricultural soils, plants, animals, aquatic life and human health. *International Journal of Environmental Research and Public Health, 17*(7), 2204.
Panwar, R. M., & Ahmed, S. (2018). Assessment of contamination of soil and groundwater due to e-waste handling. *Current Science*, 166–173.
Pourhossein, F., & Mousavi, S. M. (2023). Improvement of gold bioleaching extraction from waste telecommunication printed circuit boards using biogenic thiosulfate by Acidithiobacillus thiooxidans. *Journal of Hazardous Materials, 450*, 131073.
Prabhakar, A., Mishra, S., & Das, A. P. (2019). Isolation and identification of Lead (Pb) solubilizing bacteria from automobile waste and its potential for recovery of Lead from end of life waste batteries. *Geomicrobiological Journal, 36*(10), 894–903. https://doi.org/10.1080/01490451.2019.1654044
Priya, A., & Hait, S. (2017). Comparative assessment of metallurgical recovery of metals from electronic waste with special emphasis on bioleaching. *Environmental Science and Pollution Research, 24*, 6989–7008.
Priya, A., & Hait, S. (2020). Biometallurgical recovery of metals from waste printed circuit boards using pure and mixed strains of *Acidithiobacillus ferrooxidans and Acidiphilium acidophilum. Process Safety and Environmental Protection, 143*, 262–272.
Priyadarshini, B., Santhosh Kumar, M., & Das, A. P. (2024). Microbial degradation of synthetic plastics released from COVID-19 generated single use disposable face masks. *Environmental Science and Engineering, Part, F2827*, 1–16. https://doi.org/10.1007/978-3-031-50840-0_1
Priyadarsini, S., & Das, A. P. (2024). Analytical and structural characterization of waste lithium-ion batteries for their effective recycling strategy. *Environmental Chemistry and Ecotoxicology*. https://doi.org/10.1016/j.enceco.2024.12.004
Priyadarsini, S., & Das, A. P. (2025). Lithium bioleaching: A review on microbial-assisted sustainable technology for lithium bio-circularity. *Journal of Water Process Engineering, 69*, 106744. https://doi.org/10.1016/j.jwpe.2024.106744
Purchase, D., Abbasi, G., Bisschop, L., Chatterjee, D., Ekberg, C., Ermolin, M., et al. (2020). Global occurrence, chemical properties, and ecological impacts of e-wastes (IUPAC Technical Report). *Pure and Applied Chemistry, 92*(11), 1733–1767.
Raj, K., & Das, A. P. (2023). Lead pollution: Impact on environment and human health and approach for a sustainable solution. *Environmental Chemistry and Ecotoxicology, 5*, 79–85. https://doi.org/10.1016/J.ENCECO.2023.02.001
Rajendran, S., Kamaludeen, S. P. B., & Subramanian, A. (2024). Antioxidant defense mechanism in plants exposed to mercury toxicity: Response, tolerance and remediation. In *Mercury toxicity mitigation: Sustainable nexus approach* (pp. 225–245). Springer Nature Switzerland.
Rautela, R., Arya, S., Vishwakarma, S., Lee, J., Kim, K. H., & Kumar, S. (2021). E-waste management and its effects on the environment and human health. *Science of the Total Environment, 773*, 145623.
Ren, G., Xiao, S., Xie, M., Bing, P. A. N., Jian, C., Wang, F., & Xing, X. I. A. (2017). Recovery of valuable metals from spent lithium ion batteries by smelting reduction process based on Fe0-SiO2-Alp3 slag system. *Transactions of the Nonferrous Metals Society of China, 27*, 450–456.

Romera, E., Gonzalez, F., Ballester, A., Blázquez, M. L., & Muñoz, J. A. (2006). Biosorption with algae: A statistical review. *Critical Reviews in Biotechnology, 26*(4), 223–235.

Roshanfar, M., Golmohammadzadeh, R., & Rashchi, F. (2019). An environmentally friendly method for recovery of lithium and cobalt from spent lithium-ion batteries using gluconic and lactic acids. *Journal of Environmental Chemical Engineering, 7*, 102794.

Rouchalová, D., Rouchalová, K., & Čablík, V. (2024). Bioleaching of mine tailings by mesophilic: Acidithiobacillus spp., Leptospirillum ferrooxidans, and thermophilic: Sulfobacillus thermosulfidooxidans cultures with the addition of Ag+ additive. *Minerals, 14*(3), 255.

Sadegh Safarzadeh, M., Bafghi, M. S., Moradkhani, D., & OjaghiIlkhchi, M. (2007). A review on hydrometallurgical extraction and recovery of cadmium from various resources. *Minerals Engineering, 20*, 211–220.

Sag, Y., & Kutsal, T. (2001). Recent trends in the biosorption of heavy metals: A review. *Biotechnology and Bioprocess Engineering, 6*, 376–385.

Saha, L., Kumar, V., Tiwari, J., Rawat, S., Singh, J., & Bauddh, K. (2021). Electronic waste and their leachates impact on human health and environment: Global ecological threat and management. *Environmental Technology & Innovation, 24*, 102049.

Sand, W., Rohde, K., Sobotke, B., & Zenneck, C. (1992). Evaluation of Leptospirillum ferrooxidans for leaching. *Applied and Environmental Microbiology, 58*(1), 85–92.

Sandrasaigaran, P., Palanisamy, T., Sukalingam, R., & Murugan, S. (2022). Extended spectrum beta-lactamase expressing Chromobacterium sp. Dyh27s2016 strain is capable of metal leaching in electronic waste. *Key Engineering Materials, 937*, 181–191.

Schippers, A., & Sand, W. (1999). Bacterial leaching of metal sulfides proceeds by two indirect mechanisms via thiosulfate or via polysulfides and sulfur. *Applied and Environmental Microbiology, 65*(1), 319–321.

Seh-Bardan, B. J., Othman, R., Wahid, S. A., Husin, A., & Sadegh-Zadeh, F. (2012). Bioleaching of heavy metals from mine tailings by Aspergillus fumigatus. *Bioremediation Journal, 16*, 57–65.

Sharma, S., Wakode, S., Sharma, A., Nair, N., Dhobi, M., Wani, M. A., & Pottoo, F. H. (2020). Effect of environmental toxicants on neuronal functions. *Environmental Science and Pollution Research, 27*, 44906–44921.

Sheel, A., & Pant, D. (2018). Recovery of gold from electronic waste using chemical assisted microbial biosorption (hybrid) technique. *Bioresource Technology, 247*, 1189–1192.

Si, S. (2024). 16 Microorganism-based. In *Advanced bioseparation of industrial wastes: Sustainable recovery of high-value metal ions*, 337.

Silveira, A. V. M., Fuchs, M. S., Pinheiro, D. K., Tanabe, E. H., & Bertuol, D. A. (2015). Recovery of indium from LCD screens of discarded cell phones. *Waste Management, 45*, 334–342.

Silverman, M. P. (1967). Mechanism of bacterial pyrite oxidation. *Journal of Bacteriology, 94*(4), 1046–1051.

Singh Jadaun, J., Bansal, S., Sonthalia, A., Rai, A. K., & Singh, S. P. (2022). Biodegradation of plastics for sustainable environment. *Bioresource Technology, 347*, 126697. https://doi.org/10.1016/J.BIORTECH.2022.126697

Sinha, R., Chauhan, G., Singh, A., Kumar, A., & Acharya, S. (2018). A novel eco-friendly hybrid approach for recovery and reuse of copper from electronic waste. *Journal of Environmental Chemical Engineering, 6*(1), 1053–1061.

Sonawane, P. M. (2023). Understanding electronic waste and decoding its ambiguous nature. *Research and Development*, 345.

Srichandan, H., Mohapatra, R. K., Parhi, P. K., & Mishra, S. (2019). Bioleaching approach for extraction of metal values from secondary solid wastes: A critical review. *Hydrometallurgy, 189*, 105122.

Suganthi, R., Avudainayagam, S., Davamani, V., Banu, K. S. P., Chandrasekar, C. N., & Sivakumar, U. (2021). Indian mustard and Boston fern exhibits growth tolerance to increased dose of soil spiked mercury.

Tang, X., Qiao, J., Chen, C., Chen, L., Yu, C., Shen, C., & Chen, Y. (2013). Bacterial communities of polychlorinated biphenyls polluted soil around an e-waste recycling workshop. *Soil and Sediment Contamination: An International Journal, 22*(5), 562–573.

Tay, S. B., Natarajan, G., Rahim, M. N. B. A., Tan, H. T., Chung, M. C. M., Ting, Y. P., & Yew, W. S. (2013). Enhancing gold recovery from electronic waste via lixiviant metabolic engineering in Chromobacterium violaceum. *Scientific Reports, 3*(1), 2236.

Tembhare, S., & Bhanvase, B. A. (2023). Technological development in metal extraction processes from e-waste. In *360-Degree waste management* (Vol. 2, pp. 221–253). Elsevier.

Thakur, P., & Kumar, S. (2021). Pretreatment of low-grade shredded dust e-waste to enhance silver recovery through biocyanidation by Pseudomonas balearica SAE1. *3 Biotech, 11*(11), 454.

Tong, L., Zhao, Q., Kamali, A. R., Sand, W., & Yang, H. (2020). Effect of graphite on copper bioleaching from waste printed circuit boards. *Minerals, 10*(1), 79.

Trivedi, A., & Hait, S. (2020). Bioleaching of selected metals from e-waste using pure and mixed cultures of Aspergillus species. In *Measurement, analysis and remediation of environmental pollutants* (pp. 271–280).

Trudinger, P. A., Walter, M. R., Ralph, B. J., Olsen, T. M., Ashman, P. R., Torma, A. E., & Murr, L. E. (1980). Desulfurization of coal by Thiobacillus ferrooxidans. In *Biogeochemistry of ancient and modern environments: Proceedings of the fourth international symposium on environmental biogeochemistry (ISEB) and, conference on biogeochemistry in relation to the mining industry and environmental pollution (leaching conference), held in Canberra, Australia, 26 August–4 September 1979* (pp. 693–703). Springer Berlin Heidelberg.

Tuncuk, A., Stazi, V., Akcil, A., Yazici, E. Y., & Deveci, H. (2012). Aqueous metal recovery techniques from e-scrap: Hydrometallurgy in recycling. *Minerals Engineering, 25*, 28–37.

Vakilchap, F., Mousavi, S. M., & Shojaosadati, S. A. (2016). Role of Aspergillus niger in recovery enhancement of valuable metals from produced red mud in Bayer process. *Bioresource Technology, 218*, 991–998.

Vera, M., Schippers, A., & Sand, W. (2013). Progress in bioleaching: Fundamentals and mechanisms of bacterial metal sulfide oxidation—Part A. *Applied Microbiology and Biotechnology, 97*, 7529–7541.

Wang, J., & Chen, C. (2006). Biosorption of heavy metals by Saccharomyces cerevisiae: A review. *Biotechnology Advances, 24*(5), 427–451.

Wang, X., Chen, J., Yan, X., Wang, X., Zhang, J., Huang, J., & Zhao, J. (2015). Heavy metal chemical extraction from industrial and municipal mixed sludge by ultrasound-assisted citric acid. *Journal of Industrial and Engineering Chemistry, 27*, 368–372.

Wang, S., Dong, D., Li, P., Hua, X., Zheng, N., Sun, S., et al. (2020). Mercury concentration and fatty acid composition in muscle tissue of marine fish species harvested from Liaodong Gulf: An intelligence quotient and coronary heart disease risk assessment. *Science of the Total Environment, 726*, 138586.

Widmer, R., Oswald-Krapf, H., Sinha-Khetriwal, D., Schnellmann, M., & Boni, H. (2005). Global perspectives one-waste. *Environmental Impact Assessment Review, 25*, 436–458.

Woo, S. H., Lee, D. S., & Lim, S. R. (2015). Potential resource and toxicity impacts from metals in waste electronic devices. *Integrated Environmental Assessment and Management, 12*(2), 364–370.

World Health Organization. (2021). *Children and digital dumpsites: E-waste exposure and child health*. World Health Organization. https://apps.who.int/iris/handle/10665/341718

Wu, Z., Yuan, W., Li, J., Wang, X., Liu, L., & Wang, J. (2017). A critical review on the recycling of copper and precious metals from waste printed circuit boards using hydrometallurgy. *Frontiers of Environmental Science & Engineering, 11*, 8.

Wu, W., Liu, X., Zhang, X., Zhu, M., & Tan, W. (2018). Bioleaching of copper from waste printed circuit boards by bacteria-free cultural supernatant of iron–sulfur-oxidizing bacteria. *Bioresources and Bioprocessing, 5*(1), 1–13.

Xiang, Y., Wu, P., Zhu, N., Zhang, T., Liu, W., Wu, J., & Li, P. (2010). Bioleaching of copper from waste printed circuit boards by bacterial consortium enriched from acid mine drainage. *Journal of Hazardous Materials, 184*(1–3), 812–818.

Yachkula, A., Rozova, O., Abashina, T., Vainshtein, M., Grouzdev, D., & Bulaev, A. (2022). Attempts to stimulate leaching activity of Acidithiobacillusferrooxidans strain TFBk. *Minerals, 12*(8), 1051.

Yan, G., Guo, J., Zhu, G., Zhang, Z., Zhao, P., Xiangnan, Z., & Zhang, B. (2020). Liberation enhancement and copper enrichment improvement for waste printed circuit boards by heating pretreatment. *Waste Management, 106*, 145–154.

Yeboah, A., Lu, J., Gu, S., Liu, H., Shi, Y., Amoanimaa-Dede, H., et al. (2021). Evaluation of two wild castor (Ricinus communis L.) accessions for cadmium tolerance in relation to antioxidant systems and lipid peroxidation. *Environmental Science and Pollution Research, 28*(39), 55634–55642.

Youcai, Z. (2018). Leachate generation and characteristics. In *Pollution control technology for leachate from municipal solid waste* (pp. 1–30).

Zeng, X., Xu, X., Zheng, X., Reponen, T., Chen, A., & Huo, X. (2016). Heavy metals in PM2.5 and in blood, and children's respiratory symptoms and asthma from an e-waste recycling area. *Environmental Pollution, 210*, 346–353.

Recovery of Precious Metals from E-Waste Using Cyanogenic Microbes

D P Krishna Samal and Lala Behari Sukla

Abstract The rapid expansion of the electronics industry has profoundly reshaped modern life. However, it has simultaneously given rise to a pressing environmental challenge like a surge of electronic waste (e-waste). E-waste contains precious metals like gold, palladium, silver, platinum, copper, and rare earth elements and also has hazardous substances such as cadmium, lead, arsenic, chromium and mercury. E-waste is often discarded improperly, bypassing essential environmental protocols. Thus, these contaminants can leach into the environment, posing significant health risks to ecosystems along with humans. Utilizing microorganisms to facilitate precious metals recovery emerges as a more environmentally friendly recycling approach of e-waste. Various microbes are used in the recycling of e-wastes and recovering precious metals. Cyanogenic microbes such as *Chromobacterium*, *Bacillus, Pseudomonas, Acidithiobacillus, Penicillium*, *Aspergillus*, etc., play a critical role in recovering precious metals like gold, silver etc. These microbes release cyanide as a secondary metabolic byproduct. Cyanides form a soluble complex with metals enhancing the metal solubility. Additionally, these microorganisms can produce organic acids that further dissolve metals. This review underscores the e-waste pollution and the mechanisms by which cyanogenic microorganisms can recover precious metals from e-waste.

Keywords E-waste · Metal recovery · Cyanogenic microbes · Cyanide

D. P. K. Samal · L. B. Sukla (✉)
Biofuels and Bioprocessing Research Center, Institute of Technical Education and Research, Siksha 'O' Anusandhan (Deemed to be University), Bhubaneswar, India
e-mail: dpkrishnasamal@soa.ac.in; lalabeharisukla@soa.ac.in

A. P. Das, S. Priyadarsini (eds.), *Electronic Waste and Environmental Pollution*, Sustainable Environmental Waste Management Strategies,
https://doi.org/10.1007/978-3-031-90287-1_10

1 Introduction

The historic turnaround of the electronic industry has changed modern life for good. It boosts technological innovation along with economic development. Many electrical devices now intertwine with our everyday lives. Computers, televisions, cell phones, and other gadgets have become essential tools for communication, work, and entertainment (Haleem et al., 2022). There has been growth in manufacturing and consuming electronic devices leading to never-before-seen advancements, with technology now growing in accessibility and ingrained in our lives (Priyadarsini & Das, 2025). Outdated models quickly reach obsolete status as new electronic devices keep coming into the markets, creating demand for the newer technologies. However, the other side of this benefits is the increase in electronic waste, which is motorizing an enormous green problem all over the world (Ghulam & Abushammala, 2023). E-waste includes abandoned obsolete non-working or broken electronic devices such as mobile phones, computers, televisions, hard disks, batteries and their accessories (Mishra et al., 2023). Electronic hazardous materials are considered themselves toxic metals like lead and mercury, other hazardous organic compounds (Abalansa et al., 2021; Raj & Das, 2023). They pose a serious threat to the environment if not managed properly. Unfortunately, many of these devices enter the waste stream without the protection of proper protocols. The high toxicity of these metals leads to contamination of air, soil and water posing a serious long term environmental challenges (Twagirayezu et al., 2022; Das & Mishra, 2008; Mishra et al., 2009; Mishra & Das, 2008).

As the worldwide e-waste generation continues to increase, efficient management of e-waste has turned out to be an important concern in recent years. Developing effective recovery techniques that are minimize environmental damage while maximizing resource recovery is imperative. Research on e-waste management has gained momentum, focusing on minimizing environmental damage while reclaiming valuable materials. Due to its high metal content, e-waste is a primary carrier of precious and harmful metals. Various precious metals can be recovered from e-waste, particularly copper, silver, gold, palladium, etc. (Fornalczyk et al., 2013). Traditional metal extraction methods such as mechanical, hydrometallurgical, and pyrometallurgical processes were employed. But, due to their high costs, pollutant emissions, and energy consumption, these methods are not eco-friendly (Izydorczyk et al., 2021; Priyadarsini & Das, 2024). Bio-metallurgical processes are promising because they are eco-friendly, cost-effective, and produce minimal secondary waste. It uses microbes to recover metals from e-wastes (Tezyapar Kara et al., 2023; Das et al., 2011; Ghosh et al., 2016a).

Cyanogenic microbes are the most efficient microbes that help recover precious metals from e-waste. They are producing cyanide ions as a secondary metabolite during their metabolic processes (Adetunji et al., 2023). These microorganisms, including bacteria, fungi, and algae, have been identified as cyanide producers. Cyanidation has been used to extract valuable metals like gold, silver, copper, nickel, etc. Diverse cyanogenic microbes generate HCN. Cyanogenic microbes such as *Chromobacterium violaceum*, *Pseudomonas fluorescens*, *Penicillium simplicissimum* and *Aspergillus niger*, have studied the ability to facilitate metal dissolution from e-waste (Samal et al., 2019; Ji et al., 2022; Adetunji et al., 2023). Although this process is slower than other methods, its reduced ecological impact makes it a potential field for further research.

This chapter explores the challenges of e-waste and its pollution, highlighting its hazardous effects on both the environment and human health. A considerable focus is placed on role of cyanogenic microbes in recovering precious metals from e-waste, particularly cyanogenic microbes. It also focuses on the mechanisms these microbes employ to extract valuable metals efficiently and sustainably (Das et al., 2015b; Ghosh et al., 2016b; Ghosh & Das, 2018).

2 E-Waste

E-waste refers to discarded electronic equipment (EEE) that are non-functional or redundant. Such e-waste includes A broad range of items, including household appliances, telecommunication devices, light fixtures, batteries, electronic tools, circuit boards, toys, cathode-ray tubes, sports and leisure equipment, medical devices, monitoring equipment, and vending machines (Kaya, 2016). The printed circuit board forms a core component in many EEEs containing, among other hazardous substances, a multitude of metals. When streamed into a landfill, most e-waste can release dangerous gases and cause soil and groundwater toxicity by traditional waste incineration (Oke & Potgieter, 2024). Contaminants from e-waste can persist in the environment and be absorbed by the plants and can affect life dependent on resources from those plants.

3 E-Waste Exposure Roots and Pollution

E-waste exposure occurs through various routes. E-waste exposure is a complex process involving several routes of exposure, sources of exposure, and exposures of variable duration (Perkins et al., 2014). Sources of e-waste exposure have included discarded electronic goods, improper disposal, and recycling activities (Annamalai, 2015; Abalansa et al., 2021; Münzel et al., 2023). Some sources of exposure can be classified as informal recycling, formal recycling, and environmental poisoning. Informal recycling is where an unlicensed agent collects discarded electronics for purposes of separating reusable elements with self-invented or primitive techniques (Annamalai, 2015). Formal recycling is where licensed facilities operate equipped with appropriate machinery to disassemble obsolete equipment and sorts salvageable materials. Environmental poisoning is one through which air, soil, and water is polluted to the extent of fatal contact with these particular waste products (Münzel et al., 2023). Depending on their type and duration, exposures can be classed as acute-short-term or long-term chronic (Parvez et al., 2021). Inhalation, ingestion, or dermal contacts are the most likely avenues of exposure to dangerous components of e-waste. Poorly managed e-waste has substantial environmental impacts, including air, soil, water pollution and beach degradation (Das & Mishra, 2009; Paul et al., 2023). In burning, e-waste releases gases and heavy metals that would eventually return toxins to the environment around it, that is water reservoirs for surface and groundwater (Chakraborty et al., 2022). Lead is released into the atmosphere during the incineration of e-waste, along with cadmium and mercury (Prabhakar et al., 2019). When dumped in a landfill, e-waste can leach into the soil and ground water and threaten the topsoil and crop viability (Jain et al., 2023). Water pollution occurs when dumped e-waste seeps into the soil, contaminating groundwater with poisons like arsenic. Another hazard posed to the environment is beach degradation. `E-waste components can be broken down into cathode-ray tubes, broken computer chips, and other electronic components. The release of pollutants such as glass dust, tin, lead, brominated dioxins, beryllium, cadmium, mercury, and chemicals from the stripping and burning of electronic components into the air and rivers can lead to the acidification of aquatic ecosystems, affecting fish and other marine organisms (Ankit et al., 2021).

4 Environmental Toxicity and Health Hazards

Electronic devices consist of toxic metal contaminants such as lead, cadmium, beryllium, and brominated flame retardants. Numerous workers, including children, are involved in e-waste dismantling processes, exposing them to these hazardous materials. Nickel-cadmium from rechargeable batteries can trigger skin and lung allergies (Monika & Kishore, 2010). Arsenic causes skin diseases, lung cancer, and disrupted nerve signaling. Brominated flame retardants interfere with endocrine system functions (Kumar et al., 2020). Lead is being the most prevalent metal in e-waste. Lead can enter living systems mainly through food, air, water, and soil. Primarily, children are more at risk to poisoning of lead. E-waste recycling activities have increased blood lead levels in children (Desye et al., 2023). Exposure to beryllium may result in berylliosis, lung cancer, and skin diseases (Cooper & Harrison, 2009). Mercury can damage the brain, reproductive system, kidneys and blood disorders (Fernandes Azevedo et al., 2012).

5 Precious Metals Present in E-Wastes

E-waste contains precious metals like silver, gold, palladium, platinum, copper, zinc, tin, aluminum, nickel, lead, and iron, as well as base metals like nickel, cobalt, zinc, lead, and iron (Gulliani et al., 2023) (Table 1). These metals are lower than base metals in waste printed circuit boards but higher than other common metals.

Table 1 Different precious metals and additional metals present in e-wastes

E-waste component	Precious metals	Additional valuable materials
Printed circuit boards (PCBs)	Gold, silver, palladium	Copper, lead, nickel, tin, zinc
Battery	Lithium, cobalt	Nickel, cadmium, aluminum
Charger	Gold, silver	Copper, brominated flame retardants
Television	Gold, silver, rare earth elements like europium (Eu), terbium (Tb), yttrium (Y)	Lead, indium
Mobile phone (overall)	Gold, silver, palladium	Copper, nickel, tin, lead
DVD player	Gold, silver	Copper, nickel
Hard disk drives	Gold, silver, platinum, neodymium (Nd), dysprosium (Dy)	Copper, cobalt

Traditionally, precious metals are extracted from minerals through smelting and refining. Mobile phones and personal computers use 3% of the world's gold and silver, 13% for palladium, and 15% for cobalt (Manikandan et al., 2023).

6 Microbes Used in E-Waste Recovery

Microbes have been used for decades to extract metals from various sources, including mine resources and e-waste (Han et al., 2022, (Das et al., 2015a; Das & Ghosh, 2018; Ghosh et al., 2023). Most promising method for recovering metals from ores and e-waste is bioleaching. It involves two main stages: the reaction of organic sugars and acids with the waste materials and the growth of microorganisms (Adetunji et al., 2023). This process typically occurs in an acidic environment. It includes the oxidation of sulfur or its reduced compounds to form sulfuric acid and the production of organic acids through oxygen respiration or carbohydrate fermentation.

Microbes for metal recovery include chemolithotrophic, heterotrophic bacteria, and fungi. Microbes produce many organic complexing agents for instance citric acid, oxalic acid, carboxylic acids and tartaric acid (Das et al., 2012a; Das & Ghosh, 2023). These acids help solubilize metals from e-waste by regulating the redox potential and acidity during the bioleaching process (Das et al., 2012b). Cyanogenic bacteria include *Pseudomonas fluorescens*, *Pseudomonas aeruginosa*, *Chromobacterium violaceum*, *Pseudomonas plecoglossicida*, and *Pseudomonas aeruginosa*. Fungal species like *Aspergillus niger*, *Penicillium chrysogenum*, and yeast have been widely studied for bioleaching metals from industrial e-wastes (Pradhan & Kumar, 2012; Samal et al., 2019; Adetunji et al., 2023). Cyanide is generated by these organisms responsible for precious metal recovery.

7 Cyanogenic Microbes

Cyanogenic microbes typically thrive under moderately thermophilic and alkaline conditions. Certain microorganisms produce cyanide ions during their metabolic processes. Cyanide is produced as a secondary metabolite by the microbes (Zdor, 2015). It produced particularly during the early stationary phase of microbial growth when specific conditions are met. Cyanogenic microbes release cyanide, which dissolves metals by forming soluble complexes (Olaya-Abril et al., 2024).

Several microorganisms, including bacteria, fungi, and algae, have been identified as cyanide producers (Table 2). Cyanogenic bacteria include *Bacillus megaterium, Pseudomonas aeruginosa, Chromobacterium violaceum, Escherichia coli, Pseudomonas plecoglossicida, Rhizobium leguminosarum, Pseudomonas polycolor, Pseudomonas fluorescens, Pseudomonas syringae, Pseudomonas putida,* and *Anacystis nidulans* (Noumi et al., 2018; Zdor, 2015; Samal et al., 2019). Fungus includes *Penicillium chrysogenum*, *Aspergillus niger*, *Marasmius oreades*, *Gloeocercospora sorghi*, *Stemphylium loti*, and *Clitocybe* sp. (Zhai et al., 2023). Some algal species have also been reported to produce cyanide. *Oscillatoria tenuis*, *Chlorella vulgaris*, *Lyngbya majuscule* are most common algal species.

Table 2 Using several cyanogenic species in recovering precious metals

Microorganisms	Metals recovery	Conditions	Recovery efficiency	Reference
Chromobacterium violaceum ATCC 12471	Gold	pH 8.0–11.0, 30 °C, 150 rpm, 15 g/L pulp density	70.6%	Li et al. (2015)
Aspergillus niger	Indium	pH 7.0, 30 °C, 200 rpm, 1% (w/v) indium tin oxide, 15 days	100%	Cui et al. (2021)
Pseudomonas balearica SAE1	Gold, Silver	pH 9.0, 30 °C, 10 g/L pulp density, 5 g/L glycine	Gold: 68.5%, Silver: 33.8%	Kumar et al. (2018a, 2018b)
Pseudomonas putida WSC361	Gold	pH 7.2–8.4, 10 g/L pulp density, 150 rpm, 3 h, 10 g/L glycine	48%	Marra et al. (2018)
Bacillus megaterium PTCC 1656	Platinum, Rhenium	Glycine concentration 12.8 g/L, 4% (w/v) pulp density	Platinum: 15.7%, Rhenium: 98%	Motaghed et al. (2014)
Chromobacterium violaceum	Gold, Copper	pH 8.0–11.0, 30 °C, 150 rpm, 15 g/L pulp density	Gold: 10.8%, Copper: 11.4%	Chi et al. (2011)
Pseudomonas fluorescens CICC21620	Gold	pH 9.0, 0.33% (w/v) pulp density, glycine–methionine ratio 1:10	54%	Li et al. (2020)
Penicillium simplicissimum	Copper, Nickel	pH 7.0 (Cu), pH 2.0 (Ni), 30 °C, 150 rpm	Copper: 90%, Nickel: 89%	Arshadi et al. (2019)

The metabolic activity of cyanide-producing microbes is influenced by several physiological factors, including pH, pulp density, nutrient availability, and the presence of other metal ions, which may act as catalysts (Kumar et al., 2023; Malmir et al., 2022). Sulfide and sulfate compounds are known to react with and deplete cyanide. To enhance precious metal leaching, researchers focus on optimizing microbial growth conditions and maximizing cyanide production. Although cyanogenic microbes are well-documented, the detailed mechanism of hydrogen cyanide (HCN) formation remained elusive for many years.

8 Cyanide Leaching Mechanism by Cyanogenic Microbes

Before the recovery process of metals from e-wastes by bioleaching, pretreatment is required to obtain PCBs that will interact with microorganisms (Argumedo-Delira et al., 2020) (Fig. 1). This usually entails disassembling computers, hard disks, DVD players and mobile phones and reducing their particle size. Microorganisms such as bacteria, fungi, and microalgae play a crucial role in converting contaminants into simpler, less hazardous forms. The interaction between these microorganisms and contaminants involves two key processes: (1) the mobilization of metals from printed circuit boards (PCBs) and (2) the immobilization of these metals after they are solubilized. Bioleaching, a common microbial process, includes mechanisms like acidolysis, redoxolysis, and complexolysis for metal extraction (Naseri et al., 2023). Cyanogenic bacteria and fungi are commonly used for metallic

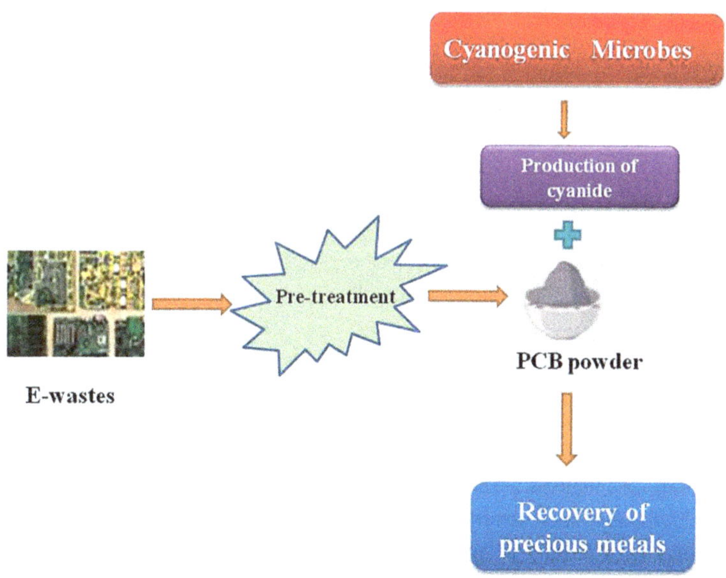

Fig. 1 Precious metals recovery by cyanogenic microbes from e-wastes

recovery of PCBs. Cyanogenic bacteria produce cyanide from glycine. More investigations have carried out by Trivedi et al. (2022) on metal recovery by microbial processes for the treatment of WEEE using a mixture of printed circuit boards (PCBs) from various types of electronic devices.

Biohydrometallurgy has been a promising innovation for recovering precious metals in the last decade. This treatment involves microbial synthesis of the metabolites that enhance metal solubility through soluble metallic complexes. Cyanide has been a cheap reagent in the concentration of precious metals such as silver and gold. Cyanide serves as a more efficient lixiviant of metals (Medina & Anderson, 2020). It is one of the best methods for extracting Gold or Silver from PCBs. Bioleaching of recovering precious metals is completed using cyanogenic microorganisms. Cyanogenic microbes carry out bioleaching of recovering precious metals. Cyanide is produced as secondary metabolite by cyanogenic microbes. Cyanide in the form of hydrocyanic acid is generated by cyanogenic microbes (Faramarzi et al., 2020). Some bacteria produce secondary metabolites through the oxidative decarboxylation of glycine with the HCN synthase enzyme. The most commonly used microbes for this process are *Bacillus megaterium*, *Chromobacterium violaceum*, *Pseudomonas* sp. etc. (Kumar et al., 2023; Adetunji et al., 2023). The concentration cyanide depends upon bacterial multiplication and incubation duration. The concentration reaches a maximum during the late logarithmic and early stationary phases. For optimal bacterial culture and production of cyanide, a chemically defined medium with an ideal carbon source, lipids, vitamins, and amino acids is necessary (Karamba et al., 2018).

Bacteria have been effectively utilized for bioleaching metals from PCBs found in laptop and phones through various methods. These methods include adjusting culture conditions, introducing PCBs in diverse sizes and pulp densities, adding substances to the bioleaching process, pretreating PCBs before bioleaching, and applying physicochemical treatments post-bioleaching (Sikander et al., 2022). To maintain optimal conditions, the culture medium can be refreshed to reinstate the early volume required for bioleaching.

PCBs from electronic devices are usually applied in the form of fragments, powder, or at various pulp densities. The lack of standardized particle size has repeatedly been shown to improve metal solubilization by bacteria, thus providing an area for further research and optimization. Placement of hydrogen peroxide into the bioleaching process has also been proposed as providing metal oxidation and thus improving metal recovery efficiency.

Fungi such as *Aspergillus niger*, *Penicillium simplicissimum*, and *Aspergillus niveus* are able to grow in an environment of low pH, with a low temperature; exposing metal contaminants. These fungi also produce organic acids such as citric acids, oxalic acids, malic acids, and gluconic acids that act as lixiviants for the solubilization of metals in electronic waste (Krishnamoorthy et al., 2021). Moreover, fungi can degrade mineral surfaces by producing protons, which is referred to as acidolysis, allowing the leaching of metals at a lesser acidic pH which, in turn, reduces environmental impact (Nkuna & Matambo, 2024). The application of organic acids during the metal mobilization phase, in the presence of iron-oxidizing bacteria, can increase solubilization of metals while also maintaining dissolved ferric iron, producing an even better result for extracting metals from e-waste.

9 Recovery of Gold and Silver

The process of cyanidation is used for the recovery of both silver and gold. It involves the use of cyanide for its ability to form water-soluble complexes along with metals and thereby facilitate the process of extraction (Kumar et al., 2018a, 2018b). An environmentally safer alternative to time-tested conventional cyanidation techniques is biogenic cyanide, or alkaline bioleaching. This makes use of the natural capabilities of microorganisms to generate hydrocyanic acid.

Cyanide, produced by certain bacteria, serves as a lixiviant for the mobilization of both Silver (Ag) and gold (Au) from waste PCBs (WPCBs). Under alkaline conditions, typically at a pH range of 8–10, cyanide forms water-soluble complexes with the precious metals in WPCBs, facilitating their extraction(Li et al., 2020). At a pH of around 7–8, some cyanide may be lost due to volatilization, such as hydrogen cyanide (Table 2). However, this pH range is considered optimal for both bacteria culture and cyanide generation, balancing the need for effective metal solubilization with the biological requirements of the microorganisms.

Some heterotrophic bacteria, such as *Chromobacterium violaceum*, *Pseudomonas fluorescens*, *Pseudomonas plecoglossicida*, and *Pseudomonas aeruginosa*, are known to produce HCN (Table 2). The enzyme that produces HCN in these organisms is named HCN synthase, which is central to bioleaching in gold recovery (Liu et al., 2016). Chemical dissolution of gold in cyanide solution is described by anodic and cathodic reactions forming Elsner's eq. A pretreatment step is needed to optimize the use of cyanide-producing bacteria for gold extraction from electronic wastes, like printed circuit boards (PCBs). This step involves the application of iron and sulfur oxidizing bacteria to remove interfering metals like copper, typically HCN consumers thereby reducing its gold extraction efficiency. Gold dissolution requires oxygen very much because it is consumed during the microbial growth phase, which may inhibit gold solubilization. For effective conduct of the process, oxygen supply must be managed in bioleaching environment (Khaing et al., 2019).

10 Conclusion

The e-waste industry is a rapidly burgeoning sector of the global economies driven by the high rise in the global use of electronic devices. E-waste can potentially release highly toxic substances into the environment during disposal, posing a considerable health risk to human beings and the environment. Making use of recycling techniques like bioleaching can be an effective solution to counter these challenges. Earth carries an important sustainable way for the reformers, especially bioleaching cyanogenic microorganisms: specific bacteria are known to produce cyanide and organic acids as metabolic byproducts that would enhance the solubilization of metals from electronic devices. This is a green method when compared with the convention extraction technique, as it inhibits environmental damage and enhances the resource recovery rate in consideration.

The growing demand for precious metals, coupled with their limited natural resources, shows the urgency for effective recycling from end-of-life products like e-waste and spent catalysts. These materials accredit for more than 90% of the industrially precious metals and have become valuable secondary resources. Technology and innovation in recycling over the past few years have led to significant improvement in recovery rates while minimizing environmental damages. However, sustainable high-recovery processing and alleviation of informal disposal practices still pose challenges. Thus, efforts in future endeavors should focus on developing eco-friendly, resource-efficient, and less pollutant recycling technologies in support of sustainable development.

Acknowledgments The authors are grateful to Prof. (Dr.) Manojranjan Nayak, President of Siksha 'O' Anusandhan University, for providing infrastructure and encouragement throughout the research process.

Declaration of Competing Interest The authors declare that they have no known competing financial interests or personal relationships that could have appeared to influence the work reported in this paper.

Credit Authorship Contribution Statement D P Krishna Samal: Conceptualization, Writing—original draft, review. Lala Behari Sukla: Supervision, editing.

References

Abalansa, S., El Mahrad, B., Icely, J., & Newton, A. (2021). Electronic waste, an environmental problem exported to developing countries: The good, the bad, and the ugly. *Sustainability, 13*(9), 5302. https://doi.org/10.3390/su13095302

Adetunji, A. I., Oberholster, P. J., & Erasmus, M. (2023). Bioleaching of metals from e-waste using microorganisms: A review. *Minerals, 13*(6), 828. https://doi.org/10.3390/min13060828

Ankit, S., Saha, L., Kumar, V., Tiwari, J., Sweta, Rawat, S., Singh, J., & Bauddh, K. (2021). Electronic waste and their leachates impact on human health and environment: Global ecological threat and management. *Environmental Technology & Innovation, 24*, 102049. https://doi.org/10.1016/j.eti.2021.102049

Annamalai, J. (2015). Occupational health hazards related to informal recycling of e-waste in India: An overview. *Indian Journal Of Occupational And Environmental Medicine, 19*(1), 61–65. https://doi.org/10.4103/0019-5278.157013

Argumedo-Delira, R., Díaz-Martínez, M. E., & Gómez-Martínez, M. J. (2020). Microorganisms and plants in the recovery of metals from the printed circuit boards of computers and cell phones: A mini review. *Metals, 10*(9), 1120. https://doi.org/10.3390/met10091120

Arshadi, M., Yaghmaei, S., & Mousavi, S. M. (2019). Study of plastics elimination in bioleaching of electronic waste using *Acidithiobacillus ferrooxidans*. *International journal of Environmental Science and Technology, 16*, 7113–7126.

Chakraborty, S. C., Qamruzzaman, M., Zaman, M. W. U., Alam, M. M., Hossain, M. D., Pramanik, B. K., Nguyen, L. N., Nghiem, L. D., Ahmed, M. F., Zhou, J. L., Mondal, M. I. H., Hossain, M. A., Johir, M. A. H., Ahmed, M. B., Sithi, J. A., Zargar, M., & Moni, M. A. (2022). Metals in e-waste: Occurrence, fate, impacts and remediation technologies. *Process Safety and Environmental Protection, 162*, 230–252. https://doi.org/10.1016/j.psep.2022.04.011

Chi, T. D., Lee, J. C., Pandey, B. D., Yoo, K., & Jeong, J. (2011). Bioleaching of gold and copper from waste mobile phone PCBs by using a cyanogenic bacterium. *Minerals Engineering, 24*, 1219–1222.

Cooper, R. G., & Harrison, A. P. (2009). The uses and adverse effects of beryllium on health. *Indian Journal Of Occupational And Environmental Medicine, 13*(2), 65–76. https://doi.org/10.4103/0019-5278.55122

Cui, J., Zhu, N., Mao, F., Wu, P., & Dang, Z. (2021). Bioleaching of indium from waste LCD panels by Aspergillus niger: Method optimization and mechanism analysis. *Science of the Total Environment, 790*, 148151.

Das, A. P., & Ghosh, S. (2018). Bioleaching of manganese from mining waste materials. *Materials Today: Proceedings, 5*(1), 2381–2390. https://doi.org/10.1016/J.MATPR.2017.11.459

Das, A. P., & Ghosh, S. (2023). Role of metabolites and metabolomics for understanding of manganese bioleaching. In *Manganese mining microorganisms* (pp. 117–127). https://doi.org/10.1016/B978-0-12-822147-1.00011-X

Das, A. P., & Mishra, S. (2008). Hexavalent chromium (VI): Health hazards and environmental pollutant. *Journal of Environmental Research and Development, 2*(3), 386–392.

Das, A. P., & Mishra, S. (2009). Hexavalent chromium [Cr (VI)]: Yellow water pollution and its remediation. *Sarovar Saurabh ENVIS Newsletter on Wetland Ecosystems, 5*(2).

Das, A. P., Sukla, L. B., Pradhan, N., & Nayak, S. (2011). Manganese biomining: A review. *Bioresource Technology, 102*(16), 7381–7387. https://doi.org/10.1016/J.BIORTECH.2011.05.018

Das, A. P., Sukla, L. B., & Pradhan, N. (2012a). Microbial recovery of manganese using *Staphylococcus epidermidis*. *International Journal of Nonferrous Metallurgy, 2012*(02), 9–12. https://doi.org/10.4236/IJNM.2012.12002

Das, A. P., Swain, S., Panda, S., Pradhan, N., & Sukla, L. B. (2012b). Reductive acid leaching of low grade manganese ores. *Geomaterials, 2012*(04), 70–72. https://doi.org/10.4236/GM.2012.24011

Das, A. P., Ghosh, S., Mohanty, S., & Sukla, L. B. (2015a). Advances in manganese pollution and its bioremediation. In L. B. Sukla, N. Pradhan, S. Panda, & B. K. Mishra (Eds.), *Environmental microbe biotechnology* (1st ed., pp. 313–328). Springer. https://doi.org/10.1007/978-3-319-19018-1_16

Das, A. P., Ghosh, S., Mohanty, S., & Sukla, L. B. (2015b). Consequences of manganese compounds: A review. *Toxicological & Environmental Chemistry, 96*(7), 981–997. https://doi.org/10.1080/02772248.2015.1005428

Desye, B., Tesfaye, A. H., Berihun, G., Ademas, A., & Sewunet, B. (2023). A systematic review of the health effects of lead exposure from electronic waste in children. *Frontiers in Public Health, 11*, 1113561. https://doi.org/10.3389/fpubh.2023.1113561

Faramarzi, M. A., Mogharabi-Manzari, M., & Brandl, H. (2020). Bioleaching of metals from wastes and low-grade sources by HCN-forming microorganisms. *Hydrometallurgy, 191*, 105228. https://doi.org/10.1016/j.hydromet.2019.105228

Fernandes Azevedo, B., Barros Furieri, L., Peçanha, F. M., Wiggers, G. A., Frizera Vassallo, P., Ronacher Simões, M., Fiorim, J., Rossi de Batista, P., Fioresi, M., Rossoni, L., Stefanon, I., Alonso, M. J., Salaices, M., & Valentim Vassallo, D. (2012). Toxic effects of mercury on the cardiovascular and central nervous systems. *Journal of Biomedical Biotechnology, 2012*, 949048. https://doi.org/10.1155/2012/949048

Fornalczyk, A., Willner, J., Francuz, K., & Cebulski, J. (2013). E-waste as a source of valuable metals. *Archives Of Materials Science And Engineering, 63*, 87–92.

Ghosh, S., & Das, A. P. (2018). Metagenomic insights into the microbial diversity in manganese-contaminated mine tailings and their role in biogeochemical cycling of manganese. *Scientific Reports, 8*(1), 1–12. https://doi.org/10.1038/s41598-018-26311-w

Ghosh, S., Mohanty, S., Akcil, A., Sukla, L. B., & Das, A. P. (2016a). A greener approach for resource recycling: Manganese bioleaching. *Chemosphere, 154*, 628–639. https://doi.org/10.1016/J.CHEMOSPHERE.2016.04.028

Ghosh, S., Mohanty, S., Nayak, S., Sukla, L. B., & Das, A. P. (2016b). Molecular identification of indigenous manganese solubilising bacterial biodiversity from manganese mining deposits. *Journal of Basic Microbiology, 56*(3), 254–262. https://doi.org/10.1002/JOBM.201500477

Ghosh, S., Tripathy, B., Dey, S., & Das, A. P. (2023). Comparative investigation of fungal and bacterial manganese biomining mechanisms. In *Manganese mining microorganisms* (pp. 99–115). https://doi.org/10.1016/b978-0-12-822147-1.00016-9

Ghulam, S. T., & Abushammala, H. (2023). Challenges and opportunities in the management of electronic waste and its impact on human health and environment. *Sustainability, 15*(3), 1837. https://doi.org/10.3390/su15031837

Gulliani, S., Volpe, M., Messineo, A., & Volpe, R. (2023). Recovery of metals and valuable chemicals from waste electric and electronic materials: A critical review of existing technologies. *RSC Sustainability, 1*, 1085–1108.

Haleem, A., Javaid, M., Qadri, M. A., & Suman, R. (2022). Understanding the role of digital technologies in education: A review. *Sustainable Operations and Computers, 3*, 275–285. https://doi.org/10.1016/j.susoc.2022.05.004

Han, P., Teo, W. Z., & Yew, W. S. (2022). Biologically engineered microbes for bioremediation of electronic waste: Wayposts, challenges, and future directions. *Engineering Biology, 6*(1), 23–34. https://doi.org/10.1049/enb2.12023

Izydorczyk, G., Mikula, K., Skrzypczak, D., Moustakas, K., Witek-Krowiak, A., & Chojnacka, K. (2021). Potential environmental pollution from copper metallurgy and methods of management. *Environmental Research, 197*, 111050. https://doi.org/10.1016/j.envres.2021.111050

Jain, M., Kumar, D., Chaudhary, J., Kumar, S., Sharma, S., & Verma, A. S. (2023). Review on e-waste management and its impact on the environment and society. *Waste Management Bulletin, 1*(3), 34–44. https://doi.org/10.1016/j.wmb.2023.06.004

Ji, X., Yang, M., Wan, A., Yu, S., & Yao, Z. (2022). Bioleaching of typical electronic waste-printed circuit boards (WPCBs): A short review. *International Journal of Environmental Research and Public Health, 19*(12), 7508. https://doi.org/10.3390/ijerph19127508

Karamba, K. I., Ahmad, S. A., Zulkharnain, A., Yasid, N. A., Ibrahim, S., & Shukor, M. Y. (2018). Batch growth kinetic studies of locally isolated cyanide-degrading Serratia marcescens strain AQ07. *3 Biotech, 8*(1), 11. https://doi.org/10.1007/s13205-017-1025-x

Kaya, M. (2016). Recovery of metals and nonmetals from electronic waste by physical and chemical recycling processes. *Waste Management, 57*, 64–90. https://doi.org/10.1016/j.wasman.2016.08.004

Khaing, S. Y., Sugai, Y., & Sasaki, K. (2019). Gold dissolution from ore with iodide-oxidising bacteria. *Scientific Reports, 9*, 4178. https://doi.org/10.1038/s41598-019-41004-8

Krishnamoorthy, S., Ramakrishnan, G., & Dhandapani, B. (2021). Recovery of valuable metals from waste printed circuit boards using organic acids synthesised by Aspergillus niveus. *IET Nanobiotechnology, 15*(2), 212–220. https://doi.org/10.1049/nbt2.12001

Kumar, A., Saini, H. S., & Kumar, S. (2018a). Bioleaching of gold and silver from waste printed circuit boards by Pseudomonas balearica SAE1 isolated from an e-waste recycling facility. *Current Microbiology, 75*, 194–201.

Kumar, A., Saini, H. S., & Kumar, S. (2018b). Enhancement of gold and silver recovery from discarded computer printed circuit boards by *Pseudomonas balearica* SAE1 using response surface methodology (RSM). *3 Biotech, 8*(2), 100. https://doi.org/10.1007/s13205-018-1129-y

Kumar, M., Sarma, D. K., Shubham, S., Kumawat, M., Verma, V., Prakash, A., & Tiwari, R. (2020). Environmental endocrine-disrupting chemical exposure: Role in non-communicable diseases. *Frontiers in Public Health, 8*, 553850. https://doi.org/10.3389/fpubh.2020.553850

Kumar, A., Shemi, A., Chipise, L., Moodley, S., Yah, C.S. & Ndlovu, S. (2023). Can microbial Bio-CN be a sustainable alternative to the chemical cyanidation of precious metals? An update and way forward. *Renewable and Sustainable Energy Reviews, 188*, 113892. https://doi.org/10.1016/j.rser.2023.113892

Li, J., Liang, C., & Ma, C. (2015). Bioleaching of gold from waste printed circuit boards by Chromobacterium violaceum. *Journal of Material Cycles and Waste Management, 17*, 529–539.

Li, J., Wen, J., Guo, Y., An, N., Liang, C., & Ge, Z. (2020). Bioleaching of gold from waste printed circuit boards by alkali-tolerant Pseudomonas fluorescens. *Hydrometallurgy, 194*, 105260. https://doi.org/10.1016/j.hydromet.2020.105260

Liu, R., Li, J., & Ge, Z. (2016). Review on *Chromobacterium violaceum* for gold bioleaching from e-waste. *Procedia Environmental Sciences, 31*, 947–953.

Malmir, N., Fard, N. A., Aminzadeh, S., Moghaddassi-Jahromi, Z., & Mekuto, L. (2022). An overview of emerging cyanide bioremediation methods. *Processes, 10*(9), 1724. https://doi.org/10.3390/pr10091724

Manikandan, S., Inbakandan, D., Nachiyar, C. V., & Raja Namasivayam, S. K. (2023). Towards sustainable metal recovery from e-waste: A mini review. *Sustainable Chemistry for the Environment, 2*, 100001. https://doi.org/10.1016/j.scenv.2023.100001

Marra, A., Cesaro, A., Rene, E. R., Belgiorno, V., & Lens, P. N. L. (2018). Bioleaching of metals from WEEE shredding dust. *Journal of Environmental Management, 210*, 180–190.

Medina, D., & Anderson, C. G. (2020). A review of the cyanidation treatment of copper-gold ores and concentrates. *Metals, 10*(7), 897. https://doi.org/10.3390/met10070897

Mishra, S., & Das, A. P. (2008). Hexavalent chromium reduction and 16S rDNA identification of bacteria isolated from a Cr (VI) contaminated site. *The Internet Journal of Microbiology, 7*(1).

Mishra, S., Das, A. P., & Seragadam, P. (2009). Microbial remediation of hexavalent chromium from chromite contaminated mines of Sukinda Valley, Orissa (India). *Journal of Environmental Research and Development*, 1122–1127.

Mishra, S., Ghosh, S., van Hullebusch, E. D., Singh, S., & Das, A. P. (2023). A critical review on the recovery of base and critical elements from electronic waste-contaminated streams using microbial biotechnology. *Applied Biochemistry and Biotechnology, 2023*, 1–30. https://doi.org/10.1007/S12010-023-04440-X

Monika, & Kishore, J. (2010). E-waste management: As a challenge to public health in India. *Indian Journal of Community Medicine, 35*(3), 382–385. https://doi.org/10.4103/0970-0218.69251

Motaghed, M., Mousavi, S. M., Rastegar, S. O., & Shojaosadati, S. A. (2014). Platinum and rhenium extraction from a spent refinery catalyst using Bacillus megaterium as a cyanogenic bacterium: Statistical modeling and process optimization. *Bioresource Technology, 171*, 401–409.

Münzel, T., Hahad, O., Daiber, A., & Landrigan, P. J. (2023). Soil and water pollution and human health: What should cardiologists worry about? *Cardiovascular Research, 119*(2), 440–449. https://doi.org/10.1093/cvr/cvac082

Naseri, T., Beiki, V., Mousavi, S. M., & Farnaud, S. (2023). A comprehensive review of bioleaching optimization by statistical approaches: Recycling mechanisms, factors affecting, challenges, and sustainability. *RSC Advances, 13*(34), 23570–23589. https://doi.org/10.1039/d3ra03498d

Nkuna, R., & Matambo, T. S. (2024). Determining the metabolic processes of metal-tolerant fungi isolated from mine tailings for bioleaching. *Minerals, 14*(3), 235. https://doi.org/10.3390/min14030235

Noumi, E., Merghni, A., Alreshidi, M., Haddad, O., Akmadar, G., De Martino, L., Mastouri, M., Ceylan, O., Snoussi, M., Al-Sieni, A., & De Feo, V. (2018). *Chromobacterium violaceum* and *Pseudomonas aeruginosa* PAO1: Models for evaluating anti-quorum sensing activity of *Melaleuca alternifolia* essential oil and its main component terpinen-4-ol. *Molecules, 23*(10), 2672. https://doi.org/10.3390/molecules23102672

Oke, E., & Potgieter, J. H. (2024). Discarded e-waste/printed circuit boards: A review of their recent methods of disassembly, sorting, and environmental implications. *Journal of Material Cycles and Waste Management, 26*, 1277. https://doi.org/10.1007/s10163-024-01917-7

Olaya-Abril, A., Biełło, K., Rodríguez-Caballero, G., Cabello, P., Sáez, L. P., Moreno-Vivián, C., Luque-Almagro, V. M., & Roldán, M. D. (2024). Bacterial tolerance and detoxification of cyanide, arsenic, and heavy metals: Holistic approaches applied to bioremediation of industrial complex wastes. *Microbial Biotechnology, 17*.

Parvez, S. M., Jahan, F., Brune, M. N., Gorman, J. F., Rahman, M. J., Carpenter, D., Islam, Z., Rahman, M., Aich, N., Knibbs, L. D., & Sly, P. D. (2021). Health consequences of exposure to e-waste: An updated systematic review. *Lancet Planetary Health, 5*(12), 905–920. https://doi.org/10.1016/S2542-5196(21)00263-1

Paul, A., Dey, S., Ram, D. K., & Das, A. P. (2023). Hexavalent chromium pollution and its sustainable management through bioremediation. *Geomicrobiology Journal, 41*, 1–11. https://doi.org/10.1080/01490451.2023.2218377

Perkins, D. N., Brune Drisse, M. N., Nxele, T., & Sly, P. D. (2014). E-waste: A global hazard. *Annals of Global Health, 80*(4), 286–295. https://doi.org/10.1016/j.aogh.2014.10.001

Prabhakar, A., Mishra, S., & Das, A. P. (2019). Isolation and identification of lead (Pb) solubilizing bacteria from automobile waste and its potential for recovery of lead from end of life waste batteries. *Geomicrobiological Journal, 36*(10), 894–903. https://doi.org/10.1080/01490451.2019.1654044

Pradhan, J. K., & Kumar, S. (2012). Metals bioleaching from electronic waste by *Chromobacterium violaceum* and *Pseudomonads* sp. *Waste Management & Research, 30*(11), 1151–1159. https://doi.org/10.1177/0734242X12437565

Priyadarsini, S., & Das, A. P. (2024). Analytical and structural characterization of waste lithium-ion batteries for their effective recycling strategy. *Environmental Chemistry and Ecotoxicology, 7*, 182. https://doi.org/10.1016/j.enceco.2024.12.004

Priyadarsini, S., & Das, A. P. (2025). Lithium bioleaching: A review on microbial-assisted sustainable technology for lithium bio-circularity. *Journal of Water Process Engineering, 69*, 106744. https://doi.org/10.1016/j.jwpe.2024.106744

Raj, K., & Das, A. P. (2023). Lead pollution: Impact on environment and human health and approach for a sustainable solution. *Environmental Chemistry and Ecotoxicology, 5*, 79–85. https://doi.org/10.1016/J.ENCECO.2023.02.001

Samal, D. P. K., Sukla, L., Pattanaik, A., & Pradhan, D. (2019). Extraction of gold from electronic scraps: A biohydrometallurgical process, overview. *Biointerface Research in Applied Chemistry, 9*, 362–367. https://doi.org/10.33263/BRIAC95.362367

Sikander, A., Kelly, S., Kuchta, K., Sievers, A., Willner, T., & Hursthouse, A. S. (2022). Chemical and microbial leaching of valuable metals from PCBs and tantalum capacitors of spent mobile phones. *International Journal of Environmental Research and Public Health, 19*(16), 10006. https://doi.org/10.3390/ijerph191610006

Tezyapar Kara, I., Kremser, K., Wagland, S. T., et al. (2023). Bioleaching metal-bearing wastes and by-products for resource recovery: A review. *Environmental Chemistry Letters, 21*, 3329–3350. https://doi.org/10.1007/s10311-023-01611-4

Trivedi, A., Vishwakarma, A., Saawarn, B., Mahanty, B., & Hait, S. (2022). Fungal biotechnology for urban mining of metals from waste printed circuit boards: A review. *Journal of Environmental Management, 323*, 116133.

Twagirayezu, G., Irumva, O., Huang, K., Xia, H., Uwimana, A., Nizeyimana, J. C., Manzi, H. P., Nambajemariya, F., & Itangishaka, A. C. (2022). Environmental effects of electrical and electronic waste on water and soil: A review. *Polish Journal of Environmental Studies, 31*(3), 2507–2525. https://doi.org/10.15244/pjoes/144194

Zdor, R. E. (2015). Bacterial cyanogenesis: Impact on biotic interactions. *Journal of Applied Microbiology, 118*(2), 267–274. https://doi.org/10.1111/jam.12697

Zhai, J., Wang, B., Sun, Y., Yang, J., Zhou, J., Wang, T., Zhang, W., Qi, C., & Guo, Y. (2023). Effects of *Aspergillus niger* on cyanogenic glycosides removal and fermentation qualities of ratooning sorghum. *Frontiers in Microbiology, 14*, 1128057. https://doi.org/10.3389/fmicb.2023.1128057

Bioleaching of Heavy Metals from Electronic Waste

Santoshi Mohanta, Lipika Parida, Apurba Sarkar, and Don Adrian Meshan Dassanayake

Abstract Electronic waste has become a rapidly growing environmental issue due to its hazardous content, particularly heavy metals. Current waste disposal practices of landfilling and incineration generate apparent environmental impacts since they emit poisonous emissions that contaminate the surroundings. The chapter describes how bioleaching provides an ecologically beneficial way for recovering metals from discarded electronic items. Several species of microbes like *Chromobacterium violaceum, Acidithiobacillus ferrooxidans, Leptospirillum ferrooxidans,* and *Aspergillus niger* employ their metabolic processes to transform heavy metals into water-soluble elements. The chapter analyzes how microorganisms operate to release metals and compares how chemolithoautotrophic and heterotrophic mechanisms affect metal extraction alongside fungal applications. Research focuses on how certain bioleaching parameters such as pH levels, temperature, and nutrition delivery influence metal extraction effectiveness. This chapter addresses all environmental dangers of e-waste exposure throughout the bioleaching process and recommends solutions to decrease such hazards. It also illustrates how bioleaching might assist circular economy projects by converting trash into useful products.

S. Mohanta
School of Chemical Engineering, KIIT Deemed University, Bhubaneswar, Odisha, India
e-mail: santoshi.mohanta@kiitbiotech.ac.in

L. Parida (✉)
Department of Chemical Engineering, Veer Surendra Sai University of Technology, Sambalpur, Odisha, India
e-mail: lparida_chemical@vssut.ac.in

A. Sarkar
Apurba Sarkar, School of Chemical Engineering, KIIT Deemed University, Bhubaneswar, Odisha, India
e-mail: 232410@biotech.kiit.ac.in

D. A. M. Dassanayake
Don Adrian Meshan Dassanayake, School of Chemical Engineering, KIIT Deemed University, Bhubaneswar, Odisha, India
e-mail: 2324008@biotech.kiit.ac.in

A. P. Das, S. Priyadarsini (eds.), *Electronic Waste and Environmental Pollution*, Sustainable Environmental Waste Management Strategies,
https://doi.org/10.1007/978-3-031-90287-1_11

This chapter displays recent research findings to indicate the way bioleaching may become a feasible alternative for e-waste management and heavy metal pollution mitigation.

Keywords Electronic waste · Heavy metals · Bioleaching · Microbial mechanisms · Sustainable recycling

1 Introduction

New technological breakthroughs generate demand for computers, cellphones, energy appliances, and industrial equipment which boost living quality today. The fast technological changes and brief lifetime make most gadgets step into the e-waste category due to their usable service (Priyadarsini & Das, 2024). These discarded electronic parts, known as electronic waste (e-waste), have arisen as a global problem due to their rising number and the health risks associated with inappropriate management, disposal, and recycling (Forti & Freitas, 2020; Priyadarsini & Das, 2025). The Global E-Waste Monitor estimates that 53.6 million metric tons of e-waste were produced worldwide in 2019, and projections indicate that this number might increase to 74 million metric tons by 2030. Nevertheless, only around 9.3 metric tons of e-waste—or 17.4% of the total amount generated—have been collected and processed (Al-Salem et al., 2022).

E-waste includes hazardous and/or nonhazardous things such as plastic, glass, flame retardants, and other ferrous and nonferrous metals (Kudpeng et al., 2020; Mishra et al., 2023). According to the presence of metal components, these are classified into six subgroups as shown in Fig. 1 (Balde et al., 2017). Additionally, compared to certain ores, e-waste has higher concentrations of base metals, rare earth metals (REMs), and other valuable metals. The metals Au, Ag, Pb, and Pt are also crucial (Prabhakar et al., 2019; Raj & Das, 2023). Because of their excellent conductivity and chemical resistance, these metals are extensively used in many different sectors, including electronics, automotive, jewelry, dentistry, and many more.

Among other components, printed circuit boards (PCBs) are most common in almost all devices, and they contain a very high percentage of precious metals, such as the central processing unit (CPU) of computers. As reported by Nasiri et al., composition of mobile PCB is shown in Fig. 2. While some of the metals are hazardous and toxic to the environment, requiring expert and careful disposal and handling, most of the metals used are very valuable and can be recycled, which includes gold, copper, and silver (Das et al., 2011; Mishra et al., 2023). Among the metals, cadmium, mercury, and lead can be considered as most toxic components and need proper treatment before safe disposal. Similarly, cathode ray tubes and solders used in most old monitors and TVs contain lead, while batteries and semiconductors contain cadmium. Switches and lighting devices often contain mercury metal or mercury vapor. Inappropriate treatment of this e-waste has been reported to provide several dangers, including neurotoxicity, carcinogenicity, and bioaccumulation in

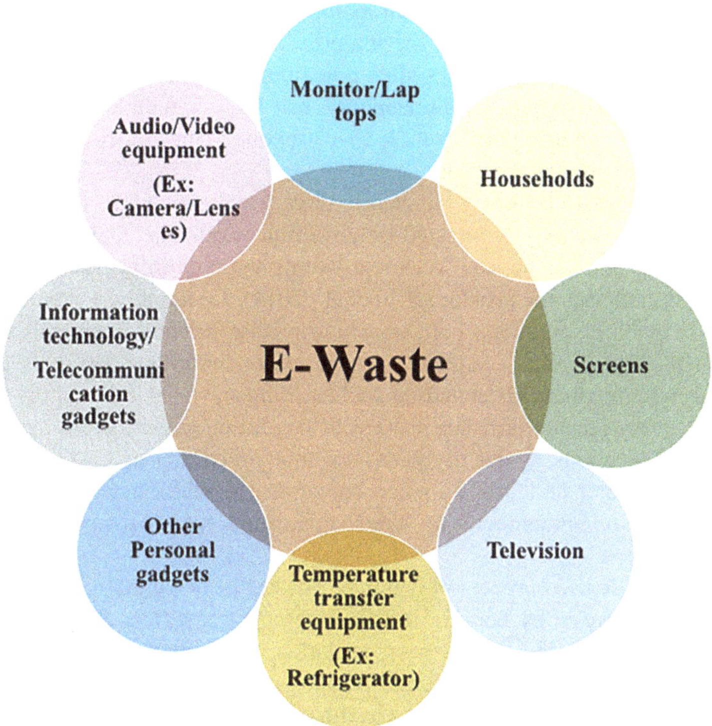

Fig. 1 Different types of e-waste

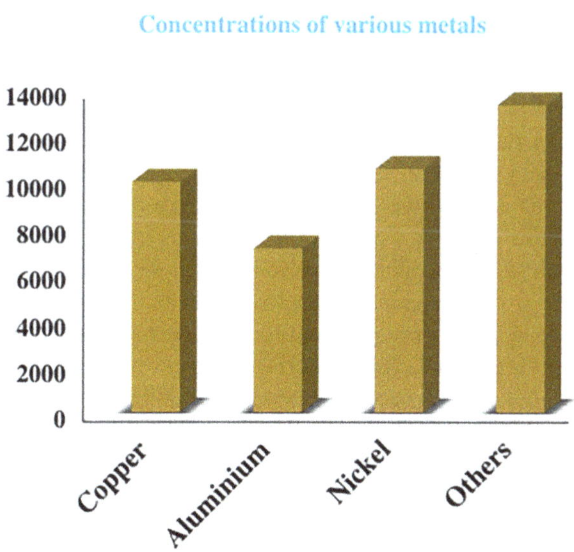

Fig. 2 Concentration of metals in ppm found in mobile phone WPCBs (Nasiri et al., 2022)

food chains. Batteries and semiconductors have cadmium, while solders and cathode ray tubes, which are found in the majority of old TVs and monitors, include lead. Switches and lighting devices often contain mercury metal or mercury vapor. Risks such as neurotoxicity, carcinogenicity, and bioaccumulation in food chains often result from inappropriately handled e-waste due to the ability of the metals to get absorbed into air, water, or soil.

Household e-waste is mostly handled in three ways worldwide: (i) by throwing it in a bin; (ii) by having it collected by government officials; and (iii) by private garbage dealers and businesses. Weak regulations and poor infrastructure in developing nations enhance the problem (Chi et al., 2014). Commonly, e-waste is stored in landfills or burnt outdoors, both of which release harmful chemicals into the environment. Furthermore, workers are exposed to hazardous materials through unofficial recycling methods including acid leaching and hand disassembly without the proper safety precautions. The process of landfilling entails disposing of waste in designated locations, frequently with poor safety measures against leachate pollution. Groundwater and soil are often contaminated with hazardous substances from e-waste, making them unfit for human consumption and causing life threats to animal life.

Another widely used process of handling e-waste is incineration, which reduces the amount of e-waste by burning it at high temperatures (Das & Ghosh, 2023a, 2023b). Despite reducing landfill overpopulation, this process has several serious disadvantages that include dioxins, furans, and heavy metal particles released during burning that seriously endanger the health of those in the proximity and contribute to air pollution Dey et al., 2023; Ghosh et al., 2016a, 2016b; Ghosh & Das, 2017; Mohanty et al., 2017). Moreover, burning causes the valuable metals, which could eventually be recycled, to be lost due to combustion.

Bioleaching has developed into a potential alternative to sustainable e-waste handling considering the current circumstances and challenges. Bioleaching stood out to be an environmentally acceptable way of extracting valuable metals from e-waste by using bioprocesses like manipulating the metabolic activities of microorganisms such as fungi, bacteria, etc. (Abdel Azim et al., 2023). As shown in Fig. 3, microbes help solubilize metals present in e-waste. According to Dutta et al. (2023), this method uses microbial activity to transform insoluble metallic compounds found in ores into soluble form.

Bioleaching methods can extract numerous metals like gold, copper, and nickel from complicated waste materials. The adoption of bioleaching technology helps shift e-waste management toward closed-loop economic principles. By applying biotechnology to harvest and reuse materials from e-waste, bioleaching helps cut down on new resource exploitation while also minimizing environmental effect. Leaching processes based on organisms function effectively at various scale levels and adapt to the evolving trends of e-waste materials.

Here, we explore the ways in which bioleaching offers an environmentally sustainable method of effectively handling e-waste. We begin by tackling both the health and environmental concerns linked to e-waste before contemplating

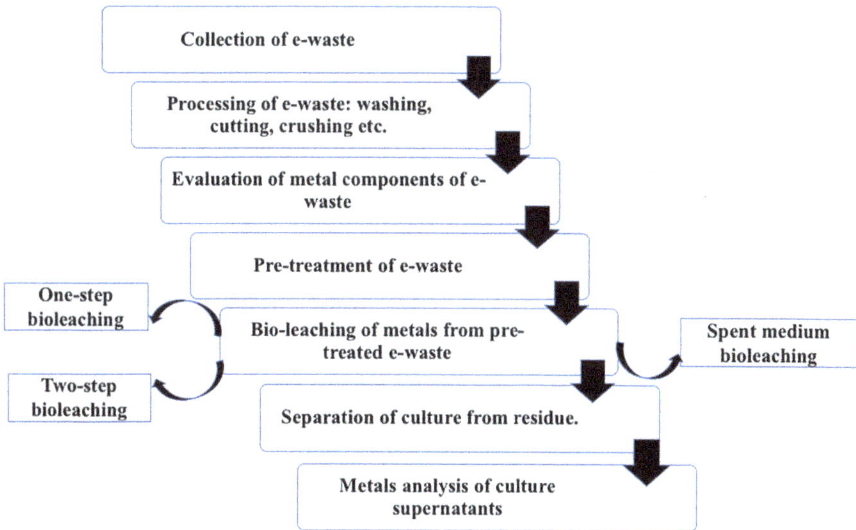

Fig. 3 Procedure of bioleaching of metals from e-waste (Abdel Azim et al., 2023)

bioleaching. This chapter outlines the potential integration of bioleaching into global e-waste management systems to support sustainability and future existence.

2 Health and Environmental Risks of E-Waste

About 2–5% of all solid waste products generated by volume is made up of e-waste. This contributes to an extremely high level of toxicity—more than 70%. This is primarily due to inappropriate e-waste handling, which has a variety of negative consequences on the environment, including soil pollution, groundwater contamination, and air pollution caused by toxic fumes. This has immediate implications for the health of living things (Fig. 4) (Rautela et al., 2021; Ahirwar & Tripathi, 2021). Through their interactions with soil, air, and water, toxins such as palladium, beryllium, chromium, lead, and others contaminate our food chain. They cause serious health issues in the human body on direct consumption. Furthermore, phthalates, ester benzenes, dioxins, hydrocarbons, etc., are harmful to humans as well as to nearby plants, soil, and microorganisms (Zhang et al., 2015).

Recent studies have highlighted the effects of e-waste leachates on soil microorganisms (such as bacteria and fungi) as well as other soil-dwelling species such as aquatic life, plants, and even humans (Purchase et al., 2020; Bao et al., 2020). Mercury, for instance, plays a crucial role in components like electronic switches, batteries, and display screens. Soil microbes can convert mercury into methylmercury, a neurotoxic compound that can accumulate in the fatty tissues of humans and other animals, posing significant health risks (Wang et al., 2020). Children are

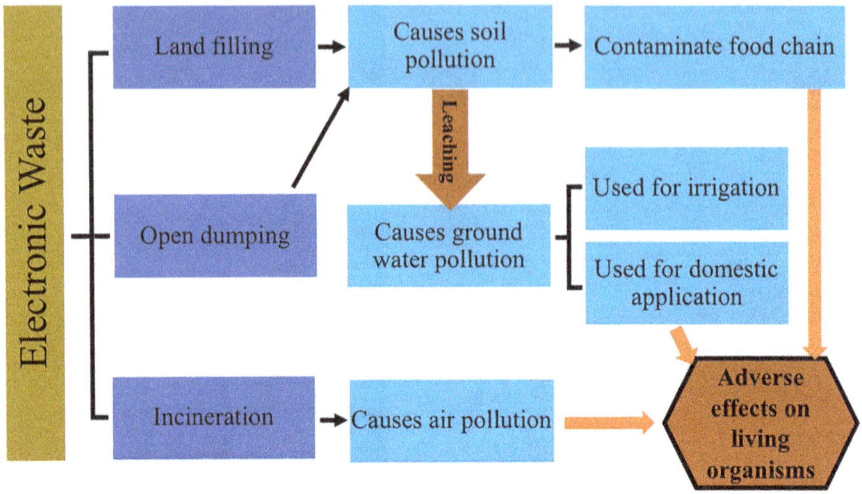

Fig. 4 Harmful effects of e-waste on living beings

getting exposed to the pollutants from a very young age, even during pregnancy also the fetus might get affected significantly. This may result in the generation of asthma, poor immune response, and on prolonged exposure, it may eventually result in neurological disease, and multiple sclerosis muscular dystrophy (Li et al., 2017; Mohod & Dhote, 2013).

E-waste contaminants can simply leach into water bodies, posing a threat to aquatic organisms by disrupting growth and reproduction. Interestingly, the abundance of fungi appears to be inversely related to e-waste pollution, showing a clear link between pollution levels and the health of these organisms. This relationship highlights how environmental contamination directly impacts ecosystem dynamics (Sanusi, 2016). Additionally, water quality varies greatly due to environmental pollution, which affects COD, BOD, and pH levels. According to Javed and Usmani (2011), heavy metals introduced into water systems via pollution accumulate and increase their concentration in aquatic species.

3 Bioleaching: A Sustainable Alternative

The fundamental metal recovery processes of hydrometallurgy and pyrometallurgy have major consequences for both the environment and the economy. According to Hocheng et al. (2017), pyrometallurgical high-temperature processing uses a significant amount of energy and creates hazardous emissions that pollute the environment. Hydrometallurgy enables the recovery of metals from solid waste, but it has negative environmental repercussions and produces a large amount of waste that must be properly treated before disposal (Chauhan et al., 2020). These solutions need more energy and are more expensive, while also being harmful to the environment. Bioleaching provides low-cost and environmentally friendly metal extraction technologies. Bioleaching makes use of bacteria and fungi, which are basic microorganisms that dissolve metals by breaking down waste materials or assisting with metal transformations (Gavrilescu, 2022). The process uses little power and works at low temperatures, requiring just small investments to produce metals in safer, more environmentally friendly ways (Nayak et al., 2024).

The effectiveness of bioleaching is affected by several operating conditions, including the mixture of solids to liquid at different pH levels and oxygen supply. Bioleaching works in a variety of systems of various sizes in laboratory research and industrial operations, such as heap, landfill, and vat reactors. According to Vítová and Mezricky (2024), bioleaching helps separate metals from complex waste products, such as fly ash, sewage sludge, galvanic sludge, e-waste, old batteries, medical waste, and low-quality ores, as demonstrated by Vítová and Mezricky (2024), which contain trace amounts of metals or pollutants that make standard treatment methods ineffective. Using bioleaching processes proves a practical approach to extract metals that is better for the environment than traditional heat and water processes. Industries are progressively adopting it as a more ecologically and economically feasible technology of metal recovery, especially as concerns about environmental pollution and energy consumption rise (Nayak et al., 2024).

4 Microorganisms Involved in Bioleaching

4.1 *Thiobacillus*

Thiobacillus bacteria play an important role in the bioleaching process. These bacteria are Gram-negative, nonspore-forming rods that live in highly acidic and aerobic conditions. According to Dey et al. (2023), most thiobacilli are chemolithoautotrophs, which means they get energy by oxidizing sulfur compounds while absorbing carbon from atmospheric CO_2. According to Bosecker (1997), Thiobacilli bacteria consume sulfides and thiosulfate to produce elemental sulfur before metabolic oxidation yields sulfate. These organisms thrive in environments that keep metal ions dissolved when exposed to acidic conditions, making them important for bioleaching activities.

The bacterium *Thiobacillus ferrooxidans* was initially discovered in 1947 and acts as a vital aspect of bioleaching because it leverages the presence of ferrous iron to supply energy while growing without oxygen on sulfur-containing materials. It breaks down rocks to release acid-soluble metal compounds in the form of sulfate according to Bosecker (1997). A recent scientific study identified two novel acidophilic thiobacilli strains that excel at specialized tasks: *Thiobacillus prosperus*, which flourishes in salty settings and unlocks metals there, and *Thiobacillus cuprinus*, which removes copper from chalcopyrite specifically.

Thiooxidans and *acidithiobacillus ferrooxidans* are of great interest to scientists due to their potential as chemolithoautotrophic bacteria for the basic recovery of metals. These bacteria are crucial to modern biotechnological procedures that recover natural resources because of their unique metabolic pathways, which allow them to live in very acidic environments.

4.2 Leptospirillum

The acid-resistant bacteria *Leptospirillum ferrooxidans* work as the main oxidant of ferrous iron in bioleaching. Scientists initially identified *L. ferrooxidans* in Armenian mine streams; thus, it thrives at low pH and withstands increased uranium, molybdenum, and silver concentrations. Although it thrives in harsh conditions, *L. ferrooxidans* decomposes quickly when exposed to copper. Due to its inability to decompose sulfur and sulfur-based components, *L. ferrooxidans* has limited ability to dissolve mineral sulfides. Combining with sulfur-oxidizing bacteria like *T. ferrooxidans* or *T. thiooxidans* helps *L. ferrooxidans* dissolve metal sulfides more effectively in bioleaching processes (Kumari & Natarajan, 2002). *L. ferrooxidans, T. ferrooxidans, and T. thiooxidans* are all mesophilic bacteria that thrive at temperatures between 25 and 35 °C, making them appropriate for a range of bioleaching settings (Dey, 2023). This synergy and adaptability highlight the usefulness of *L. ferrooxidans* in bio-hydrometallurgical processes.

4.3 Thermophilic Bacteria

Thermophilic and very thermophilic bacteria are required for bioleaching, which allows metal extraction at high temperatures. Thiobacillus-like bacteria are moderately thermophilic, thriving at roughly 50 °C and oxidizing ferrous iron while growing on minerals such as pyrite, pentlandite, and chalcopyrite, albeit yeast extract is required for growth. In contrast, very thermophilic species such as *Acidianusbrierleyi*, *Sulfolobus*, and *Sulfobacillus thermosulfidooxidans* may function at temperatures beyond 60 °C. *Acidianusbrierleyi* is a chemolithoautotrophic archaeon that oxidizes metal sulfides, elemental sulfur, and ferrous iron. Under anaerobic circumstances, it converts sulfur to hydrogen sulfide (H_2S). Similarly, sulfolobus species, both aerobic and facultatively chemolithotrophic, oxidize sulfide minerals, sulfur and ferrous iron. A spore-forming bacterium, *Sulfobacillus thermosulfidooxidans*, also oxidizes these chemicals but, like Th-bacteria, needs yeast extract to thrive (Bosecker, 1997; Dey, 2023). Collectively, these microorganisms drive the biochemical oxidation of insoluble metal sulfides into water-soluble metal sulfates, which are an important step in bioleaching (Nayak et al., 2024). Their capacity to flourish in high-temperature, harsh conditions broadens the scope of bioleaching to geothermally active and deep subterranean mining operations, making them critical for long-term and effective metal recovery.

4.4 *Heterotrophic Microorganisms*

Heterotrophic microorganisms, such as bacteria and fungi, have a significant role in metal leaching by consuming organic nutrients for development and energy. Unlike autotrophs, these organisms get no direct benefit from the leaching process. Metal solubilization by heterotrophs is accomplished by methods such as enzymatic reduction of oxidized metal compounds or the generation of organic acids such as gluconic, citric, oxalic, and lactic acids. These acids, along with phenolic derivatives and other chemicals with hydrophilic reactive groups, dissolve metals by removing metal ions from the ore matrix and producing soluble metal complexes and chelates (Anjum et al., 2012). Members of Bacillus are highly effective, whereas the most significant contributions from fungi include *Penicillium* and *Aspergillus*. Heterotrophic microorganisms have shown promise in e-waste biomining, providing benefits such as improved metal dissolution under less-oxidative conditions, elimination of the need for sulfur or iron supplements, utilization of organic waste for economic and environmental benefits, production of biodegradable and nontoxic leaching agents, overproduction of organic acids in the presence of heavy metals, and reduced corrosiveness of bioleaching media. However, limits include susceptibility to contamination at near-neutral pH and the lack of regenerating oxidants such as the Fe^{2+}/Fe^{3+} cycle, which are seen in autotrophs. These characteristics make heterotrophic bioleaching a desirable, if complicated, method of metal extraction, notably in sustainable e-waste recycling.

5 Microbial Metal Mobilization Mechanisms

Bioleaching microorganisms include both chemolithotrophs that break down inorganic materials and chemoorganotrophs that metabolize reduced organic compounds. Living at low pH conditions helps acidophiles grow as chemolithotrophs. The microbes grow best at specified temperatures, which allows scientists to define them as mesophilic, moderate thermophilic, or thermophilic. Chemolithotrophs are categorized into two types: the group chemo-lithotrophic autotrophs derive energy from inorganic materials, whereas chemo-lithotrophic heterotrophs require organic carbon as their source of energy. In bioleaching processes, the majority of the microorganisms are acidophiles. They are more often used by scientists than cyanogenic bacteria or fungi.

5.1 Acidophilic Bioleaching

Acidophilic bioleaching utilizes acidophilic microorganisms, mostly bacteria and some fungi, to remove metals from e-waste. These bacteria flourish in acidic settings and accelerate mechanisms that solubilize metals for recovery. Key microorganisms participating in bioleaching include *Acidithiobacillus ferrooxidans*, *Leptospirillum ferrooxidans*, *Leptospirillum ferriphilum*, *Thiobacillus thioparus*, *Aspergillus niger*, and *Penicillium* species. The two principal pathways for solubilizing metals are the thiosulfate pathway (e.g., for pyrite) and the polysulfide pathway (e.g., for chalcopyrite, arsenopyrite, sphalerite) (Rawlings, 2004; Kara et al., 2023).

5.1.1 Thiosulfate Pathway

1. Oxidation of ferrous iron to ferric iron:

$$2Fe_2^{+} + 0.5O_2 + 2H^{+} \rightarrow 2Fe^{3+} + H_2O \tag{1}$$

2. Pyrite oxidation to thiosulfate:

$$FeS_2 + 6Fe^{3+} + 3H_2O \rightarrow S_2O_3^{2-} + 7Fe^{2+} + 6H^{+} \tag{2}$$

3. Thiosulfate oxidation:

$$S_2O_3^{2-} + 8Fe^{3+} + 5H_2O \rightarrow 2SO_4^{2-} + 8Fe^{2+} + 10H^{+} \tag{3}$$

5.1.2 Polysulfide Pathway

1. Reaction with ferric iron and acid:

$$MS + Fe^{3+} + H^{+} \rightarrow M^{2+} + 0.5H_2S_n + 2Fe^{2+} \quad (n \geq 2) \tag{4}$$

2. Polysulfide oxidation:

$$0.5H_2S_n + Fe^{3+} \rightarrow 0.125S_8 + Fe^{2+} + H^{+} \tag{5}$$

3. Elemental sulfur oxidation:

$$0.125S_8 + 1.5O_2 + H_2O \rightarrow SO_4^{2-} + 2H^{+} \tag{6}$$

Ferrooxidans have been utilized in different experiments to recover metals from waste materials. Benzal et al. (2020) found that copper bioleaching from PCBs from discarded mobile phones resulted in a 95–100% recovery rate in 48 h. Chen et al.

(2015) also reported 95% copper recovery after 28 days. The bioleaching mechanism may be expressed as follows:

$$Cu^0 + 2Fe^{3+} \rightarrow Cu^{2+} + 2Fe^{2+} \quad (7)$$

$$4Fe^{2+} + O_2 + 4H^+ \rightarrow 4Fe^{3+} + H_2O \quad (8)$$

$$2Cu^0 + 4H^+ + O_2 \rightarrow 2Cu^{2+} + 2H_2O \quad (9)$$

Gholami et al. (2010) employed *A. ferrooxidans* to evaluate the recovery of nickel, cobalt, molybdenum, and aluminum from a used processing catalyst. According to Gholami et al. (2010), after 30 days, 63% Al, 96% Co, 84% Mo, and 99% Ni dissolution were accomplished. Using *A. ferrooxidans*, Wang et al. (2009) examined the mobilization of copper, lead, and zinc from PCBs. After 9 days, 99% of the Cu was recovered, and after 5 days of leaching, over 88.9% of the Zn and Pb were soluble (Wang et al., 2009).

$$2Fe^{3+} + Zn^0 \rightarrow 2Fe^{2+} + Zn^{2+} \quad (10)$$

$$2Fe^{3+} + Pb^0 \rightarrow 2Fe^{2+} + Pb^{2+} \quad (11)$$

5.2 *Fungal Bioleaching*

Fungal bioleaching serves as a cost-effective and environmentally sustainable method for extracting heavy metals from various waste sources, including electronic trash. Fungi generate organic acids during metabolic processes, which possess remarkable metal solubilization properties, making them suitable for bioleaching applications (Hocheng et al., 2017). Fungal bioleaching operates via four main pathways: acidolysis, complexolysis, redoxolysis, and bioaccumulation, which occurs due to the concurrent influence of organic acid synthesis and enzyme expression. The metal solubilization process using *Penicillium simplicissimum* and *Aspergillus niger* requires the use of glucose or sucrose as a carbon source to produce vital organic acids. The organic acids used in fungal bioleaching include glutonic acids, malic and tartaric acids, oxalic and citric acids, and others (Nayak et al., 2024).

5.2.1 Acidolysis

- Fungi secrete organic acids that provide protons to react with metals on the surface of waste materials, leading to metal ion solubilization.

$$NiO + 2H^+ \rightarrow Ni^{2+} + H_2O \quad (12)$$

5.2.2 Complexolysis

- Metal ions released during acidolysis are stabilized through complexation with organic acids.

$$Ni^{2+} + C_6H_8O_7 \rightarrow Ni(C_6H_8O_7)^- + 2H^+ \tag{13}$$

5.2.3 Redoxolysis

- Involves the oxidation–reduction reactions where fungal metabolites facilitate electron.

$$MnO_2 + 2e^- + 4H^+ \rightarrow Mn^{2+} + H_2O \tag{14}$$

5.2.4 Bioaccumulation

- Metals are actively transferred across the fungal cell membrane and accumulate inside cells, disrupting equilibrium and continuously solubilizing metals.

Applications of fungal bioleaching include recovery from PCBs, bioleaching of rare earth metals (REMs), and cleaning waste batteries. *Aspergillus niger* obtained up to 90% copper recovery and 70% zinc recovery from PCBs within 10–14 days. *Penicillium simplicissimum* solubilizes REMs such as yttrium and cerium from e-waste. Fungal species also recover cobalt, lithium, and nickel from wasted batteries by secreting organic acids.

6 Key Factors Affecting the Bioleaching Process

Several interdependent factors influence the effectiveness of the bioleaching process, including target metal dissolution, microbial activity, and process sustainability (Roy et al., 2021). Enhancing bioleaching efficiency across a range of ore types and operating scales requires an understanding of and ability to modify these factors.

6.1 Microbial Type and Acclimatization, and Inoculum Size

The process of bioleaching is largely dependent on microbiological factors, notably regarding acclimatization, inoculum size, and microbial species. The acclimatization process is a crucial step that improves microbial resistance to metal contamination seen in e-waste and ores containing toxic heavy metals and chemical inhibitors.

Over time, microorganisms that are exposed to metal-containing surfaces improve their capacity to remove metals from them. Acclimatized *Acidithiobacillus ferrooxidans* strains exhibit enhanced heavy metal resistance, thereby producing higher metal recovery rates, notably for copper and gold. The efficiency of leaching relies considerably on inoculum size since inadequate inoculum leads to poor leaching performance, while raised inoculum levels induce nutrient competition that affects efficacy. Research reveals that inoculum concentrations lying between 5% and 15% (v/v) often give the greatest outcomes, although the selected inoculum concentration relies on both the substrate type and process equipment design.

The bioleaching method needs microorganisms to undergo acclimatization to metal-bearing materials so that they may establish resistance to metal toxicity. Microbes adapt to metal-bearing material using a specialized method of subculture in liquid solution while adding increasing quantities of solid material until the requisite solids-per-liquids ratio is attained. Laboratory study reveals that following adaptation, bacteria display higher performance compared to nonacclimatized counterparts while dissolving metals (Kara et al., 2023). Lab testing demonstrated that acidic circumstances permitted acclimatized strains of both *A. ferrooxidans* and *Leptospirillum ferrooxidans* to get 57% and 66% uranium solubilization from apatite-rich low-grade uraninite ore. Through the adaptation of *A. ferrooxidans* to waste PCBs, researchers acquired 94.8% copper recovery during 28 days of culture. The combination of *Sulfobacillus thermosulfidooxidans* with *Thermoplasma acidophilum* in acclimatized mixed cultures allowed for bioleaching of electronic scrap and led to substantial zinc (80%), aluminum (64%), copper (86%), and nickel (74%) metal recoveries over extended periods (Kara et al., 2023).

Bioleaching effectiveness together with leaching time is regulated greatly by inoculum concentration (ZareTavakoli et al., 2017). Research reveals bioleaching performance improves from increasing inoculum concentrations; however, having optimum concentrations assures the avoidance of metal ion precipitation difficulties. High inoculum concentrations above their ideal value reduce bioleaching efficacy when dealing with low-grade ores and e-waste materials. New optimization methods, such as factorial designs and multivariable approaches, have revealed that inoculum concentration stands as a major determinant (Arshadi et al., 2019). Higher inoculum levels result in improved bioleaching performance but need more operating costs to maintain their sustained concentration. Thus, inoculum concentration should be optimized to balance bioleaching efficiency and operational costs in future studies.

6.2 pH

Bioleaching effectiveness depends fundamentally on medium pH adjustment, as this value regulates metal ion solubility and microbiological activities and inhibits jarosite precipitation layer development. The ideal acidic range for bioleaching operations runs from pH 2.0 to 2.5 because it promotes microbial activity while conserving metal ion solubility. Maintaining a pH below 2.5 promotes the production of protons for target metal breakdown and prevents metal precipitation, which improves bioleaching performance. Bioleaching performance may suffer when ores include acid-neutralizing minerals such as calcium and magnesium and potassium, and sodium because these components generate high pH conditions that limit acidophilic microbe activity (Kara et al., 2023). Studies for optimization coupled with changes using sulfuric acid, HCl, or NaOH are required to resolve this, depending on the chosen microorganisms. With the use of pH sensors and automated pH control systems, modern metal recovery processes are enhanced by controlling conditions and maintaining microbial activity.

6.3 Temperature

Temperature has a key influence in determining growth rates among microorganisms and both enzyme activities and bioleaching system efficiency. Acidophilic bacteria have the highest efficiency at temperatures between 25 and 30 °C, but thermophilic microorganisms demand thermal conditions equivalent to or greater than 40–45 °C. *Aspergillus niger* experiment findings indicate that restricting the operating temperature range between 25 and 30 °C is necessary for volume-dependent iron bioleaching to be successful. Bioleaching systems may endure temperature adjustments via numerous means that rely on their physical configuration. Stirred tank reactors rely on precise temperature control systems to manage the exothermic properties of the process, which enables steady microbial activity. The regulation of temperature during column bioleaching relies on water jackets and water baths, while heap bioleaching relies on irrigation coupled with aeration rates. The exothermic nature of bioleaching and the heterogeneity of indigenous microbial communities in ores may lead to shifts in the dominant microbial population over time in heap systems (Natarajan, 2018). By changing temperature based on the microorganisms employed, bioleaching systems may boost metal recovery rates and maintain stable performance.

6.4 Aeration and Dissolved Oxygen

Aeration and oxygen availability are crucial for microbial respiration and the oxidative processes involved in bioleaching. Sufficient oxygen supply prevents anaerobic circumstances that would hinder microbial activity, while maintaining a proper redox potential, often between 400 and 800 mV for ores such as chalcopyrite, is crucial for enhancing bioleaching efficiency. Aeration enhances oxygen transfer and microbial bio-oxidation, yet excessive aeration may generate shear stress, which kills microbial cells. The suggested aeration rates for most systems range from 0.07 to 0.5 L/min, with oxygen-enriched air increasing metal recovery (Natarajan, 2018).

6.5 Agitation Rate

Agitation rate is also a significant metric, depending on the system type. Agitation in shake flasks and stirred reactors facilitates efficient gas–liquid–solid transfer and heat dispersion (Natarajan, 2018). The appropriate agitation rate is dictated by the reactor type and the target metal, with effective ranges of 120–240 rpm for shake flasks and 150–600 rpm for stirred tanks. High metal dissolving rates (>90%) have been reached under similar conditions, with particular studies stressing the relevance of agitation speed for certain metals. For example, the optimal speed for cobalt (Co) leaching was determined at 250 rpm, whereas nickel (Ni) and copper (Cu) required 300 and 350 rpm, respectively. Another investigation indicated that an agitation speed of 180 rpm was optimum for *S. thermosulfidooxidans*, resulting in substantial metal dissolution rates for zinc (72%), cobalt (68%), copper (78%), nickel (81%), and iron (70%) (Ilyas et al., 2010). Furthermore, agitation is required in bacterial growth; a rotation rate of 120–145 rpm is recommended for maximum microbial development. Working conditions must be carefully adjusted to enhance microbial activity and bioleaching performance. In contrast, excessive agitation may generate friction and abrasion stress.

6.6 Particle Size

Particle size is crucial in bioleaching, impacting both microbial activity and metal dissolving rates. Smaller particle sizes increase surface area, improve microbial adhesion, and speed up metal breakdown. Excessively tiny particles (<1 mm) may clog systems and reduce oxygen transport, requiring a balance of efficiency and operational issues. Optimal particle sizes for bioleaching are 75–200 μm, which balance dissolving efficiency and system performance. Particle size is classified as small (<1 mm) or coarse (1–25 mm or bigger), with applications in vat (1–10 mm), heap (10–25 mm), and dump leaching (>25 mm) (Natarajan, 2018). Fine particles have faster dissolving rates but may more effectively buffer system pH, while coarser particles decrease buffering but may restrict mass transfer, resulting in slower dissolution. Acid agglomeration methods efficiently decrease the clogging concerns associated with tiny particles, resulting in efficient bioleaching processes across a wide range of ore types and system designs.

6.7 Pulp Density

Pulp density, which represents the solid content in the bioleaching slurry, has a substantial impact on microbial activity, nutrient diffusion, and metal dissolving efficiency. Optimal pulp densities for shake flasks and stirred tanks generally range from 5% to 20% (w/v), balancing dissolving rates and operating expenses. Although lower pulp densities (<1%) improve dissolution efficiency, they are not cost-effective for large-scale applications. Higher pulp densities, on the other hand, might buffer the pH, necessitating the injection of acid to sustain microbial activity.

According to research, controlling pulp density is critical for various bioleaching techniques and materials. For example, low pulp densities (1–5%) have been useful in extracting metals such as La, Ce, and W utilizing two-step techniques, with dissolving rates surpassing 90% in certain instances. However, greater pulp densities often impair recovery rates owing to restricted oxygen transport, the release of hazardous chemicals, and inadequate agitation. Higher pulp densities (>20% w/v) are feasible in column and heap bioleaching, since commercial operations have shown effective large-scale bioleaching of refractory ores at such concentrations.

Noncontact bioleaching enables greater pulp densities since microorganisms are not present in the leaching solution. However, during contact bioleaching, high pulp densities may impair microbial efficiency, despite the increased availability of critical metal ions in the leachate. Overall, optimizing pulp density is crucial for increasing bioleaching efficiency and assuring economic feasibility across a wide range of applications and reactor types.

6.8 *Energy Sources*

Bioleaching bacteria rely on energy sources like ferrous ions (Fe^{2+}) and sulfur compounds to maintain metabolic activity and improve metal recovery. Optimizing the concentration of these energy sources is critical for promoting microbial growth while avoiding inhibitory effects. Ferrous ions (Fe^{2+}) enhance metal recovery from certain ores. The sort of energy source used is determined by the microorganism and the substance being digested. Incineration ashes and slags need both ferric ions (Fe^{3+}) and sulfuric acid (H_2SO_4), whereas wasted lithium-ion batteries and discarded PCBs depend mainly on Fe^{2+}, Fe^{3+}, or H_2SO_4. Studies have proposed that energy source concentrations be optimized between 0.1% and 20% (w/v) to produce high bioleaching yields while maintaining economic viability. In the bioleaching of Pb/Zn smelting slag, indium (In) dissolution was enhanced by a mix of acid dissolution via biogenic H_2SO_4 and oxidation–reduction processes mediated by Fe^{2+}/Fe^{3+}, whereas Zn, Cd, and Pb followed an acid dissolution mechanism. Similarly, in bioleaching sewage sludge, ferrous sulfate showed greater metal recovery rates (e.g., 64% Cu and 76% Zn) than ammonium ferrous sulfate (56% Cu and 68% Zn) (Kara et al., 2023). The kind and concentration of energy sources must be tailored to the individual bacteria and materials to maximize efficiency and achieve long-term bioleaching results.

6.9 Pretreatment Techniques

Pretreatment approaches improve bioleaching performance by altering the substrate's physical and chemical characteristics, resulting in better microbial interaction and metal recovery. Common pretreatment processes include acid washing, which eliminates surface pollutants and stabilizes pH at microorganism-friendly levels, heat treatment, mechanical pulverization, sterilization, and water washing. Acid washing eliminates surface impurities and inhibits chemicals, particularly for alkaline materials. Thermal treatment promotes substrate bioavailability by breaking down complex structures and removing organic waste, suppressing acidophilic bacteria. Mechanical pulverization lowers particle size, increasing surface area for microbial activity. Autoclaving is used in laboratory settings to remove indigenous microorganisms. Water washing eliminates water-soluble salts from substrates, boosting metal recovery. While bioleaching is effective on substrates having up to 10% organic matter, some materials with high organic content or particular metal compositions may necessitate specialized pretreatment procedures. Thermal pretreatment is especially beneficial for selective metal recovery, but it must be carefully tuned to limit undesired side effects, such as the production of insoluble compounds.

6.10 Catalysts

The inclusion of catalysts to the bioleaching medium has been demonstrated to improve bioleaching performance, with multiple studies showing considerable increases in metal solubilization. Researchers at Zeng et al. (2013) evaluated the *Acidithiobacillus ferrooxidans* bioleaching process using silver ions (Ag+) on spent lithium batteries and found 98% cobalt recovery after 7 days, whereas the control achieved only 43% recovery. The inclusion of biochar into bioleaching medium allowed Wang et al. (2016) to observe full copper (Cu) solubilization from e-waste samples in just 2 days whereas pure bioleaching without biochar needed 4 days to attain 80% Cu recovery. Gu et al. (2017) evaluated graphene's catalytic behavior in copper extraction from discarded PCBs using *A. ferrooxidans* treatment, which indicated that graphene enhanced copper dissolving rates by 10% relative to an untreated sample. These investigations indicate that catalysts may greatly increase bioleaching yields by increasing metal solubility. However, the effect of these extra components on bioleaching kinetics is complicated and warrants more investigation. To increase bioleaching efficiency, catalyst addition should be considered after adjusting the key operating parameters.

7 Environmental Implications of Bioleaching

Bioleaching's key benefit results from its capacity to minimize e-waste toxicity levels. E-waste disposal techniques like incineration coupled with landfilling generate harmful byproducts that contaminate the environment. Bioleaching acts as a full recycling loop that employs microorganisms to extract metals with no secondary waste creation. The bioleaching technique saves energy consumption and chemical waste since high temperatures and toxic chemicals are unnecessary for its operation (Harumi & Santosa, 2019). According to Narayanasamy et al. (2021), bacteria, fungi, and other microorganisms use bioleaching techniques to acquire metals via biochemical reduction and oxidation strategies. Awasthi and Li (2019) have shown in many recent publications that fungi, bacteria, and biomass from agricultural waste have a wide range of potential for HM biosorption. The bioleaching process makes it simple to recover rare earth elements and precious metals like copper and gold that are often lost in conventional waste disposal systems.

It also reduces the need to mine virgin ores, eventually protecting the environment and encouraging resource conservation. According to studies, under ideal circumstances, bioleaching can recover metals like copper and gold with recovery efficiencies of up to 95%. These results have a major impact on the circular economy, which minimizes waste formation by continuously reusing and recycling products (Saha et al., 2021).

7.1 Role of Bioleaching in Long-Term Sustainability

The overall effectiveness of recycling e-waste can be improved by including bioleaching into current waste management systems. For example, metals can be efficiently extracted while nonmetallic components are separated for additional processing by combining bioleaching with mechanical separation techniques. These hybrid strategies support a circular economy by maximizing resource recovery and minimizing waste (Okorhi et al., 2019).

Furthermore, the present limits of bioleaching may be overcome with the help of developments in microbial engineering and bioprocess optimization. By increasing microorganisms' metal tolerance and activity, genetic alterations can shorten recovery times and increase scalability. Similar to this, bioleaching can become more environmentally friendly and economically feasible with the introduction of bioreactor systems that offer better control over process parameters (Li & Achal, 2020).

8 Bioleaching in Support of the Circular Economy

A circular economy describes "A form of production and consumption working via material and product sharing and reusing and repairing and reusing existing pieces for maximum duration." The active product life cycle is prolonged using this strategy. Bioleaching plays a crucial function in recycling procedures and is most often used in product restoration and repair operations.

8.1 Metal Recovery from PCBs

Metal-like Cu recovery is done using microorganisms like *Acidithiobacillus ferrooxidans*. They do not directly separate copper but rather react with Fe^{3+} ions, which start a series of reactions that ultimately convert copper into a water-soluble form. In this series of reactions, the bacteria react with ferrous (Fe^{2+}) ions and oxidize them to ferric (Fe^{3+}) ions, which then leach Cu from PCBs and other e-waste. Cu^{2+} is soluble in water and can be separated as solutions like $CuSO_4$ (Kaur et al., 2022). For Ni recovery, fungi like *Aspergillus* spp. produce organic acids (e.g., citric acid) that solubilize nickel from PCBs or other e-waste (Kaur et al., 2022). Similar to Cu recovery in Ni recovery, a microorganism (fungi) is used in the process. In this process, Ni is being recovered from nickel oxide, unlike Cu which was recovered from its pure form. NiO is mostly found in Ni-MH batteries, semiconductors, and capacitors.

The bioleaching processes discussed were about recovering metals. Microbes that produce enzymes like Laccase (*Pleurotus florida* and *Pseudomonas* spp.) can be used to degrade organic materials like resins and plastics that bind metals. The result of this also facilitates metal recovery but differently compared to the previous ones (Kaur et al., 2022). Just like these metals, various others such as cobalt, cadmium, zinc, and lead can be recovered using bioleaching processes. These metals can then be reused in production, which will result in less demand for virgin materials. Apart from reducing raw material demand, bioleaching simultaneously reduces waste during the recovery process, thus supporting the circular economy.

Bioleaching seems to have a promising future in e-waste management. With more and more electronics being produced, there is always going to be more waste to manage. Bioleaching addresses challenges that are faced by conventional methods. Unlike chemical leaching, bioleaching does not produce environmentally harmful substances, making it eco-friendly. Traditional methods also leave more waste in the recovery process whereas in bioleaching, used materials can be discarded easily, leaving very little or no waste behind. These reasons create a potentially solid future for bioleaching in e-waste management.

9 Challenges and Future Directions

Like other modern eco-friendly processes, one of the main challenges faced when implementing bioleaching is its efficiency compared to conventional leaching and thermal methods such as hydrometallurgy and pyrometallurgy. For instance, recovery efficiencies for copper in bioleaching range from 10.5% to 18% (Kaur et al., 2022).

Scaling up bioleaching processes from laboratories to industrial levels is another challenge to be overcome. In lab conditions, maintaining parameters such as temperature, pH, and nutrient supply is easy. However, in mass scales, maintaining these parameters can be challenging and minor changes in these parameters may interrupt the process since some of these microorganisms are highly sensitive to these conditions (Gopikrishnan et al., 2020).

Bioleaching is slower compared to chemical leaching and smelting processes. For example, laboratory studies report maximum laccase enzyme activity (critical for bioleaching by fungi) only after 20 days of incubation. This might be unfeasible for industries requiring rapid processing (Kaur et al., 2022).

In most cases, e-waste does not provide a suitable environment for bio-organisms. Meaning the microorganisms used in bioleaching have to go through a harsh toxic environment. Finding and developing microorganisms that can withstand these conditions is a major obstacle. For example, high concentrations of lead or mercury may render microbes ineffective (Gopikrishnan et al., 2020).

9.1 Research Gaps and Areas for Further Study

Being a relatively new and unexplored area in the scientific world bioleaching has its own gaps in research which are areas for further study.

- One of these is the limited research done on microbial consortium. Although research has been done on individual microbes in bioleaching, less focus has been given on combining microbes with complementary capabilities for enhanced performance (Kaur et al., 2022).
- Another area for further study is the biochemical and enzymic mechanisms of bioleaching. Like scientific studies and experiments show that enzymes like laccase play a role in metal solubilization, but the interaction between microbial enzymes and metals is poorly understood (Kaur et al., 2022).
- Handling nonmetal waste is another area left to explore. Most of the research is done on recovering metals but using bioleaching to recover other nonmetal materials like plastic and glass has been little explored (Arya & Kumar, 2020).
- To commercialize bioleaching in the e-waste management field trials, large-scale experiments need to be conducted to test the feasibility. This is also an area left for further development, which if developed can optimize bioleaching processes and make them economically viable. While mining operations successfully use bioleaching for ores, the application to e-waste has seen minimal real-world testing (Arya & Kumar, 2020; Gopikrishnan et al., 2020).

10 Conclusion

The increased demand for electronic gadgets has led to a growth in e-waste, which comprises crucial metals and valuable rare earth elements. E-waste is financially important but also environmentally damaging when incorrectly disposed of. Bioleaching, a promising, sustainable, and ecologically friendly process for recovering metal from e-waste, has shown larger recovery rates than existing methods. However, efficiency is governed by several physicochemical characteristics that must be improved. New technologies, such as catalysts, may boost bio-recovery kinetics. Genetically engineered microorganisms may also be utilized to increase metal recovery. Alternative carbon sources, such as agro-industrial waste, may be employed to culture bacteria in the bioleaching medium. Efficient separation procedures for harmful compounds are vital for decreasing e-waste toxicity. Bioleaching, which employs acidophilic bacteria and fungi, is a low-cost and ecologically benign approach for extracting metals from complicated materials. However, slow dissolution kinetics remain an issue. Multivariable optimization methodologies, such as response surface analysis, Plackett-Burman, and orthogonal array design may boost bioleaching yields. Further pilot studies and interdisciplinary collaboration are required to scale up bioleaching for commercial uses. Life cycle and technoeconomic assessments may assist in analyzing the environmental and economic consequences of bioleaching.

References

Abdel Azim, A., Bellini, R., Vizzarro, A., Bassani, I., Piri, C. F., & Menin, B. (2023). Highlighting the role of archaea in urban mine waste exploitation and valorisation. *Recycling, 8*, 20.

Ahirwar, R., & Tripathi, A. K. (2021). E-waste management: A review of recycling process, environmental and occupational health hazards, and potential solutions. *Environmental Nanotechnology, Monitoring and Management, 15*, 100409.

Al-Salem, S. M., Leeke, G. A., El-Eskandarany, M. S., Van Haute, M., Constantinou, A., Dewil, R., & Baeyens, J. (2022). On the implementation of the circular economy route for e-waste management: A critical review and an analysis for the case of the State of Kuwait. *Journal of Environmental Management, 323*, 116181.

Anjum, F., Shahid, M., & Akcil, A. (2012). Biohydrometallurgy techniques of low grade ores: A review on black shale. *Hydrometallurgy, 117–118*, 1–12.

Arshadi, M., Nili, S., & Yaghmaei, S. (2019). Ni and Cu recovery by bioleaching from the printed circuit boards of mobile phones in non-conventional medium. *Journal of Environmental Management, 250*, 109502. https://doi.org/10.1016/j.jenvman.2019.109502

Arya, S., & Kumar, S. (2020). Bioleaching: Urban mining option to curb the menace of E-waste challenge. *Bioengineered, 11*(1), 640–660. https://doi.org/10.1080/21655979.2020.1775988

Awasthi, A. K., & Li, J. (2019). Sustainable bioprospecting of electronic waste. *Trends in Biotechnology, 37*(7), 677–680.

Balde, C. P., Forti, V., Gray, V., Kuehr, R., & Stegmann, P. (2017). *The global E-waste monitor*. United Nations University (UNU)/International Telecommunication Union (ITU)/International Solid Waste Association.

Bao, S., Pan, B., Wang, L., Cheng, Z., Liu, X., Zhou, Z., & Nie, X. (2020). Adverse effects in Daphnia magna exposed to e-waste leachate: Assessment based on life trait changes and responses of detoxification-related genes. *Environmental Research, 188*, 109821.

Benzal, E., Solé, M., Lao, C., Gamisans, X., & Dorado, A. D. (2020). Elemental copper recovery from e-wastes mediated with a two-step bioleaching process. *Waste and Biomass Valorization, 11*(10), 5457–5465. https://doi.org/10.1007/s12649-020-01040-2

Bosecker, K. (1997). Bioleaching: Metal solubilization by microorganisms. *FEMS Microbiology Reviews, 20*(3–4), 591–604. https://doi.org/10.1111/j.1574-6976.1997.tb00340.x

Chauhan, G., Kaur, P. J., Pant, K. K., & Nigam, K. (2020). *Sustainable metal extraction from waste streams*. Wiley. https://doi.org/10.1002/9783527826704

Chen, S., Yang, Y., Liu, C., Dong, F., & Liu, B. (2015). Column bioleaching copper and its kinetics of waste printed circuit boards (WPCBs) by Acidithiobacillus ferrooxidans. *Chemosphere, 141*, 162–168.

Chi, X., Wang, M. Y. L., & Reuter, M. A. (2014). E-waste collection channels and household recycling behaviors in Taizhou of China. *Journal of Cleaner Production, 80*, 87–95.

Das, A. P., & Ghosh, S. (2023a). Factors affecting bioleaching. In *Manganese mining microorganisms* (pp. 77–86). https://doi.org/10.1016/B978-0-12-822147-1.00004-2

Das, A. P., & Ghosh, S. (2023b). Mixed culture bioleaching: An insight into manganese biomining process and efficacy. In *Manganese mining microorganisms* (pp. 129–141). https://doi.org/10.1016/B978-0-12-822147-1.00001-7

Das, A. P., Sukla, L. B., Pradhan, N., & Nayak, S. (2011). Manganese biomining: A review. *Bioresource Technology, 102*(16), 7381–7387. https://doi.org/10.1016/J.BIORTECH.2011.05.018

Dey, S. (2023). *Biohydrometallurgical processes: Metal recovery and remediation*. CRC Press. https://doi.org/10.1201/9781003451457

Dey, S., Das, A., Mallick, K., Sahu, A., & Das, A. P. (2023). *Environmental petroleum waste: Pollution, toxicity, sustainable remediation* (pp. 159–175). https://doi.org/10.1007/978-3-031-48220-5_7

Dutta, D., Rautela, R., Gujjala, L. K. S., Kundu, D., Sharma, P., Tembhare, M., & Kumar, S. (2023). A review on recovery processes of metals from E-waste: A green perspective. *Science of the Total Environment, 859*, 160391.

Forti, C. A. B., & Freitas, K. S. (2020). Relationship between dividends and the rating classification of Brazilian companies. *Contemporary Accounting Journal, 17*(42), 120–137. https://doi.org/10.5007/2175-8069.2020v17n42p120

Gavrilescu, M. (2022). Microbial recovery of critical metals from secondary sources. *Bioresource Technology, 344*, 126208. https://doi.org/10.1016/j.biortech.2021.126208

Gholami, R. M., Borghei, S. M., & Mousavi, S. M. (2010). Bacterial leaching of a spent Mo–Co–Ni refinery catalyst using Acidithiobacillusferrooxidans and Acidithiobacillusthiooxidans. *Hydrometallurgy, 106*(1–2), 26–31. https://doi.org/10.1016/j.hydromet.2010.11.011

Ghosh, S., & Das, A. P. (2017). Bioleaching of manganese from mining waste residues using Acinetobacter sp. *Geology, Ecology, and Landscapes, 1*(2), 77–83. https://doi.org/10.1080/24749508.2017.1332847

Ghosh, S., Mohanty, S., Akcil, A., Sukla, L. B., & Das, A. P. (2016a). A greener approach for resource recycling: Manganese bioleaching. *Chemosphere, 154*, 628–639. https://doi.org/10.1016/J.CHEMOSPHERE.2016.04.028

Ghosh, S., Mohanty, S., Nayak, S., Sukla, L. B., & Das, A. P. (2016b). Molecular identification of indigenous manganese solubilising bacterial biodiversity from manganese mining deposits. *Journal of Basic Microbiology, 56*(3), 254–262. https://doi.org/10.1002/JOBM.201500477

Gopikrishnan, V., Vignesh, A., Radhakrishnan, M., Joseph, J., Shanmugasundaram, T., Doble, M., & Balagurunathan, R. (2020). Microbial leaching of heavy metals from e-waste: Opportunities and challenges. In V. K. Sharma & G. C. Pandey (Eds.), *Biovalorisation of wastes to renewable chemicals and biofuels* (pp. 189–216). Elsevier. https://doi.org/10.1016/B978-0-12-817951-2.00010-9

Gu, W., Bai, J., Dong, B., Zhuang, X., Zhao, J., Zhang, C., Wang, J., & Shih, K. (2017). Catalytic effect of graphene in bioleaching copper from waste printed circuit boards by

Acidithiobacillusferrooxidans. *Hydrometallurgy, 171*, 172–178. https://doi.org/10.1016/j.hydromet.2017.05.012

Harumi, M., & Santosa, S. J. (2019). Recovery of Au (III) from printed circuit board waste by chitosan/SiO_2 coated on iron sand magnetic material. *Materials Today, 19*, 1101–1110.

Hocheng, H., Chakankar, M., & Jadhav, U. (2017). *Biohydrometallurgical recycling of metals from industrial wastes*. CRC Press. https://doi.org/10.1201/9781315195544

Ilyas, S., Bhatti, H. N., Bhatti, I. A., Sheikh, M. A., & Ghauri, M. A. (2010). Bioleaching of metal ions from low grade sulphide ore: Process optimization by using orthogonal experimental array design. *African Journal of Biotechnology, 9*, 2801–2810.

Javed, M., & Usmani, N. (2011). Accumulation of heavy metals in fishes: A human health concern. *International Journal of Environmental Sciences, 2*.

Kara, I. T., Kremser, K., Wagland, S. T., & Coulon, F. (2023). Bioleaching metal-bearing wastes and by-products for resource recovery: A review. *Environmental Chemistry Letters, 21*(6), 3329–3350. https://doi.org/10.1007/s10311-023-01611-4

Kaur, P., Sharma, S., Albarakaty, F. M., Kalia, A., Hassan, M. M., & Abd-Elsalam, K. A. (2022). Biosorption and bioleaching of heavy metals from electronic waste varied with microbial genera. *Sustainability, 14*(2), 935. https://doi.org/10.3390/su14020935

Kudpeng, K., et al. (2020). Bioleaching of gold from sulfidic gold ore concentrate and electronic waste by Roseovariustolerans and Roseovariusmucosus. *Microorganisms, 8*(11), 1783.

Kumari, A., & Natarajan, K. (2002). Electrochemical aspects of leaching of ocean nodules in the presence and absence of microorganisms. *International Journal of Mineral Processing, 66*(1–4), 29–47. https://doi.org/10.1016/s0301-7516(02)00010-8

Li, W., & Achal, V. (2020). Environmental and health impacts due to e-waste disposal in China—A review. *Science of the Total Environment, 737*, 139745.

Li, X., Tian, Y., Zhang, Y., Ben, Y., & Lv, Q. (2017). Accumulation of polybrominated diphenyl ethers in breast milk of women from an e-waste recycling centre in China. *Journal of Environmental Sciences, 52*, 305–313.

Mishra, S., Ghosh, S., van Hullebusch, E. D., Singh, S., & Das, A. P. (2023). A critical review on the recovery of base and critical elements from electronic waste-contaminated streams using microbial biotechnology. *Applied Biochemistry and Biotechnology, 2023*, 1–30. https://doi.org/10.1007/S12010-023-04440-X

Mohanty, S., Ghosh, S., Nayak, S., & Das, A. P. (2017). Bioleaching of manganese by Aspergillus sp. isolated from mining deposits. *Chemosphere, 172*, 302–309. https://doi.org/10.1016/J.CHEMOSPHERE.2016.12.136

Mohod, C., & Dhote, J. (2013). Review of heavy metals in drinking water and their effect on human health. *International Journal of Innovative Research in Science, Engineering and Technology, 2*, 2992–2996.

Narayanasamy, M., Dhanasekaran, D., & Thajuddin, N. (2021). Bioremediation of noxious metals from e-waste printed circuit boards by Frankia. *Microbiological Research, 245*, 126707.

Nasiri, T., Mokhtari, M., Teimouri, F., & Abouee, E. (2022). Remediation of metals and plastic from e-waste by iron mine indigenous acidophilic bacteria. *Waste Management & Research the Journal for a Sustainable Circular Economy, 41*(4), 894–902. https://doi.org/10.1177/0734242x221126418

Natarajan, K. A. (2018). *Biotechnology of metals: Principles, recovery methods, and environmental concerns* (1st ed.). Elsevier. 488 pages. https://doi.org/10.1016/C2015-0-00161-7

Nayak, J., Chakrabortty, S., Tripathy, S. K., & Shah, M. P. (2024). *Advanced bioseparation of industrial wastes: Sustainable recovery of high-value metal ions*. CRC Press. https://doi.org/10.1201/9781003415541

Okorhi, J. O., Amadi-Echendu, J. E., Aderemi, H. O., Uhunmwangho, R., & Okwubunne, A. C. (2019). Disconnect between policy and practice in developing countries: Evidence of managing e-waste from Nigeria. *African Journal of Science, Technology, Innovation and Development, 11*(4), 523–531.

Prabhakar, A., Mishra, S., & Das, A. P. (2019). Isolation and identification of lead (Pb) solubilizing bacteria from automobile waste and its potential for recovery of lead from end of life waste batteries. *Geomicrobiological Journal, 36*(10), 894–903. https://doi.org/10.1080/01490451.2019.1654044

Priyadarsini, S., & Das, A. P. (2024). Analytical and structural characterization of waste lithium-ion batteries for their effective recycling strategy. *Environmental Chemistry and Ecotoxicology., 7*, 182. https://doi.org/10.1016/j.enceco.2024.12.004

Priyadarsini, S., & Das, A. P. (2025). Lithium bioleaching: A review on microbial-assisted sustainable technology for lithium bio-circularity. *Journal of Water Process Engineering, 69*, 106744. https://doi.org/10.1016/j.jwpe.2024.106744

Purchase, D., Abbasi, G., Bisschop, L., Chatterjee, D., Ekberg, C., Ermolin, M., & Wong, M. H. (2020). Global occurrence, chemical properties, and ecological impacts of E-wastes (IUPAC Technical Report). *Pure and Applied Chemistry, 92*(11), 1733–1767.

Raj, K., & Das, A. P. (2023). Lead pollution: Impact on environment and human health and approach for a sustainable solution. *Environmental Chemistry and Ecotoxicology, 5*, 79–85. https://doi.org/10.1016/J.ENCECO.2023.02.001

Rautela, R., Arya, S., Vishwakarma, S., Lee, J., Kim, K. H., & Kumar, S. (2021). E-waste management and its effects on the environment and human health. *Science of the Total Environment, 773*, 145623.

Rawlings, D. E. (2004). Microbially assisted dissolution of minerals and its use in the mining industry. *Pure and Applied Chemistry, 76*(4), 847–859.

Roy, J. J., Cao, B., & Madhavi, S. (2021). A review on the recycling of spent lithium-ion batteries (LIBs) by the bioleaching approach. *Chemosphere, 282*, 130944. https://doi.org/10.1016/j.chemosphere.2021.130944

Saha, L., Kumar, V., Tiwari, J., Rawat, S., Singh, J., & Bauddh, K. (2021). Electronic waste and their leachates impact on human health and environment: Global ecological threat and management. *Environmental Technology & Innovation, 24*, 102049.

Sanusi, A. I. (2016). Heavy metal profile of oreochromisniloticus harvested from E-waste polluted vials and associated fungi. *Advances in Microbiology, 6*(08), 555.

Vítová, M., & Mezricky, D. (2024). Microbial recovery of rare earth elements from various waste sources: A mini review with emphasis on microalgae. *World Journal of Microbiology and Biotechnology, 40*(6). https://doi.org/10.1007/s11274-024-03974-4

Wang, J., Bai, J., Xu, J., & Liang, B. (2009). Bioleaching of metals from printed wire boards by Acidithiobacillusferrooxidans and Acidithiobacillusthiooxidans and their mixture. *Journal of Hazardous Materials, 172*(2–3), 1100–1105. https://doi.org/10.1016/j.jhazmat.2009.07.102

Wang, S., Zheng, Y., Yan, W., Chen, L., DummiMahadevan, G., & Zhao, F. (2016). Enhanced bioleaching efficiency of metals from E-wastes driven by biochar. *Journal of Hazardous Materials, 320*, 393–400. https://doi.org/10.1016/j.jhazmat.2016.08.054

Wang, S., Dong, D., Li, P., Hua, X., Zheng, N., Sun, S., & Li, X. (2020). Mercury concentration and fatty acid composition in muscle tissue of marine fish species harvested from Liaodong Gulf: An intelligence quotient and coronary heart disease risk assessment. *Science of the Total Environment, 726*, 138586.

ZareTavakoli, H., Abdollahy, M., Ahmadi, S. J., & Darban, A. K. (2017). Enhancing recovery of uranium column bioleaching by process optimization and kinetic modeling. *Transactions of Nonferrous Metals Society of China (English Edition), 27*, 2691–2703. https://doi.org/10.1016/S1003-6326(17)60298-X

Zeng, G., Luo, S., Deng, X., Li, L., & Au, C. (2013). Influence of silver ions on bioleaching of cobalt from spent lithium batteries. *Minerals Engineering, 49*, 40–44. https://doi.org/10.1016/j.mineng.2013.04.021

Zhang, Y., Luo, X. J., Mo, L., Wu, J. P., Mai, B. X., & Peng, Y. H. (2015). Bioaccumulation and translocation of polyhalogenated compounds in rice (Oryza sativa L.) planted in paddy soil collected from an electronic waste recycling site, South China. *Chemosphere, 137*, 25–32.

Lithium Bioleaching: A Prospective Technology for Lithium Recovery from Spent Mobile Battery

Sailaja Priyadarsini and Alok Prasad Das

Abstract Lithium, an essential element for energy storage technologies, is experiencing surging demand due to its extensive application in lithium-ion batteries (LIBs). Traditional extraction methods, such as mining and chemical leaching, pose substantial environmental challenges, including high energy consumption, substantial waste production, and excessive water usage. In this framework, lithium bioleaching has arisen as a promising, eco-friendly, and sustainable substitute for lithium recovery, which leverages the metabolic activities of microbes to solubilize lithium from primary ores, secondary sources like waste LIBs, and other lithium-containing materials. Microbials have demonstrated potential in bioleaching applications because they can yield organic acids and other metabolites that enhance lithium solubilization. Additionally, bioleaching offers the advantage of operating at ambient conditions, reducing the carbon footprint and operational costs accompanying traditional methods. Recent advancements in molecular biology and metabolic engineering have further optimized microbial efficiency, enhanced lithium recovery rates, and made the process more viable for industrial applications. This review explores the sources of lithium, the analytical structure of LIBs, and the extraction of lithium from spent LIBs. It also discusses the mechanisms and methods of lithium bioleaching, and different factors affecting bioleaching. The integration of the bio-circular economy perspective with lithium bioleaching into closed-loop systems for LIB recycling to minimize waste and recover valuable materials.

Keywords Lithium · Lithium-ion battery · Bioleaching · Recovery

S. Priyadarsini · A. P. Das (✉)
Department of Life Sciences, Rama Devi Women's University, Bhubaneswar, Odisha, India

A. P. Das, S. Priyadarsini (eds.), *Electronic Waste and Environmental Pollution*, Sustainable Environmental Waste Management Strategies,
https://doi.org/10.1007/978-3-031-90287-1_12

1 Introduction

The necessity for rechargeable batteries in many industries such as mobile, computers, automobiles, gadgets, and storage devices is incessantly rising. Lithium-ion batteries (LIBs) represent most rechargeable batteries, and their usage is developing every day. (Priyadarsini & Das, 2024; Roy et al., 2021). The expansion of industrial sectors leads to an increase in the amounts of pollutants like metals (Das et al., 2015b; Dey et al., 2023b; Ghosh et al., 2017, 2020; Ghosh & Das, 2018; Mohanty et al., 2018), plastics (Das et al., 2024b; Mishra et al., 2019, 2020, 2021; Paul et al., 2024; Singh et al., 2020), biomedical waste (Khatun et al., 2023; Sadasivuni et al., 2019), and chemicals (Das & Swain, 2013) which generally discharge into the water and pollute the environment (Mishra et al., 2022, 2024; Mishra & Das, 2021, 2022, 2023). The global lithium marketplace is estimated to expand by nearly 88% by 2028 due to the anticipated expansion of LIBs (Razmjou et al., 2019). In the same period, consumer demand will increase to 900,000 tons per year, resulting in a price increase. (Baniasadi et al., 2019). LIB's appealing properties such as a high content of energy, compact size, elevated voltage, broad working temperature range prolonged lifespan, decreased self-discharge, and weight, and good electrochemical features make them preferable among other types of batteries (Fergus, 2010). Lithium's unique physicochemical properties, such as its high energy density and low weight, make it a critical element in modern energy storage solutions. LIBs have revolutionized energy, enabling longer-lasting, more efficient, and lighter batteries. Beyond LIBs, lithium finds applications in aerospace technologies, pharmaceuticals, and nuclear fusion, further underscoring its strategic importance. Meeting this multifaceted demand necessitates innovative approaches to lithium recovery (Priyadarsini & Das, 2024). Continuous exploitation depletes these primary resources and causes toxicity to the environment (Das & Mishra, 2010; Das & Bissoyi, 2011; Das & Mishra, 2008, 2009; Mallick et al., 2024; Mishra & Das, 2008). However, it is expected that the lithium demand will be five times more supplementary than the capacity of its mineral sources during the next 20 years. The worldwide extraction capacity of lithium brines and ores must be exceeded by market claim via lithium recovery from other secondary resources. Traditional lithium extraction methods primarily rely on hard rock mining and brine evaporation. Hard rock mining, involving minerals such as spodumene and lepidolite, is resource-intensive and generates significant waste. Brine extraction, while less invasive, is associated with extensive water usage and ecological disruption, particularly in arid regions. Both methods are energy-intensive, costly, and contribute to greenhouse gas emissions, raising concerns about their long-term viability in a decarbonizing world (Priyadarsini & Das, 2025). Bioleaching, an environmentally friendly and cost-effective technology, leverages the metabolic capabilities of microbes to recover metals from ores and waste materials (Das et al., 2011; Ghosh et al., 2016b). While usually bioleaching employed for extracting base metals such as copper and gold, bioleaching is gaining popularity for lithium recovery due to its low environmental effect and ability to treat low-grade and unconventional deposits. Bioleaching, which employs the inherent abilities of bacteria and fungus, represents a paradigm shift in lithium recovery strategies (Das & Ghosh, 2018, 2023a, 2023f; Ghosh et al., 2023).The bioleaching process for LIBs typically consists of multiple phases. First, spent LIBs are disassembled, and components like cathode material are separated. Pretreatment procedures such as shredding, grinding, and thermal treatment are frequently used to increase the surface area and release metals for microbial activity. The microbes are then cultivated under optimal conditions to enhance acid production and metal solubilization. The prepared cathode material is combined with the microbial culture or bioleaching medium. Microorganisms degrade metals over time, generating soluble metal ions in the leachate (Das et al., 2024a; Mohanty et al., 2024). Finally, solubilized lithium is extracted from the leachate using procedures such as precipitation, solvent extraction, or ion exchange. Recent research has concentrated on enlightening the effectiveness and scalability of bioleaching using a variety of techniques (Das & Ghosh, 2023b, 2023c; Ghosh & Das, 2017). Microorganisms with improved acid production and metal tolerance have been developed through genetic engineering. Microbial consortia are used to improve metal recovery by leveraging synergies (Bal et al., 2018; Das et al., 2012b; Das & Ghosh, 2023d; Mishra et al., 2009). To optimize the process, advanced bioreactor designs that improve aeration, agitation, and temperature control have been created. Bioleaching parameters are also refined and efficiency is increased using statistical and computational methodologies (Prabhakar et al., 2019; Raj & Das, 2023).

Li et al. (2013) perform bioleaching using *A. ferrooxidans* at different values of pH and ferrous ions (Fe^{2+}) concentration, indicated that 7.6% of the Co was dissolved from wasted LIBs. Recent research has focused on increasing the efficiency and scalability of bioleaching using several approaches. Genetic engineering has resulted in the development of microorganisms with increased acid output and metal tolerance. Microbial consortiums are utilized to increase metal recovery by utilizing synergies. Advanced bioreactor designs have been developed to increase aeration, agitation, and temperature control. Statistical and computational approaches are also used to fine-tune bioleaching parameters and boost efficiency (Naseri et al., 2019). Bioleaching is a promising method for the long-term recovery of lithium from wasted LIBs. By utilizing the power of microbes, it provides a cost-effective substitute for standard recycling methods (Das et al., 2015a; Kanungo et al., 2024; Paul et al., 2023; Pradhan et al., 2023). While hurdles persist, continued research and technological developments are paving the road for its practical application (Bhattacharjee et al., 2021; Dey et al., 2024; Nag et al., 2023). As the demand for lithium rises, bioleaching will become gradually significant in assuring the long-term management of LIB waste and the conservation of vital resources.

2 Structural Model of Lithium-Ion Batteries (LIBs)

The broad usage of LIBs has facilitated the expansion of microelectronics, which has evolved into the preferred power supply for portable electronic devices. This choice arises from their superior performance in gravimetric and volumetric energy densities compared to other rechargeable systems. Compared to systems utilizing aqueous electrolytes, the higher operating voltages produced with water-free, non-aqueous electrolytes are responsible for the greater energy density (Manthiram, 2017). In addition, renewable energy sources such as wind and solar energy are expected to play a key part in the progressive replacement of fossil fuel-based power generation, resulting in a cleaner, more sustainable society. A LIB cell typically consists of two electrodes: graphite for the anode and lithium iron phosphate ($LiFePO_4$), lithium manganese oxide ($LiMnO_4$), and lithium cobalt oxide ($LiCoO_2$) for the cathode, as well as an electrolyte and a separator (Nitta et al., 2015). Figure 1 describes a schematic diagrammatic depiction of both the internal and external structure of LIBs.

Through an electrochemical reaction, the anode oxidizes and releases electrons to the external circuit. Lithiated graphite, or $LixC_6$, is a widely used anode material that is made of graphite sheets that have been intercalated with lithium. Researchers are investigating novel materials, including combinations of elements such as silicon. Electrons are lost to the external circuit by the cathode, also referred to as the positive or oxidizing electrode (Grey & Hall, 2020). For the cathode to be effective, it must be inexpensive, be a good conductor and diffuser of Li ions, hold a large amount of Li without experiencing major structural changes, and have strong chemical and electrochemical stability when combined with electrolytes (Nitta et al., 2015). One crucial component of nonelectroconductive parts is the electrolyte. It may affect the interfacial chemistry, rate capability, energy density, and other important aspects of battery performance (Osiak et al., 2014). A permeable membrane positioned between the anode and cathode of a battery is called a separator. To prevent brief electrical contacts, a separator's primary function is to maintain the distance between the two electrodes. It also promotes the flow of ionic charge carriers, which is necessary for an electrochemical cell to close the circuit when current passes through it (Li et al., 2021).

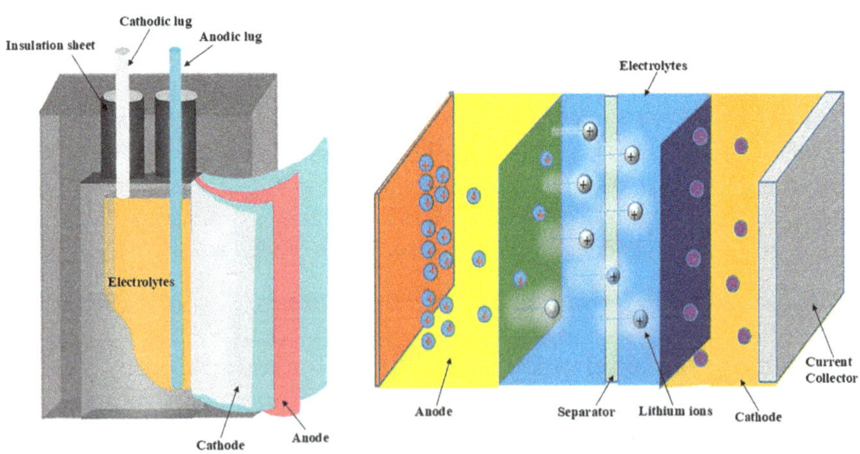

Fig. 1 Represent a schematic representation of the external and internal structural models of spent LIBs

3 Pretreatment Stages for Recycling of LIBs

Before the metal is extracted, there are a few stages in the recycling of LIBs known as pretreatment. The reprocessing of wasted batteries is a critical component of LIB recycling. These batteries contain toxic chemicals and valuable metals, demanding recycling for environmental and sustainability benefits. The major pretreatment steps for LIBs include the following steps:

3.1 Discharging

Before the metal is extracted, there are a few stages in the recycling of LIBs known as pretreatment. The reprocessing of wasted batteries is a critical component of LIB recycling. These batteries contain toxic chemicals and valuable metals, demanding recycling for environmental and sustainability benefits (Kim et al., 2021). There are different methods for discharging spent LIBs based on the battery type and structure. Whenever LIBs are connected to a load, such as a resistor or halogen light, they can be discharged to a safe voltage (Widijatmoko et al., 2020). The most practical and advised method for scaling up is self-discharging based on immersion in brine (a salt solution). For LIB discharge, various electrolyte solutions, including NaCl, KCl, $NaNO_3$, $NaSO_4$, $MnSO_4$, $MgSO_4$, $FeSO_4$, and $ZnSO_4$, have been investigated. Of these, NaCl is the most effective (Kaya, 2022). On an industrial scale, it was reported that shredding LIBs might not require a discharge phase. However, it is crucial to carry out this procedure in a system with adequate ventilation to collect any emitted gases if the LIBs are shredded without venting. For activities conducted on an industrial scale, eliminating the discharge stage is also a feasible alternative. After being completely depleted, lithium batteries are moved to a controlled location where they can be opened securely because some of the chemicals in the batteries may react with moisture in the air and oxygen (Ou & Wang, 2022).

3.2 Dismantling

The dismantling of separators, plastics, partial adhesives, metal shells, and electrolytes following the discharging step improved the separation efficiency in the subsequent battery recycling stages. Manually disassembling individual battery cells is routine in laboratory investigations, but it is costly for industrial processing (Zhou et al., 2021). Although this approach is difficult due to the wide range of battery designs, automatic or mechanical disassembly can be devised for operations on industrial scales. Therefore, shredding is a common step in processing flowsheets for tiny LIBs, and before materials can be recovered, various components must be separated (Kaya, 2022). When the spent LIBs were disassembled, the emission of volatile organic chemicals was examined.

3.3 Crushing

Crushing is a vital initial treatment process that employs a substantial quantity of discarded LIBs generated after sorting and disassembly. Its primary objectives are to diminish the volume and quintessence of valuable components. Crushing significantly diminishes the dimensions of spent LIBs, thereby resulting in a notable reduction in overall volume. This technique reveals all the ingredients of the LIBs, including electrodes, electrolytes, and other materials. Crushing increases surface area, which is advantageous for subsequent procedures such as leaching or chemical treatments, hence improving overall material recovery efficiency. During the crushing procedure, all LIB components are exposed, which increases the risk of micro short circuits caused by contact between the cathode and anode materials. Impact crushers, granulators, cutting mills, and shredders are among the most often investigated crushers in research, with a focus on their usefulness in recycling spent LIBs using cathodes. The crushing process causes tremendous friction and high-speed impact, resulting in a temperature increase of up to 300 °C. This elevated temperature might cause the electrolyte to disintegrate and volatilize, resulting in the release of toxic gasses like HF. Crushing LIBs in liquid N_2 or inert N_2 is an alternate method for reducing hazardous gas emissions (Zhang et al., 2013).

3.4 Particle Size Classification

The final processing procedure includes sieving to distinguish and intensify the metallic section of discarded LIBs. This stage depicts the overall distribution of precious metals across different particle sizes in the crushed sample. Metals can be divided into five size fractions: ultrafine (<0.5 mm), fine (0.5–1 mm), mid (1–2.5 mm), coarse (2.5–6 mm), and biggest (>6 mm) using a vibrating machine system and several sorting sieves (Roy et al., 2021). The process is mainly carried out via sieving or screening, which can be mechanized with an automatic screening system. Larger components, such as organic membranes, metallic shells, copper/aluminum foils, and different polymers, are distinguished from tiny particles, including graphite and fine $LiCoO_2$ particles, also known as black masses. To achieve separation, several approaches for size and concentration are used, including magnetic separation, sieving, electrostatic separation, air separation, electrodynamic separation, and methods based on the physical qualities of the product (Yu et al., 2021).

4 Extraction of Lithium from Spent LIBs

To recover important metals pretreated spent LIB materials—which are primarily cathodic—move on to the following recycling cycle. Retrieval of valuable metals from abandoned LIBs is often accomplished using the following methods: bioleaching, pyrometallurgy, or biohydrometallurgy

4.1 Pyrometallurgy

Pyrometallurgy constitutes a metallurgical discipline centered on employing elevated temperatures to extract and treat metals derived from ores (Das & Ghosh, 2023e, 2023g; Mohanty et al., 2016). Within the realm of pyrometallurgical procedures, the emphasis lies on subjecting raw materials, such as ores and concentrates, to thermal treatment. This treatment induces physical and chemical changes, leading to the extraction of valuable metals from spent LIBs. Key activities in pyrometallurgy include smelting, roasting, and various heating or melting techniques, utilizing heat to segregate metals from their ores and attain specific metallurgical objectives. In contrast, pyrometallurgy differs from hydrometallurgy in that it involves the use of aqueous solutions for the extraction of metals (Ali et al., 2022). After pretreatment, the active ingredients are ground into a powder and calcined. Li_2CO_3 and metal oxides are produced when the anode and cathode's lithium metal oxides react at temperatures above 700 °C. The calcined powder is leached afterward to dissolve Li_2CO_3 in water, as metal oxides do not undergo dissolution in water. Filtration was employed for water leaching, leading to the parting of the aqueous solution and the metal oxide residue, resulting in a Li_2CO_3 solution. Subsequently, water evaporation was carried out to produce Li_2CO_3 (Bae & Kim, 2021).

Li et al. (2016) developed a pyrometallurgical process to extract Li, Co, and graphite from wasted LIB. This unique and environmentally beneficial method incorporated an oxygen-free wet magnetic separation method. Co, Li, and graphite had recovery rates of 95.72%, 98.93%, and 91.05%, respectively. There are various disadvantages to pyrometallurgy, including the difficulties in reusing the electrolyte and the ease with which Li can be misplaced in the slag. Furthermore, pyrometallurgy releases pollutants into the environment during the smelting process, such as sulfur dioxide and other gasses.

4.2 Hydrometallurgy

Metals that constitute cathodic materials are liquified at low temperatures during the hydrometallurgical process. This aqueous approach makes it easier to recover metals after separation and is followed by purification. Acid and biological leaching are the main types of leaching and reduction that hydrometallurgy typically entails (Das et al., 2012a; Das & Ghosh, 2023g; Ghosh et al., 2016a; Mohanty et al., 2017). A group of scientists recently conducted an experiment that resulted in the total extraction of Co, Li, and Ni, as well as a 93% extraction of Mn. This was proficient in the absence of a reducing agent, using 1 mol/L H_2SO_4 at 90 °C and a solid-to-liquid ratio of 1:10 from spent Lithium Nickel Manganese Cobalt battery (NMC-type batteries) (Biswal & Balasubramanian, 2023). The hydrometallurgical process is thought to provide greater advantages than the pyrometallurgical method since it produces less greenhouse emissions and requires less energy (Ali et al., 2022).

4.3 Bioleaching

Bioleaching employs three primary processes. The first method is acidolysis, in which bacteria produce organic acids that dissolve lithium compounds. The second method is complexation, in which bacteria produce chelating molecules that bind to metal ions, increasing their solubility. The third type is redox reactions, in which specific microbes facilitate redox processes that transform metals into soluble forms. Iron-oxidizing bacteria convert ferrous Fe^{2+} to Fe^{3+}, allowing lithium to be extracted from cathodes. Several microorganisms have been explored for lithium bioleaching, including bacteria like *Acidithiobacillus ferrooxidans*, identified for its capability to oxidize sulfides and facilitate the leaching of lithium and cobalt, and *Leptospirillum ferrooxidans*, which works synergistically with *A. ferrooxidans* to enhance metal. Fungi are also efficient, creating organic acids like citric acid, that can dissolve lithium compounds. Numerous factors affect the efficiency of lithium bioleaching. Microorganisms require certain pH ranges for optimal activity, with most acidophilic bacteria working best in acidic circumstances (pH 1.5–3.0). Temperature also plays an important influence, as most bacteria thrive between 25 and 37 °C. The solid-to-liquid ratio, or pulp density, influences microbial activity and metal solubilization because high pulp densities can restrict microbial growth. High levels of hazardous metals, such as cobalt, can also impede microbial activity, needing pretreatment or dilution. Furthermore, appropriate oxygenation and mixing promote microbial activity and acid generation. Bioleaching has various advantages over traditional recycling processes. It is environmentally friendly, resulting in lower greenhouse gas emissions and hazardous byproducts. The procedure is energy efficient, working at ambient temperatures and consuming less energy. Bioleaching is also cost-effective because it reduces the need for costly chemicals (Priyadarsini & Das, 2025). Furthermore, it enables the selective recovery of specific metals, lowering contamination and increasing recovery rates. Despite its benefits, bioleaching confronts numerous hurdles. The procedure is slower than hydrometallurgical methods, requiring more processing time. The transition from laboratory-scale to industrial-scale bioleaching presents substantial technological and economic obstacles. Microorganisms are susceptible to environmental factors such as hazardous metal concentrations, which can impair their efficiency. Furthermore, effective bioleaching frequently necessitates LIB pretreatment, which increases the process's complexity and cost.

4.3.1 Methods of Bioleaching

As shown in Fig. 2, bioleaching experiments can be conducted using one-step, two-step, and cell-free spent medium. Spent LIBs are inserted with microbes in the culture medium, allowing for metal leaching (Moazzam et al., 2021). The impact of lithium from LIB waste on microbial growth could hinder the efficiency of the bioleaching. Conducting fermentation and bioleaching concurrently in a single step has

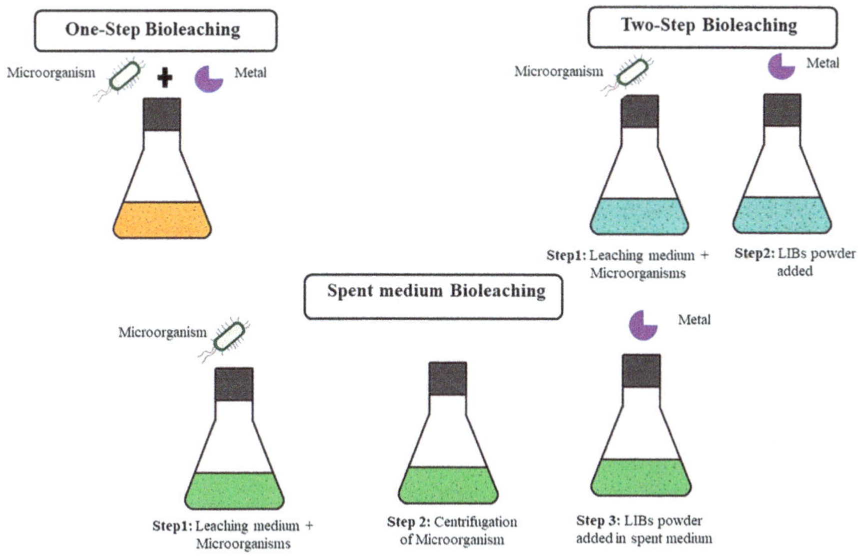

Fig. 2 Describes the schematic representation of three primary bioleaching strategies for treating spent LIBs: (a) single-step bioleaching, (b) two-step bioleaching, and (c) spent medium bioleaching

the potential to decrease operational costs during one-step leaching due to the presence of microbes in the culture (Biswal & Balasubramanian, 2023). The bacteria proliferate in the source medium during the two-step bioleaching procedure, eventually attaining the logarithmic growth phase. Subsequently, LIB waste is supplementary to start the leaching process of metal. Two-step bioleaching develops an acidic solution with the assistance of microorganisms that have already added spent LIBs to the medium, making it possibly more appropriate for industrial applications than one-step bioleaching (Mishra & Goyal, 2014). Although the two-step methodology surpasses the one-step recovery produced, method efficacy can be improved by removing microorganisms from the culture and mixing the spent medium with LIBs waste. A method known as "spent medium leaching" involves growing bacteria in growth media till they yield metabolites. The microbial solution was filtered and centrifuged in a liquid medium to provide a cell-free medium for bioleaching treatments. In the penultimate stage of bioleaching, LIB waste is put into a cell-free medium containing metabolites produced during biogenesis (Heydarian et al., 2018). In comparison to the other two methods, the spent-medium bioleaching method requires less time and a simpler operating procedure. Separating the leaching and metabolite synthesis phases permits the autonomous optimization of chemical and biological processes in the absence of microorganisms (Moosakazemi et al., 2023). Due to the toxicity, which causes microbial cell breakdown and cell wall degradation, this method does not restrict the usage of high pulp densities (Das & Singh, 2011; Dey et al., 2023a; Kanungo et al., 2023). However, spent-medium technology involves significant operating costs due to the necessity of additional bioreactor tanks. This is because microbes ferment, and the metabolites produced by them are utilized in two stages.

Horeh et al. (2016) used *Aspergillus niger* to extract metals from LIBs using three different bioleaching techniques. The spore suspension was inoculated into the source medium, together with the LIBs powder, in one step using a 30-day incubation period. Ni and Co significantly leached during the procedure, whereas Li, Al, Cu, and Mn leached only 8%, 100%, 58%, and 11%, respectively. After the two steps of the bioleaching process, involving pH reduction and organic acid generation during a 3-day logarithmic growth, the fungi were precultivated in culture media. Bioleaching commenced by introducing battery powder into the source medium. Through this approach, after 27 days of incubation, Li, Al, Mn, Cu, Co, and Ni exhibited efficiencies of 100%, 61%, 10%, 6%, and 1%, respectively. Utilizing the spent media approach for 14 days to culture fungi in the medium resulted in the fungi reaching the logarithmic state. Analysis revealed that Co, Li, Mn, Al, Co, and Ni achieved the highest metal recoveries, with efficiencies of 100%, 95%, 70%, 65%, 45%, and 38%, respectively.

Biswal et al. (2018) investigated the one-step and spent-medium methods used to recover Li and Co from spent LIBs utilizing *Aspergillus niger* strains (MM1 and SG1) and *Acidithiobacillus thiooxidans*. Additional advantages over other methods were obtained from the cultivation of microorganisms and their production of organic acids in the absence of spent LIBs, followed by the initiation of bioleaching by cell-free waste media. When the metal extraction in the spent medium technique was compared utilizing bacterial and fungal leaching, it was shown that *Acidithiobacillus thiooxidans* removed only 66% of the Li and 23% of the Co, but *Aspergillus niger* strains recovered 100% of the Li and 82% of the Co.

4.3.2 Factors Affecting Lithium Bioleaching

Sanitizing Li bioleaching methods requires comprehensive knowledge and cautious management of several parameters such as temperature, pH, pulp density, aeration, and growth media. Customizing bioleaching parameters according to the attributes of the Li source material and the specific microorganisms utilized is fundamental to accomplishing effective and economically sound Li extraction from spent LIBs. Optimizing process parameters during bioleaching is imperative for achieving maximum metal recovery, increasing microbial activity, minimizing reaction time, and effectively regulating the formation of byproducts. Several factors can significantly affect the efficiency of Li bioleaching processes.

Aerobic Environment (Dissolved Oxygen Level)

During the bioleaching process, microbes require a sufficient supply of oxygen to grow and be extremely productive. Agitation through shaking, stirring, or aeration serves as the means to achieve an aerobic environment. The choice of aerating conditions is primarily dictated by the type of microbes laboring in a bioleaching process. Typically, bioleaching bacteria thrive and remain active under aerobic conditions (Moazzam et al., 2021). The oxidation of Fe^{2+} as an electron donor in the substrate or the conversion of Fe^{2+} to Fe^{3+} provides *A. ferrooxidans* with the energy needed to proliferate. Dissolved oxygen (O_2) serves as the last electron acceptor in this process (Li et al., 2010). Some bioleaching bacteria can thrive and operate not only under aerobic conditions but also in anaerobic environments. *Acidithiobacillus* sp. is a bacterium frequently employed in both oxygen-rich and deficit environments for bioleaching investigations. *Acidithiobacillus ferrooxidans* is largely an anaerobic bacterium, but it also finds energy via the oxidation of S^0 or Fe^{2+} molecules. Yet, it can also initiate anaerobic metabolism and change Fe^{3+} to Fe^{2+} in anaerobic environments (Tezyapar Kara et al., 2023).

Effect of Pulp Density

The pulp density, which is determined by dividing the mass of solid matter by the volume of liquid, is essential for bioleaching procedures. High pulp densities negatively affect the development and productivity of microorganisms. An increase in pulp density typically diminishes the effectiveness of bioleaching due to the increase in the concentration of harmful heavy metals that exist in the leaching sources. The pulp density that is employed must match its maximum value to progress the economic efficiency of the bioleaching process (Wang et al., 2014). Through both contact and noncontact mechanisms, increases in pulp density or the ratio of bioleaching media to LIBs usually decrease bioleaching efficiency. This is because higher pulp density results in higher concentrations of potentially harmful substances (including electrolytes), higher pH, lower ORP Oxidation-reduction potential, and more metals that can be leached, all of which increase the concentration of metals in the solution (Sethurajan & Gaydardzhiev, 2021). Increased pulp density from 1% to 4% resulted in a significant drop in both metals' leaching efficiency (Niu et al., 2014).

Temperature

Temperature is an important element in bioleaching since it affects the development and metabolism of related microbes as well as bioleaching processes (Rasoulnia et al., 2021). The pace at which bacteria grow is strongly impacted by temperature fluctuations, even if they are only 2 °C. Nonetheless, certain strains might adjust to the temperature shift and begin growing with a marginally longer lag phase (Sethurajan & Gaydardzhiev, 2021). Normally, fungi thrive and exhibit activity within the temperature range of 20–40 °C. When temperatures exceed this optimal range, the protein, and enzymes crucial for microbial processes begin to undergo denaturation. Therefore, one of the key factors influencing the excellent of the appropriate temperature for bioleaching is the specific type of microorganism employed (Moazzam et al., 2021).

pH

Considering pH influences microbial growth and activity, it is critical to adjust this variable during bioleaching experiments. The optimal pH range for a given bioleaching technique is determined by both the microorganisms used and the operating system (Moazzam et al., 2021). *Acidithiobacillus ferrooxidans* and *Acidithiobacillus thiooxidans* require an acidic environment to thrive and operate effectively as bioleaching agents. A pH of 2–2.5 is optimal for these organisms, and exceeding this range dramatically lowers bioleaching yield (Gu et al., 2018). Fungal-based bioleaching can be carried out in a pH range of 3.0–7.0. According to one study, an initial pH rise from 1.5 to 4.0 had little effect on *A. ferrooxidans*' bioleaching of Co from LIBs, but leaching efficiency fell when the pH was reduced further to 1.0 (Li et al., 2013). A fungal bioleaching experiment produces a varied spectrum of metabolites, such as gluconic acid, citric acid, and malic acid, with an initial pH of 5.44, a sucrose concentration of 116.90 g/L, and an inoculum size of 3.45% (v/v) (Bahaloo-Horeh & Mousavi, 2017).

Medium and Cell Nutrients

The rate and extent of microbial growth, as well as the production of metabolites crucial for the solubilization of lipophilic acids in metal processes, are influenced by the type and quantity of nutrients employed. Significant chemical components, including the necessary elements for microbial growth, biosynthesis, and activity, must be present in a suitable culture medium. Among the ingredients required in an adequate medium are magnesium, phosphorus, and nitrogen. Ammonium, potassium, and phosphate make up 9K media, the most used media in bioleaching processes (Moazzam et al., 2021). As an essential component of sulfur metabolism, S^0 is an electron donor that provides bioleaching microorganisms with energy.

5 Conclusion

In comparison to conventional hydrometallurgical or pyrometallurgical procedures, lithium bioleaching is a viable, sustainable, and cost-effective way of recovering lithium from discarded mobile batteries. Bioleaching, which uses microbial processes to dissolve lithium and other precious metals under mild circumstances, is an environmentally conscious choice. The higher possibility of employing bacteria and fungi, particularly acidophilic and metal-tolerant species, offers substantial promise in increasing lithium yields while reducing environmental pollution. Considering its benefits, studies involve problems such as extended processing times, bacterial adaption, and performance adjustments. Enhancing microbial engineering procedures, adjusting various parameters, and creating biotechnology can enhance extraction yields and ultimately scale up the process on a large scale. Because of a decreased demand for primary resources and its impact on the environment, bioleaching can help to create a more sustainable lithium supply chain. Additional studies should be directed at refining the microbial consortia, improving the circumstances for bioleaching procedures, and developing combination bio-hydrometallurgical ways for obtaining more lithium. Through additional research, lithium bioleaching has the potential to transform LIBs recycling, resulting in better resource management and a shift toward sustainable energy supply.

References

Ali, H., Khan, H. A., & Pecht, M. (2022). Preprocessing of spent lithium-ion batteries for recycling: Need, methods, and trends. *Renewable and Sustainable Energy Reviews, 168*, 112809. https://doi.org/10.1016/J.RSER.2022.112809

Bae, H., & Kim, Y. (2021). Technologies of lithium recycling from waste lithium ion batteries: A review. *Materials Advances, 2*(10), 3234–3250. https://doi.org/10.1039/d1ma00216c

Bahaloo-Horeh, N., & Mousavi, S. M. (2017). Enhanced recovery of valuable metals from spent lithium-ion batteries through optimization of organic acids produced by Aspergillus niger. *Waste Management, 60*, 666–679. https://doi.org/10.1016/j.wasman.2016.10.034

Bal, B., Ghosh, S., & Das, A. P. (2018). Microbial recovery and recycling of manganese waste and their future application: A review. *Geomicrobiological Journal, 36*(1), 85–96. https://doi.org/10.1080/01490451.2018.1497731

Baniasadi, M., Vakilchap, F., Bahaloo-Horeh, N., Mousavi, S. M., & Farnaud, S. (2019). Advances in bioleaching as a sustainable method for metal recovery from e-waste: A review. *Journal of Industrial and Engineering Chemistry, 76*, 75–90. https://doi.org/10.1016/J.JIEC.2019.03.047

Bhattacharjee, J., Mishra, S., & Das, A. P. (2021). Recent advances in sensor-based detection of toxic dyes for bioremediation application: A review. *Applied Biochemistry and Biotechnology, 2021*, 1–20. https://doi.org/10.1007/S12010-021-03767-7

Biswal, B. K., & Balasubramanian, R. (2023). Recovery of valuable metals from spent lithium-ion batteries using microbial agents for bioleaching: A review. In *Frontiers in microbiology* (Vol. 14). Frontiers Media S.A. https://doi.org/10.3389/fmicb.2023.1197081

Biswal, B. K., Jadhav, U. U., Madhaiyan, M., Ji, L., Yang, E. H., & Cao, B. (2018). Biological leaching and chemical precipitation methods for recovery of Co and Li from spent lithium-ion batteries. *ACS Sustainable Chemistry and Engineering, 6*(9), 12343–12352. https://doi.org/10.1021/acssuschemeng.8b02810

Das, A. P., & Bissoyi, A. (2011). Computational approach for comparative phylogenetic analysis of isolated chromium resistant strain Brevibacterium casei. *Journal of Engineering and Technology Research, 3*(3), 82–87. https://doi.org/10.5897/JETR.9000020

Das, A. P., & Ghosh, S. (2018). Bioleaching of manganese from mining waste materials. *Materials Today: Proceedings, 5*(1), 2381–2390. https://doi.org/10.1016/J.MATPR.2017.11.459

Das, A. P., & Ghosh, S. (2023a). Cost-effectiveness of the biomining process. In *Manganese mining microorganisms* (pp. 143–151). https://doi.org/10.1016/B978-0-12-822147-1.00008-X

Das, A. P., & Ghosh, S. (2023b). Factors affecting bioleaching. In *Manganese mining microorganisms* (pp. 77–86). https://doi.org/10.1016/B978-0-12-822147-1.00004-2

Das, A. P., & Ghosh, S. (2023c). Mechanism of manganese bioleaching. In *Manganese mining microorganisms* (pp. 87–98). https://doi.org/10.1016/B978-0-12-822147-1.00014-5

Das, A. P., & Ghosh, S. (2023d). Metagenomic insights into unculturable microbial diversity. In *Manganese mining microorganisms* (pp. 63–76). https://doi.org/10.1016/B978-0-12-822147-1.00002-9

Das, A. P., & Ghosh, S. (2023e). Microorganisms in manganese biomining processes. In *Manganese mining microorganisms* (pp. 55–61). https://doi.org/10.1016/B978-0-12-822147-1.00013-3

Das, A. P., & Ghosh, S. (2023f). Role of metabolites and metabolomics for understanding of manganese bioleaching. In *Manganese mining microorganisms* (pp. 117–127). https://doi.org/10.1016/B978-0-12-822147-1.00011-X

Das, A. P., & Ghosh, S. (2023g). *Manganese mining microorganisms*. Elsevier.

Das, A. P., & Mishra, S. (2008). Hexavalent chromium (VI): Health hazards and environmental pollutant. *Journal of Environmental Research and Development, 2*(3), 386–392.

Das, A. P., & Mishra, S. (2009). Hexavalent chromium [Cr (VI)]: Yellow water pollution and its remediation. *Sarovar Saurabh ENVIS Newsletter on Wetland Ecosystems, 5*(2).

Das, A., & Mishra, S. (2010). Biodegradation of the metallic carcinogen hexavalent chromium Cr(VI) by an indigenously isolated bacterial strain. *Journal of Carcinogenesis, 9*, 6. https://doi.org/10.4103/1477-3163.63584

Das, A. P., & Singh, S. (2011). Occupational health assessment of chromite toxicity among Indian miners. *Indian Journal Of Occupational And Environmental Medicine, 15*(1), 6. https://doi.org/10.4103/0019-5278.82998

Das, A. P., & Swain, S. (2013). Algal biosorption of toxic dye methylene blue. In *A potential source of food, feed, biochemicals, biofuels and biofertilizers. International Conference on Algal Biorefinery, Indian Institute of Technology* (p. 13).

Das, A. P., Sukla, L. B., Pradhan, N., & Nayak, S. (2011). Manganese biomining: A review. *Bioresource Technology, 102*(16), 7381–7387. https://doi.org/10.1016/J.BIORTECH.2011.05.018

Das, A. P., Sukla, L. B., & Pradhan, N. (2012b). Microbial recovery of manganese using *Staphylococcus epidermidis*. *International Journal of Nonferrous Metallurgy, 2012*(02), 9–12. https://doi.org/10.4236/IJNM.2012.12002

Das, A. P., Swain, S., Panda, S., Pradhan, N., & Sukla, L. B. (2012a). Reductive acid leaching of low grade manganese ores. *Geomaterials, 2012*(04), 70–72. https://doi.org/10.4236/GM.2012.24011

Das, A. P., Ghosh, S., Mohanty, S., & Sukla, L. B. (2015a). Advances in manganese pollution and its bioremediation. In L. B. Sukla, N. Pradhan, S. Panda, & B. K. Mishra (Eds.), *Environmental microbe biotechnology* (1st ed., pp. 313–328). Springer. https://doi.org/10.1007/978-3-319-19018-1_16

Das, A. P., Ghosh, S., Mohanty, S., & Sukla, L. B. (2015b). Consequences of manganese compounds: A review. *Toxicological & Environmental Chemistry, 96*(7), 981–997. https://doi.org/10.1080/02772248.2015.1005428

Das, A., Dey, S., Bala, K., Gnanasekaran, R., & Das, A. P. (2024a). Microbial bioremediation technology for sustainable treatment and management of synthetic microfiber waste. In *Sustainable microbial technology for synthetic and cellulosic microfiber bioremediation* (pp. 221–238). https://doi.org/10.1007/978-3-031-62600-5_12

Das, A., Dey, S., & Das, A. P. (2024b). Ecotoxicological impacts of synthetic microfiber pollutants and development of sustainable mitigation strategies. *Environmental Chemistry and Ecotoxicology, 7*, 201. https://doi.org/10.1016/J.ENCECO.2024.12.003

Dey, S., Das, A., Mallick, K., Sahu, A., & Das, A. P. (2023a). Environmental petroleum waste: Pollution, toxicity, sustainable remediation (pp. 159–175). https://doi.org/10.1007/978-3-031-48220-5_7

Dey, S., Tripathy, B., Kumar, M. S., & Das, A. P. (2023b). Ecotoxicological consequences of manganese mining pollutants and their biological remediation. *Environmental Chemistry and Ecotoxicology, 5*, 55. https://doi.org/10.1016/J.ENCECO.2023.01.001

Dey, S., Dash, A., & Das, A. P. (2024). Bioremediation as a sustainable and circular solution for mining waste management and bioenergy generation. In *Sustainable manage-*

ment of mining waste and tailings: A circular economy approach (pp. 107–119). https://doi.org/10.1201/9781003442455-6/BIOREMEDIATION-SUSTAINABLE-CIRCULAR-SOLUTION-MINING-WASTE-MANAGEMENT-BIOENERGY-GENERATION-SUDESHNA-DEY-AKANSHYA-DASH-ALOK-PRASAD-DAS

Fergus, J. W. (2010). Recent developments in cathode materials for lithium ion batteries. *Journal of Power Sources, 195*(4), 939–954. https://doi.org/10.1016/J.JPOWSOUR.2009.08.089

Ghosh, S., & Das, A. P. (2017). Bioleaching of manganese from mining waste residues using Acinetobacter sp. *Geology, Ecology, and Landscapes, 1*(2), 77–83. https://doi.org/10.1080/24749508.2017.1332847

Ghosh, S., & Das, A. P. (2018). Metagenomic insights into the microbial diversity in manganese-contaminated mine tailings and their role in biogeochemical cycling of manganese. *Scientific Reports, 8*(1), 1–12. https://doi.org/10.1038/s41598-018-26311-w

Ghosh, S., Mohanty, S., Akcil, A., Sukla, L. B., & Das, A. P. (2016a). A greener approach for resource recycling: Manganese bioleaching. *Chemosphere, 154*, 628–639. https://doi.org/10.1016/J.CHEMOSPHERE.2016.04.028

Ghosh, S., Mohanty, S., Nayak, S., Sukla, L. B., & Das, A. P. (2016b). Molecular identification of indigenous manganese solubilising bacterial biodiversity from manganese mining deposits. *Journal of Basic Microbiology, 56*(3), 254–262. https://doi.org/10.1002/JOBM.201500477

Ghosh, S., Bal, B., & Das, A. P. (2017). Enhancing manganese recovery from low-grade ores by using mixed culture of indigenously isolated bacterial strains. *Geomicrobiology, 35*(3), 242–246. https://doi.org/10.1080/01490451.2017.1362080

Ghosh, S., Gandhi, M., van Hullebusch, E. D., & Das, A. P. (2020). Proteomic insights into Lysinibacillus sp.-mediated biosolubilization of manganese. *Environmental Science and Pollution Research, 28*(30), 40249–40263. https://doi.org/10.1007/S11356-020-10863-4

Ghosh, S., Tripathy, B., Dey, S., & Das, A. P. (2023). Comparative investigation of fungal and bacterial manganese biomining mechanisms. In *Manganese mining microorganisms* (pp. 99–115). https://doi.org/10.1016/b978-0-12-822147-1.00016-9

Grey, C. P., & Hall, D. S. (2020). Prospects for lithium-ion batteries and beyond—A 2030 vision. *Nature Communications, 11*(1). https://doi.org/10.1038/s41467-020-19991-4

Gu, T., Rastegar, S. O., Mousavi, S. M., Li, M., & Zhou, M. (2018). Advances in bioleaching for recovery of metals and bioremediation of fuel ash and sewage sludge. *Bioresource Technology, 261*, 428–440. https://doi.org/10.1016/j.biortech.2018.04.033

Heydarian, A., Mousavi, S. M., Vakilchap, F., & Baniasadi, M. (2018). Application of a mixed culture of adapted acidophilic bacteria in two-step bioleaching of spent lithium-ion laptop batteries. *Journal of Power Sources, 378*, 19–30. https://doi.org/10.1016/j.jpowsour.2017.12.009

Horeh, N. B., Mousavi, S. M., & Shojaosadati, S. A. (2016). Bioleaching of valuable metals from spent lithium-ion mobile phone batteries using Aspergillus niger. *Journal of Power Sources, 320*, 257–266. https://doi.org/10.1016/j.jpowsour.2016.04.104

Kanungo, J., Sahoo, T., Swain, L. P., & Behera, I. D. (2023). Toxicity of persistent hydrocarbon pollutants, sources and sustainable remediation process (pp. 39–65). https://doi.org/10.1007/978-3-031-48220-5_2.

Kanungo, J., Sahoo, T., Bal, M., & Behera, I. D. (2024). Performance of bioremediation strategy in waste lubricating oil pollutants: A review. *Geomicrobiology Journal, 41*(4), 360–373. https://doi.org/10.1080/01490451.2023.2245395

Kaya, M. (2022). State-of-the-art lithium-ion battery recycling technologies. *Circular Economy, 1*(2), 100015. https://doi.org/10.1016/j.cec.2022.100015

Khatun, R., Singh, S., Dubey, N. K., & Prasad Das, A. (2023). A review on marine-based phytochemicals and their application in biomedical research. In *Recent frontiers of phytochemicals: Applications in food, pharmacy, cosmetics, and biotechnology* (pp. 383–395). Elsevier. https://doi.org/10.1016/B978-0-443-19143-5.00013-X

Kim, S., Bang, J., Yoo, J., Shin, Y., Bae, J., Jeong, J., Kim, K., Dong, P., & Kwon, K. (2021). A comprehensive review on the pretreatment process in lithium-ion battery recycling. *Journal of Cleaner Production, 294*. https://doi.org/10.1016/j.jclepro.2021.126329

Li, L., Ge, J., Chen, R., Wu, F., Chen, S., & Zhang, X. (2010). Environmental friendly leaching reagent for cobalt and lithium recovery from spent lithium-ion batteries. *Waste Management, 30*(12), 2615–2621. https://doi.org/10.1016/J.WASMAN.2010.08.008
Li, L., Zeng, G. S., Luo, S. L., Deng, X. R., & Xie, Q. J. (2013). Influences of solution pH and redox potential on the bioleaching of $LiCoO_2$ from spent lithium-ion batteries. *Journal of the Korean Society for Applied Biological Chemistry, 56*(2), 187–192. https://doi.org/10.1007/s13765-013-3016-x
Li, J., Wang, G., & Xu, Z. (2016). Environmentally-friendly oxygen-free roasting/wet magnetic separation technology for in situ recycling cobalt, lithium carbonate and graphite from spent $LiCoO_2$/graphite lithium batteries. *Journal of Hazardous Materials, 302*, 97–104. https://doi.org/10.1016/j.jhazmat.2015.09.050
Li, A., Yuen, A. C. Y., Wang, W., De Cachinho Cordeiro, I. M., Wang, C., Chen, T. B. Y., Zhang, J., Chan, Q. N., & Yeoh, G. H. (2021). A review on lithium-ion battery separators towards enhanced safety performances and modelling approaches. *Molecules, 26*(2). https://doi.org/10.3390/molecules26020478
Mallick, K., Sahu, S., Sahu, A., Dey, S., & Das, A. P. (2024). A critical review of marine microfiber pollution routes, toxicity, and its sustainable remediation. *Environmental Science and Engineering, Part, F2333*, 189–211. https://doi.org/10.1007/978-3-031-51792-1_10
Manthiram, A. (2017). An outlook on lithium ion battery technology. *ACS Central Science, 3*(10), 1063–1069. https://doi.org/10.1021/acscentsci.7b00288
Mishra, S., & Das, A. P. (2008). Hexavalent chromium reduction and 16S rDNA identification of bacteria isolated from a Cr (VI) contaminated site. *The Internet Journal of Microbiology, 7*(1).
Mishra, S., & Das, A. P. (2021). Current treatment technologies for removal of microplastic and microfiber pollutants from wastewater. In *Wastewater treatment: Cutting-edge molecular tools, techniques and applied aspects* (pp. 237–251). https://doi.org/10.1016/B978-0-12-821881-5.00011-8
Mishra, S., & Das, A. P. (2022). Treatment of the wastewater polluted with synthetic microfiber released from washing machine. In B. B. Das, H. Hettiarachchi, P. K. Sahu, & S. Nanda (Eds.), *Lecture notes in civil engineering* (Vol. 207, pp. 109–117). Springer. https://doi.org/10.1007/978-981-16-7509-6_9/COVER
Mishra, S., & Das, A. P. (2023). Microbial remediation of synthetic microfiber contaminated wastewater. In M. P. Shah (Ed.), *Microbial technologies in industrial wastewater treatment* (pp. 337–350). Springer. https://doi.org/10.1007/978-981-99-2435-6_17
Mishra, D., & Goyal, P. (2014). Estimation of vehicular emissions using dynamic emission factors: A case study of Delhi, India. *Atmospheric Environment, 98*, 1–7. https://doi.org/10.1016/J.ATMOSENV.2014.08.047
Mishra, S., Das, A. P., & Seragadam, P. (2009). Microbial remediation of hexavalent chromium from chromite contaminated mines of Sukinda Valley, Orissa (India). *Journal of Environmental Research and Development*, 1122–1127.
Mishra, S., Rath, C. C., & Das, A. P. (2019). Marine microfiber pollution: A review on present status and future challenges. *Marine Pollution Bulletin, 140*, 188–197. https://doi.org/10.1016/J.MARPOLBUL.2019.01.039
Mishra, S., Singh, R. P., Rath, C. C., & Das, A. P. (2020). Synthetic microfibers: Source, transport and their remediation. *Journal of Water Process Engineering, 38*, 101612. https://doi.org/10.1016/J.JWPE.2020.101612
Mishra, S., Rout, P. K., & Das, A. P. (2021). Emerging microfiber pollution and its remediation. In R. Prasad (Ed.), *Environmental and microbial biotechnology* (pp. 247–266). Springer. https://doi.org/10.1007/978-981-15-5499-5_9
Mishra, S., Singh, R. P., Rout, P. K., & Das, A. P. (2022). Membrane bioreactor (MBR) as an advanced wastewater treatment technology for removal of synthetic microplastics. In *Development in wastewater treatment research and processes* (pp. 45–60). https://doi.org/10.1016/B978-0-323-85583-9.00022-3
Mishra, S., Dash, D., & Das, A. P. (2024). Isolation and molecular characterization of polyester-degrading bacterial strains from municipal wastewater and their application in microplastic

degradation in contaminated wastewater. *Geomicrobiology Journal, 41*, 732. https://doi.org/10.1080/01490451.2024.2376101

Moazzam, P., Boroumand, Y., Rabiei, P., Baghbaderani, S. S., Mokarian, P., Mohagheghian, F., Mohammed, L. J., & Razmjou, A. (2021). Lithium bioleaching: An emerging approach for the recovery of Li from spent lithium ion batteries. *Chemosphere, 277*, 130196. https://doi.org/10.1016/j.chemosphere.2021.130196

Mohanty, S., Ghosh, S., Nayak, S., & Das, A. P. (2016). Isolation, identification and screening of manganese solubilizing fungi from low-grade manganese ore deposits. *Geomicrobiological Journal, 34*(4), 309–316. https://doi.org/10.1080/01490451.2016.1189016

Mohanty, S., Ghosh, S., Nayak, S., & Das, A. P. (2017). Bioleaching of manganese by Aspergillus sp. isolated from mining deposits. *Chemosphere, 172*, 302–309. https://doi.org/10.1016/J.CHEMOSPHERE.2016.12.136

Mohanty, S., Ghosh, S., Bal, B., & Das, A. P. (2018). A review of biotechnology processes applied for manganese recovery from wastes. *Reviews in Environmental Science and Bio/Technology, 17*(4), 791–811. https://doi.org/10.1007/S11157-018-9482-1

Mohanty, M., Tripathy, B., Koteswara Reddy, G., & Das, A. P. (2024). Greener technologies for establishing restraint over microplastic pollution. In *Sustainable microbial technology for synthetic and cellulosic microfiber bioremediation* (pp. 239–258). https://doi.org/10.1007/978-3-031-62600-5_13

Moosakazemi, F., Ghassa, S., Jafari, M., & Chelgani, S. C. (2023). Bioleaching for recovery of metals from spent batteries—A review. *Mineral Processing and Extractive Metallurgy Review, 44*(7), 511–521. https://doi.org/10.1080/08827508.2022.2095376

Nag, M., Lahiri, D., Ghosh, S., Sarkar, T., Pati, S., Das, A. P., Ram, D. K., Bhattacharya, D., & Ray, R. R. (2023). Application of microorganisms in biotransformation and bioremediation of environmental contaminant: A review. *Geomicrobiology Journal, 41*, 374. https://doi.org/10.1080/01490451.2023.2261443

Naseri, T., Bahaloo-Horeh, N., & Mousavi, S. M. (2019). Bacterial leaching as a green approach for typical metals recovery from end-of-life coin cells batteries. *Journal of Cleaner Production, 220*, 483–492. https://doi.org/10.1016/j.jclepro.2019.02.177

Nitta, N., Wu, F., Lee, J. T., & Yushin, G. (2015). Li-ion battery materials: Present and future. *Materials Today, 18*(5), 252–264. https://doi.org/10.1016/j.mattod.2014.10.040

Niu, Z., Zou, Y., Xin, B., Chen, S., Liu, C., & Li, Y. (2014). Process controls for improving bioleaching performance of both Li and Co from spent lithium ion batteries at high pulp density and its thermodynamics and kinetics exploration. *Chemosphere, 109*, 92–98. https://doi.org/10.1016/J.CHEMOSPHERE.2014.02.059

Osiak, M., Geaney, H., Armstrong, E., & O'Dwyer, C. (2014). Structuring materials for lithium-ion batteries: Advancements in nanomaterial structure, composition, and defined assembly on cell performance. *Journal of Materials Chemistry A, 2*(25), 9433–9460. https://doi.org/10.1039/c4ta00534a

Ou, X., & Wang, W. (2022). Recycling battery cathode materials. In *Nano technology for battery recycling, remanufacturing, and reusing* (pp. 303–319). https://doi.org/10.1016/B978-0-323-91134-4.00017-0

Paul, A., Dey, S., Ram, D. K., & Das, A. P. (2023). Hexavalent chromium pollution and its sustainable management through bioremediation. *Geomicrobiology Journal, 1–11*, 324. https://doi.org/10.1080/01490451.2023.2218377

Paul, A., Priyadarsini, S., Karakuş, S., Behera, J. K., Gnanasekaran, R., & Das, A. P. (2024). Fate and accumulation of synthetic microfibers in the environment. In *Sustainable microbial technology for synthetic and cellulosic microfiber bioremediation* (pp. 61–76). https://doi.org/10.1007/978-3-031-62600-5_4

Prabhakar, A., Mishra, S., & Das, A. P. (2019). Isolation and identification of lead (Pb) solubilizing bacteria from automobile waste and its potential for recovery of lead from end of life waste batteries. *Geomicrobiological Journal, 36*(10), 894–903. https://doi.org/10.1080/01490451.2019.1654044

Pradhan, G., Tripathy, B., Ram, D. K., Digal, A. K., & Das, A. P. (2023). Bauxite mining waste pollution and its sustainable management through bioremediation. *Geomicrobiology Journal, 41*, 1–10. https://doi.org/10.1080/01490451.2023.2235353

Priyadarsini, S., & Das, A. P. (2024). Analytical and structural characterization of waste lithium-ion batteries for their effective recycling strategy. *Environmental Chemistry and Ecotoxicology, 7*, 182. https://doi.org/10.1016/j.enceco.2024.12.004

Priyadarsini, S., & Das, A. P. (2025). Isolation and Identification of Native Lithium-Resistant Bacteria from Electronic Waste Contaminated sites: A focused Investigation on Tolerance Assessment with Potential Lithium Bio-recovery Application. *Geomicrobiology Journal*, 1–15. https://doi.org/10.1080/01490451.2025.2539509

Raj, K., & Das, A. P. (2023). Lead pollution: Impact on environment and human health and approach for a sustainable solution. *Environmental Chemistry and Ecotoxicology, 5*, 79–85. https://doi.org/10.1016/J.ENCECO.2023.02.001

Rasoulnia, P., Barthen, R., & Lakaniemi, A. M. (2021). A critical review of bioleaching of rare earth elements: The mechanisms and effect of process parameters. *Critical Reviews in Environmental Science and Technology, 51*(4), 378–427. https://doi.org/10.1080/10643389.2020.1727718

Razmjou, A., Asadnia, M., Hosseini, E., Habibnejad Korayem, A., & Chen, V. (2019). Design principles of ion selective nanostructured membranes for the extraction of lithium ions. *Nature Communications, 10*(1). https://doi.org/10.1038/s41467-019-13648-7

Roy, J. J., Cao, B., & Madhavi, S. (2021). A review on the recycling of spent lithium-ion batteries (LIBs) by the bioleaching approach. *Chemosphere, 282*, 130944. https://doi.org/10.1016/J.CHEMOSPHERE.2021.130944

Sadasivuni, K. K., Rattan, S., Waseem, S., Brahme, S. K., Kondawar, S. B., Ghosh, S., Das, A. P., Chakraborty, P. K., Adhikari, J., Saha, P., & Mazumdar, P. (2019). Silver nanoparticles and its polymer nanocomposites—Synthesis, optimization, biomedical usage, and its various applications. In *Lecture notes in bioengineering* (pp. 331–373). https://doi.org/10.1007/978-3-030-04741-2_11

Sethurajan, M., & Gaydardzhiev, S. (2021). Bioprocessing of spent lithium ion batteries for critical metals recovery—A review. *Resources, Conservation and Recycling, 165*, 105225. https://doi.org/10.1016/j.resconrec.2020.105225

Singh, R. P., Mishra, S., & Das, A. P. (2020). Synthetic microfibers: Pollution toxicity and remediation. *Chemosphere, 257*, 127199. https://doi.org/10.1016/j.chemosphere.2020.127199

Tezyapar Kara, I., Kremser, K., Wagland, S. T., & Coulon, F. (2023). Bioleaching metal-bearing wastes and by-products for resource recovery: A review. *Environmental Chemistry Letters., 21*, 3329–3350. https://doi.org/10.1007/s10311-023-01611-4

Wang, Y., Song, Z., Everaert, N., De Ketelaere, B., Willemsen, H., Decuypere, E., & Buyse, J. (2014). The anorectic effects of alpha-lipoicacid are mediated by central AMPK and are not due to taste aversion in chicken (Gallus gallus). *Physiology & Behavior, 132*, 66–72. https://doi.org/10.1016/J.PHYSBEH.2014.04.047

Widijatmoko, S. D., Fu, G., Wang, Z., & Hall, P. (2020). Recovering lithium cobalt oxide, aluminium, and copper from spent lithium-ion battery via attrition scrubbing. *Journal of Cleaner Production, 260*, 120869. https://doi.org/10.1016/J.JCLEPRO.2020.120869

Yu, D., Huang, Z., Makuza, B., Guo, X., & Tian, Q. (2021). Pretreatment options for the recycling of spent lithium-ion batteries: A comprehensive review. *Minerals Engineering, 173*, 107218. https://doi.org/10.1016/j.mineng.2021.107218

Zhang, T., He, Y., Ge, L., Fu, R., Zhang, X., & Huang, Y. (2013). Characteristics of wet and dry crushing methods in the recycling process of spent lithium-ion batteries. *Journal of Power Sources, 240*, 766–771. https://doi.org/10.1016/J.JPOWSOUR.2013.05.009

Zhou, M., Liaw, S., Sun, Q., & He, R. (2021). *Pretreatment, recycling, and regeneration strategies of cathode active materials from spent lithium ion batteries*.

An Overview of Occurrence, Application, and Recovery of Rare Earth Metals from E-Waste

Berat Batuhan Kaplangı and Mehmet Ali Küçüker

Abstract The concept of urban mining is a circular economy application that treats the waste electric and electronic equipment (WEEE or e-waste) as a secondary resource for valuable elements. It reduces dependence on raw materials by recovering these valuable elements, including rare-earth metals (REMs). WEEE comprises discarded electronic devices such as smartphones, cameras, computers, and household appliances. E-waste contains significant quantities of REMs like neodymium (Nd), dysprosium (Dy), lanthanum (La), and cerium (Ce). Recovering these kinds of metals is challenging due to the complex composition of e-waste. Environmental challenges also arise associated with improper disposal. In this chapter, recovery methods of REMs from e-waste are focused on. Conventional recovery methods including hydrometallurgical and pyrometallurgical techniques have shown remarkable recovery rates for REMs from WEEE components such as printed circuit boards, and fluorescent lamps. Other alternative methods, such as bioleaching and ionic liquid techniques, offer more environmentally friendly and sustainable solutions by reducing harsh chemical consumption during recovery operations. In addition, emerging pretreatment methods including alkaline roasting offer eco-friendly potential to increase the REM extraction efficiency. These recovery techniques have economic and environmental advantages rather than REM mining. Life cycle assessment studies in the literature highlights the significant reduction in greenhouse gas emissions, energy consumption, and waste generation-especially hazardous wastes—due to the urban mining efforts, including REM recovery techniques, compared to traditional mining.

B. B. Kaplangı · M. A. Küçüker (✉)
Department of Environmental Engineering, İzmir Institute of Technology, İzmir, Türkiye

A. P. Das, S. Priyadarsini (eds.), *Electronic Waste and Environmental Pollution*, Sustainable Environmental Waste Management Strategies,
https://doi.org/10.1007/978-3-031-90287-1_13

1 Introduction

E-waste, or waste electric and electronic equipment (WEEE), is one of the waste streams that grows rapidly on a global scale. WEEEs have both environmental and economic effects. Improper disposal and treatment techniques of these types of wastes contribute to the depletion of the resources and environmental impacts (Murthy & Ramakrishna, 2022; Mishra et al., 2023). The growing demand for electric and electronic devices increases the need for rare-earth metals (REMs). This reality underlines the importance of the innovative resource recovery practices from e-waste (Xavier et al., 2021).

Nd, La, and Ce can be classified as REMs associated with the production of modern electric and electronic devices including smartphones, cameras, computers, etc. (Vuppaladadiyam et al., 2024). REMs are crucial for the operation of advanced technological equipment including integrated circuits, superconductors, lasers, and batteries. Liquid crystal displays (LCDs) and cathode ray tubes (CRTs) are the other devices in that REMs play a key role during their operations.

In addition to their usage in electric and electronic equipment, REMs enhance metallurgical processes. They improve the strength of the alloys and optimize the purification of materials. The widespread use of REMs in diverse sectors and their ability to provide enhanced magnetic, optical, and catalytic properties make them essential for modern technology (Ambaye et al., 2020). In terms of green transition, the mentioned properties make the applications of REMs very important for clean energy structures and eco-friendly devices. For instance, Nd and Pr are widely used in the production of magnets for wind turbines and electric vehicles. La and Ce are other REMs that are used in emerging battery technology (Ramprasad et al., 2022). Therefore, ensuring REMs availability is crucial for green technology and proper e-waste recycling methods serve as a circular economy concept (Liang et al., 2024). In Table 1, applications of the REMs in electric and electronic devices are summarized. This table highlights the wide range of usage and availability of REMs in the modern world.

In the natural reserves, the availability of REMs is becoming limited day by day. To overcome this problem, e-waste represents a valuable alternative source of these essential elements. Mitigating resource scarcity and reducing the environmental impacts of traditional REM mining are two main benefits of innovative and advanced recovery processes (Ambaye et al., 2020). Within the scope of the urban mining concept, extraction of REMs from WEEE is one of the objectives that should be achieved for a circular economy and these efforts reduce the dependence on primary resources, foster resource consumption efficiency, and support global sustainable development (Ouro-Salim, 2024; Erdiaw-Kwasie et al., 2024).

On the other hand, it should be noted that recovering valuable materials, especially REMs, from e-waste has significant challenges. The complex and heterogeneous composition of e-waste is one of the challenges. For instance, printed circuit boards (PCBs) contain over 1000 different ingredients including hazardous elements (Pb, Cd, brominated flame retardants), and these ingredients make the extraction process much more complicated (Vuppaladadiyam et al., 2024). In addition, the

Table 1 Rare-earth elements and their applications

Rare-earth metal (REM)	Primary components found in	Applications for electronic devices
Neodymium (Nd)	Permanent magnets (e.g., NdFeB magnets)	High-performance magnets are used in hard disk drives, electric vehicle motors, and wind turbine generators
Dysprosium (Dy)	Permanent magnets	Enhances thermal resistance in magnets for electric vehicle motors and wind turbines
Yttrium (Y)	Phosphors in fluorescent lamps	Red and green phosphors for displays and fluorescent lamps
Europium (Eu)	Phosphors in fluorescent lamps	Red phosphor in display screens, fluorescent lamps, and LED technologies
Lanthanum (La)	Nickel-metal hydride batteries (NiMH)	Electrodes in batteries, optical lenses, and catalytic converters
Cerium (Ce)	Catalysts and display technologies	Polishing powders for glass, additives in catalytic converters, and phosphors in televisions and displays
Praseodymium (Pr)	Magnets, display phosphors	Magnets in electronic devices and green phosphors in display technologies
Terbium (Tb)	Phosphors	Green phosphors for screens, lighting systems, and LED displays
Samarium (Sm)	Magnets, capacitors	High-temperature-resistant magnets and components in capacitors for electronics
Gadolinium (Gd)	Alloys, imaging technologies	Contrast agents in medical imaging and magnetic refrigeration systems

REM concentration in e-waste is less than 0.1% on average. This necessitates advanced separation techniques for the isolation of REMs from other ingredients (Balaram, 2024). To solve these problems, conventional techniques should be improved, and emerging techniques, including new pretreatment methods, should be widely used.

The most used recovery methods for REMs from e-waste are hydrometallurgical and pyrometallurgical techniques (Liang et al., 2024). During the hydrometallurgical processes, acids and solvents are used to dissolve and extract REMs like Y and Eu from fluorescent lamps and PCBs (Ramprasad et al., 2022). This intense chemical usage causes the generation of hazardous chemical wastes compared to other conventional methods. In contrast, chemical consumption in pyrometallurgical techniques is less than the hydrometallurgical techniques. However, due to the high-temperature smelting, pyrometallurgical techniques consume more energy. This high energy demand and associated CO_2 emissions cause environmental challenges (Liang et al., 2024). Hydrometallurgical techniques are more selective; on the other hand, pyrometallurgical methods recover a broader range of metals (Tan & Li, 2019).

In recent years, several emerging eco-friendly techniques are being developed for the recovery of REMs. Bioleaching and ionic liquid techniques are one of the remarkable emerging techniques and they have demonstrated high potential for REMs recovery. In bioleaching techniques, microorganisms are employed to extract REMs, especially Nd and Dy (Ambaye et al., 2020). Similarly, recyclable and nonvolatile ionic liquids can achieve selective separation of REMs during the applications of ionic liquid techniques. Generation of less chemical waste rather than the conventional techniques makes them sustainable option (Kaim et al., 2023).

Besides these conventional and emerging techniques, pretreatment techniques can play a vital role in improving recovery efficiencies. Alkaline roasting is one of these methods that shows improvements in REM recovery compared to untreated waste (Li et al., 2019). Membrane solvent extraction can also achieve high REM purity levels and less energy consumption relative to the conventional methods (Deshmane et al., 2019).

In Fig. 1, an overview of the main topics discussed in this book chapter is presented. The figure outlines the environmental and economic advantages of recovering REMs. It underlines the best practices for sustainable WEEE management in terms of REMs recovery. This figure serves as a visual representation of the content of the chapter.

2 E-Waste Recycling and REM Recovery Challenges

La, Lu, Y, Sc, and the other REMs found in e-waste are critical due to their unique properties, as discussed in the previous part. In addition to REMs, WEEEs contain other valuable metals including Co, Ni, Li, Ge, Te, Pt, etc., which are essential for the operations of electric and electronic devices (Balaram, 2024).

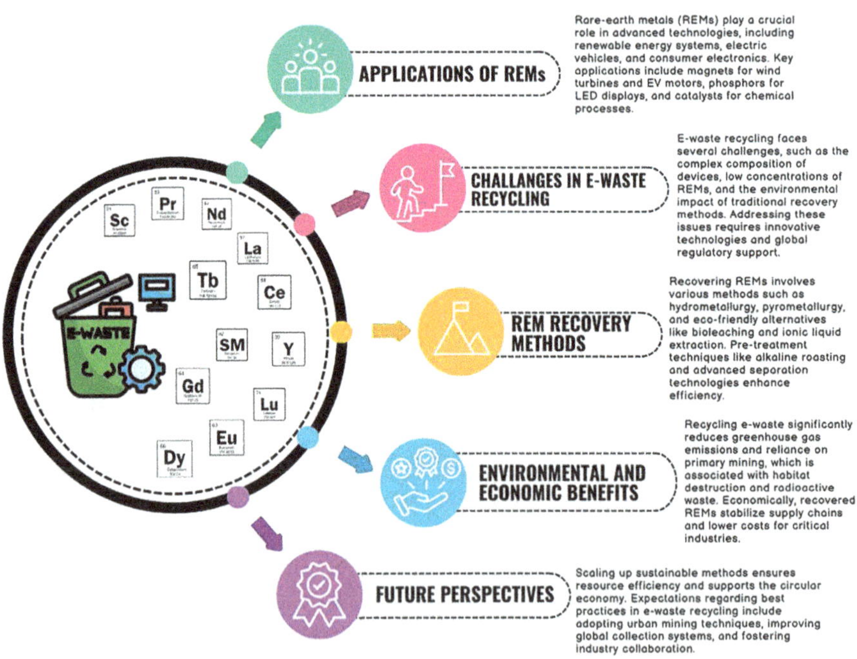

Fig. 1 Overview of occurrence, application, and recovery of REMs from e-waste

PCBs account for 3–6% of the total amount of electric and electronic equipment in the world and this type of material contains REMs higher than the natural ores. Magnets like NdFeBs are rich in Nd, Pr, and Dy. Optimized recovery techniques can achieve 98% efficiency from these magnets. Similarly, fluorescent lamps contain high concentrations of Y and Eu. Hydrometallurgical methods enable recovery rates of these metals to exceed 85%. Nickel-metal hydride (NiMH) batteries also contain REMs, especially La. Advanced recovery techniques, such as leaching processes, can achieve recovery efficiencies of up to 93% (Vuppaladadiyam et al., 2024).

Besides these high recovery ratios, the challenges associated with REMs recovery from e-waste are complex and multifaceted. One of the major challenges is the diverse composition of e-waste. These types of waste contain mixes of metals (Mishra et al., 2022), plastics (Priyadarshini et al., 2024; Singh Jadaun et al., 2022), polymers complicating the application of standardized recovering processes (Vuppaladadiyam et al., 2024).

In addition, low REMs concentration levels within WEEE require advanced and energy-intensive treatment technologies for efficient recovery. Pyrometallurgical techniques need significant energy input, and this causes high operational costs and emissions. Like pyrometallurgical techniques, hydrometallurgical methods have adverse environmental impacts due to the intense chemical consumption during the treatment. New treatment processes, including bioleaching, have some advantages, but slow reaction kinetics and expensive reagents can be classified as common barriers for the operations. Moreover, price fluctuations in REMs in the market directly affect the feasibility of the recovery operations (Li et al., 2019).

The lack of recovery infrastructure and policies can be another obstacle, especially for developing countries. Improper recovery facilities may lead to informal processing techniques. This kind of facility poses significant occupational health and safety as well as environmental risks. Regulatory and policy gaps, especially in developing countries, cause insufficient waste collection and slow the e-waste treatment applications. These challenges directly influence e-waste treatment and REMs recovery (Sharma et al., 2021; Ramprasad et al., 2022).

3 REM Recovery Methods from E-Waste

3.1 Hydrometallurgical Techniques

Leaching with hydrochloric acid (HCl), sulfuric acid (H_2SO_4), or nitric acid (HNO_3), REMs can be dissolved and then extracted and purified by solvent extraction and ion exchange methods. Hydrometallurgical techniques involve processes such as leaching, solvent extraction, ion exchange, and precipitation. During these steps, acidic or basic solutions are consumed to selectively dissolve and separate valuable materials from ore and waste.

Leaching is the first step of the hydrometallurgical technique and acidic solutions are effective for dissolving REMs from the e-waste matrix. The factors that affect the efficiency of the leaching process are type of acid and its concentration, temperature, and pressure. Leaching processes can be enhanced by employing microwave and this method can reduce the processing time and increase efficiency (Danouche et al., 2024; Kaim et al., 2023).

After the leaching step, impurities should be removed through selective precipitation. In this step, chemical agents such as oxalic acid are used to concentrate REMs to separate the other elements from REMs. After selective precipitation, precipitated REMs are separated by solvent extraction, ion exchange, and advance membrane techniques. Solvent extraction is the most common method where REMs are transferred into an organic solvent phase (Danouche et al., 2024). In the final stage, crystallization or additional membrane applications are used for converting REMs into a form suitable for industrial applications. At the end of the operations, chemical waste should carefully be managed to enhance the overall environmental performance (Kaim et al., 2023). The hydrometallurgical steps are summarized in Fig. 2.

3.2 *Pyrometallurgical Techniques*

The main factor of pyrometallurgical techniques is the use of high temperatures to extract metals from ore or waste materials (Priyadarsini & Das, 2025). Pyrometallurgical techniques consist of smelting, incineration or pyrolysis (thermal degradation without oxygen). These steps isolate hazardous substances from bulk materials or wastes (Priyadarsini & Das, 2024). Pyrometallurgy is effective in segregating precious metals (PMs) including gold, silver, and platinum group metals (PGMs) as well as REMs. E-wastes are processed in smelters, where metals are recovered in molten baths while oxides form a separate slag phase. This technique is suitable for complex materials like PCBs and does not require extensive pretreatment. Combustible fractions of e-waste can also be utilized for energy recovery (Khaliq et al., 2014).

Initial phase of the pyrometallurgy is the separation of the hazardous substances such as batteries and capacitors among the e-waste and size reduction of the feed.

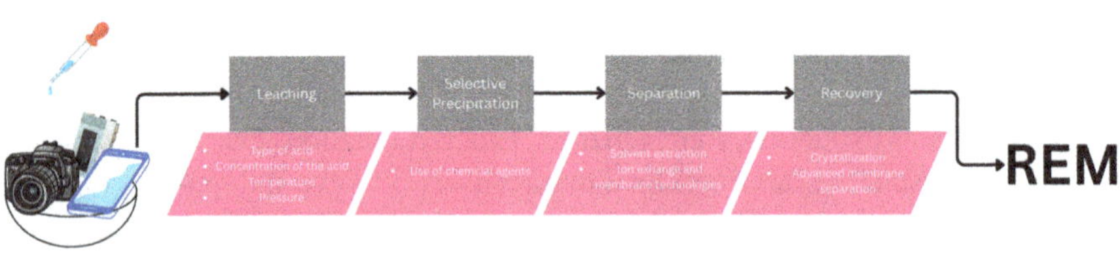

Fig. 2 Flow chart of hydrometallurgical process

This step prepares the input for further treatment and increases the efficiency of the recovery (Faraji et al., 2022).

The next and the main step of this technique is high-temperature processing. This step is held in smelting furnaces. Metals are extracted from bulk material into a molten metal phase, while some others like Zn and Pb are volatilized and condensed. Other components including nonmetallic materials (plastics etc.) are combusted. Residues are incorporated into the slag phase and can be used as a byproduct. Slag phase is essential for isolating valuable metals from nonmetallic impurities (Khaliq et al., 2014).

Copper plays a key role in metal recovery. Gold and silver partition into the molten copper phase and, therefore, metal recovery is achieved. Other components such as aluminum and iron are generally found in the slag phase. Slag phase is essential for extracting targeted elements (Dutta et al., 2023). Slag generated during the smelting process should be stabilized for disposal or subjected to further extractions for additional recovery of remaining metals (Faraji et al., 2022).

Figure 3 represents the steps of the pyrometallurgical technique.

3.3 *Bioleaching*

Microorganisms are utilized to provide the dissolution and extraction of metals by biological-based steps including acidolysis, redoxolysis, and complexolysis. *Acidithiobacillus ferrooxidans* and *Pseudomonas fluorescens* are commonly used microbial species because they have organic acids (Das et al., 2011; Dey et al., 2023; Ghosh et al., 2016; Mohanty et al., 2017). Organic acids aid in leaching metals from solid matrices into soluble forms. Bioleaching is an eco-friendly and sustainable solution for REM recovery (Dev et al., 2020).

Crushed and sieved materials combined with specific microorganisms in a leaching medium. The medium consists of carbon sources such as glucose (Das & Ghosh, 2018, 2023b). The typical pH level is pH 4–5 for fungi such as *Aspergillus niger* (Andersen et al., 2009). Microorganisms utilize carbon sources and produce organic acids typically sulfuric, citric, and oxalic acids. These organic acids act as lixiviants to solubilize metals from the substrate through acidolysis or complexolysis (Adetunji et al., 2023).

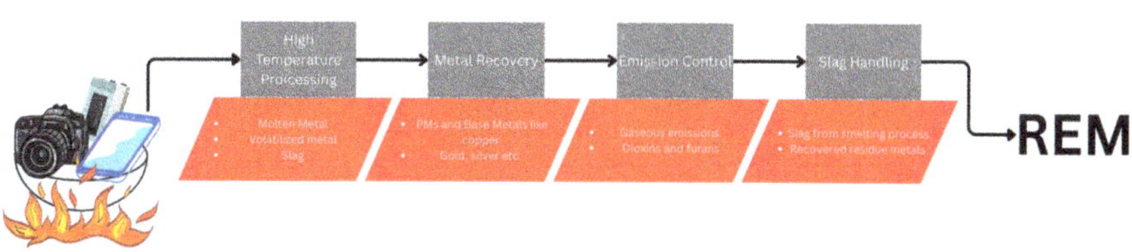

Fig. 3 Flow chart of a pyrometallurgical process

During the bioleaching, metals break down into metal oxides and soluble metal organic complexes are formed (Das & Ghosh, 2023c; Ghosh & Das, 2017). This is due to the reaction of organic acids with the metals in e-waste. Types of microorganisms, concentration of the lixiviants, and contact time between the e-waste and microorganisms are critical parameters for bioleaching (Das & Ghosh, 2023a; Nag et al., 2023). For instance, in bioleaching experiments using *A. niger*, citric acid production plays a critical role in dissolving metals like cerium and lanthanum under optimal conditions. At peak efficiency, the leachate is separated from the solid residue by filtration or centrifugation. Precipitation or solvent extraction techniques are applied for the REMs recovery after centrifugation (Hosseinzadeh et al., 2021; Brewer et al., 2019). Figure 4 summarizes the bioleaching steps.

3.4 *Ionic Liquid Technique*

This technique involves using solvents and extractants for recovery. For this purpose, room-temperature ionic liquids (RTILs) are utilized. These are salts and they remain in a liquid state at relatively low temperatures below 100 °C (Benedetto, 2017). This property creates significant physicochemical advantages including negligible volatility and high thermal stability. In addition, ionic liquids are highly selective in extracting metals such as Nd, Sm, and Eu in the means of ion exchange and neutral complexation mechanisms. These liquids enhance separation efficiency and reduce the environmental impacts (Prusty et al., 2021).

Initially, grounded WEEE is mixed with a solution containing extractant ionic liquids, which are tailored organic salts capable of selectively binding to REM ions (Qi, 2018). Metal ions are transferred to the ionic liquid phase through ionic exchange. Affinity of the ionic liquids for REMs can be increased by using specific ligands or donor atoms (Zhang et al., 2021). Temperature, pH, and the presence of cosolvents result in high selectivity and recovery rates.

When the phases are separated, metal-loaded ionic liquid is ready for subsequent recovery. REMs are then precipitated or stripped from the ionic liquid solution. The stripped metals are purified through electrowinning applications and high-purity REM are obtained. Ionic liquid can be regenerated by using acid or alkali solutions for reuse (Inman et al., 2022). Ionic liquid technique is summarized in Fig. 5.

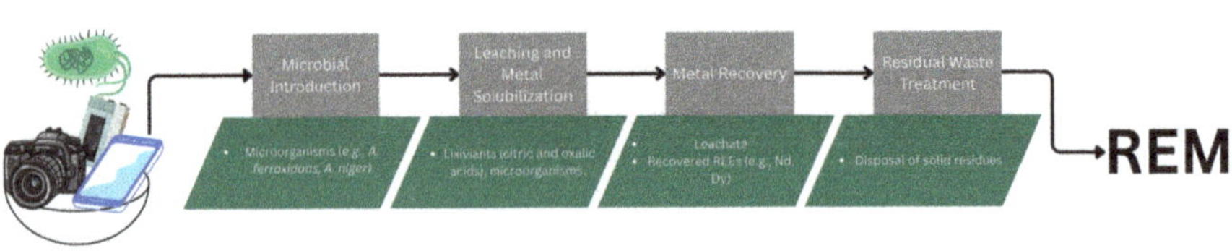

Fig. 4 Flow chart of bioleaching with microorganisms process

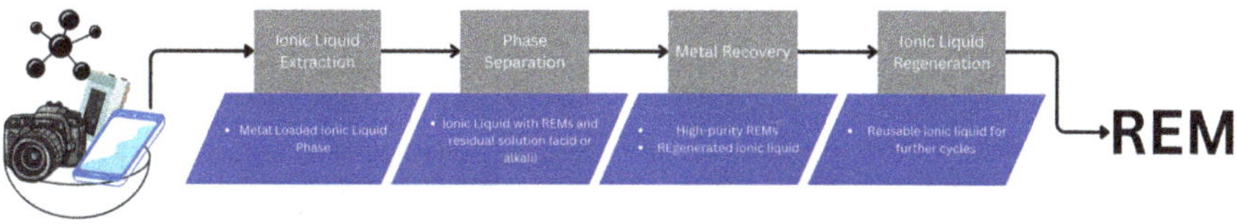

Fig. 5 Flow chart of ionic liquid technique process

3.5 Alkaline Roasting

Alkaline roasting is a widely used method for recovering REMs into oxides, aluminates, or carbonates. These types of structures can then be extracted using acid or water leaching (Perez et al., 2019; Liang et al., 2016). The e-waste or mineral feed is ground and mixed with an alkaline agent. Typically, sodium hydroxide (NaOH) or sodium carbonate (Na_2CO_3) is used as an alkaline agent (Perez et al., 2019). The mixture is heated to 500 °C–900 °C, breaking down complex structures such as aluminum spinels in a controlled environment. At the end of the roasting, the slag is formed, and this formation is washed with either water or acidic solutions to remove soluble compounds such as sodium aluminate. At the end of the washing step, medium rich in terms of REMs is obtained. REMs are extracted from the slag using HCl and a reductant like ferrous chloride ($FeCl_2$) to increase the solubility and recovery.

Unlike pyrometallurgical methods that require >1000°, this process operates at lower temperatures over an extended period, ensuring the efficient dissolution of REMs (Hamza et al., 2022). Studies have reported impressive leaching efficiencies for REMs such as yttrium, europium, cerium, and terbium reaching up to 99%, 99.4%, 98.6%, and 98.8%, respectively (Xie et al., 2023).

The type of alkaline agent used, roasting temperature, duration, and leaching conditions affect the results. Using NaOH at ~900 °C for 120 min has shown to yield high recovery rates, particularly for yttrium (Y) and europium (Eu) (Xie et al., 2023; Panda et al., 2020).

Alkaline roasting process is summarized in Fig. 6.

3.6 Membrane Extraction

This technique begins with the preparation of the membrane system. Impregnation of an extractant into the porous membrane material is included in this stage (Mwewa et al., 2022). Between the stripping solution for recovery and the feed solution containing the target REMs, this impregnated membrane acts as a selective barrier. Membranes serve as an interface between an aqueous solution and an organic extractant (Swain & Mishra, 2019). Aqueous solution contains the target REMs. Organic phase within the pores of the membrane serves as a medium to bind and transport the target metal ions across the membrane interface.

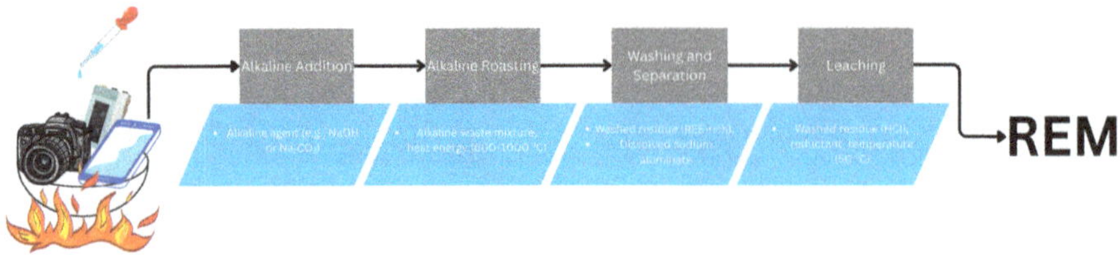

Fig. 6 Flow chart of alkaline technique process

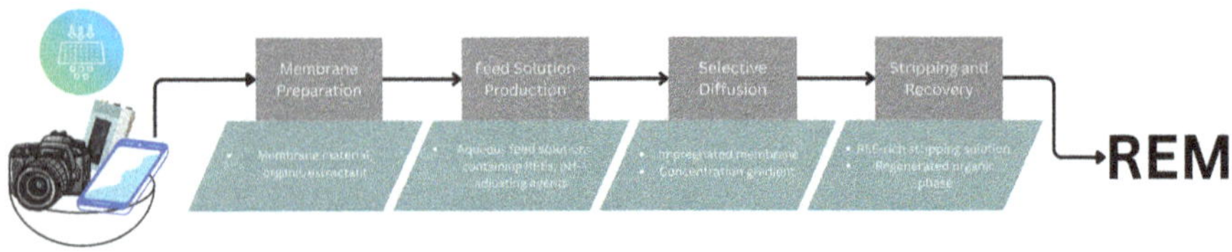

Fig. 7 Flow chart of membrane extraction technique

When one side of the membrane contacts with the feed solution, the concentration gradient causes the target REMs to diffuse through the membrane. The extractant and the REMs interact within the membrane to produce complexes that have the ability to pass to the opposite side. The stripping solution releases the REMs into a solution for collection after breaking these complexes on the other side of the membrane (Arrachart et al., 2021).

Flow rate, temperature, extractant used in the organic phase, pH of the feed, and stripping solutions are the operational parameters of membrane extraction. Membranes have low stability in aggressive media like acidic or alkali (Elbashier et al., 2021). Under optimal conditions, studies have demonstrated that the recovery of REEs like neodymium (Nd) and dysprosium (Dy) can achieve efficiencies greater than 95% (Arrachart et al., 2021; Khanzada et al., 2024). Figure 7 delves into a detailed flowchart describing the membrane extraction techniques.

4 Advantages and Disadvantages of REM Recycling Techniques in Terms of Environmental and Economic Aspects

REM recovery techniques address critical issues of resource scarcity and ecological degradation (Kazancoglu et al., 2020). The recovering of REMs from e-waste can reduce the environmental effects compared to virgin production and this is validated by the recent life cycle assessment (LCA) studies (Li et al., 2019). When the numbers are investigated, producing 1 ton of REM generates approximately 60,000 m^3 of gas emissions containing sulfuric acid and hydrofluoric acid, 200 m^3 of acid water, and 1.4 tons of radioactive waste (Walters & Lusty, 2011). Besides, leaching processes require about 5.2 tons of ammonium sulfate, 2.2 tons of ammonium carbonate, 0.5 tons of sulfuric acid, and 400 kilowatt-hours of electricity per ton of rare-earth oxides produced. Annual emissions from operations include 67.67 tons of CO_2, 20.66 tons of volatile organic compounds (VOCs), 282.3 tons of nitrogen oxides, and 2.23 tons of sulfur oxides (Meyer & Bras, 2011). In addition, beneficiation processes achieve an overall recovery rate of 61% for rare-earth oxides, highlighting the substantial resource and energy demands inherent in production (Navarro & Zhao, 2014).

On the other hand, e-waste recycling efforts can reduce energy consumption by up to 60% and greenhouse gas (GHG) emissions by 70% during REM recovery (Vuppaladadiyam et al., 2024; Li et al., 2019). By this way, resource scarcity can be decreased, and supply chains can be stabilized. It is estimated that recycling REMs could save up to $3.5 billion annually in extraction costs, particularly for critical elements essential for renewable energy technologies (Sanchez Moran et al., 2024). When the e-wastes are considered specifically, recovering 1 kg of REE (comprising 95% Y and 5% Eu) from fluorescent lamp waste results in net carbon emissions of 74 kg CO_2 equivalent and consumes 61,982 MJ of fossil energy (Liu, 2020). Similarly, the recovery of 1 kg of gallium (Ga) from LED waste generates 3687 kg CO_2 equivalent emissions and uses 63,352 MJ of fossil energy. The high environmental impact of Ga recovery is due to its low concentration in LED chips (0.234% w/w). Nonetheless, the process achieves high efficiencies, with REE recovery from phosphor fractions reaching 91% and Ga recovery using hydrometallurgical methods attaining a leaching efficiency of 97% and overall recovery efficiency of 89% (Liu & Keoleian, 2020).

4.1 Advantages and Disadvantages of Hydrometallurgical Techniques

Life cycle assessment (LCA) achieved for hydrometallurgical operations indicates that the process reaches REMs recovery (from fluorescent lamps and hard disk magnets) rates of up to 98% for neodymium and 95% for dysprosium under optimized conditions (Kumari & Sahu, 2023; Jha et al., 2016). This method can process low-grade ore and secondary resources, which are not economically viable for other techniques. The controlled nature of chemical reactions in hydrometallurgy allows for precise separation of metals, resulting in products suitable for industrial applications (Danouche et al., 2024). Examples of optimized conditions included a sulfuric acid concentration of 3.0 M, a leaching duration of 4 h at 70 °C, and a pulp density of 110.8 g/L, yielding highly efficient recovery rates (Kaya, 2024; Kumari & Sahu, 2023). Using oxalic acid for precipitation, individual REEs like Nd and Dy recovered with 95% purity. However, the process resulted in material losses of up to 15% due to the codissolution of iron and other transition metals.

Another study, Gkika et al. (2024) report that using hydrometallurgical techniques, recovery rates were achieved at 97% for Eu and 92% for Y from waste computer monitors utilizing sulfuric and hydrochloric acids. Leaching efficiencies for Dy, Nd, and Pr from waste permanent magnets reached 76%, 78%, and 83%, respectively, using 5 M HCl. Stripping efficiencies in certain hydrometallurgical processes were recorded at 98.60%, with purities of recovered REEs reaching 99.85%.

Hydrometallurgical methods also generate less dust and airborne emissions than pyrometallurgical approaches, making them relatively environmentally friendly regarding air pollution (Yuksekdag et al., 2022; Kanungo et al., 2023).

However, hydrometallurgical techniques generate significant wastewater and sludge that require additional treatment (Kaya, 2024). They are heavily reliant on acids and other chemicals, which can lead to significant hazardous waste generation. For instance, sulfuric and nitric acid leaching processes produce large volumes of acidic effluents requiring careful treatment and disposal, often increasing operational costs by 20–30% compared to alternative methods (Kaim et al., 2023; Danouche et al., 2024).

The process is also energy intensive, with solvent extraction stages consuming substantial energy resources. Moreover, hydrometallurgical approaches demand large quantities of water, further limiting their sustainability in regions with water scarcity (Jha et al., 2016).

4.2 Advantages and Disadvantages of Pyrometallurgical Techniques

In pyrometallurgical applications, smelting processes can recover approximately 95% of precious metals while simultaneously extracting base metals (Faraji et al., 2022). In addition to the high recovery efficiency, some modifications of the processes are energy efficient. For instance, the combustion of plastic fractions in e-waste as a partial energy source reduces the external energy input (Khaliq et al., 2014).

Since they do not rely on liquid chemical reagents, pyrometallurgical methods generate less water pollution compared to hydrometallurgical techniques. Furthermore, the slag produced in these processes can be used as a secondary raw material in construction, thus promoting circular economy practices (Gkika et al., 2024; Godbole et al., 2025).

Besides the advantages, pyrometallurgical methods have drawbacks. Smelting processes require temperatures above 1200 °C and therefore contribute to high CO_2 emissions. Studies in the literature underline that these processes can generate up to 2 tons of CO_2 per one ton of e-waste processed (Dutta et al., 2023). Additionally, dioxins and furans generated due to the thermal operations should be treated. These techniques require advanced and expensive filtration systems.

Another limitation is the loss of certain metals, particularly those with low boiling points like Zn, which volatilize during high-temperature processing and efficient recovery is challenging. The complex structure of e-waste also complicates the recovery of metals present in trace amounts (Faraji et al., 2022). Pyrometallurgical recycling facilities are capital intensive, requiring substantial investments in infrastructure and emission control technologies. These costs make smaller-scale operations unfeasible, especially in regions with limited regulatory oversight or market incentives for e-waste recycling (Khaliq et al., 2014).

4.3 Advantages and Disadvantages of Bioleaching Method

Bioleaching is a cost-effective and environmentally friendly method for recovering REMs as it significantly reduces the use of hazardous chemicals compared to other techniques (Hosseinzadeh et al., 2021; Auerbach et al., 2019). Microorganisms are utilized such as *Acidithiobacillus ferrooxidans* and *Pseudomonas fluorescens* (Das et al., 2015). These types of microorganisms produce organic acids and oxidizing agents to dissolve REMs from e-waste. As a result, bioleaching minimizes chemical waste as well as helps reduce environmental pollution.

Studies show that generally 70–80% extraction efficiencies can be achieved by bioleaching of e-waste such as fluorescent lamps and phosphors (Dev et al., 2020; Brewer et al., 2019). However, studies have shown that the type of microorganism greatly impacts the recovery efficiency. For example, *Leptospirillium ferrooxidans* has been shown to recover ratios up to 99% of Pr, Dy, and 91% Nd by maintaining stable pH levels, reducing the need for additional chemicals (Zhang et al, 2020; Sarkodie et al., 2022).

On the other hand, bioleaching has some limitations. Under nonideal conditions, recovery rates for REMs can drop significantly (Dev et al., 2020). In addition, the duration of the process is long up to 14 days for complete metal extraction, making it slower than the other conventional methods. Over such an extended period, maintaining optimal temperature and pressure also increases energy costs (AliAkbari et al., 2020; Pathapati et al., 2023).

4.4 Advantages and Disadvantages of Ionic Liquid Techniques

Ionic liquid applications provide highly selective and efficient recovery. Tailored ionic liquids for targeted elements have been shown to achieve extraction efficiencies of up to 95% for certain REMs founded in complex waste matrices (Prusty et al., 2021). Ionic liquids are nonvolatile and stable over a wide temperature range (Rothe et al., 2021) and this property reduces the release of volatile organic compounds (VOCs). Room temperature is generally sufficient for ionic liquid extractions. LCA studies have demonstrated that processes using ionic liquids can reduce GHG emissions by as much as 30% in comparison with traditional solvent extraction methods (Jovell et al., 2021). Studies report energy savings of 25–30% in such operations (AliAkbari et al., 2020). In addition, ionic liquids can be reused multiple times with minimal loss in extraction efficiency. Multiple usage helps reduce chemical waste and operational costs (AliAkbari et al., 2020).

In contrast, synthesizing and utilizing ionic liquids are expensive and this is a significant obstacle to their broader industrial adoption (Greer et al., 2020). Achieving optimal performance also requires precise control over operational parameters such as pH and temperature. When complex feed materials are considered, this can be challenging. (Pathapati et al., 2023).

In addition, due to the high viscosity of ionic liquids, difficulties during mixing and phase separation can be faced. To overcome these difficulties, diluents are sometimes added, but this compromises the process's environmental benefits (AliAkbari et al., 2020). Although ionic liquids are generally less toxic during the operations, their synthesis can involve energy-intensive processes and hazardous chemicals. Therefore, production processes of the ionic liquids might offset some of the environmental advantages gained during their application in metal recovery (Prusty et al., 2021).

4.5 Advantages and Disadvantages of Alkaline Roasting

Alkaline roasting is commonly applied in the treatment of phosphor powders and other REM-rich waste streams to extract elements like Y, Eu, and Ce (Perez et al., 2019). One of the major advantages of alkaline roasting for REM recovery is its ability to process materials with complex compositions and produce high-purity outputs. Especially followed by acid leaching, 85–95% recovery efficiency can be achieved (Thomas et al., 2024; Shahbaz, 2022). A decrease of 40–50% in acid consumption during the leaching can also be achieved by alkaline roasting (Bektas, 2023). The use of alkali agents like potassium hydroxide (KOH) during alkaline roasting for REM extraction not only enhances metal recovery but also enables the production of valuable byproducts such as tripotassium phosphate, cryolite, which are used in various industrial applications improving the process's economic viability (Galvin & Safarzadeh, 2018; Shahbaz, 2022).

One of the major disadvantages of alkaline roasting is the generation of fly ash due to the metal ions in the pregnant leach solution. It makes further separation steps complicated and expensive (Thomas et al., 2024). Additionally, the process generates alkaline waste, causing additional disposal efforts and creating environmental impact. On the other hand, high energy consumption required to maintain elevated temperatures is another drawback. Elevated temperatures up to 800° increase energy consumption compared to standard acid–base leaching systems (Sinha et al., 2023).

4.6 Advantages and Disadvantages of Membrane Solvent Extraction

According to studies, recovery ratios can reach up to 95% efficiency. Target metals, REMs, can be recovered with excellent selectivity using membrane solvent extraction, even if the complexity of the feed (Agrahari et al., 2024). In addition, this technique is known for its energy efficiency. Less organic solvent is used than the other conventional recovery techniques and this results in less chemical waste. Furthermore, its modular architecture makes it appropriate for a wide range of industrial uses (Yadav et al., 2024).

One of the major problems of this technique is membrane fouling when feed solutions consist of high levels of impurities or particulates. Membrane replacement or frequent cleaning is often required causing additional costs. Precise separation is ensured by the high selectivity of the method. On the other hand, it can slow down the recovery rate, making it difficult to meet the demands of large-scale processing facilities. Another disadvantage of this technique is extractants and membranes can be affected by extreme conditions such as high temperatures and harsh chemical environments. These factors can compromise system performance and longevity, posing additional challenges for industrial applications (Shamsul Kahar et al., 2023).

5 Future Perspectives in E-Waste Management Policies

The rapid rise in WEEE has turned e-waste management into a global challenge. In 2019, the world generated approximately 53.6 MT of e-waste, and experts predict this number could reach 75 MT by 2030 (Liu et al., 2023; Shittu et al., 2021). Despite this growing issue, only a small fraction—around 16–17%—of WEEE is properly handled. As a result, valuable materials including REMs are often unrecovered, leading to significant resource loss (Liu et al., 2023; Halim & Suharyanti, 2019). The situation is even more severe in developing countries. In developing countries, outdated e-devices accumulate rapidly, worsening the problem. Furthermore, informal recycling practices used in developing countries pose serious risks to both human health and the environment. The lack of proper regulations and sustainable recovering solutions, e-waste will continue to grow at an alarming rate, exacerbating environmental and socioeconomic challenges worldwide (Shittu et al., 2021).

Addressing the challenges of e-waste management requires a multifaceted approach that focuses on three key areas: integrating advanced technologies, adopting sustainable practices, and enforcing strong legislative frameworks.

Conventional recovery methods have demonstrated high efficiencies achieving REM recovery up to 99.95%. At the same time, innovative approaches like bioleaching offer a more environmentally friendly solution by using microorganisms to extract valuable material despite their slower processing speed being a limitation

(Jha et al., 2023). Incorporating green design principles including extended producer responsibility (EPR), and innovative recovery methods such as carbon based nanocomposites and porous metal-organic frameworks can further enhance the recovery process and decrease the environmental hazards (Halim & Suharyanti, 2019; Liu et al., 2023).

As of 2019, China led the world in REM production. The country has supplied 80% of the global REM market. However, REM recovery efforts remain low, with the rates at just 25, a huge contrast to the over 90% recycling rates for materials like iron and steel (Opare et al., 2021; Jha et al., 2023). As demand for REMs rises—driven largely by green technologies—it is projected to reach 51.9 thousand MT by 2030 (Bhattarai et al., 2024). In order to meet this demand, advancements in recovery methods are required. The innovations including sophisticated chemical and biological separation techniques play a crucial role in the production of high-performance magnets for electric vehicles and wind turbines, as well as energy-efficient lighting and catalytic converters (Bhattarai et al., 2024; Jha et al., 2023).

Strong policy frameworks are essential for the integration of the circular economy principles into WEEE management. The European Union's WEEE Directive has established ambitious recycling targets of 65%. This leads to notable improvements in material recovery and reducing landfill dependence (Murthy & Ramakrishna, 2022). Similarly, countries like Japan and South Korea have implemented EPR programs. These programs require manufacturers to take back and recover e-waste, effectively closing the loop on resource use (Xavier et al., 2021).

Economic analyses highlight the revenue potential of e-waste recycling. When only smartphone recycling efforts are considered, it is estimated to gain €2 billion annually by 2030 (Halim & Suharyanti, 2019). However, realizing this potential is challenging due to inconsistent regulations and the absence of standardized reporting systems. Only 78 nations worldwide had e-waste legislation in place as of 2019. Tif number emphasizes the necessity of coordinated international policies to support effective and long-lasting recovery initiatives (Shittu et al., 2021).

The foundation for incorporating the concepts of the circular economy into the management of WEEE is urban mining. Urban mining creates solutions for growing issue of e-waste while lowering the demand for primary extraction by utilizing discarded electronics as a resource. In comparison to conventional techniques, case studies have demonstrated that urban mining may recover up to 90% of REMs from devices that are used widely in daily life like laptops and smartphones while lowering processing costs by up to 30% (Sharma et al., 2021).

The increasing academic focus on e-waste management further underscores its significance. In the years between 1981 and 2018, approximately 2800 research publications addressed various aspects of the issue. Among these studies, China leads the way by contributing 40.93% of the total studies (Gao et al., 2019). Topics such as recycling management in developing countries, the degradation of waste materials, and the health risks from heavy metals are the key research areas for this field, reflecting an increasing focus on environmental and human health impacts (Gao et al., 2019).

6 Concluding Remarks

Advantages and disadvantages of the techniques mentioned in this book chapter are summarized in Table 2. Hydrometallurgical methods are commonly used all over the world. In spite of this, hydrometallurgical processes produce significant chemical waste and consume large amounts of water. Pyrometallurgical techniques, on the other hand, are particularly effective in recovering precious metals and managing complex e-waste streams. However, their high energy requirements and substantial CO_2 emissions present notable environmental challenges.

Biological methodology, bioleaching, emerges as an environmentally friendly and cost-efficient alternative among the REM recovery techniques but slow reaction kinetics and its sensitivity to pH and temperature fluctuations should be considered during the applications. Advanced methods, such as ionic liquid extraction and membrane solvent extraction, offer exceptional efficiency and selectivity, achieving recovery rates exceeding 95%. Yet, these techniques are constrained by high operational costs and complexity.

Alkaline roasting generates alkaline waste and necessitates precise operational control, but it is one of the recovery techniques that high recovery efficiencies can be achieved. Collectively, these techniques represent a diverse toolkit for tackling

Table 2 Advantages and disadvantages of techniques

Technique	Advantages	Disadvantages
Hydrometallurgical techniques	High recovery rates (up to 90%), suitable for low-grade ores, controlled reactions, and scalability.	Generates toxic chemical waste, requires significant water usage.
Pyrometallurgical techniques	Handles complex and heterogeneous e-waste, high recovery of precious metals (95%), uses combustibles for energy efficiency.	High energy consumption, high CO_2 emissions, loss of volatile metals, capital-intensive.
Bioleaching techniques	Eco-friendly, cost-effective, uses microorganisms, recovery efficiencies up to 80% for REMs.	Slow reaction kinetics, sensitivity to pH and temperature, scalability challenges.
Ionic liquid extraction	High selectivity, efficiency (up to 95%), reusable, reduced chemical waste, low energy requirements.	High cost of ionic liquids, complex mechanisms, potential environmental impact of IL production.
Alkaline roasting	Efficient for breaking down complex materials, high recovery rates (up to 99% for specific REMs), effective for mixed waste streams.	Energy-intensive, alkaline waste generation requires precise operational control.
Membrane solvent extraction	High selectivity, energy efficiency, minimal solvent losses, scalable recovery rates exceed 95% for certain REEs.	Membrane fouling, high manufacturing and maintenance costs, lower throughput rates.

the challenges of sustainable REM recovery, each with distinct strengths and limitations.

Figure 8 compares the energy demands of different rare earth metal (REM) recycling techniques. Pyrometallurgy exhibits the highest energy requirement, exceeding 3000 kJ, due to its reliance on high-temperature smelting processes.

Techniques like pyrometallurgy and alkaline roasting consume significantly higher energy, making them energy-intensive applications. For bioleaching, energy consumption includes the additional energy required to maintain stable temperature conditions, as the process is highly sensitive to temperature variations, which are crucial for optimal microbial activity and recovery efficiency. Hydrometallurgical methods, on the other hand, show moderate energy consumption, while advanced methods like ionic liquid extraction and membrane solvent extraction demonstrate the lowest energy requirements, highlighting their potential for energy-efficient recycling.

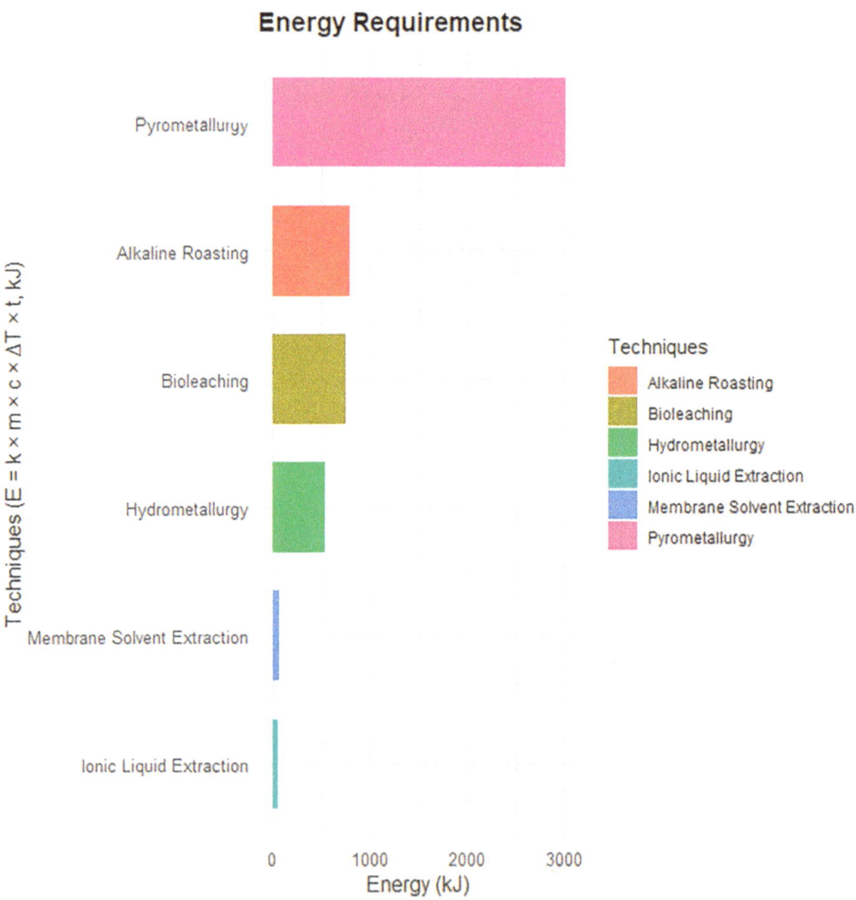

Fig. 8 Energy requirements for the recovery techniques

Figure 9 illustrates the recycling efficiencies of various REM recovery techniques. Alkaline roasting achieves the highest efficiency, approaching 99%, followed closely by ionic liquid extraction, membrane solvent extraction, and pyrometallurgical methods, all exceeding 90%.

Qiu and Suh (2019) detail the techniques and economic outcomes associated with REM from end-of-life lighting technologies. The cost of recycling was estimated at $7200 per ton of REM phosphors for a recycling plant with an annual capacity of 100 tons. By increasing the plant capacity to 1500 tons per year, the cost was reduced to $2500 per ton. However, the market prices are another factor that affects the feasibility of the recovery of REMs. Same study showed that recycling REMs was not economically feasible under 2018 market conditions, regardless of plant scale. It is obvious that when the plant scale increases, the feasibility of the recovery operations will also increase; however, market prices all over the world are fluctuating and affects the stability of the investments.

To achieve profitability, REM market prices need to increase by factors of 6.3, 2.8, and 2.2 for plants with capacities of 100, 800, and 1500 tons per year, respectively. This analysis underscores that while scaling up operations can lead to

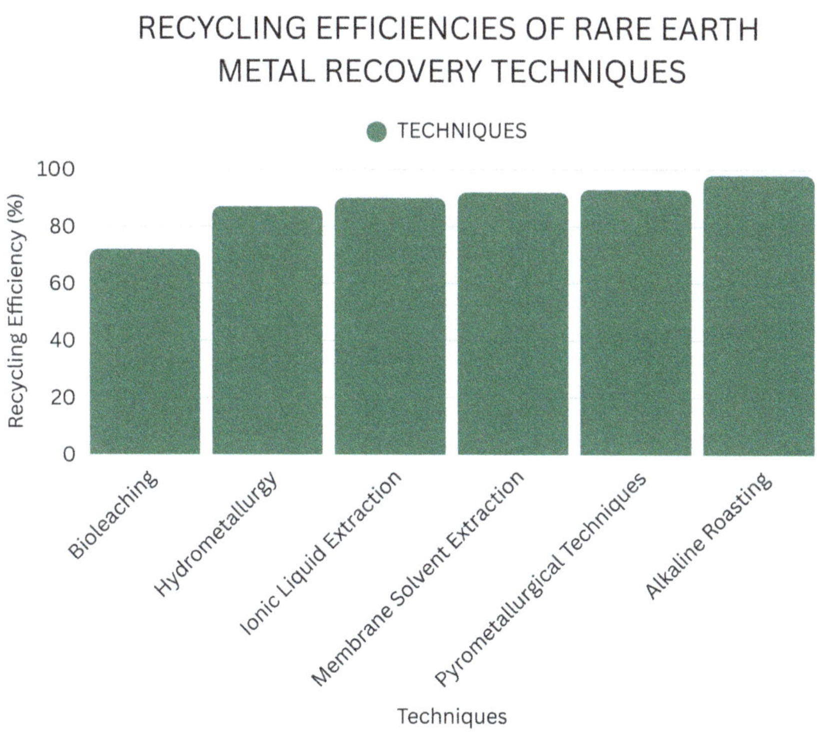

Fig. 9 Recycling efficiencies of REM recovery techniques

significant cost reductions; the economic viability of recycling REM remains heavily reliant on market conditions and REM values.

In conclusion, results in the literature underline how urgent policy changes, including financial incentives or subsidies, are needed to close the economic gap. Furthermore, to ensure that REM recovering turns into a sustainable and profitable choice, initiatives to stabilize or raise REM market prices are essential. Further developments in recovery technologies lower operational costs, as well. To address growing waste amounts and environmental risks, the management of WEEE in the future requires combining technological innovation, improved recycling systems and cooperative international initiatives. Streamlining recovery techniques for REMs and filling legal and reporting gaps will make REM recovery sustainable and feasible (Liu et al., 2023; Shittu et al., 2021; Bhattarai et al., 2024; Jha et al., 2023).

References

Adetunji, A. I., Oberholster, P. J., & Erasmus, M. (2023). Bioleaching of metals from E-waste using microorganisms: A review. *Minerals, 13*(6), 828. https://doi.org/10.3390/min13060828

Agrahari, G. K., Vignesh, M. S., & Nigam, K. D. P. (2024). Novel devices for the extraction and recovery of rare-earth metals through recycling of waste. *Journal of Material Cycles and Waste Management, 26*(1), 109–137.

AliAkbari, R., Marfavi, Y., Kowsari, E., & Ramakrishna, S. (2020). Recent studies on ionic liquids in metal recovery from E-waste and secondary sources by liquid-liquid extraction and electrodeposition: A review. *Materials Circular Economy, 2*, 1–27.

Ambaye, T. G., Vaccari, M., Castro, F. D., & Prasad, S. (2020). Emerging technologies for the recovery of rare earth elements (REEs) from end-of-life electronic wastes: A review on progress, challenges, and perspectives. *Environmental Science and Pollution Research, 27*, 36052–36074.

Andersen, M. R., Lehmann, L., & Nielsen, J. (2009). Systemic analysis of the response of *Aspergillus niger* to ambient pH. *Genome Biology, 10*, R47. https://doi.org/10.1186/gb-2009-10-5-r47

Arrachart, G., Couturier, J., Dourdain, S., Levard, C., & Pellet-Rostaing, S. (2021). Recovery of Rare Earth Elements (REEs) using ionic solvents. *PRO, 9*(7), 1202. https://doi.org/10.3390/pr9071202

Auerbach, R., Bokelmann, K., Stauber, R., Gutfleisch, O., Schnell, S., & Ratering, S. (2019). Recycling of rare earth metals out of end-of-life magnets by bioleaching with various bacteria as an example of an intelligent recycling strategy. *Minerals Engineering, 134*, 104–117.

Balaram, V. (2024). Sustainable recovery of rare earth elements by recycling of E-waste for a circular economy: perspectives and recent advances. *Environmental Materials and Waste*, 499–544.

Bektas, T. U. (2023). *Extraction of rare earth elements from aqueous solutions by using hydrometallurgical techniques* (Doctoral dissertation, University of Saskatchewan Saskatoon).

Benedetto, A. (2017). Room-temperature ionic liquids meet bio-membranes: The state-of-the-art. *Biophysical Reviews, 9*(4), 309–320. https://doi.org/10.1007/s12551-017-0279-1

Bhattarai, A., Gardas, R. L., Yadav, S. N., Chatterjee, S. K., & Dev, R. K. (2024). *Recovery of Rare Earth Elements (REEs) from different sources of E-waste and their potential applications: A critical review*. Preprints.org.

Brewer, A., Dohnalkova, A., Shutthanandan, V., Kovarik, L., Chang, E., Sawvel, A. M., Mason, H. E., Reed, D., Ye, C., Hynes, W. F., Lammers, L. N., Park, D. M., & Jiao, Y. (2019). Microbe Encapsulation for Selective Rare-Earth Recovery from Electronic Waste Leachates. *Environmental Science & Technology, 53*(23), 13888–13897.

Danouche, M., Bounaga, A., Oulkhir, A., Boulif, R., Zeroual, Y., Benhida, R., & Lyamlouli, K. (2024). Advances in bio/chemical approaches for sustainable recycling and recovery of rare earth elements from secondary resources. *Science of the Total Environment, 912*, Article 168811.

Das, A. P., & Ghosh, S. (2018). Bioleaching of manganese from mining waste materials. *Materials Today: Proceedings, 5*(1), 2381–2390. https://doi.org/10.1016/J.MATPR.2017.11.459

Das, A. P., & Ghosh, S. (2023a). Factors affecting bioleaching. In *Manganese mining microorganisms* (pp. 77–86). https://doi.org/10.1016/B978-0-12-822147-1.00004-2

Das, A. P., & Ghosh, S. (2023b). Mixed culture bioleaching: An insight into manganese biomining process and efficacy. In *Manganese mining microorganisms* (pp. 129–141). https://doi.org/10.1016/B978-0-12-822147-1.00001-7

Das, A. P., & Ghosh, S. (2023c). Role of metabolites and metabolomics for understanding of manganese bioleaching. In *Manganese mining microorganisms* (pp. 117–127). https://doi.org/10.1016/B978-0-12-822147-1.00011-X

Das, A. P., Sukla, L. B., Pradhan, N., & Nayak, S. (2011). Manganese biomining: A review. *Bioresource Technology, 102*(16), 7381–7387. https://doi.org/10.1016/J.BIORTECH.2011.05.018

Das, A. P., Ghosh, S., Mohanty, S., & Sukla, L. B. (2015). Advances in manganese pollution and its bioremediation. In L. B. Sukla, N. Pradhan, S. Panda, & B. K. Mishra (Eds.), *Environmental microbe biotechnology* (1st ed., pp. 313–328). Springer. https://doi.org/10.1007/978-3-319-19018-1_16

Deshmane, V. G., Islam, S. Z., & Bhave, R. R. (2019). Selective recovery of rare earth elements from a wide range of e-waste and process scalability of membrane solvent extraction. *Environmental Science & Technology, 54*(1).

Dev, S., Sachan, A., Dehghani, F., Ghosh, T., Briggs, B. R., & Aggarwal, S. (2020). Mechanisms of biological recovery of rare-earth elements from industrial and electronic wastes: A review. *Chemical Engineering Journal, 397*, 124596.

Dey, S., Tripathy, B., Kumar, M. S., & Das, A. P. (2023). Ecotoxicological consequences of manganese mining pollutants and their biological remediation. *Environmental Chemistry and Ecotoxicology*. https://doi.org/10.1016/J.ENCECO.2023.01.001

Dutta, D., Rautela, R., Gujjala, L. K. S., Kundu, D., Sharma, P., Tembhare, M., & Kumar, S. (2023). A review on recovery processes of metals from E-waste: A green perspective. *Science of the Total Environment, 859*, 160391.

Elbashier, E., Mussa, A., Hafiz, M., & Hawari, A. H. (2021). Recovery of rare earth elements from waste streams using membrane processes: An overview. *Hydrometallurgy, 204*, 105706.

Erdiaw-Kwasie, M. O., Abunyewah, M., & Baah, C. (2024). A systematic review of the factors–Barriers, drivers, and technologies–Affecting e-waste urban mining: On the circular economy future of developing countries. *Journal of Cleaner Production, 436*, 140645.

Faraji, F., Golmohammadzadeh, R., & Pickles, C. A. (2022). Potential and current practices of recycling waste printed circuit boards: A review of the recent progress in pyrometallurgy. *Journal of Environmental Management, 316*, 115242.

Galvin, J., & Safarzadeh, M. S. (2018). Decomposition of monazite concentrate in potassium hydroxide solution. *Journal of Environmental Chemical Engineering, 6*(1), 1353–1363.

Gao, Y., Ge, L., Shi, S., Sun, Y., Liu, M., Wang, B., Shang, Y., Wu, J., & Tian, J. (2019). Global trends and future prospects of e-waste research: A bibliometric analysis. *Environmental Science and Pollution Research, 26*(17), 17809–17820.

Ghosh, S., & Das, A. P. (2017). Bioleaching of manganese from mining waste residues using Acinetobacter sp. *Geology, Ecology, and Landscapes, 1*(2), 77–83. https://doi.org/10.1080/24749508.2017.1332847

Ghosh, S., Mohanty, S., Akcil, A., Sukla, L. B., & Das, A. P. (2016). A greener approach for resource recycling: Manganese bioleaching. *Chemosphere, 154*, 628–639. https://doi.org/10.1016/J.CHEMOSPHERE.2016.04.028

Gkika, D. A., Chalaris, M., & Kyzas, G. Z. (2024). Review of methods for obtaining rare earth elements from recycling and their impact on the environment and human health. *Processes, 12*(6), 1235. https://doi.org/10.3390/pr12061235

Godbole, P., Meshram, P., Jawadand, S. et al. (2025). A critical analysis of industrial slags, their hazard potential and remediation with reference to Sustainable Development Goals (SDGs). *Discov Civ Eng 2*, 117. https://doi.org/10.1007/s44290-025-00242-8

Greer, A. J., Jacquemin, J., & Hardacre, C. (2020). Industrial applications of ionic liquids. *Molecules, 25*(21), 5207. https://doi.org/10.3390/molecules25215207

Halim, L., & Suharyanti, Y. (2019). E-waste: Current research and future perspective on developing countries. *International Journal of Industrial Engineering and Engineering Management, 1*(2), 24–39.

Hamza, H., Eterigho-Ikelegbe, O., Jibril, A., & Bada, S. O. (2022). Application of the response surface methodology to optimise the leaching process and recovery of rare earth elements from discard and run of mine coal. *Minerals, 12*(8), 938. https://doi.org/10.3390/min12080938

Hosseinzadeh, F., Rastegar, S. O., & Ashengroph, M. (2021). Bioleaching of rare earth elements from spent automobile catalyst as pretreatment method to improve Pt and Pd recovery: Process optimization and kinetic study. *Process Biochemistry, 105*, 1–7.

Inman, G., Nlebedim, I. C., & Prodius, D. (2022). Application of ionic liquids for the recycling and recovery of technologically critical and valuable metals. *Energies, 15*(2), 628. https://doi.org/10.3390/en15020628

Jha, M. K., et al. (2016). Review on hydrometallurgical recovery of rare earth metals. *Hydrometallurgy, 165*, 2–26.

Jha, H., Sanskrit, U. S. V., & Dubey, B. K. (2023). Recovery of precious metals and rare earth elements from e-waste. In *Waste-to-wealth: Resource recovery and value-added products for sustainable development*. CRC Press.

Jovell, D., Pou, J. O., Llovell, F., & Gonzalez-Olmos, R. (2021). Life cycle assessment of the separation and recycling of fluorinated gases using ionic liquids in a circular economy framework. *ACS Sustainable Chemistry & Engineering, 10*(1), 71–80.

Kaim, V., Rintala, J., & He, C. (2023). Selective recovery of rare earth elements from e-waste via ionic liquid extraction: A review. *Separation and Purification Technology, 306*, 122699.

Kanungo, J., Sahoo, T., Swain, L. P., & Behera, I. D. (2023). Toxicity of persistent hydrocarbon pollutants, sources and sustainable remediation process. 39–65. https://doi.org/10.1007/978-3-031-48220-5_2.

Kaya, M. (2024). An overview of NdFeB magnets recycling technologies. *Current Opinion in Green and Sustainable Chemistry, 46*, 100884.

Kazancoglu, Y., Ozkan-Ozen, Y. D., Mangla, S. K., & Ram, M. (2020). Risk assessment for sustainability in e-waste recycling in circular economy. *Clean Technologies and Environmental Policy*, 1–13.

Khaliq, A., Rhamdhani, M. A., Brooks, G., & Masood, S. (2014). Metal extraction processes for electronic waste and existing industrial routes: A review and Australian perspective. *Resources, 3*(1), 152–179.

Khanzada, N. K., Al-Juboori, R. A., Khatri, M., Ahmed, F. E., Ibrahim, Y., & Hilal, N. (2024). Sustainability in membrane technology: Membrane recycling and fabrication using recycled waste. *Membranes, 14*(2), 52.

Kumari, A., & Sahu, S. K. (2023). A comprehensive review on recycling of critical raw materials from spent neodymium iron boron (NdFeB) magnets. *Separation and Purification Technology, 317*, 123527.

Li, Z., Diaz, L. A., Yang, Z., Jin, H., Lister, T. E., & Vahidi, E. (2019). Comparative life cycle analysis for value recovery of precious metals and rare earth elements from electronic waste. *Resources, Conservation & Recycling, 149*, 20–30.

Liang, Y., Liu, Y., Lin, R., Guo, D., & Liao, C. (2016). Leaching of rare earth elements from waste lamp phosphor mixtures by reduced alkali fusion followed by acid leaching. *Hydrometallurgy, 163*, 99–103.

Liang, B., Gu, J., Zeng, X., Yuan, W., Rao, M., Xiao, B., & Hu, H. (2024). A review of the occurrence and recovery of rare earth elements from electronic waste. *Molecules, 29*(19), 4624.

Liu, L. (2020). *Improving building sustainability: Lighting life cycle optimization and management, and HVAC demand response* (Doctoral dissertation).

Liu, L., & Keoleian, G. A. (2020). LCA of rare earth and critical metal recovery and replacement decisions for commercial lighting waste management. *Resources, Conservation & Recycling, 159*, 104846.

Liu, K., Tan, Q., Yu, J., & Wang, M. (2023). A global perspective on e-waste recycling. *Circular Economy, 2*, 100028.

Meyer, L., & Bras, B. (2011). Rare earth metal recycling. In *Sustainable Systems and Technology (ISSST), 2011 IEEE International Symposium* (pp. 1–6). Institute of Electrical and Electronics Engineers.

Mishra, S., Dash, D., Al-Tawaha, A. R. M. S., & Das, A. P. (2022). A review on heavy metal ion adsorption on synthetic microfiber surface in aquatic environments. *Applied Biochemistry and Biotechnology, 194*(10), 4639–4654. https://doi.org/10.1007/S12010-022-04029-W/METRICS

Mishra, S., Ghosh, S., van Hullebusch, E. D., Singh, S., & Das, A. P. (2023). A critical review on the recovery of base and critical elements from electronic waste-contaminated streams using microbial biotechnology. *Applied Biochemistry and Biotechnology, 2023*, 1–30. https://doi.org/10.1007/S12010-023-04440-X

Mohanty, S., Ghosh, S., Nayak, S., & Das, A. P. (2017). Bioleaching of manganese by Aspergillus sp. isolated from mining deposits. *Chemosphere, 172*, 302–309. https://doi.org/10.1016/J.CHEMOSPHERE.2016.12.136

Murthy, V., & Ramakrishna, S. (2022). A review on global e-waste management: Urban mining towards a sustainable future and circular economy. *Sustainability, 14*(2), 647.

Mwewa, B., Tadie, M., Ndlovu, S., Simate, G. S., & Matinde, E. (2022). Recovery of rare earth elements from acid mine drainage: A review of the extraction methods. *Journal of Environmental Chemical Engineering, 10*(3), 107704.

Nag, M., Lahiri, D., Ghosh, S., Sarkar, T., Pati, S., Das, A. P., Ram, D. K., Bhattacharya, D., & Ray, R. R. (2023). Application of microorganisms in biotransformation and bioremediation of environmental contaminant: A review. *Geomicrobiology Journal*. https://doi.org/10.1080/01490451.2023.2261443

Navarro, J., & Zhao, F. (2014). Life-cycle assessment of the production of rare-earth elements for energy applications: A review. *Frontiers in Energy Research, 2*, Article 45.

Opare, E. O., Struhs, E., & Mirkouei, A. (2021). A comparative state-of-technology review and future directions for rare earth element separation. *Renewable and Sustainable Energy Reviews, 143*, 110917.

Ouro-Salim, O. (2024). Urban mining of e-waste management globally: Literature review. *Journal of Cleaner Production, 9*, 100162.

Panda, R., Jadhao, P. R., Pant, K. K., Naik, S. N., & Bhaskar, T. (2020). Eco-friendly recovery of metals from waste mobile printed circuit boards using low temperature roasting. *Journal of Hazardous Materials, 395*, 122642.

Pathapati, S. V. S. H., Free, M. L., & Sarswat, P. K. (2023). A comparative study on recent developments for individual rare earth elements separation. *Processes, 11*(7), 2070.

Perez, J. P. H., Folens, K., Leus, K., Vanhaecke, F., Van Der Voort, P., & Du Laing, G. (2019). Progress in hydrometallurgical technologies to recover critical raw materials and precious metals from low-concentrated streams. *Resources, Conservation and Recycling, 142*, 177–188.

Priyadarshini, B., Santhosh Kumar, M., & Das, A. P. (2024). Microbial degradation of synthetic plastics released from COVID-19 generated single use disposable face masks. *Environmental Science and Engineering, Part, F2827*, 1–16. https://doi.org/10.1007/978-3-031-50840-0_1

Priyadarsini, S., & Das, A. P. (2024). Analytical and structural characterization of waste lithium-ion batteries for their effective recycling strategy. *Environmental Chemistry and Ecotoxicology*. https://doi.org/10.1016/j.enceco.2024.12.004

Priyadarsini, S., & Das, A. P. (2025). Lithium bioleaching: A review on microbial-assisted sustainable technology for lithium bio-circularity. *Journal of Water Process Engineering, 69*, 106744. https://doi.org/10.1016/j.jwpe.2024.106744

Prusty, S., Pradhan, S., & Mishra, S. (2021). Ionic liquid as an emerging alternative for the separation and recovery of Nd, Sm and Eu using solvent extraction technique-A review. *Sustainable Chemistry and Pharmacy, 21*, 100434.

Qi, D. (2018). Extractants used in solvent extraction-separation of rare earths: Extraction mechanism, properties, and features. In D. Qi (Ed.), *Hydrometallurgy of rare earths* (pp. 187–389) Elsevier.

Qiu, Y., & Suh, S. (2019). Economic feasibility of recycling rare earth oxides from end-of-life lighting technologies. *Resources, Conservation & Recycling, 150*, 104432.

Ramprasad, C., Gwenzi, W., & Chaukura, N. (2022). Strategies and options for the sustainable recovery of rare earth elements from electrical and electronic waste. *Chemical Engineering Journal*.

Rothe, M., Müller, E., Denk, P., & Kunz, W. (2021). Ionic liquids based on the concept of melting point lowering due to ethoxylation. *Molecules, 26*(13), 4034. https://doi.org/10.3390/molecules26134034

Sanchez Moran, E., Prodius, D., & Nlebedim, I. C. (2024). Rare-earth elements recovery from electronic waste: Techno-economic and life cycle analysis. *ACS Sustainable Chemistry & Engineering, 12*(38).

Sarkodie, E. K., Jiang, L., Li, K., Yang, J., Guo, Z., Shi, J., Deng, Y,. Liu, H., Jiang, H., Liang, Y., Yin, H., & Liu, X. (2022). A review on the bioleaching of toxic metal(loid)s from contaminated soil: Insight into the mechanism of action and the role of influencing factors. *Front. Microbiol. 13*:1049277. https://doi.org/10.3389/fmicb.2022.1049277

Shahbaz, A. (2022). A systematic review on leaching of rare earth metals from primary and secondary sources. *Minerals Engineering, 184*, 107632.

Shamsul Kahar, I. N., Othman, N., Mohamed Noah, N. F., & Suliman, S. S. (2023). Recovery of copper and silver from industrial e-waste leached solutions using sustainable liquid membrane technology: A review. *Environmental Science and Pollution Research, 30*(25), 66445–66472.

Sharma, M., Joshi, S., & Govindan, K. (2021). Issues and solutions of electronic waste urban mining for circular economy transition: An Indian context. *Journal of Environmental Management, 290*, 112373.

Shittu, O. S., Williams, I. D., & Shaw, P. J. (2021). Global e-waste management: Can WEEE make a difference? A review of e-waste trends, legislation, contemporary issues, and future challenges. *Waste Management, 120*, 549–563.

Singh Jadaun, J., Bansal, S., Sonthalia, A., Rai, A. K., & Singh, S. P. (2022). Biodegradation of plastics for a sustainable environment. *Bioresource Technology, 347*, 126697. https://doi.org/10.1016/J.BIORTECH.2022.126697

Sinha, M. K., Tanvar, H., Sahu, S. K., & Mishra, B. (2023). A review on recovery of terbium from primary and secondary resources: Current state and future perspective. *Mineral Processing and Extractive Metallurgy Review, 45*(7), 743–766. https://doi.org/10.1080/08827508.2023.2253490

Swain, N., & Mishra, S. (2019). A review on the recovery and separation of rare earths and transition metals from secondary resources. *Journal of Cleaner Production, 220*, 884–898.

Tan, Q., & Li, J. (2019). Rare earth metal recovery from typical e-waste. *Resource Recycling*.

Thomas, B. S., Dimitriadis, P., Kundu, C., Vuppaladadiyam, S. S. V., Raman, R. S., & Bhattacharya, S. (2024). Extraction and separation of rare earth elements from coal and coal fly ash: A review on fundamental understanding and on-going engineering advancements. *Journal of Environmental Chemical Engineering, 112769*.

Vuppaladadiyam, S. S., Thomas, B. S., & Kundu, C. (2024). Can e-waste recycling provide a solution to the scarcity of rare earth metals? An overview of e-waste recycling methods. *Science of the Total Environment, 924*, 171453.

Walters, A., & Lusty, P. (2011). *Rare earth elements*. British Geological Survey.

Xavier, L. H., Giese, E. C., Ribeiro-Duthie, A. C., & Lins, F. A. F. (2021). Sustainability and the circular economy: A theoretical approach focused on e-waste urban mining. *Energy Policy*.

Xie, B., Liu, C., Wei, B., Wang, R., & Ren, R. (2023). Recovery of rare earth elements from waste phosphors via alkali fusion roasting and controlled potential reduction leaching. *Waste Management, 163*, 43–51.

Yadav, J., Sarker, S. K., Bruckard, W., Jegatheesan, V., Haque, N., Singh, N., & Pramanik, B. K. (2024). Greening the supply chain: Sustainable approaches for rare earth element recovery from neodymium iron boron magnet waste. *Journal of Environmental Chemical Engineering, 113169.*

Yuksekdag, A., Kose-Mutlu, B., Wiesner, M. R., & Koyuncu, I. (2022). Effect of pre-concentration on membrane solvent extraction process for the recovery of rare earth elements from dilute acidic leachate. *Process Safety and Environmental Protection, 161*, 210–220.

Zhang, Y., Cao, Y., & Wang, H. (2021). Multi-interactions in ionic liquids for natural product extraction. *Molecules, 26*(1), 98. https://doi.org/10.3390/molecules26010098

Zhang, Y., Gu, F., Su, Z., Liu, S., Anderson, C., & Jiang, T. (2020). Hydrometallurgical Recovery of Rare Earth Elements from NdFeB Permanent Magnet Scrap: *A Review. Metals, 10*(6), 841. https://doi.org/10.3390/met10060841

Advancements in the Recovery of Rare-Earth Metals from Electronic Waste

Rituparna Saha

Abstract Recent breakthroughs in scientific and techno-industrial domains have made the development of smart and clean technologies possible. This has created an extensive demand for the separation and extraction of rare-earth metals (REMs) as they form one of the essential raw materials. However, recovery of REMs from ore matrices using traditional methods is complex, energy-intensive, and can cause serious environmental damage. In this regard, recovery of REMs from secondary sources like electronic waste (E-waste) using physical and chemical methods has gained prominence as they can meet the emerging demands by segregating the rare elements from the complex electronic materials. However, issues like high expenses and the generation of toxic waste streams have made the industry stakeholders look into biological recovery methods. The biorecovery method uses microorganisms to separate REMs from E-waste, followed by extraction and purification. The recovery process employs microbial processing methods such as bioleaching, biosorption, bioaccumulation, and bioprecipitation. The biological recovery of REMs offers many advantages like cost-effectiveness, reduction of chemical and energy input, and minimal environmental footprint.

In this chapter, the different physical and chemical methods utilized for separation of REMs from E-waste are discussed. The various strains of microorganisms and the diverse techniques used for rare metals extraction are discussed in detail. The current advancements in biorecovery methods are also summarized. Lastly, the challenges and future progress required for advancements in the recovery techniques of REMs from E-waste are also mentioned.

Keywords Rare-earth metals · Electronic waste · Biological recovery · Microorganisms · Biomaterials

R. Saha (✉)
Department of Life Sciences, Kristu Jayanti College (Autonomous), Bengaluru, Karnataka, India
e-mail: rituparna.s@kristujayanti.com

A. P. Das, S. Priyadarsini (eds.), *Electronic Waste and Environmental Pollution*, Sustainable Environmental Waste Management Strategies,
https://doi.org/10.1007/978-3-031-90287-1_14

1 Introduction

Rapid scientific and technological advancements in the past decade have boosted the quality of life globally. This progress has increased the use and production of electronic devices which has fueled the proliferation of the largest and fastest-growing industrial sector (Seif et al., 2024). As a result, electronics have become heavily incorporated into human life as they are used in telecommunication, healthcare, transportation, and security. However, fast-paced changes in the features and capabilities of electronic devices, and a consumption-driven society have significantly decreased their lifespan making them quickly obsolete. Electronic devices are getting replaced and discarded within a few years as waste with no intention of reuse (Danouche et al., 2024; Priyadarsini & Das, 2024). This contributes to the proliferation of electronic waste (E-waste), which has become an issue of concern for environmental policymakers and legislators as it is the fastest growing waste stream in the world (Mishra et al., 2023). According to a United Nations report, the total E-waste generated globally in 2019 amounted to 53.6 metric tonnes (Mt) worth 57 billion USD, among which only 17% were adequately recycled and the rest accumulated in landfills or was inadequately treated. With the existing generating rate, the report has estimated that the annual production of E-waste could reach 110 Mt globally by the year 2050 (Baldé et al., 2022). However, waste of this magnitude contains valuable as well as harmful materials. Still, due to the absence of any stringent rules and regulations, E-waste management is done improperly leading to contamination of different ecosystems and has proved dangerous to human health (Pimassoni et al., 2023).

Generally, E-waste includes all nonfunctional and nearly obsolete devices and has been classified into ten categories. Of these, the major contributors include large household items (42.1%), telecommunication and IT devices (33.9%), consumer devices (13.7%), and small household items (4.7%). And types like biomedical, electrical, lighting, and electronic tools account for <2% whereas toys, monitoring, and control devices along with automatic dispensers contribute <1% of E-waste (Shahabuddin et al., 2023) (Fig. 1). The components that make up any electronic waste depends on factors like manufacturer, date of manufacturing, model, type, and age. E-waste comprises an expensive cache of metals, like gold, silver, palladium, copper, iron, aluminum, and rare-earth metals (REMs). Among these, precious metals like gold and silver are commonly used in high-tech devices because of their high chemical stability and excellent electrical conductivity. In contrast, copper is the most commonly used followed by aluminum and iron (Hossain & Sahajwalla, 2024). Most importantly, REMs have found diverse applications in electronic devices, especially in clean energy technologies. Coupled with the exponential market expansion and poor distribution of REMs in nature, they have been categorized as critical raw materials by the US Department of Energy and the European Commission due to their supply risk (Nnorom & Osibanjo, 2008). Consequently, the recycling industry has taken a keen interest in recovering the metals as the voluminous amount of E-waste provides an important source with the

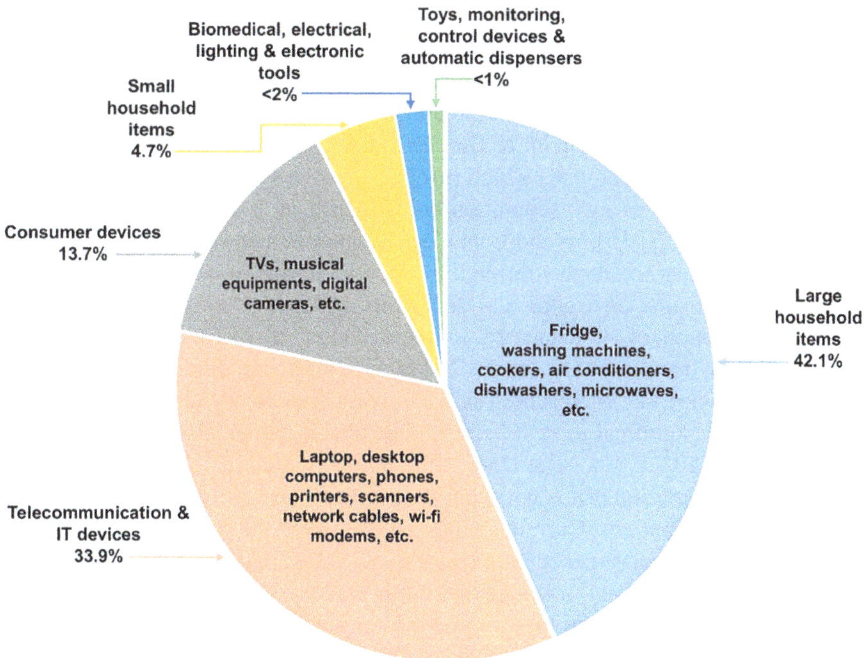

Fig. 1 Common categories of E-waste and their examples

grade of the rare elements noticeably higher than the primary ores (Xia & Ghahreman, 2024).

Recycling of REMs present in E-waste is extremely significant as they are one of the most important components present in modern devices and are therefore crucial for sustaining the global electronic industrial market. The International Union of Pure and Applied Chemistry (IUPAC) categorized the REMs into 17 elements involving scandium, yttrium, and lanthanides. They came to prominence in recent decades because of their distinctive optical, magnetic, and electronic properties, making their recovery from E-waste extremely valuable (Vuppaladadiyam et al., 2024). Not only does it help to reduce the dependency on primary mining and its associated impact on the environment, but also the recovery contributes to a circular economy as it minimizes waste disposal, reduces loss of natural resources, and promotes sustainable development. Exploitation of REMs from secondary sources is becoming vital with regard to its effect on the economy and the environment. But, selective recovery though promising poses challenges due to the heterogeneous composition of electronic materials and the absence of any systematic recycling techniques (Kaim et al., 2023; Liang et al., 2024).

Researchers have attempted to extract REMs sustainably from E-waste which is frequently designated as "urban mines." The extraction techniques that have been developed range from mechanical separation to physical and chemical processes. These recovery techniques involve multiple steps—starting from pretreatment to enrichment, followed by separation using pyrometallurgical, hydrometallurgical, or electrochemical methods, and finally extraction and purification of the critical elements (Bhavan et al., 2023). The most common industrial practice used for recovering REMs is pyrometallurgy (thermal treatment) which requires a high-temperature furnace for melting (Priyadarsini & Das, 2025). An alternative technique is hydrometallurgy (aqueous treatment) which involves dissolving metals from E-waste by leaching along with various separation and purification methods (Das & Ghosh, 2018; Ghosh et al., 2016). In addition, electrochemical methods reduce the use of reagents and extract the metals through the electroplating process. However, these methods have major environmental footprints involving exorbitant operational costs, consumption of high energy, low specificity, production of heavy metal sludge, and low recovery (Dev et al., 2020). In contrast, research into alternative green technologies like bioleaching is of growing interest as they are more cost-effective, require minimal to zero harmful chemicals, and are also energy efficient (Das & Ghosh, 2023f; Ghosh & Das, 2017; Mohanty et al., 2017). Bioleaching is carried out by microorganisms with higher specificity and efficacy in REM extraction at low concentrations. Current studies have revealed the promising potential of biometallurgy but challenges remain (Baez et al., 2024). However, overcoming the problems can not only help to reduce the supply chain risk and industry dependency for REMs but can also contribute to making microbial recovery techniques more mainstream.

This book chapter focuses on the physical, chemical, and biorecovery techniques employed for separation of REMs from E-waste. It especially addresses different methods employed by microbes, and the modifications required to enhance the recovery of the critical elements. The challenges and economic perspectives of the various recovery systems are highlighted along with the future scopes of developing viable recovery technology to extract REMs from E-waste in an industrial scale.

2 Rare-Earth Metals in Electronic Waste

REMs are primarily 17 metallic elements which include 15 lanthanides in addition to scandium and yttrium. On the basis of their atomic weight, they are further categorized as light rare earth elements (LREEs), and the less common and thus more valuable heavy rare earth elements (HREEs) (Chen et al., 2024a) (Fig. 2). REMs exhibit unique physico-chemical characteristics due to their distinct electronic structure, making them useful for optical, electrical, and magnetic applications (Yan et al., 2024). Praseodymium (Pr) and neodymium (Nd) are extremely useful in the production of permanent magnets for electric motors, whereas europium (Eu) and terbium (Tb) are vital for the lighting industry. Erbium (Er) and ytterbium (Yb) play significant roles in the telecommunications sector as they are incorporated frequently into fiber optic cables that enable high-speed data transmission. In addition, cerium (Ce) is used commonly as a catalyst in the manufacturing of chemicals, oil refining, and pollution control (Danouche et al., 2024). In the past decade, their use

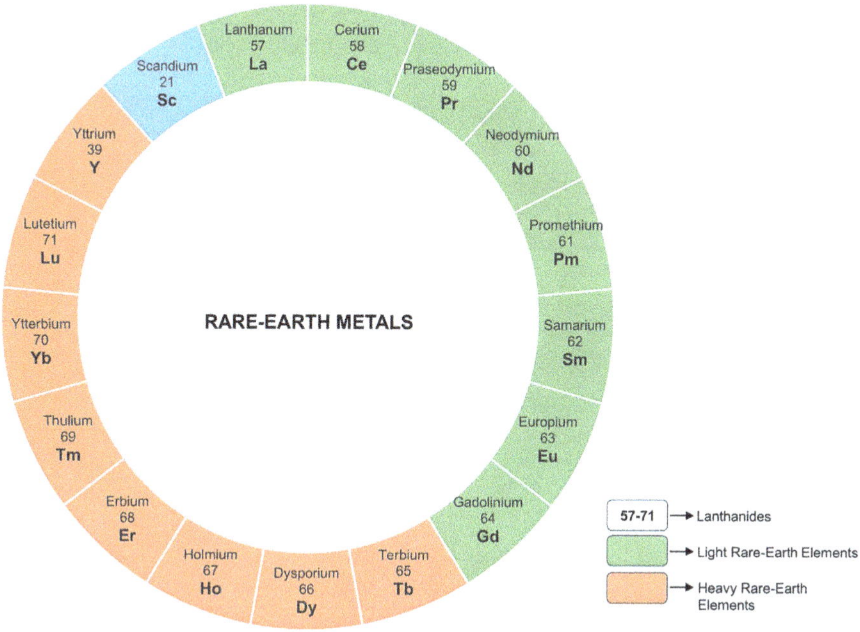

Fig. 2 Rare-earth metals and its types

has exploded as high-tech industries have become more dependent on REMs. The earth's crust is rich in rare metals but mineable concentrations are scarce and are mostly limited to China, Brazil, and Vietnam. As demand outweighs supply, along with significant challenges related to their purification at an industrial scale, and threat to their availability because of uneven distribution and unstable geopolitics, REMs have become particularly expensive that by 2026 valuation is estimated to reach USD 9.6 billion (Vuppaladadiyam et al., 2024).

REMs are predominantly extracted from natural resources where they are distributed either in the form of rare earth ores or ionic clay deposits, with the HREEs making up the major component in ionic rare earth deposits (Brown et al., 2024; Azimi, 2024). Except for promethium (Pm) which is produced by radioactive decay, almost all the other REMs are found in the form of mineral deposits like bastnasite, allanite, xenotime, monazite, and gadolinite. Irrespective of the technological advancements, few of these minerals have been used for the extraction of REMs on the industrial scale so far, thus making their supply from primary resources very unpredictable (Cheng et al., 2024). Scientists and policymakers have been looking into secondary sources as a sustainable replacement to meet the projected global demand for REMs. The regulatory framework proposed for REM extraction focuses on worker safety, protection for the environment, and resource conservation. In this regard, E-waste has been identified as a prominent secondary resource for REMs owing to their critical concentrations (Gkika et al., 2024). The process of recycling the components of E-waste not only eliminates the separation of low-value

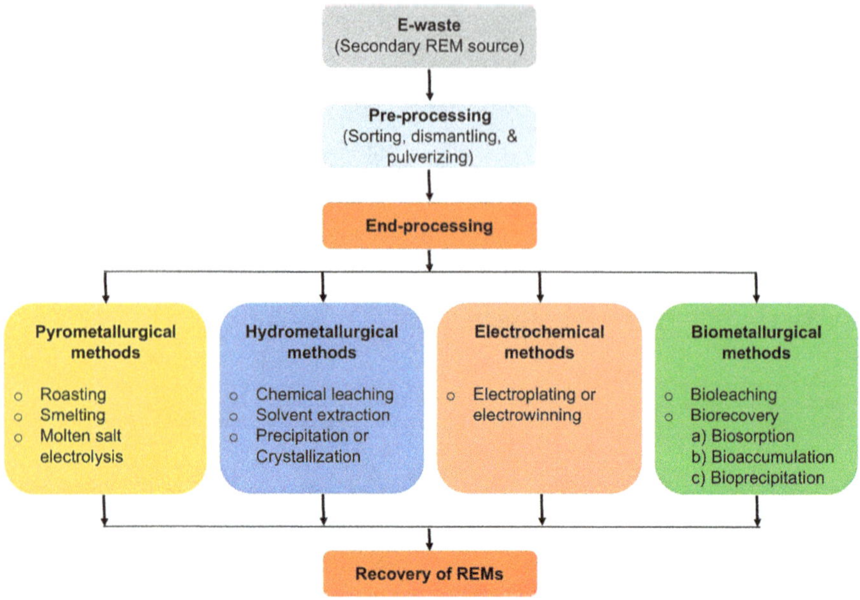

Fig. 3 Strategies involved in the extraction of rare-earth metals from E-waste

radioactive elements but also minimizes land disturbance and reduces environmental impact (Mishra et al., 2024). The recycling process is focused on maximizing resource recovery and includes two fundamental steps—preprocessing which involves sorting, dismantling, and pulverizing the electronic waste materials, and end-processing that consists of metallurgical processes that promote the extraction and recovery of the critical elements (Lazo, 2019) (Fig. 3). However, the major problem associated with recovering REMs from electronics are their heterogeneous mixture of components and sophisticated design. Extreme caution must be taken during the operation and processing of E-waste because it contains over 60 toxic and hazardous compounds which if released into the environment can contaminate the ecosystem and cause severe human health issues (Kumar et al., 2024). Thus, it has become extremely important to develop a REM recovery process that is not only cost-effective but also eco-friendly.

3 Common Methods of REMs Extraction from E-Waste

Challenges involved with primary mining have made it crucial to develop sustainable strategies for the recycling of REMs from secondary sources like E-waste. Various recycling techniques have been devised to separate and extract the metals but each has its own advantages and challenges. Although many recovery methods are currently used, three types are the most common which are discussed below.

3.1 Pyrometallurgical Methods

Pyrometallurgy is a traditional process that involves heating the metals and is the most sought-after method for the separation of REMs from secondary sources. This process separates the elements by melting the electronic items at an extremely high temperature of 900–3000 °C (Fazari et al., 2024). Various pyrometallurgical methods used for REM extraction include roasting, smelting, and molten salt electrolysis. The process of roasting the metals is performed in the presence of additives like sodium hydroxide (NaOH) or sodium carbonate (Na_2CO_3). It is a temperature-dependent process that converts the REM-containing compounds into oxides, enhancing both the conversion efficiency as well as selectivity. In the case of smelting, it concentrates the REMs in a molten phase by melting the E-waste at a high temperature, which makes it easy for the elements to be separated from the slag. To minimize energy consumption and maximize recovery, the process is optimized by adjusting parameters like temperature, reducing agents, and flux composition. And lastly, molten salt electrolysis is an electrochemical process that utilizes molten salts as an electrolyte at a high temperature, and is suitable for extracting REMs from calcium and phosphate matrix. The molten electrolyte usually consists of lithium fluoride in which the E-waste materials are dissolved following which the REMs are deposited at the cathode. These methods are frequently used for recovery because they separate the rare metallic elements altogether in a bulk from the complex electronic wastes (Zhang et al., 2024; Kwon et al., 2024). However, these methods face major limitations as they are very complicated and most of the time require further separation and purification of the REMs. Another issue pertains to huge investments and utilization of vast amounts of energy. Also, a major drawback is the environmental impact that is caused due to the release of harmful substances, toxic gases, and production of dust particles (Xu et al., 2024a; Das & Ghosh, 2023c, 2023e). All of these issues curb the use of pyrometallurgical methods widely due to the stringent government policies and environmental regulations all across the world.

3.2 Hydrometallurgical Methods

Another widely applied recovery technique is the hydrometallurgical method. It is a chemical approach that is used when impurities like plastics and adhesives, and metals like Co, Ni, and Zn, interfere with the use of the pyrometallurgy process. In hydrometallurgy, chemicals such as organic solvents, alkalis, and acids are used to extract and purify the REMs from E-waste (Diallo et al., 2024). The key steps in the recovery method involve: (i) leaching, (ii) solvent extraction, and (iii) precipitation or crystallization. Initially, the rare metals containing parts are separated from the E-waste and crushed into a fine powder, after which the metals are converted to their respective oxides by roasting them at 950 °C. Then, selective leaching is carried out using sulfuric acid (H_2SO_4), nitric acid (HNO_3), and hydrochloric acid (HCl). The leaching agent is selected based on the REM-containing material and the transition metals that may interfere with the recovery process. Leaching is followed by the solvent extraction process, which separates and purifies the REMs from the leachate solution. Solvent extraction actively involves organic solvents that bind easily with rare elements and help in their separation from other dissolved ions. Finally, the separated REM-containing solutions are then exposed to precipitation and crystallization to obtain the solid metallic products. This is generated either by changing the pH of the solution or by using precipitating agents like ammonium bicarbonate or oxalic acid. The precipitated rare metals are then filtered, washed, and calcined to derive high-purity oxides of REMs (Tuncay et al., 2024; Yadav et al., 2024).

Hydrometallurgy plays an essential role in the separation and recovery of REMs from electronic materials. Although the process necessitates less expenditure and requires less energy input compared to pyrometallurgy, its daily operating costs are high and the use of chemical substances produces toxic sludge and poisonous fumes which are harmful to the environment and subsequently to humans (Fazari et al., 2024; Das & Mishra, 2009; Paul et al., 2023).

3.3 *Electrochemical Recovery Methods*

The electrochemical method is a selective recovery process used for the separation of rare metallic elements. This approach uses fewer chemical reagents, requires less energy, is easily controlled, and is more eco-friendly than the previous two methods. The electrochemical method is also more sustainable as it avoids the use of toxic chemicals, does not require any harsh pH conditions, and generates no additional waste (Vuppaladadiyam et al., 2024).

The electrochemical technique includes the process of electrowinning in which the rare element extraction is achieved by an electroplating process. The electrochemical cell consists of an inert anode from which the current is passed through an electrolyte that is a metal-containing liquid leach-solution and deposits the REMs onto the cathode, and electrorefining is where an acidic electrolyte is used to pass the current from the anode to the cathode—the anode consists of recycled material which finally corrodes and releases the rare elements as ions into the solution, following which electrowinning takes place (Guo et al., 2024). Irrespective of the advantages associated, electrochemical recovery applications of REMs have been hardly described in the literature. Also, more research is required to overcome the issues regarding low recovery rate, high operational costs, and overall feasibility of the electrochemical recovery method.

In conclusion, as the demand for REMs grows exponentially, more advanced and environment-friendly methods are required so that the electronic waste recycling industry can efficiently extract and purify these rare elements, and contribute to a more resilient and sustainable circular economy.

4 Biological Strategies for REMs Extraction from E-Waste

Issues like high expenses, use of harmful chemicals, excessive energy input, and generation of toxic waste streams have restricted the use of pyrometallurgy and hydrometallurgy as recovery methods for REMs' extraction from E-waste. Similarly, less is known about the electrochemical process as extensive studies regarding their application in REM extraction are few and far between. This has led to the advancement in alternative extraction technologies which are efficient, inexpensive, and eco-friendly, and are guaranteed to meet the increasing demands for REMs.

Biometallurgy is a novel recovery process that utilizes the specificity of microorganisms to distinctly separate rare earth metals from electronic materials to produce concentrated products (Das et al., 2011; Das & Ghosh, 2023b; Ghosh et al., 2023). The process uses oxidation, reduction, absorption, and dissolution of the critical elements by microbes or their secondary metabolites to accomplish extraction and enrichment (Balaram, 2024). The biological recovery technique is quickly gaining interest among industry stakeholders due to its high selectivity, low energy consumption, and production of a much lower carbon footprint than the previous recovery methods (Das & Ghosh, 2023a, 2023d). However, the major limiting factors for biometallurgy are its longer life cycle, requirement of ample area for extraction at an industrial scale, and relatively insufficient recovery efficiency (Thacker et al., 2024).

Microorganisms have become a favorable choice due to their ability to quickly adapt to harsh environments (Das et al., 2015). Different types of bacteria, fungi, and algae can extract REMs through different biological approaches (Kiran et al., 2024). These microorganisms separate the rare metals by mobilizing them from solid to liquid phase using different leaching mechanisms (Rasoulnia et al., 2021). Microorganisms also use their natural metabolic systems to recover REMs through processes like biosorption, bioaccumulation, and bioprecipitation (Balaram, 2024). The following section highlights the different biological strategies and the microorganisms employed for separation and extraction of REMs from E-waste in detail.

4.1 Bioleaching

Bioleaching is a novel technique that utilizes the metabolic products of microorganisms to solubilize rare metallic elements and extract them from electronic materials. The process of bioleaching is primarily executed by heterotrophic microorganisms which solubilizes REMs by using iron/sulfur oxidation, hydrogen cyanide pathways, and organic acids (Zhou et al., 2024). The biochemical reactions involved in bioleaching are complex and various reaction mechanisms are involved in the mobilization of the REMs into the leachate solution, namely acidolysis, redoxolysis, and complexolysis.

The process of acidolysis is performed by iron and sulfur-oxidizing bacteria which oxidizes ferrous ion (Fe^{2+}) and sulfide (S) to produce ferric ion (Fe^{3+}) and sulfuric acid (H_2SO_4), which further oxidize the REMs and subsequently releases them into the culture medium. This process is driven by the protons which interact with the metals ultimately leading to their solubilization (Golzar-Ahmadi et al., 2024).

Microorganisms carry out redoxolysis by catalyzing the oxidation or reduction of rare metals and interacting with the source material by two different modes. In case of contact mode, the microorganism produces extracellular polymeric substances (EPS) which become the location for the biochemical reaction (Baez et al., 2024). There are two variations of the contact mode: a one-step process, where both the microbial culture and the REM-containing material are added to the same culture medium simultaneously, and the two-step process, in which the microbial culture is pregrown after which the REM-containing material gets added. The one-step and the two-step process is favored for low and high metal concentrations, respectively. In the noncontact mode, the waste material is not directly exposed to the microbial culture, but the REM-containing material is added to a culture medium that contains only the leaching agents secreted by a pregrown microbe (Cesaro, 2024).

In the process of complexolysis, microorganisms form complexes with rare metals with the help of organic acids produced as a metabolite and cause their dissolution. In this process, microorganisms do not interact directly with the source material but instead, they synthesize metabolic products that chelate the metals by forming soluble complexes (Golzar-Ahmadi et al., 2024). In the case of heterotrophic microorganisms, they chelate the REMs by forming soluble complexes with the organic acids produced by the microbes from the exogenous carbon sources during growth (Baez et al., 2024). Similarly, hydrogen cyanide produced by cyanogenic bacteria forms cyanide complexes with rare elements that are water-soluble (Mishra et al., 2024). Some microorganisms also produce siderophores that facilitate the dissolution of the metallic elements by forming complexes with them. The REMs containing bioleachate are then further subjected to biological or chemical methods to separate and harvest the metals.

4.2 Biorecovery of REMs

4.2.1 Biosorption

Biosorption is the removal of REMs from solutions through interaction with living microbial cells like bacteria, fungi, and algae. Both living and dead cells can be used to extract the REMs from the leachate using a combination of electrostatic interaction, ion exchange, precipitation, and surface complexation (Vítová & Mezricky, 2024). The metals are removed through binding with the cell surface by physical and chemical action. Biosorption is a metabolism-dependent process in the case of living microbial cells as the rare elements are adsorbed onto the cell surface and then transported across the membrane. Whereas, it is a metabolism-independent process for dead microbial cells where the functional groups that are present on the cell surface bind with the REMs and separate them from the solution (Liang et al., 2024). Amino, acyl, hydroxyl, carboxyl, phosphate, sulfhydryl, and hydrophobic groups of lipids, polysaccharides, and glycoproteins present on the cell surface act as the binding sites for the REMs. For example, negatively charged carboxyl groups use the ion-exchange mechanism to bind with metal cations. Amino groups can easily chelate cations and also adsorb anions either through hydrogen bonds or electrostatic interactions (Zhang et al., 2024). Some major advantages of the biosorption process include no production of chemical waste streams, ease of operation, and efficiency in removing rare metals from dilute solutions (Danouche et al., 2024).

4.2.2 Bioaccumulation

Another metabolic-dependent mechanism is bioaccumulation which helps in the uptake of REMs from solutions and transports the same intracellularly from the cell surface. The process of bioaccumulation can only be implemented by living microbial cells. Once all metallic elements accumulate in the cell cytoplasm, they get sequestered by a storage system which is mostly made up of proteins and peptide ligands (Balaram, 2024). The REMs are generally accumulated by channels like α-helical proteins which facilitate the movement of the metals by passive diffusion. In the case of gram-negative bacteria, REMs are sequestered by the porins present on the outer membrane. The rare metals get transported through the β-channels to the periplasmic space, after which they are transported to the cytoplasm for storage. Other protein carriers involved in the process of bioaccumulation are symporters, uniporters, and antiporters. These are energy-dependent carriers, where symporters use the proton motive force as a cosubstrate which creates a charge difference to translocate the target metal. Similarly, uniporters also use the charge difference to transfer cationic metals (Zhang et al., 2024).

4.2.3 Bioprecipitation

The process of bioprecipitation is initiated by the translocation of the phosphatase enzyme to the extracellular polymeric matrix. Phosphatase then starts producing inorganic phosphate from organic phosphate. This inorganic phosphate then causes precipitation by binding with the REMs present in solution. Bioprecipitation is also used for selective extraction of both LREEs and HREEs (Zhang et al., 2024).

4.3 Types of Microorganisms

4.3.1 Bacteria

Microorganisms that are extensively used for bioleaching are acidophilic bacteria as they can grow under extremely acidic conditions. Among them, *Acidithiobacillus ferooxidans* and *Leptospirillum ferooxidans* are most widely used for oxidative dissolution of REMs as they can leach metals by oxidizing Fe^{2+} to Fe^{3+}. At the same time, *A. ferrooxidans* and *A. thiooxidans* can extract the metals by generating sulfuric acid from the oxidation of sulfide (Chen et al., 2024b). Nevertheless, the REM recovery rates remain quite low and often require considerable adjustments in highly acidic conditions along with a period of 2 weeks minimum for effective extraction of REMs. Recent studies with the bacterium *Gluconobacter oxydans* reported that replacement of KH_2PO_4 with yeast extract in the growth medium resulted in a high conversion yield of gluconic acid which led to the extraction of 19.5% of the total REMs in 1 day that is 11.0 times more than with the phosphate medium, enhancing the overall recovery rate (Rasoulnia et al., 2022). Another type of bacteria used for bioleaching are cyanogenic microorganisms which produce CN^- as a secondary metabolite by utilizing C and N elements present in the growth medium. *Chromobacterium violaceum* is the most common CN^- producer but its use is limited in the biorecovery process as it can only survive in tropical and subtropical climates (Sandrasaigaran et al., 2022).

4.3.2 Fungi

Fungi perform the leaching process through the synthesis of organic acids as secondary metabolites. *Aspergillus* sp. is the most commonly used to extract REMs from different electronic materials. Citric acid, lactic acid, gluconic acid, and oxalic acid, produced by *Aspergillus niger*, *Penicillium simplex*, and *Penicillium chrysogenum* make them highly effective to leach the metallic elements (Lhamo & Mahanty, 2022). Some marine fungi, like *Dipodasaceae*, *Gymnoascaceae*, *Microascaceae*, and *Trichocomaceae*, have also been studied for their bioleaching and bioaccumulation capability of REMs from electronic waste materials (Zhang et al., 2024). The major advantage of using fungi is their ability to grow over a wide pH range making extraction from alkaline substances more productive, and hence reducing the overall metal leaching period from E-waste (Chen et al., 2024b). A recent study with the gluconic acid-producing *Penicillium expansum* demonstrated a bioleaching efficiency of ~70% for Pr, Nd, Gd, 50% for Tb, and 40% for La, under 24 h (Baez et al., 2024).

4.3.3 Algae

Different species of algae extract REMs mostly through the biosorption and bioaccumulation processes. Some of the most extensively studied algae that use the biosorption process are *Turbinaria conoides* (Yb, Eu, Ce, La), *Posidonia oceanica* (Sc), and *Sargassum* sp. (Nd, La), whereas *Desmodesmus quadricaudata*, *Chlamydomonas reinhardtii*, and *Chlorella kessleri* have been demonstrated to have the ability to accumulate La from leachate solutions (Lhamo & Mahanty, 2022; Vítová & Mezricky, 2024).

5 Modifications for Enhancing Biological Recovery of REMs

Although biological strategies for extraction have been universally acknowledged as a sustainable approach for the extraction of critical elements from E-waste, substantial research is lacking on how to overcome limitations like specificity, rate of recovery, and leaching time. Recently, studies have been focused on the biosorption mechanism for selective recovery of rare metals using engineered biomaterials (Giese, 2020). Protein/peptide-based biosorption has emerged as a promising method because of its fast adsorption kinetics, high selectivity and affinity, and low chemical and energy inputs. It has been identified that methylotrophic bacteria isolated from different environments use lanthanides to carry out their essential functions. They can selectively acquire the elements by utilizing highly specific REM-binding proteins. Researchers have exploited state-of-the-art techniques like proteomics, rational design, phage display, and homology modeling to genetically engineer proteins and peptides with highly improved REM-binding affinity and selectivity. For example, REM-binding peptides like FL 464 and FL606 identified through phage display were observed to bind specifically REM-containing phosphor materials including LaPO4:Ce^{3+}, Tb^{3+} (Ye et al., 2024). Another peptide DM39 was also identified to have a strong affinity for Y_2O_3 with its efficiency being more than the wild-type phage by 531.6-fold (Maass et al., 2024). In addition, artificial peptides synthesized through rational design is another approach. For REM-binding artificial peptides, the EF-hand motif of the calcium-binding protein calmodulin has been used; however, the low affinity and selectivity of these peptides have limited their practical applications in the recovery of REMs (Ye et al., 2024).

Irrespective of the significant efforts made in the direction of using engineered biomaterials, some crucial challenges exist that need to be addressed. First, the robustness of the protein/peptide-based biomaterial under application conditions should be ensured. Second, the biomaterial should be designed for reusability and scalability so that the metallic elements can be recovered at an industrial scale. And lastly, more innovative and sustainable technologies are still required to effectively separate the REMs because the existing processes require high energy and chemical input (Chakankar et al., 2024). Moving forward, more efforts are required to understand the mechanism of action of different biorecovery methods of REMs to explore and innovate the extraction strategies from E-waste.

6 Existing Challenges and Future Perspective

Regardless of the advancements made in the recovery technologies to extract REMs from E-waste, there is still a pressing need to refine the recovery systems. The current recycling industry does not have the proper infrastructure and regulations to manage the soaring amount and sophisticated REM containing E-waste (Xu et al.,

2024b). Also, the inadequate collection, sorting, and processing methods make the extraction highly inefficient and therefore limit the large-scale recovery of REMs.

To fix these issues, collaboration between the policymakers and the industry stakeholders is necessary to develop strategies for electronic waste management which prioritizes the recycling of rare metallic elements. This includes holding the manufacturers responsible for their product disposal process as well as providing incentives to the public for proper disposal of their E-waste. Also, efforts should be made to increase public awareness which can help in the proper collection and sorting of E-waste (Liang et al., 2024).

It is also essential to optimize the different recovery technologies to enhance the overall extraction efficiency. Every extraction and enrichment process has its advantages and drawbacks, and innovative synergistic combinations of them can lead to increased efficacy, affordability, and sustainable techniques (Gkika et al., 2024). For instance, combining hydrometallurgy with bioleaching can minimize energy consumption, reduce the input of chemicals, and impact on the environment while providing high rates of recovery. Researchers and industries should explore hybrid recovery processes as they can contribute to more efficient and sustainable recovery techniques (Sophia & Vidya, 2023).

By advancing efficient utilization of E-waste and sustainable management strategies, governments and industries can contribute significantly to the REMs recycling and recovery industry by creating a more resilient circular economy.

7 Conclusions

REMs have become indispensable due to our current technology-dependent lifestyle and expansion to use alternative clean technologies. As such, recycling of REMs from secondary sources has assumed considerable significance because recycling will allow countries to procure REMs even without any natural REM reserves and will play an important role in fulfilling the market demands. Further investigations into REM recovery are critical to establish innovative technologies for efficient extraction and selective dissolution processes that are sustainable and eco-friendly. Thus, through the collaborative efforts of the stakeholders, it will be possible to expand on the sustainable recovery of rare metals and ensure their availability to future generations.

References

Azimi, G. (2024). Technological advancements in rare earth elements recovery from ionic clays: A comprehensive review. *Hydrometallurgy, 231*, 106414. https://doi.org/10.1016/j.hydromet.2024.106414

Baez, A. G., Muñoz, L. P., Timmermans, M. J., Garelick, H., & Purchase, D. (2024). Molding the future: Optimization of bioleaching of rare earth elements from electronic waste by *Penicillium expansum* and insights into its mechanism. *Bioresource Technology, 402*, 130750. https://doi.org/10.1016/j.biortech.2024.130750

Balaram, V. (2024). Sustainable recovery of rare earth elements by recycling of E-waste for a circular economy: Perspectives and recent advances. In *Environmental materials and waste* (pp. 499–544). https://doi.org/10.1016/B978-0-443-22069-2.00023-1

Baldé, C. P., D'Angelo, E., Deubzer, V. L. O., & Kuehr, R. (2022). *Global transboundary E-waste flows monitor—2022*. United Nations Institute for Training and Research (UNITAR).

Bhavan, J. S., Joy, J., & Pazhani, A. (2023). Identification and recovery of rare earth elements from electronic waste: Material characterization and recovery strategies. *Materials Today Communications, 36*, 106921. https://doi.org/10.1016/j.mtcomm.2023.106921

Brown, D., Zhou, R., & Sadan, M. (2024). Critical minerals and: Rare earth elements in a planetary just transition: An interdisciplinary perspective. *The Extractive Industries and Society, 19*, 101510. https://doi.org/10.1016/j.exis.2024.101510

Cesaro, A. (2024). Microbial leaching strategies for extraction of rare earth elements from primary and secondary resources. In *Biotechnological innovations in the mineral-metal industry* (pp. 67–77). Springer. https://doi.org/10.1007/978-3-031-43625-3_4

Chakankar, M., Lederer, F., Jain, R., Matys, S., Kutschke, S., & Pollmann, K. (2024). State-of-the-art biotechnological recycling processes. In *Management of electronic waste: Resource recovery, technology and regulation* (pp. 375–405). https://doi.org/10.1002/9781119894360.ch15

Chen, P., Ilton, E. S., Wang, Z., Rosso, K. M., & Zhang, X. (2024a). Global rare earth element resources: A concise review. *Applied Geochemistry, 175*, 106158. https://doi.org/10.1016/j.apgeochem.2024.106158

Chen, Z., Han, Z., Gao, B., Zhao, H., Qiu, G., & Shen, L. (2024b). Bioleaching of rare earth elements from ores and waste materials: Current status, economic viability and future prospects. *Journal of Environmental Management, 371*, 123217. https://doi.org/10.1016/j.jenvman.2024.123217

Cheng, S., Li, W., Han, Y., Sun, Y., Gao, P., & Zhang, X. (2024). Recent process developments in beneficiation and metallurgy of rare earths: A review. *Journal of Rare Earths, 42*(4), 629–642. https://doi.org/10.1016/j.jre.2023.03.017

Danouche, M., Bounaga, A., Oulkhir, A., Boulif, R., Zeroual, Y., Benhida, R., & Lyamlouli, K. (2024). Advances in bio/chemical approaches for sustainable recycling and recovery of rare earth elements from secondary resources. *Science of the Total Environment, 912*, 168811. https://doi.org/10.1016/j.scitotenv.2023.168811

Das, A. P., & Ghosh, S. (2018). Bioleaching of manganese from mining waste materials. *Materials Today: Proceedings, 5*(1), 2381–2390. https://doi.org/10.1016/J.MATPR.2017.11.459

Das, A. P., & Ghosh, S. (2023a). Cost-effectiveness of the biomining process. In *Manganese mining microorganisms* (pp. 143–151). https://doi.org/10.1016/B978-0-12-822147-1.00008-X

Das, A. P., & Ghosh, S. (2023b). Evolution of biomining technology. In *Manganese mining microorganisms* (pp. 5–19). https://doi.org/10.1016/B978-0-12-822147-1.00012-1

Das, A. P., & Ghosh, S. (2023c). Mechanism of manganese bioleaching. In *Manganese mining microorganisms* (pp. 87–98). https://doi.org/10.1016/B978-0-12-822147-1.00014-5

Das, A. P., & Ghosh, S. (2023d). Microorganisms in manganese biomining processes. In *Manganese mining microorganisms* (pp. 55–61). https://doi.org/10.1016/B978-0-12-822147-1.00013-3

Das, A. P., & Ghosh, S. (2023e). Mixed culture bioleaching: An insight into manganese biomining process and efficacy. In *Manganese mining microorganisms* (pp. 129–141). https://doi.org/10.1016/B978-0-12-822147-1.00001-7

Das, A. P., & Ghosh, S. (2023f). Role of metabolites and metabolomics for understanding of manganese bioleaching. In *Manganese mining microorganisms* (pp. 117–127). https://doi.org/10.1016/B978-0-12-822147-1.00011-X

Das, A. P., & Mishra, S. (2009). Hexavalent chromium [Cr (VI)]: Yellow water pollution and its remediation. *Sarovar Saurabh ENVIS Newsletter on Wetland Ecosystems, 5*(2).

Das, A. P., Sukla, L. B., Pradhan, N., & Nayak, S. (2011). Manganese biomining: A review. *Bioresource Technology, 102*(16), 7381–7387. https://doi.org/10.1016/J.BIORTECH.2011.05.018

Das, A. P., Ghosh, S., Mohanty, S., & Sukla, L. B. (2015). Advances in manganese pollution and its bioremediation. In L. B. Sukla, N. Pradhan, S. Panda, & B. K. Mishra (Eds.), *Environmental microbe biotechnology* (1st ed., pp. 313–328). Springer. https://doi.org/10.1007/978-3-319-19018-1_16

Dev, S., Sachan, A., Dehghani, F., Ghosh, T., Briggs, B. R., & Aggarwal, S. (2020). Mechanisms of biological recovery of rare-earth elements from industrial and electronic wastes: A review. *Chemical Engineering Journal, 397*, 124596. https://doi.org/10.1016/j.cej.2020.124596

Diallo, S., Tran, L. H., Larivière, D., & Blais, J. F. (2024). Selective leaching of rare earths, base metals and precious metals from used smartphones. *Environmental Technology, 46*, 1–16. https://doi.org/10.1080/09593330.2024.2423906

Fazari, J., Hossain, M. Z., & Charpentier, P. (2024). A review on metal extraction from waste printed circuit boards (wPCBs). *Journal of Materials Science, 59*(27), 12257–12284. https://doi.org/10.1007/s10853-024-09941-6

Ghosh, S., & Das, A. P. (2017). Bioleaching of manganese from mining waste residues using Acinetobacter sp. *Geology, Ecology, and Landscapes, 1*(2), 77–83. https://doi.org/10.1080/24749508.2017.1332847

Ghosh, S., Mohanty, S., Akcil, A., Sukla, L. B., & Das, A. P. (2016). A greener approach for resource recycling: Manganese bioleaching. *Chemosphere, 154*, 628–639. https://doi.org/10.1016/J.CHEMOSPHERE.2016.04.028

Ghosh, S., Tripathy, B., Dey, S., & Das, A. P. (2023). Comparative investigation of fungal and bacterial manganese biomining mechanisms. In *Manganese mining microorganisms* (pp. 99–115). https://doi.org/10.1016/b978-0-12-822147-1.00016-9

Giese, E. C. (2020). Biosorption as green technology for the recovery and separation of rare earth elements. *World Journal of Microbiology and Biotechnology, 36*(4), 52. https://doi.org/10.1007/s11274-020-02821-6

Gkika, D. A., Chalaris, M., & Kyzas, G. Z. (2024). Review of methods for obtaining rare earth elements from recycling and their impact on the environment and human health. *Processes, 12*(6), 1235. https://doi.org/10.3390/pr12061235

Golzar-Ahmadi, M., Bahaloo-Horeh, N., Pourhossein, F., Norouzi, F., Schoenberger, N., Hintersatz, C., Chakankar, M., Holuszko, M., & Kaksonen, A. H. (2024). Pathway to industrial application of heterotrophic organisms in critical metals recycling from e-waste. *Biotechnology Advances, 77*, 108438. https://doi.org/10.1016/j.biotechadv.2024.108438

Guo, M., Deng, R., Gao, M., Xu, C., & Zhang, Q. (2024). Sustainable recovery of metals from e-waste using deep eutectic solvents: Advances, challenges, and perspectives. *Current Opinion in Green and Sustainable Chemistry, 47*, 100913. https://doi.org/10.1016/j.cogsc.2024.100913

Hossain, R., & Sahajwalla, V. (2024). Current recycling innovations to utilize e-waste in sustainable green metal manufacturing. *Philosophical Transactions of the Royal Society A, 382*, 20230239. https://doi.org/10.1098/rsta.2023.0239

Kaim, V., Rintala, J., & He, C. (2023). Selective recovery of rare earth elements from e-waste via ionic liquid extraction: A review. *Separation and Purification Technology, 306*, 122699. https://doi.org/10.1016/j.seppur.2022.122699

Kiran, N. S., Yashaswini, C., Chatterjee, A., & Shah, M. P. (2024). Biotechnological approaches for metal recovery from electronic wastes. *Current Microbiology, 81*(12), 419. https://doi.org/10.1007/s00284-024-03945-w

Kumar, P., Singh, S., Gacem, A., Yadav, K. K., Bhutto, J. K., Alreshidi, M. A., Kumar, M., Kumar, A., Yadav, V. K., Soni, S., Kumar, R., Qasim, M. T., Tariq, M., & Alam, M. W. (2024). A review on e-waste contamination, toxicity, and sustainable clean-up approaches for its management. *Toxicology, 508*, 153904. https://doi.org/10.1016/j.tox.2024.153904

Kwon, G., Yoon, K., Kwon, E., Park, J., Lee, H., & Song, H. (2024). Technical advancement in valorization of electronic waste and its contribution to establishing economic value-chain. *Chemical Engineering Journal, 494*, 153154. https://doi.org/10.1016/j.cej.2024.153154

Lazo, D. E. (2019). *Selective complexation leaching of rare earth minerals: A new hydrometallurgical route inspired by geochemical process* (Doctoral dissertation, Curtin University). http://hdl.handle.net/20.500.11937/79965

Lhamo, P., & Mahanty, B. (2022). Bioleaching of rare earth elements from industrial and electronic wastes: Mechanism and process efficiency. In *Environmental technologies to treat rare earth elements pollution: Principles and engineering*. IWA Publishing.

Liang, B., Gu, J., Zeng, X., Yuan, W., Rao, M., Xiao, B., & Hu, H. (2024). A review of the occurrence and recovery of rare earth elements from electronic waste. *Molecules, 29*(19), 4624. https://doi.org/10.3390/molecules29194624

Maass, D., Boelens, P., Bloss, C., Claus, G., Harter, S., Günther, D., Pollmann, K., & Lederer, F. (2024). Identification of yttrium oxide-specific peptides for future recycling of rare earth elements from electronic scrap. *Biotechnology and Bioengineering, 121*(3), 1026–1035. https://doi.org/10.1002/bit.28629

Mishra, S., Ghosh, S., van Hullebusch, E. D., Singh, S., & Das, A. P. (2023). A critical review on the recovery of base and critical elements from electronic waste-contaminated streams using microbial biotechnology. *Applied Biochemistry and Biotechnology, 2023*, 1–30. https://doi.org/10.1007/S12010-023-04440-X

Mishra, K., Siwal, S. S., & Thakur, V. K. (2024). E-waste recycling and utilization: A review of current technologies and future perspectives. *Current Opinion in Green and Sustainable Chemistry, 47*, 100900. https://doi.org/10.1016/j.cogsc.2024.100900

Mohanty, S., Ghosh, S., Nayak, S., & Das, A. P. (2017). Bioleaching of manganese by Aspergillus sp. isolated from mining deposits. *Chemosphere, 172*, 302–309. https://doi.org/10.1016/J.CHEMOSPHERE.2016.12.136

Nnorom, I. C., & Osibanjo, O. (2008). Overview of electronic waste (e-waste) management practices and legislations, and their poor applications in the developing countries. *Resources, Conservation and Recycling, 52*(6), 843–858. https://doi.org/10.1016/j.resconrec.2008.01.004

Paul, A., Dey, S., Ram, D. K., & Das, A. P. (2023). Hexavalent chromium pollution and its sustainable management through bioremediation. *Geomicrobiology Journal, 41*, 1–11. https://doi.org/10.1080/01490451.2023.2218377

Pimassoni, Y. S., Weitzel Dias Carneiro Lima, M. T., Yamane, L. H., & Siman, R. R. (2023). The recovery of rare earth elements from waste electrical and electronic equipment: A review. *Hydrometallurgy, 222*, 106156. https://doi.org/10.1016/j.hydromet.2023.106156

Priyadarsini, S., & Das, A. P. (2024). Analytical and structural characterization of waste lithium-ion batteries for their effective recycling strategy. *Environmental Chemistry and Ecotoxicology, 7*, 182. https://doi.org/10.1016/j.enceco.2024.12.004

Priyadarsini, S., & Das, A. P. (2025). Lithium bioleaching: A review on microbial-assisted sustainable technology for lithium bio-circularity. *Journal of Water Process Engineering, 69*, 106744. https://doi.org/10.1016/j.jwpe.2024.106744

Rasoulnia, P., Barthen, R., & Lakaniemi, A. M. (2021). A critical review of bioleaching of rare earth elements: The mechanisms and effect of process parameters. *Critical Reviews in Environmental Science and Technology, 51*(4), 378–427. https://doi.org/10.1080/10643389.2020.1727718

Rasoulnia, P., Barthen, R., Lakaniemi, A. M., Ali-Löytty, H., & Puhakka, J. A. (2022). Low residual dissolved phosphate in spent medium bioleaching enables rapid and enhanced solubilization of rare earth elements from end-of-life NiMH batteries. *Minerals Engineering, 176*, 107361. https://doi.org/10.1016/j.mineng.2021.107361

Sandrasaigaran, P., Palanisamy, T., Sukalingam, R., & Murugan, S. (2022). Extended spectrum beta-lactamase expressing Chromobacterium sp. Dyh27s2016 strain is capable of metal leaching in electronic waste. *Key Engineering Materials, 937*, 181–191. https://doi.org/10.4028/p-65o2dj

Seif, R., Salem, F. Z., & Allam, N. K. (2024). E-waste recycled materials as efficient catalysts for renewable energy technologies and better environmental sustainability. *Environment, Development and Sustainability, 26*, 5473–5508. https://doi.org/10.1007/s10668-023-02925-7

Shahabuddin, M., Uddin, M. N., Chowdhury, J. I., et al. (2023). A review of the recent development, challenges, and opportunities of electronic waste (e-waste). *International jour-*

nal of Environmental Science and Technology, 20*, 4513–4520. https://doi.org/10.1007/s13762-022-04274-w

Sophia, S., & Vidya, S. K. (2023). Recovery of rare earth elements and critical metals from electronic waste. In *Electronic waste management: Policies, processes, technologies, and impact* (pp. 202–225). https://doi.org/10.1002/9781119891543.ch15

Thacker, S. C., Tipre, D. R., & Dave, S. R. (2024). Biohydrometallurgical metal recycling/recovery from e-waste: Current trend, challenges, and future perspective. In *Management of electronic waste: Resource recovery, technology and regulation* (pp. 436–464). https://doi.org/10.1002/9781119894360.ch17

Tuncay, G., Yuksekdag, A., Mutlu, B. K., & Koyuncu, I. (2024). A review of greener approaches for rare earth elements recovery from mineral wastes. *Environmental Pollution, 357*, 124379. https://doi.org/10.1016/j.envpol.2024.124379

Vítová, M., & Mezricky, D. (2024). Microbial recovery of rare earth elements from various waste sources: A mini review with emphasis on microalgae. *World Journal of Microbiology and Biotechnology, 40*(6), 1–13. https://doi.org/10.1007/s11274-024-03974-4

Vuppaladadiyam, S. S. V., Thomas, B. S., Kundu, C., Vuppaladadiyam, A. K., Duan, H., & Bhattacharya, S. (2024). Can e-waste recycling provide a solution to the scarcity of rare earth metals? An overview of e-waste recycling methods. *Science of the Total Environment, 924*, 171453. https://doi.org/10.1016/j.scitotenv.2024.171453

Xia, J., & Ghahreman, A. (2024). Sustainable technologies for the recycling and upcycling of precious metals from e-waste. *Science of the Total Environment, 916*, 170154. https://doi.org/10.1016/j.scitotenv.2024.170154

Xu, Q., Wu, D., Hu, W., Zhang, Z., Liu, X., Yang, F., & Wang, Z. (2024a). Recycling NdFeB magnets and rare earth fluorescent materials from electronic waste. *JOM, 76*(3), 1319–1328. https://doi.org/10.1007/s11837-023-06235-1

Xu, T., Zheng, X., Ji, B., Xu, Z., Bao, S., Zhang, X., Li, G., & Li, Z. (2024b). Green recovery of rare earth elements under sustainability and low carbon: A review of current challenges and opportunities. *Separation and Purification Technology, 330*, 125501. https://doi.org/10.1016/j.seppur.2023.125501

Yadav, J., Sarker, S. K., Bruckard, W., Jegatheesan, V., Haque, N., Singh, N., & Pramanik, B. K. (2024). Greening the supply chain: Sustainable approaches for rare earth element recovery from neodymium iron boron magnet waste. *Journal of Environmental Chemical Engineering, 12*, 113169. https://doi.org/10.1016/j.jece.2024.113169

Yan, L., Zhang, Y., Zhang, Y., Liu, R., Liu, C., Wu, F., & Xu, F. (2024). Progress of rare earth-transition metal alloys and their compounds for electromagnetic wave absorption. *Journal of Magnetism and Magnetic Materials, 597*, 172003. https://doi.org/10.1016/j.jmmm.2024.172003

Ye, Q., Wang, D., & Wei, N. (2024). Engineering biomaterials for the recovery of rare earth elements. *Trends in Biotechnology, 42*(5), 575–590. https://doi.org/10.1016/j.tibtech.2023.10.011

Zhang, X., Tan, N., Rastegar, S. O., & Gu, T. (2024). Advances in bioleaching of rare earth elements from electronic wastes. In *Management of electronic waste: Resource recovery, technology and regulation* (pp. 321–358). https://doi.org/10.1073/pnas.2308502120

Zhou, H., Wang, J., Yu, X., Kang, J., Qiu, G., Zhao, H., & Shen, L. (2024). Effective extraction of rare earth elements from ion-adsorption type rare earth ore by three bioleaching methods. *Separation and Purification Technology, 330*, 125305. https://doi.org/10.1016/j.seppur.2023.125305

Innovative Bioprocessing Techniques for Sustainable E-Waste Management

Noor Alam Chowdhary and L. S. Songachan

Abstract Global technological advancements and industrialization have led to the increased usage of electronic devices. The rapid accumulation of electronic waste (e-waste) presents significant threats to our environment and public health, largely due to improper disposal practices. As the need for sustainable solutions grows, innovative bioprocessing techniques offer a promising alternative to conventional e-waste recycling methods. These approaches harness the natural capabilities of microorganisms to interact with metals, degrade plastics, and manage hazardous substances found in electronic waste. By employing bio-based products derived from renewable resources, like microbial diversity and plant-based materials, bioprocessing provides an environmentally sustainable solution for e-waste disposal and management.

This chapter summarizes the latest advancements in bioprocessing approaches for sustainable management of e-waste, using biomaterials for disassembly and separation of electronic components. It emphasizes the role of microbial diversity in retrieving valuable metals, reducing waste, and addressing toxic contaminants. Additionally, the chapter offers a comprehensive analysis of bioleaching, biosorption, phytoremediation techniques, and their contributions to enhancing the efficiency of e-waste recycling systems in a sustainable manner.

Keywords Electronic waste (e-waste) · Bioleaching · Phytoremediation · Genetically modified organisms (GMOs) · Sustainable development

N. A. Chowdhary · L. S. Songachan (✉)
Department of Botany, Banaras Hindu University, Varanasi, UP, India
e-mail: chowdharynooralam@bhu.ac.in

A. P. Das, S. Priyadarsini (eds.), *Electronic Waste and Environmental Pollution*, Sustainable Environmental Waste Management Strategies,
https://doi.org/10.1007/978-3-031-90287-1_15

1 Introduction

Electronic waste, commonly known as e-waste, e-scrap, or WEEE (Waste Electrical and Electronic Equipment), applies to discarded electronic devices that are approaching or near the end of their useful life (Tecchio et al., 2018, Mishra et al., 2023). Due to the overuse and short lifespan of innovative electronics, e-waste is becoming a growing concern in both environmental and human health concerns (Fig. 1) (Rienzie et al., 2019). The increase in disposable incomes, urbanization, mobility, and industrialization in various regions is contributing to the rising amount of e-waste (Priyadarsini & Das, 2025). Recent reports indicate that global e-waste statistics estimate that the world generated 53.6 million tons (Mt) of e-waste in 2019, an increase from 41.8 Mt in 2014 to 44.7 Mt in 2016. Projections indicate that this figure will rise to 74.7 Mt by 2030, with an annual expansion rate of approximately 2 Mt (Baldé et al., 2017; Forti et al., 2020). Currently, per annum generation of e-waste is estimated to be between 40 and 50 Mt, with projections indicating a potential increase of 500 times over the next decade, especially in rapidly developing countries such as China and India (Wilson et al., 2015). A staggering 83% of the global e-waste goes unrecorded, which means it is either illegally disposed of or burned in the open, posing a threat to both human and environmental health (Dhivya & Premalatha, 2024).

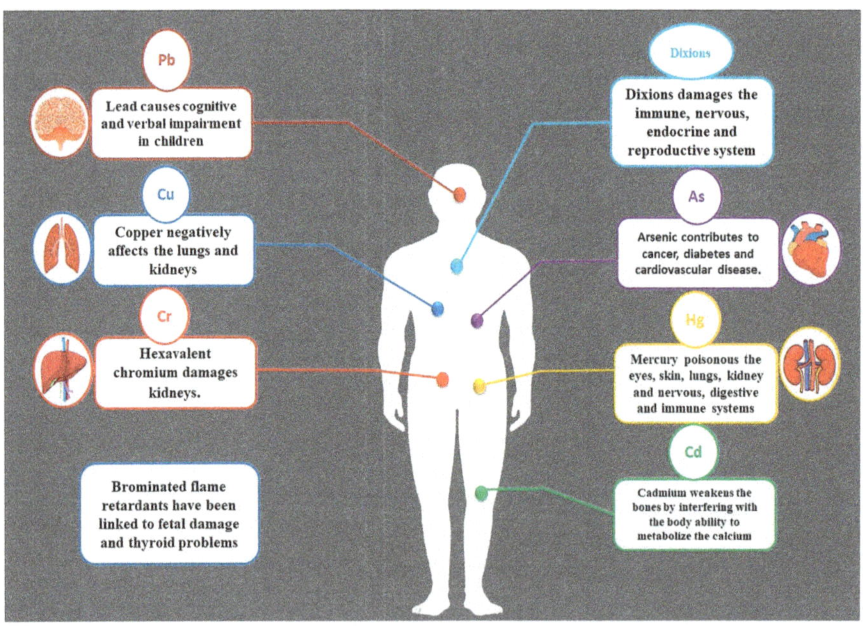

Fig. 1 Overview of toxic metals commonly found in e-waste and their associated health risks to humans

E-waste contains valuable resources alongside hazardous substances, including metals, heavy metals (Mishra et al., 2022), plastics (Priyadarshini et al., 2024; Singh Jadaun et al., 2022), brominated flame retardants (BFRs), and chlorofluorocarbons (CFCs) or hydrochlorofluorocarbons (HCFCs) (Duan et al., 2016). Among these, up to 60% of e-waste may contain metal (Ari, 2016). The recycling of materials and resources recovery from e-waste is of immense significance (Kaya, 2019; Gundupalli et al., 2018). Most common metals present in e-waste include lead (Pb), nickel (Ni), mercury (Hg), cadmium (Cd), iron (Fe), zinc (Zn), cobalt (Co), chromium (Cr), and copper (Cu) (Prabhakar et al., 2019). In addition to these, e-waste also contains rare earth metals (REEs), which consist of yttrium (Y), scandium (Sc), and the lanthanide series. The lanthanides include elements such as lanthanum (La), samarium (Sm), europium (Eu), gadolinium (Gd), terbium (Tb), dysprosium (Dy), holmium (Ho), erbium (Er), thulium (Tm), ytterbium (Yb), cerium (Ce), praseodymium (Pr), neodymium (Nd), promethium (Pm), and lutetium (Lu). Moreover, e-waste often contains elements from the platinum group (e.g., platinum (Pt) and palladium (Pd)), alkaline metals, radioactive metals, and precious metals such as gold (Au) and silver (Ag). Other notable elements include aluminum (Al), arsenic (As), manganese (Mn), and uranium (U). The presence of these diverse metals highlights the complex nature and potential value of e-waste recycling, as well as the environmental issues arising from improper disposal (Dutta et al., 2023; Şahan, 2016). The excessive release of e-waste poses a serious threat to living organisms and the environment. However, annually, 50 tons of Hg and 71 kilotons of BFR-containing plastics are detected in undocumented world e-waste flows, resulting in significant pollution and adverse effects on living organisms. Poor e-waste management drastically leads to global warming. Simultaneously, the world is coping with the adverse effects that petroleum-based products have on the environment. These products are nonbiodegradable and may persist intact for hundreds of years. The incineration of plastics releases significant amounts of CO_2 into the atmosphere, which significantly increases global warming (Naser et al., 2021). Neglecting proper e-waste management often results in its accumulation in landfills, where harmful substances leak into the surrounding ecosystem, contributing to pollution and being detrimental to both marine and terrestrial life. The rising amount of e-waste needs a sustainable and environmentally friendly approach for metal extraction from e-waste (Awasthi et al., 2017). There is a significant demand for valuable metals, particularly for the manufacturing of information and communication technology equipment. Therefore, the economical and environmentally friendly recovery of elements from e-waste can facilitate source sustainability (Zeng et al., 2018).

In the vision of ecological and economic benefits, microorganisms, such as bacteria, archaea, cyanobacteria, fungi, mycorrhiza, and plants, present a promising remedy to these concerns due to their inherent diverse metabolic capabilities (Gopikrishnan et al., 2020). The bio-based approaches present a cost-effective alternative to traditional and expensive techniques for extracting valuable metal elements and recycling toxic substances from e-waste (Narayanasamy et al., 2021; Liu et al., 2023). Moreover, certain microbes can synthesize bioplastics such as polyhydroxy butyrate (PHB), which is a biodegradable polymer belonging to the PHA family, offering an eco friendly alternative to petroleum-derived plastics (Tsang et al., 2019).

2 Current Status of E-Waste Management

Numerous techniques, such as pyrometallurgy, hydrometallurgy, and electrometallurgy, have been successfully established for the recovery of value-added products from e-waste (Kaya, 2016). Pyrometallurgy is the technique of recovering valuable metals from e-waste by incineration, smelting, and roasting. Thermal plants such as Aurubis, Noranda, Ronnskar, and Umicore process e-waste to produce pure copper blisters. However, smelting can produce harmful byproducts and uses more energy (Chauhan et al., 2018). These methods may result in environmental issues and extremely dangerous byproducts; hence, restricting their scope of use (Kuyucak & Akcil, 2013). Chemical products, including dioxins, furans, and heavy metals, are often released during e-waste recycling processes such as incineration, acid baths, open burning, and metallic acid stripping, which adversely affect the surrounding environment.

Hydrometallurgical technologies offer benefits compared to pyrometallurgical processes, including diminished hazardous gas emissions, reduced energy requirements, simplified laboratory execution, cheaper operational expenses, and enhanced efficiency rates. They also ensure substantial recovery of base metals, optimize leaching efficiency, and employ cyanide and noncyanide chemical reagents as potent lixiviants (Dutta et al., 2023). Despite sufficient information regarding the toxicity and safe handling of cyanide, environmental concerns remain significant and restrict its industrial application (Chauhan et al., 2018). The primary drawback of this procedure is the corrosive and toxic nature of the liquid utilized (Kinoshita et al., 2003).

Landfilling is one of the most commonly used e-waste discarding strategies. It is well known that all landfills exhibit leakage. Leachate frequently carries heavy metals and other hazardous compounds that may negatively impact groundwater and water resources. Even advanced landfills, designed to prevent toxins from infiltrating the soil, are not entirely impermeable over a period of time. Older landfill sites and ungovernable dumps have a significantly higher risk of leaking toxic pollutants. Hg, Cd, and Pb are among the most hazardous leachates. Hg will leach when specific electronic gadgets, such as circuit breakers, are damaged (Shahabuddin et al., 2023). Chemical recycling is the method of deploying chemical processes to convert electronic trash into valuable raw materials. This method enables the disassembly of e-waste into smaller components that can be recycled into new products. However, this approach is environmentally detrimental and may result in the emission of toxic chemicals (Chakraborty et al., 2022). Among the diverse traditional technologies, encompassing both chemical and mechanical approaches utilized for metal removal from e-waste, microbial technologies are rapidly gaining curiosity within the scientific community. Conventional processes are either expensive or

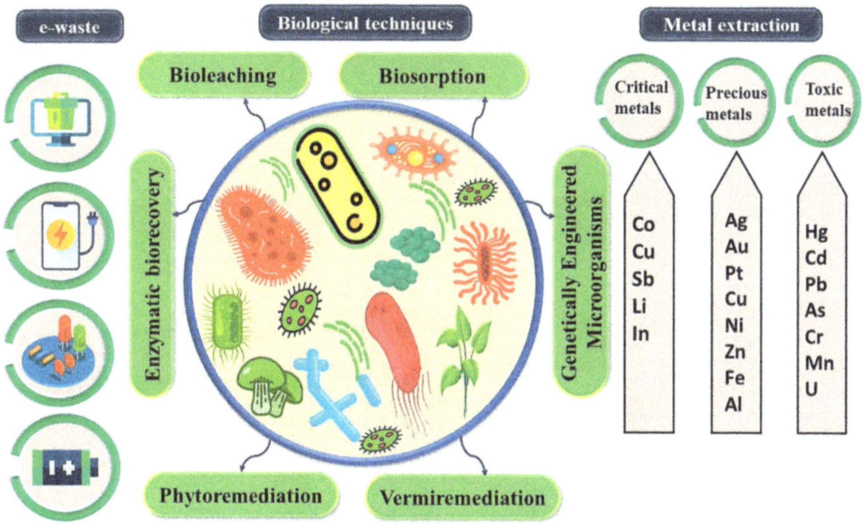

Fig. 2 An illustration of biological techniques for metal extraction from e-waste

produce further contamination needing remediation, but the biological method is environmentally sustainable, especially for metal recovery from electronic waste. This strategy uses microorganisms to transform or recover pollutants into less or nontoxic forms by microbial metabolism (Roy et al., 2022). Furthermore, researchers are investigating the ability of some bacteria and fungi to selectively bleach metals like Cu, Ag, and Au from complex e-waste compositions. By utilizing these natural processes, the field aims to develop more efficient and sustainable methods for the recovery of metals, ultimately contributing to a circular economy and reducing the effects of e-waste on the environment. Microorganisms help in the extraction of metals via numerous mechanisms, like biotransformation, biosorption, bioaccumulation, bioleaching, enzymatic bio-recovery, biomineralization, and vermiremediation are described in Fig. 2.

3 Bioprocessing Techniques for Metal Recovery

A significant challenge linked with e-waste is its metallic content. The need for metal recovery from e-waste arises from the hazardous nature of these metals, especially heavy metals, which pose major threats to both human health and the atmosphere (Das et al., 2011; Ghosh et al., 2023). This is due to their intrinsic properties, solubility across various pH levels, oxidation-reduction potentials, and complexation characteristics. Additionally, e-waste offers significant economic opportunities as a secondary source of rare and precious metals (Fig. 3) (Jagannath et al., 2017). The concept of urban mining has attracted significant attention in various

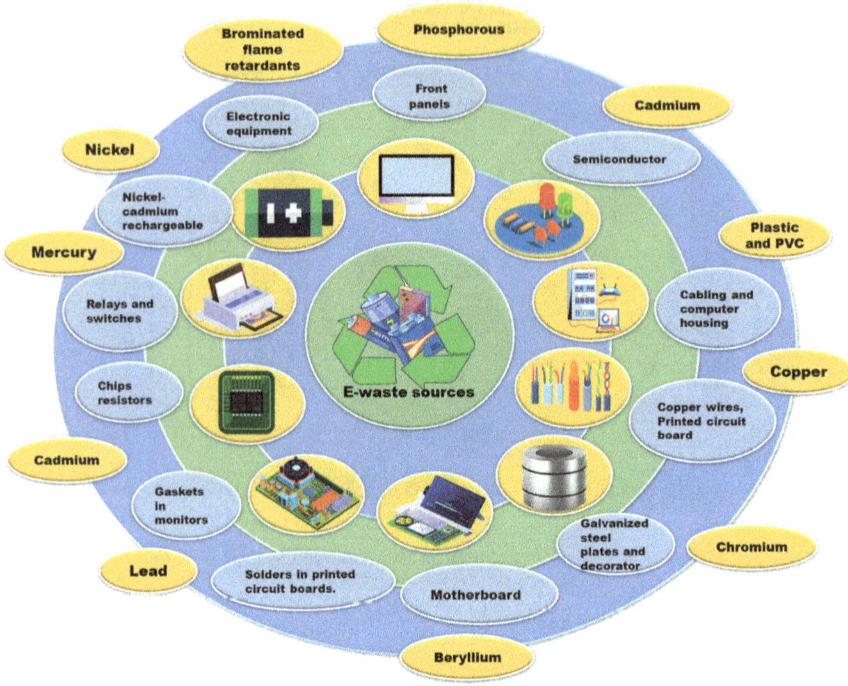

Fig. 3 An illustration of sources of e-waste and its associated metals

fields due to high concentrations of precious and rare metals found in e-waste. The availability of these metals in natural resources is declining, while their extraction from e-waste demands considerably lower energy than conventional mining methods (Wang et al., 2017). Biological techniques for metal extraction from e-waste rely on mechanisms similar to those used in hydrometallurgical processes, using microorganisms to produce reagents that facilitate metal extraction (Brierley, 2016). Specificity, cost-effectiveness and environmental sustainability are the main benefits of biological innovative processes that make them sustainable alternatives for recovering metals (Natarajan, 2018; Das & Ghosh, 2023a, d).

4 Innovative Bioprocessing Techniques in E-Waste Management

4.1 Bioleaching

Bioleaching has become a crucial technique for metal extraction from ores like pyrites, chalcopyrite, and calcite, as well as from e-waste (Das & Ghosh, 2018, 2023e; Ghosh et al., 2016; Ghosh & Das, 2017). This technique utilized the metabolic activities of microorganisms, such as fungi, cyanogenic bacteria, and iron and sulfur-oxidizing bacteria (Trivedi et al., 2022; Priyadarsini & Das, 2024). The principal bacterial groups associated with bioleaching include chemolithotrophic acidophiles, including *Acidithiobacillus ferrooxidans*, *A. thiooxidans*, and *Leptospirillum ferrooxidans*. Additionally, heterotrophic bacteria, including *Sulfolobus* species and fungi including *Aspergillus* and *Penicillium* species, play a key role in the bioleaching technique (Lee & Pandey, 2012; Nguyen et al., 2021; Liapun & Motola, 2023).

The bioleaching technique depends on mechanisms such as complexolysis, redoxolysis, and acidolysis, which enable microorganisms to produce biolixiviants that dissolve metals from solid substrates (Ilyas et al., 2014). This technology is currently employed on a large scale to extract metals such as Au, Cu, etc., and to some extent for extracting Co, Ni, and U from primary ores (Rawlings & Johnson, 2007). In recent times, more attention has been paid to the removal and recovery of metals from secondary sources, including fly ash (Xu et al., 2014), mining waste, polluted river sludge (Zhang et al., 2018), spent printed circuit boards (PCBs) (Bryan et al., 2020; Wu et al., 2018; Hubau et al., 2020), and outdated batteries (Liao et al., 2022).

4.1.1 Factors Influencing Bioleaching Efficiency

The competency of recovery of metals via bioleaching from e-waste is dependent on various factors (Das & Ghosh, 2023b). This comprises the physiology of the microorganisms and optimal physicochemical conditions that facilitate their growth and metabolite secretion (Trivedi et al., 2022). Additionally, the effectiveness of the leaching method and operational modes, which include heap leaching, column leaching, shake-flask experiments, or bioreactor systems, significantly affect bioleaching results (Dutta et al., 2023; Kaksonen et al., 2018). These factors may act independently or in combination to determine the overall bioleaching efficiency (Vardanyan et al., 2022).

4.1.2 Bioleaching Methods

Based on the interactions between biomass and waste, researchers have developed three different bioleaching techniques: the one-step method, the two-step method, and the spent-medium method (Das & Ghosh, 2023c; Mohanty et al., 2017; Priyadarsini & Das, 2025). The one-step method of bioleaching involves the direct addition of e-waste to the culture medium. However, the presence of e-waste can inhibit both microbial growth and metabolite production (Natarajan & Kumari, 2014). This method is less efficient compared to other methods.

The two-step method distinguishes biomass growth from the leaching phase. Microorganisms are first cultured without e-waste, enabling them to attain the logarithmic growth phase before the addition of e-waste in the second stage

(Natarajan & Ting, 2015; Bryan et al., 2015; Zhu et al., 2011). This method increases microbial growth, maximizes Fe^{2+} oxidation, and enables higher metal recovery rates. Research indicates that this strategy significantly lowers leaching duration and improves metal recovery efficiency (Zhu et al., 2011). For example, Işıldar et al. (2016) used a multistep method to successively extract Cu and Au from PCBs. The initial step removed Cu utilizing proteobacteria, but Au was extracted in a later phase by cyanide synthesis by cyanogenic microorganisms. This sequential approach mitigates the interference of base metals like Cu with Au cyanidation. An increase in pulp density can diminish the efficacy of bioleaching due to the noxious effects of e-waste on cyanide production (Natarajan & Ting, 2015).

In the spent-medium method, cells are isolated from the culture medium upon reaching optimal growth and production of the metabolite (Kim et al., 2016). The cell-free metabolite is subsequently utilized for bioleaching. This approach provides multiple benefits, including the prevention of direct biomass interaction with metal-containing waste, facilitating biomass recycling, and enhancing metabolite production. Additionally, higher waste concentrations can be managed without concerns regarding microbial contamination of the waste (Cui & Zhang, 2008).

Using e-waste, bioleaching has been extensively studied for the removal of metals like Al, Au, Ni, Cu, le, Pb, and Zn (Adetunji et al., 2023). The procedure often needs the addition of iron (Fe^{2+}) and a sulfur or pyrite source, as these substrates are lacking in e-waste (Harikrushnan et al., 2016). Acidophilic bacteria oxidize Fe^{2+} to Fe^{3+} and sulfur, producing protons that transform insoluble metals such as Cu^0, Ni^0, and Zn^0 into their soluble forms (Cu^{2+}, Ni^{2+}, and Zn^{2+}, respectively) in acidic conditions. This transformation is depicted in Eqs. (1) to (6) (Khatri et al., 2018). Table 1 highlights the role of microorganisms in the bioleaching technique.

$$4Fe^{2+} + O_2 \rightarrow 4Fe^{3+} + H_2O; \Delta G = 39.7 \ \text{kJ mol}^{-1} \tag{1}$$

$$2Fe^{3+} + Cu^0 \rightarrow 4Fe^{3+} + Cu^{2+}; \Delta G = -347.1 \ \text{kJ mol}^{-1} \tag{2}$$

$$2Fe^{3+} + Zn^0 \rightarrow 2Fe^{2+} + Zn^{2+}; \Delta G = -1235 \ \text{kJ mol}^{-1} \tag{3}$$

$$2Fe^{3+} + Ni^0 \rightarrow 2Fe^{2+} + Ni^{2+}; \Delta G = -822.6 \ \text{kJ mol}^{-1} \tag{4}$$

$$S^0 + 105O_2 + H_2O \rightarrow SO_4^{2-} + 2H^+; \Delta G = -822.5 \ \text{kJ mol}^{-1} \tag{5}$$

$$2Cu^0 + 2H^+ \rightarrow Cu^{2+} + 2H_2O; \Delta G = -774.5 \text{kJ mol}^{-1} \tag{6}$$

Bioleaching is a promising, green technique for metal recovery from e-waste, with various methods tailored to enhance efficiency based on specific applications. By optimizing microbial growth conditions and leaching parameters, researchers continue to improve the process for sustainable e-waste management.

Table 1 Role of microorganisms in bioleaching for e-waste and metal recovery

e-waste	Microorganisms	Biological role	Type	Recovery efficiency of metals (%)	Bioleaching conditions	Reference
Nickel batteries	*Acidithiobacillus ferrooxidans*	Sulfur and iron oxidizer	Chemolithotrophic bacteria	Co 99.9	Temperature: 30 °C, pH: 3.0, incubation time: 6 d	Zeng et al. (2012)
Mobile phone PCBs	*Acidithiobacillus ferrooxidans*	Sulfur and iron oxidizer	Chemolithotrophic bacteria	Al 93.53, Sn 95.33 Zn 92.58, Cu 95.92, Ni 65.27	Temperature: 20–35 °C, waste concentration: 5%, inoculation volume: 5% (v/v)	Fu et al. (2021)
Mobile phone PCBs	*Acidithiobacillus thiooxidans*	Sulfur oxidizer	Chemolithotrophic bacteria	Cu 98, Ni 82	Temperature: 30 °C, rpm: 160, incubation time: 72 h	Kadivar et al. (2021)
OLED screens (smartphones)	*Acidithiobacillus ferrooxidans*	Sulfur and iron oxidizer	Chemolithotrophic bacteria	In 100, Sr 5	Temperature: 29 °C, rpm: 140, pH: 1.1, incubation time: 15 d	Pourhossein et al. (2021)
Computer PCBs	*Acidiphilium acidophilum*	Facultative anaerobic bacterium, iron reducer	Chemolithotrophic bacteria	Cu 100	Concentration: 30%, incubation time: 10 d	Chandane et al. (2020)
PCBs	*Acidithiobacillus ferrooxidans and Acidithiobacillus acidophilum*	Sulfur and iron oxidizer, facultative anaerobic bacterium	Chemolithotrophic bacteria (consortium)	Cu 96, Ni 75, Zn 94.5, Pb 74.5	Pulp density: 7.5 g/L, temperature: 30 °C, rpm: 170, pH: 2.5, incubation time: 18 d	Priya and Hait (2020)
Lithium batteries	*Acidithiobacillus caldus, Leptospirillum ferriphilum, Sulfobacillus* sp.*, and Ferroplasma* sp.	Sulfur and iron oxidizer	Chemolithotrophic bacteria (consortium)	Co 99.6, Ni 99.7, Li 84	Temperature: 45 °C, rpm: 130	Ghassa et al. (2020)
Mobile phone PCBs	*Chromobacterium violaceum*	Cyanogenic bacterium	Heterotrophic bacteria	Cu 24.6, Au 11.31	Pulp density: 15 g/L, Temperature: 30 °C, rpm: 150, pH: 8.0–11.0, incubation time: 8 d	Chi et al. (2011)

(continued)

Table 1 (continued)

e-waste	Microorganisms	Biological role	Type	Recovery efficiency of metals (%)	Bioleaching conditions	Reference
AMOLED displays	*Bacillus foraminis*	Organic acid producer	Heterotrophic bacteria	Mo 56.8, Ag 100, Cu 41.4	Temperature: 40 °C, rpm: 160, incubation time: 12 d	Golzar-Ahmadi and Mousavi (2021)
PCBs	*Pseudomonas fluorescens*	Organic acid producer and metal chelator	Heterotrophic bacteria	Au 54	Pulp density: 0.33%, pH: 9.0, glycine to methionine ratio: 1:10	Li et al. (2020)
SIM card	*Chromobacterium violaceum*	Cyanogenic bacterium	Heterotrophic bacteria	Cu 13.79, Au 0.44, Ag 2.55	Pulp density: 10 g/L, pH: 9.0, incubation time: 7 d	Sahni and Kumar (2016)
PCBs	*Pseudomonas balearica*	Organic acid producer and metal solubilizer	Heterotrophic bacteria	Ag 33.8, Au 68.5	Pulp density: 10 g/L, temperature: 30 °C, pH: 9.0	Kumar et al. (2018)
Desktop PCBs	*Trichoderma viride*	Fungal metal solubilizer	Heterotrophic fungi	Pd 1, Au 10	pH: 5.0, incubation time: 30 d	Díaz-Martínez et al. (2019)
PCBs	*Phanerochaete chrysosporium*	White rot fungi, lignocellulose degrader	Heterotrophic fungi	Cu 60.96	Pulp density: 2%, pH: 5.0, temperature: 30 °C, rpm: 150, incubation time: 14 d	Liu et al. (2020)
Lithium-ion, smartphone batteries	*Aspergillus niger*	Fungal acid producer and metal solubilizer	Heterotrophic fungi	Li 100, Mn 72, Ni 45, Co 38, Cu 94, Al 62	Temperature: 30 °C, rpm: 150, incubation time: 30 d	Bahaloo-Horeh et al. (2018)
Desktop PCBs	*Aspergillus tubingensis*	Fungal acid producer and metal solubilizer	Heterotrophic fungi	Cu 34, Zn 54, Ni 8	Pulp density: 0.25–1%, pH: 5.0, incubation time: 33 d	Trivedi and Hait (2020)

4.2 Biosorption

Biosorption refers to the extraction or adhesion of substances from solutions using bio-derived materials, like material from plants, algae, fungi, and bacteria (Farooq et al., 2010). This process comprises intricate mechanisms, including hydrogen bonding, ion exchange, complexation, and electrostatic attraction (Yu et al., 2020). Recycled e-waste has emerged as a promising source of REEs, which consist of Y, Sc, and the lanthanide series: La, Pm, Sm, Eu, Gd, Nd, Ce, Ho, Pr, Tb, Er, Dy, Yb, Tm, and Lu (Cotton, 2006). Sorption mechanisms employing microbial biomass are currently established for the sustainable extraction of REEs from e-waste (Parajuly et al., 2020). Biosorption is a physical and chemical adsorption process that is similar to ion exchangers of biological origin in that it employs dead, inactive microbial biomass to capture and concentrate heavy metals from dilute aqueous solutions (Chojnacka, 2010). The cell walls of microorganisms, that contain functional groups such as amino, carboxyl, and hydroxyl groups, sequester metal ions via ion exchange, complexation, and electrostatic attraction, hence facilitating metal removal (Awasthi et al., 2019). This process does not depend on microbial metabolism, taking place on the surface of inactivated biomass. The particular metal-binding mechanisms can differ depending on the type of biomass used, but exhibit similarities to conventional adsorption methods. Biosorption does not always need expensive sorbents because biosorbents are often cheap materials made from industrial or agricultural waste (Vijayaraghavan & Yun, 2008). Common biosorbents contain algae, bacteria, fungi, farming residues, industrial byproducts, and biopolymers. This approach has been extensively and effectively investigated for wastewater treatment in recent decades. Inspired by the notable sorption capacities and cheap nature of biosorbents, researchers have extended their application for extraction of valuable metals, including Au and Pt group elements (Won et al., 2014; Pinto et al., 2023), and, more recently, the extraction of rare REEs (Gupta et al., 2019).

Biosorption has been investigated for metal recovery from e-waste utilizing multiple organisms and biomass types. For example, Kaur et al. (2022) investigate the biosorption of Cu and Fe from e-waste using *Pleurotus florida* and *Pseudomonas* spp., achieving recovery levels of 97 $mgkg^{-1}$ for Cu and 94 $mgkg^{-1}$ for Fe using *Pleurotus florida*. The results increased to 98 mg/kg and 96.2 mg/kg with *Pseudomonas*. Similarly, Sheel and Pant (2018) reported an 85% retrieval of Au from e-waste employing a bacterium *Lactobacillus acidophilus* and ammonium thiosulfate together. Some macroalgae, such as *Ulva lactuca, Ulva intestinalis, Fucus spiralis, Fucus vesiculosus, Osmundea pinnatifida*, and *Gracilaria* sp., have been utilized in biosorption for the extraction of REEs, such as Y, La, Ce, Pr, Eu, Gd, Nd, Dy, and Tb, showing an absorption capacity of 48.5 $\mu g\ g^{-1}$ (Costa et al., 2020). Náhlík et al. (2023) studied the microalgae *Galdieria sulphuraria* for its potential in the biosorption of rare REEs, for Ce, Nd, La, and Y. The investigation recorded absorption capacities of 26, 15, 11, and 11 $\mu g\ g^{-1}$ for Ce, Nd, La, and Y, respectively, utilizing a biomass concentration ranging from 3.3 to 6.1 mg L^{-1} and a sorbent dosage of 1×10^6 cells L^{-1}. Sun et al. (2022) report a study on the biosorption of La, Y, and Sm using *Galdieria sulphuraria* microalgae from solutions with concentrations between 40 and 160 mg L^{-1}. The research employed a biomass loading of 2.5% w/v, resulting in a La recovery efficiency of 7.2%, but the recovery efficiencies for Y and Sm were 96.2% and 98.9%, respectively. Fungal biomass serve as highly effective biosorbent, and progressively associated with a concept of mycosorption in context to fungal biosorption (Azin & Moghimi, 2018; Şenol et al., 2021). Yeasts, which are single cell fungi, possess a remarkable biosorption capacity that make most suitable effective biosorbent for metal ions recovery. The biomass of *Saccharomyces cerevisiae* has been extensively researched and shown to possess significant biosorption capability (Hlihor et al., 2015; Şenol et al., 2021). According to Shet et al. (2011), they reported a maximum concentration of 1699 g Zn (II) per g of yeast biomass. These studies highlight the significant potential of biosorption for metal recovery from e-waste. Despite ongoing issues, including the enhancement of biosorbents' adsorption capability and lifespan, advancements in pretreatment and modification techniques are consistently enhancing the potential of this sustainable and economical technology (Gadd, 2009; Gupta et al., 2019).

4.3 Biotransformation

Biotransformation refers to the alteration of a chemical form of a substance through various reactions. Harmful metals produced from e-waste exhibit changes in oxidation states through the addition or subtraction of electrons, resulting in alterations to their chemical properties (Narayanasamy et al., 2021). For the biotransformation technique, there are two approaches. Enzymatic direct reduction facilitated the extraction of harmful metal ions from e-waste through ion exchange mechanisms occurring outside the cellular structure. Enzymatic indirect reduction was employed to immobilize hazardous metal ions from e-waste in subsurface environments through the activities of metal-reducing microbes (Tabak et al., 2005). Numerous fungi, such as *Stachybotrys* sp., *Klebsiella oxytoca*, *Botryosphaeria rhodina*, *Phlebia* sp., and *Allescheriella* sp., exhibit significant metal-binding capability (D'Annibale et al., 2007). In addition, gram-positive bacterial strains, which include *Cellulosimicrobium* sp., exhibit resistance to xenobiotics and heavy metals, specifically Pb, Cr, Cd, and Hg (Khan et al., 2024). A research conducted by Thatoi et al. (2014) documented that the bacteria strain *Bacillus* sp. SFC 500 can reduce Cr to its less toxic form via biotransformation. Research indicates that fungi, including *Rhodobacter sphaeroides*, effectively eliminate hydrophobic toxic metals including Cd and Zn from soil (Peng et al., 2018). A research by Ayyasamy and Lee (2012) investigated the transformation of heavy metals in e-waste-contaminated soils utilizing the facultative anaerobic microbe *Shewanella* sp. HN-41. A study was carried out to explore how glucose affects the biotransformation of metals in soil. The solubilization rates for metals, including Fe, Al, As, Mn, Pb, and Cr, varied from 3 to 7664 mg per kg at different pH levels. It was observed that the rate of metal solubilization was minimum at a pH of approximately 7. However, at both acidic and basic pH levels, the solubilization rates were observed to be significantly higher. These findings enhance the understanding of metal biotransformation processes, which can be beneficial for soil improvement.

4.4 Enzymatic Bio-recovery

One promising biotechnological technique for retrieving valuable metals from e-waste is enzymatic bio-recovery. This technique utilizes enzymes, which act as biological catalysts, to specifically knock out complex metal-containing molecules that exist in e-waste, thus facilitating the release of associated metals for further recovery. Various enzymes have shown effectiveness in this approach, displaying a more environmentally friendly and particular solution to conventional metallurgical techniques (Rana et al., 2020). Enzymes including cytochrome p450, dihydrodiol dehydrogenase, nitrilases, esterases, laccases, amidases, proteases, manganese peroxidase (MnP), glyoxal oxidase, and glucose oxidase are crucial for the degradation of various classes of contaminants (Bhandari et al., 2021). The use of hydrolases, which aid in the breakdown of chemical bonds, is a prominent example of enzymatic bio-recovery. Recently, α-Amylase has been employed to degrade starch-based binders and improve metal extraction from organic compounds that exist in e-waste. Additionally, protease has demonstrated potential in dismantling proteins that can bind or enclose valuable metals in electronic gadgets (Machineni, 2023). The enzymes known as oxidoreductases are extremely important for enzymatic bio-recovery of metals. Enzymes like peroxidase and laccase efficiently oxidize organic compounds, helping in the degradation of intricate matrices found in e-waste. According to Zhang et al. (2024), these enzymes play a role in the extraction of metals by binding with organic ligands and making them accessible for further recovery procedures. Researchers have also studied engineered enzymes developed for retrieving specific metals. Scientists explore methods to increase the accuracy and proficiency of these enzymes in targeting precious metals, utilizing genetic modifications or integrating metal-binding domains to architecture. Using this approach, enzymatic bio-recovery procedures can be customized to handle e-waste with varying compositions (Zhao et al., 2021). The enzyme cyanide hydrolase (CHL) is employed in the selective removal of Au from e-waste. CHL facilitates the breakdown of toxic cyanide complexes, resulting in the release of Au ions that can be retrieved via precipitation or another separation method. This enzymatic approach provides a more environmentally friendly and sustainable option compared to conventional cyanide-based techniques employed in Au removal (Liu et al., 2022). Enzymatic bio-recovery holds potential; however, issues related to enzyme stability, scalability, and cost require improvement for broader application. Ongoing research aims to identify novel enzymes, optimize reaction conditions, and develop cost-effective methods for enzyme production. Enzymatic bio-recovery, characterized by its versatility and specificity, represents a significant aspect of biotechnological methods aimed at sustainable and effective extraction of precious metals from e-waste (Argumedo-Delira et al., 2020).

4.5 Bioaccumulation

Bioaccumulation refers to the sorption or assimilation of pollutants from e-waste by microorganisms. According to Juwarkar and Yadav (2010), contaminants are directly transported to a biomass cell inside the cellular system, where they accumulate. This process requires the active metabolism of microbes. In the cytoplasm of cells, many chemical reactions take place to transform certain organic pollutants into beneficial chemical compounds. On the other hand, metals that enter the cell cytoplasm are deposited without undergoing any reaction (Demirbaş, 2001). A comprehensive study by K Aderonke et al., 2017, extracted Cu, Zn, and Pb-tolerant microbes from soils affected by heavy metal leaching from e-waste and assessed their metal accumulation capabilities. *B. polymyxa, Bacillus licheniformis, Pseudomonas aeruginosa, Aspergillus niger*, and *A. flavus* were chosen for his bioaccumulation investigation due to their established resistance to heavy metals. *Bacillus licheniformis* demonstrated high efficiency in the extraction of Cu (71.3%) and Pb (70.1%). The deposition of Pb from *Aspergillus flavus* was quantified at 65.76%. The accumulation of Zn from *Pseudomonas aeruginosa* and *Aspergillus niger* has been determined to be 74.1% and 78.3%, respectively. Numerous metals, including Cd, Cr, Pb, and U, present in e-waste, were bioaccumulated by soil microbes such as *Aspergillus niger, Bacillus circulans, Bacillus megaterium,* and *Micrococcus luteus,* which contribute to the removal of hazardous metals Pb, Cr, and U, respectively (Narayanasamy et al., 2018). Gram-negative (*Pseudomonas aeruginosa*) and Gram-positive bacteria (*Tsukamurella paurometabola*) have been utilized in the bioaccumulation of Zn and Cd (Nnaji et al., 2023). This study examined the removal of Pb, Cd, Hg, and As by *Pseudomonas putida*, *S. cerevisiae*, and *Fusarium focciferum* (Massoud et al., 2019). Another investigation examined the bioaccumulation of Cu (II), Pb(II), and Cr (VI) by *Aspergillus niger*, revealing that *A. niger* exhibited high susceptibility to all concentrations of Cr (VI). These results highlight that *A. niger* may function as an effective living biosorbent for the isolation of heavy metal ions (Dursun et al., 2003). Bacteria, including *Bacillus megaterium, Bacillus circulans,* and *Deinococcus radiodurans*, as well as fungi like *Monodictys pelagica* and *Aspergillus niger,* have been documented to accumulate U, Cr, and Pb from e-waste (Patel & Kasture, 2014).

4.6 Biomineralization

A prominent biotechnological technique that uses the distinctive characteristics of microorganisms is biomineralization, which is employed to recover precious metals from e-waste (Gilbert et al., 2022). The ability of microorganisms, including bacteria, fungi, and archaea, to selectively absorb and take up metals in their cells leads to the formation of biominerals. This mechanism provides an ecologically friendly and sustainable substitute for conventional metal retrieving methods (Zhang et al., 2023). An example of biomineralization is observed in acidophilic bacteria such as *Acidithiobacillus thiooxidans and A. Acidithiobacillus ferrooxidans,* and they are crucial in removing metals from e-waste and flourish in acidic environments. Metal-binding proteins are formed by *Acidithiobacillus* species that assist in metal ion precipitation in microbial cells (Kiran et al., 2024). Consequently, precious metals, for example, Au and Pd, are efficiently sequestered and subsequently extracted from biomass of microbes. Among the most notable examples is the use of *Bacillus* species of bacteria. These microorganisms are capable of recovering metals from e-waste matrices by selective binding. The biomineralization mechanism is characterized by the production of extracellular polymeric substances (EPS), which serve as metal ion-binding agents. *Bacillus* species facilitate the production of microbial biofilms, which facilitate microenvironments that enhance the concentration and precipitation of valuable metals (Nadar & Khan, 2024). Fungi, including *Aspergillus* and *Penicillium* species, play a role in biomineralization technique for metal removal. In order to help in metal extraction from e-waste solubilization and precipitation as metal-containing substances, these fungi produce organic acids and exopolysaccharides. Fungi have immense potential for bioleaching and extraction applications because their mycelial structures provide a complex system for metal accumulation (Jiang et al., 2022). *Cupriavidus metallidurans*, a metallophilic bacterium, has the capacity for cellular detoxification, making it a potential candidate for the accumulation of Au (III) (Reith et al., 2009). In addition, bacterial strains including *Sporosarcina ginsengisoli*, *Bacillus fusiformis,* and *Cupriavidus metallidurans* are recognized for their involvement in biomineralization, tidily removing heavy metals namely Pb, As, and Cd (Ahmed et al., 2019). A research investigated by Achal et al., 2012 highlighted the effective biomineralization potential of As (III) by *Sporosarcina ginsengisoli.* Additionally, advancements in genetic engineering have allowed the improvement of microbial strains that boost biomineralization efficacy. Researchers have engineered microbes to produce metal-binding peptides or proteins, thereby improving uptake of specific metals and enhance extraction potential. The modified microbial strains provide precise strategy for particular metal recovery, and create a pathway to progress toward the creation of specialized biotechnological methods (Chen et al., 2022).

4.7 Vermiremediation

Earthworms enhance activity of microbes by promoting suitable environment for soil bacteria through aeration, continuing interactions, and their supply of nutrients through their casts. This method can be employed to remediate sites contaminated by e-waste. Research indicates that earthworms can bioaccumulate chemical contaminants in their tissues and may either biodegrade or biotransform these substances into nontoxic products through enzymatic processes (Sinha et al., 2010). This indicates a promising potential for the vermiremediation of soil contaminated by e-waste. According to Ahadi et al. (2020), earthworms have shown enormous promise in the restoration of soil contaminated with heavy metals. Cd and Cu concentrations in earthworms increase with increasing levels of Cd and Cu in the environment (Liu et al., 2005). The intestinal tissue of earthworms, known as chloragogenous tissue, has the capacity to accumulate large concentrations of heavy metals (Liang et al., 2011). This tissue is believed to operate as a cation exchange system, effectively absorbing and retaining heavy metals (Sivakumar, 2015). Another investigation revealed that earthworms such as *Eisenia fetida* eliminated 17.60% of Cd from Cd-contaminated soil (Wu et al., 2020). Additionally, *E. fetida* demonstrated notable decrease in metal concentrations, specifically Mn (31.8%), Zn (32%), and Cu (43.5%), from spent wash by accumulating significant levels of Mn (83.7–138.6 mg kg^{-1}), Zn (103.7–143.3 mg kg^{-1}), and Cu (16.8–25.5 mg kg^{-1}) in their tissues. The heavy metal bioaccumulation of earthworms exhibited variation depending on the species of heavy metal involved. Richardson et al. (2015) reported that earthworms significantly bioaccumulate only Zn, Hg, and Cd, as their average bioaccumulation factors (BAF) were notably higher than 1.0. Similar investigation revealed that the hierarchy of BAF values for *Allolobophora rosea* and *Nicodrilus caliginosus* was as follows: Cd >> Zn >> Cu > As = Pb = Sb. Furthermore, earthworms living within the similar ecological physiology classification have identical behaviors in the absorption and bioaccumulation of heavy metals (Nannoni et al., 2014).

4.8 Phytoremediation

Phytoremediation is an economical and eco-friendly biological remediation method that utilizes plants to extract, stabilize, and transform various compounds, including potentially toxic elements, from soil and water (Joradon et al., 2024). Numerous plant species are extensively employed in phytoremediation owing to their exceptional capacity to absorb, stabilize, or eliminate contaminants. *Brassica juncea* shows significant efficacy in the accumulation of heavy metals, namely Pb, Cd, and Se (Bortoloti & Baron, 2022). *Helianthus annuus* exhibits the capacity to extract Pb, Zn, and radioactive substances from contaminated environments (Al-Jobori & Kadhim, 2019). *Populus* spp. contribute a crucial role in soil stabilization and the volatilization of specific metals, making them appropriate for long-term remediation strategies (Li et al., 2024). *Thlaspi caerulescens* is recognized for its remarkable capacity to accumulate high levels of Zn and Cd (Zhao et al., 2003). *Eichhornia crassipes* is commonly employed in rhizofiltration to effectively eliminate heavy metals from aquatic environments. These plants are essential for sustainable environmental remediation, employing natural processes to reduce pollution (Tabinda et al., 2020). Phytoremediation encompasses the interactions between plants and their associated rhizosphere microbial diversity for the remediation of heavy metal-contaminated sediments. In plant systems, heavy metal accumulation takes place in multiple stages: mobilization for effective adsorption, uptake by the root, xylem loading, transport from root to shoot, compartmentalization, sequestration, and detoxification. Heavy metal mobilization occurs via exudates formed by plants and rhizosphere microbiota. Plant roots utilize both active and passive transport mechanisms to absorb heavy metal ions, creating complexes with organic acids like carbonates, phosphates, and sulfates, which are then immobilized in cell walls or vacuoles within the cells. Subsequently, xylem will transport the heavy metal complexes stored in vacuoles and distribute them to the shoots (Yan et al., 2020).

These mechanisms are primarily associated with phytoremediation processes, including biological degradation in the rhizosphere, phytodegradation, rhizofiltration, phytoextraction or phytoaccumulation, phytostabilization, phytovolatilization, and phytotransformation (Fig. 4) (Ansari et al., 2020). Among them, phytoextraction is a widely utilized remediation technology that typically includes three steps: cultivating appropriate metal-accumulating plants, harvesting metal-enriched plant biomass, and conducting postharvest management of the biomass to enhance commercial benefits. Phytoextraction of valuable metals, for example, Pd, Ag, Pt, and Au, presents considerable economic and environmental potential, especially in the management of e-waste for metal recovery. This technique is applicable in two main areas: phytoremediation and phytomining (Dinh et al., 2022). Phytoremediation involves heavy metals extraction from contaminated soil through their accumulation in plant shoots (Ghosh et al., 2015). Phytomining is utilized to recover precious metals from mineralized soils, low-grade ores, and mill tailings (Dinh et al., 2022).

Hyperaccumulator plants exhibit the accumulation potential of metals in their root and stem tissues at concentrations reaching 10,000 mg/kg. The plants include perennial shrubs, trees, and grasses, represented by species such as *Sedum alfredii, Trifolium alexandrinum, Thlaspi caerulescens,* and *Panicum maximus* (Kafle et al., 2022). Sheoran et al. (2013) classified plants that accumulate more than 1 mg/kg of Au as hyperaccumulators, based on the typical concentration of approximately 0.01 mg/kg in plants. The concentrations necessary for economically viable recovery vary among noble metals, including Pt, Au, Pd, and Ag. *Brassica juncea* (L.) Czern. is an important example of a hyperaccumulator plant capable of accumulating Au at concentrations surpassing 57 mg/kg dry weight under suitable substrate treatment (Anderson et al., 2001). Investigation by Chinmayee et al. (2012) found that a potentially invasive plant, *Amaranthus spinosus*, is an important phytoremediating plant in soil contaminated with trace metals. This research highlighted the plants' efficacy in bioaccumulating and translocating Cd, Cu, and Zn. In addition, Wei et al. (2018) examined three invasive species in China: *Praxelis clematidea,*

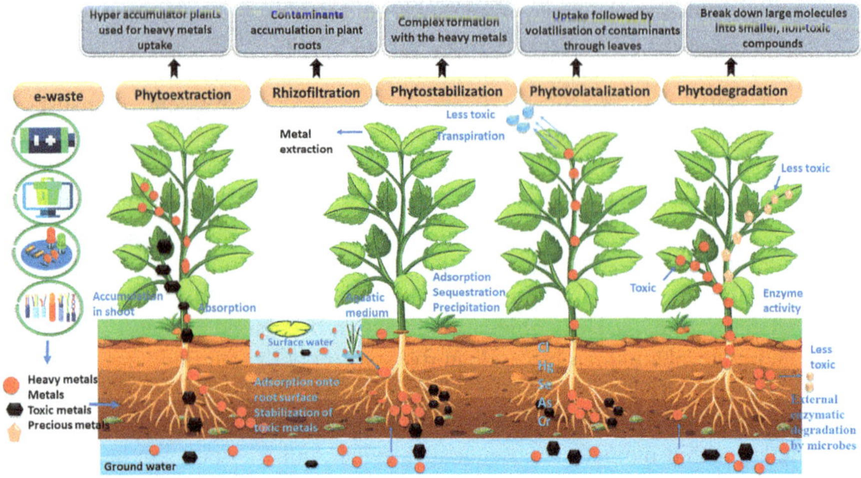

Fig. 4 An illustration of the phytoremediation mechanism for heavy metals extraction from e-waste

Bidens pilosa, and *Chromolaena odorata*, for their phytoremediation capabilities particularly for Cd extraction. This finding indicated that all three species demonstrated tolerance to Cd-contaminated soils, with *Chromolaena odorata* exhibiting the greatest bioaccumulation capacity of all three. Additionally, Pandey (2012) identified *Ipomoea carnea* as an effective phytoremediation factor based on its favorable characteristics. These factors consist of easy propagation via vegetative methods and high resilience to unfavorable conditions, including desiccation, flooding, pH fluctuations, salinity, and exposure to hazardous metals. Phytoremediation of heavy metals from e-waste has been described in Table 2.

4.9 Genetically Modified Organisms (GMOs) in Degrading E-Waste Constituent

Genetic engineering offers efficient methods for managing heavy metals and contaminants, including polyaromatic hydrocarbons (PAHs), which often show resistance to traditional bioremediation techniques (Preeti et al., 2023). GMOs offer notable benefits for bioremediation, as they are more environmentally friendly and present lower health risks than physiochemical methods, which tend to be less sustainable and potentially hazardous (Iravani & Varma, 2022).

Research on metal accumulation and recovery primarily centers on genetically engineered microorganisms to understand the molecular mechanisms of metal tolerance, oxidation, or degradation. *Delftia acidovorans* synthesizes delftibactin, a siderophore-like nonribosomal peptide that detoxifies soluble Au and favors Au biomineralization (Johnston et al., 2013). Research on the bioleaching of e-waste indicates significant advancements. For example, *Pseudomonas putida* 06909,

Table 2 Phytoremediation plants and their roles in the extraction of metals from e-waste

S. no.	Metals	Plant	Family	Type	References
1	Cu	*Aeolanthus biformifolius*	Lamiaceae	Hyperaccumulators	Van Der Ent et al. (2019)
2	Cd	*Arabidopsis halleri*	Brassicaceae	Hyperaccumulators	Ueno et al. (2008)
3	La Ce	*Dicranopteris linearis*	Gleicheniaceae	Hyperaccumulators	Jally et al. (2021)
4	As	*Pterisryukyuensis Tagawa, Eleocharis acicularis L.*	Pteridaceae, Cyperaceae	Hyperaccumulators	Srivastava et al. (2006), Sakakibara et al. (2011)
5	Hg	*Achillea millefolium L., Silene vulgaris (Moench) Garcke,*	Asteraceae, Caryophyllaceae	Hyperaccumulators	Antoniadis et al. (2021), Pérez-Sanz et al. (2012)
6	Cr	*Pteris vittata L.*	Pteridaceae	Hyperaccumulators	Kalve et al. (2011)
7	Pb	*Medicago sativa L.*	Fabaceae	Hyperaccumulators	Koptsik (2014)
8	Pb, Cd	*Pelargonium hortorum*	Geraniaceae	Phytoextraction	Gul et al. (2019)
9	Cd, Pb, Ni, Zn	*Xanthium strumarium*	Asteraceae	Phytoextraction	Khalid et al. (2019)
10	Pb, Ti	*Zea mays*	Poaceae	Phytoextraction	Huang et al. (2019)
11	Ni, Co, Fe	*Lactuca sativa*	Asteraceae	Phytoextraction	Hernández et al. (2019)
12	Pb	*Mesembryanthemum criniforum*	Aizoaceae	Phytoextraction	Manzoor et al. (2020)
13	As	*Eupatorium cannabinum*	Asteraceae	Phytoimmobilization	González et al. (2019)
14	Cu, Zn, Cd	*Solanum nigrum*	Solanaceae	Phytoimmobilization	Li et al. (2019)
15	As	*Azolla caroliniana*	Salviniaceae	Rhizofiltration	Favas et al. (2016)
16	U	*Lemna minor*	Araceae	Rhizofiltration	Pratas et al. (2012)
17	Se	*Scirpus robustus*	Cyperaceae	Phytovolatilization	Arthur et al. (2005)
19	Ofloxacin (OFX)	*Spirodela polyrhiza*	Araceae	Phytodegradation	Singh et al. (2019)
20	Ibuprofen	*Phragmites australis*	Poaceae	Phytodegradation	He et al. (2017)

modified with recombineering tools, was engineered to synthesize the high-cadmium-affinity peptide EC20. This innovation improved cadmium binding and reduced Cd-induced cellular toxicity (Wu et al., 2006). Another investigation employed an in silico genome-scale metabolic model (GEM)-based method to identify gene targets that enhance metabolic flux for cyanide synthesis, consequently improving Au and Cu recovery (Aminian-Dehkordi et al., 2020). Additionally, Zhu et al. (2020) engineered synthetic bacterial cells to display heavy-metal-capturing proteins (SynHMB) on their surfaces, along with magnetic nanoparticles, resulting in high-efficiency removal of Cd and Pb through novel biotic-abiotic coassemblies.

The importation of genetic toolkits from *E. coli*, including shuttle vectors and regulatory elements, has established a foundation for subsequent engineering initiatives aimed at enhancing bioleaching efficiencies (Gumulya et al., 2018). *Chromobacterium violaceum* is a nonmodel microorganism notable for its remarkable capacity to bio-dissolve Au (Liu et al., 2016). Cyanogenic bacteria synthesize cyanide lixiviants that form complexes with Au, thereby enhancing the bioleaching process. Cyanide production by *C. violaceum* is regulated through quorum sensing and is restricted to 20 mg/L during the stationary phase (Natarajan, 2019). Engineered strains utilizing exogenous promoters (pBAD and pTAC) have achieved a 70% increase in cyanide production, resulting in enhanced Au recovery rates (Tay et al., 2013). The introduction of an extra copy of the cyanide-producing operon increased Au extraction from 11% to 30% (Natarajan et al., 2015).

5 Conclusion

The pressing need for sustainable and eco-friendly strategies to handle the huge burden of e-waste has brought innovative bioprocessing techniques to the forefront of research and development. Among these, bio-innovative methods offer a promising alternative to traditional methods and offer an environmentally sustainable pathway for metal recovery. Innovative microbial mechanisms such as bioleaching, biosorption, biotransformation, bioaccumulation, enzymatic recovery, vermiremediation, and phytoremediation highlight the immense potential of microorganisms in e-waste remediation. Bioleaching is an environmentally friendly technique for the recovery of metals from e-waste, although it should focus on methods such as the application of OMICS approaches and the application of microbial consortia. This focus will enhance the understanding of microbial contributions to the process and facilitate the genetic manipulation of microorganisms to produce more efficient strains that improve bioleaching capabilities. Although use of microorganisms highlights metabolic ability to mitigate e-waste, significant challenges remain, particularly in optimizing process design and enhancing overall efficiency for large-scale implementation. Addressing these challenges through in-depth research and technological innovation will strengthen biotechnological approaches, enabling the efficient retrieval of toxic, precious, and valuable metals from e-waste.

In conclusion, effective e-waste management is crucial not only as a way of dealing waste, but it also provides an opportunity to recover precious metals. Biological e-waste management represent a transformative approach that set with the principles of a circular economy. By addressing existing gaps in research and optimizing biotechnologies, microorganisms can offer a sustainable, eco-friendly, and economically viable solution for mitigating the global e-waste crisis, contributing to environmental sustainability and metals recovery.

References

Achal, V., Pan, X., Fu, Q., & Zhang, D. (2012). Biomineralization based remediation of As (III) contaminated soil by Sporosarcina ginsengisoli. *Journal of Hazardous Materials, 201*, 178–184.

Aderonke, A. K., Oladimeji, O. O., Shittu, O. B., Okeyode, I. C., & Taiwo, M. O. (2017). Bioaccumulation of heavy metals using selected organisms isolated from electronic waste dumpsite of two south-western states in Nigeria. *Applied Environmental Research, 39*(2), 29–40.

Adetunji, A. I., Oberholster, P. J., & Erasmus, M. (2023). Bioleaching of metals from e-waste using microorganisms: A review. *Minerals, 13*(6), 828.

Ahadi, N., Sharifi, Z., Hossaini, S. M., Rostami, A., & Renella, G. (2020). Remediation of heavy metals and enhancement of fertilizing potential of a sewage sludge by the synergistic interaction of woodlice and earthworms. *Journal of Hazardous Materials, 385*, 121573.

Ahmed, T., Liaqat, I., Murtaza, R., & Rasheed, A. (2019). Bioremediation approaches for e-waste management: A step toward sustainable environment. In *Electronic waste pollution: Environmental occurrence and treatment technologies* (pp. 267–290).

Al-Jobori, K. M., & Kadhim, A. K. (2019). Evaluation of sunflower (Helianthus annuus L.) for phytoremediation of lead contaminated soil. *Journal of Pharmaceutical Sciences and Research, 11*(3), 847–854.

Aminian-Dehkordi, J., Mousavi, S. M., Marashi, S. A., Jafari, A., & Mijakovic, I. (2020). A systems-based approach for cyanide overproduction by Bacillus megaterium for gold bioleaching enhancement. *Frontiers in Bioengineering and Biotechnology, 8*, 528.

Anderson, C., Deram, A., Petit, D., Brooks, R., Stewart, R. B., & Simcock, R. (2001). Induced hyperaccumulation: Metal movement and problems. In *Trace elements in soil* (pp. 79–92). CRC Press.

Ansari, A. A., Naeem, M., Gill, S. S., & AlZuaibr, F. M. (2020). Phytoremediation of contaminated waters: An eco-friendly technology based on aquatic macrophytes application. *The Egyptian Journal of Aquatic Research, 46*(4), 371–376.

Antoniadis, V., Shaheen, S. M., Stärk, H. J., Wennrich, R., Levizou, E., Merbach, I., & Rinklebe, J. (2021). Phytoremediation potential of twelve wild plant species for toxic elements in a contaminated soil. *Environment International, 146*, 106233.

Argumedo-Delira, R., Gómez-Martínez, M. J., & Uribe-Kaffure, R. (2020). Fungal tolerance: An alternative for the selection of fungi with potential for the biological recovery of precious metals. *Applied Sciences, 10*(22), 8096.

Ari, V. (2016). A review of technology of metal recovery from electronic waste. In *E-waste in transition—From pollution to resource* (pp. 121–158).

Arthur, E. L., Rice, P. J., Rice, P. J., Anderson, T. A., Baladi, S. M., Henderson, K. L., & Coats, J. R. (2005). Phytoremediation—An overview. *Critical Reviews in Plant Sciences, 24*(2), 109–122.

Awasthi, A. K., Shivashankar, M., & Majumder, S. (2017, November). Plastic solid waste utilization technologies: A review. In *IOP conference series: Materials science and engineering* (Vol. 263, No. 2, p. 022024). IOP Publishing.

Awasthi, A. K., Hasan, M., Mishra, Y. K., Pandey, A. K., Tiwary, B. N., Kuhad, R. C., et al. (2019). Environmentally sound system for E-waste: Biotechnological perspectives. *Current Research in Biotechnology, 1*, 58–64.

Ayyasamy, P. M., & Lee, S. (2012). Biotransformation of heavy metals from soil in synthetic medium enriched with glucose and Shewanella sp. HN-41 at various pH. *Geomicrobiology Journal, 29*(9), 843–851.

Azin, E., & Moghimi, H. (2018). Efficient mycosorption of anionic azo dyes by Mucor circinelloides: Surface functional groups and removal mechanism study. *Journal of Environmental Chemical Engineering, 6*(4), 4114–4123.

Bahaloo-Horeh, N., Mousavi, S. M., & Baniasadi, M. (2018). Use of adapted metal tolerant Aspergillus niger to enhance bioleaching efficiency of valuable metals from spent lithium-ion mobile phone batteries. *Journal of Cleaner Production, 197*, 1546–1557.

Baldé, C. P., Forti, V., Gray, V., Kuehr, R., & Stegmann, P. (2017). *The global e-waste monitor 2017: Quantities, flows and resources*. United Nations University, International Telecommunication Union, and International Solid Waste Association.

Bhandari, S., Poudel, D. K., Marahatha, R., Dawadi, S., Khadayat, K., Phuyal, S., et al. (2021). Microbial enzymes used in bioremediation. *Journal of Chemistry, 2021*(1), 8849512.

Bortoloti, G. A., & Baron, D. (2022). Phytoremediation of toxic heavy metals by Brassica plants: A biochemical and physiological approach. *Environmental Advances, 8*, 100204.

Brierley, C. L. (2016). Biological processing: Biological processing of sulfidic ores and concentrates—Integrating innovations. In *Innovative process development in metallurgical industry: Concept to commission* (pp. 109–135).

Bryan, C. G., Watkin, E. L., McCredden, T. J., Wong, Z. R., Harrison, S. T. L., & Kaksonen, A. H. (2015). The use of pyrite as a source of lixiviant in the bioleaching of electronic waste. *Hydrometallurgy, 152*, 33–43.

Bryan, C. G., Williamson, B. J., Całus-Moszko, J., van Haute, Q., Guezennec, A. G., Gaydardzhiev, S., et al. (2020). CEReS–co-processing of coal mine & electronic wastes: Novel resources for a sustainable future. *Hydrometallurgy, 197*, 105444.

Chakraborty, S. C., Qamruzzaman, M., Zaman, M. W. U., Alam, M. M., Hossain, M. D., Pramanik, B. K., et al. (2022). Metals in e-waste: Occurrence, fate, impacts and remediation technologies. *Process Safety and Environmental Protection, 162*, 230–252.

Chandane, P., Jori, C., Chaudhari, H., Bhapkar, S., Deshmukh, S., & Jadhav, U. (2020). Bioleaching of copper from large printed circuit boards for synthesis of organic-inorganic hybrid. *Environmental Science and Pollution Research, 27*, 5797–5808.

Chauhan, G., Jadhao, P. R., Pant, K. K., & Nigam, K. D. P. (2018). Novel technologies and conventional processes for recovery of metals from waste electrical and electronic equipment: Challenges & opportunities–a review. *Journal of Environmental Chemical Engineering, 6*(1), 1288–1304.

Chen, J., Liu, B., Zhong, M., Jing, C., & Guo, B. (2022). Research status and development of microbial induced calcium carbonate mineralization technology. *PLoS One, 17*(7), e0271761.

Chi, T. D., Lee, J. C., Pandey, B. D., Yoo, K., & Jeong, J. (2011). Bioleaching of gold and copper from waste mobile phone PCBs by using a cyanogenic bacterium. *Minerals Engineering, 24*(11), 1219–1222.

Chinmayee, M. D., Mahesh, B., Pradesh, S., Mini, I., & Swapna, T. S. (2012). The assessment of phytoremediation potential of invasive weed Amaranthus spinosus L. *Applied Biochemistry and Biotechnology, 167*, 1550–1559.

Chojnacka, K. (2010). Biosorption and bioaccumulation–the prospects for practical applications. *Environment International, 36*(3), 299–307.

Costa, M., Henriques, B., Pinto, J., Fabre, E., Viana, T., Ferreira, N., et al. (2020). Influence of salinity and rare earth elements on simultaneous removal of Cd, Cr, Cu, Hg, Ni and Pb from contaminated waters by living macroalgae. *Environmental Pollution, 266*, 115374.

Cotton, S. A. (2006). Scandium, yttrium, the lanthanides and the actinides. *Annual Reports Section "A" (Inorganic Chemistry), 102*, 308–331.

Cui, J., & Zhang, L. (2008). Metallurgical recovery of metals from electronic waste: A review. *Journal of Hazardous Materials, 158*(2–3), 228–256.

D'Annibale, A., Leonardi, V., Federici, E., Baldi, F., Zecchini, F., & Petruccioli, M. (2007). Leaching and microbial treatment of a soil contaminated by sulphide ore ashes and aromatic hydrocarbons. *Applied Microbiology and Biotechnology, 74*, 1135–1144.

Das, A. P., & Ghosh, S. (2018). Bioleaching of manganese from mining waste materials. *Materials Today: Proceedings, 5*(1), 2381–2390. https://doi.org/10.1016/J.MATPR.2017.11.459

Das, A. P., & Ghosh, S. (2023a). Cost-effectiveness of the biomining process. In *Manganese mining microorganisms* (pp. 143–151). https://doi.org/10.1016/B978-0-12-822147-1.00008-X

Das, A. P., & Ghosh, S. (2023b). Factors affecting bioleaching. In *Manganese mining microorganisms* (pp. 77–86). https://doi.org/10.1016/B978-0-12-822147-1.00004-2

Das, A. P., & Ghosh, S. (2023c). Mechanism of manganese bioleaching. In *Manganese mining microorganisms* (pp. 87–98). https://doi.org/10.1016/B978-0-12-822147-1.00014-5

Das, A. P., & Ghosh, S. (2023d). Mixed culture bioleaching: An insight into manganese biomining process and efficacy. In *Manganese mining microorganisms* (pp. 129–141). https://doi.org/10.1016/B978-0-12-822147-1.00001-7

Das, A. P., & Ghosh, S. (2023e). Role of metabolites and metabolomics for understanding of manganese bioleaching. In *Manganese Mining Microorganisms* (pp. 117–127). https://doi.org/10.1016/B978-0-12-822147-1.00011-X

Das, A. P., Sukla, L. B., Pradhan, N., & Nayak, S. (2011). Manganese biomining: A review. *Bioresource Technology, 102*(16), 7381–7387. https://doi.org/10.1016/J.BIORTECH.2011.05.018

Demirbaş, A. (2001). Heavy metal bioaccumulation by mushrooms from artificially fortified soils. *Food Chemistry, 74*(3), 293–301.

Dhivya, K., & Premalatha, G. (2024). E-waste challenges & solutions. In *Sustainable management of electronic waste* (pp. 255–275).

Díaz-Martínez, M. E., Argumedo-Delira, R., Sánchez-Viveros, G., Alarcón, A., & Mendoza-López, M. R. (2019). Microbial bioleaching of Ag, Au and Cu from printed circuit boards of mobile phones. *Current Microbiology, 76*, 536–544.

Dinh, T., Dobo, Z., & Kovacs, H. (2022). Phytomining of noble metals–A review. *Chemosphere, 286*, 131805.

Duan, H., Hu, J., Tan, Q., Liu, L., Wang, Y., & Li, J. (2016). Systematic characterization of generation and management of e-waste in China. *Environmental Science and Pollution Research, 23*, 1929–1943.

Dursun, A. Y., Uslu, G., Tepe, O., Cuci, Y., & Ekiz, H. I. (2003). A comparative investigation on the bioaccumulation of heavy metal ions by growing Rhizopus arrhizus and Aspergillus niger. *Biochemical Engineering Journal, 15*(2), 87–92.

Dutta, D., Rautela, R., Gujjala, L. K. S., Kundu, D., Sharma, P., Tembhare, M., & Kumar, S. (2023). A review on recovery processes of metals from E-waste: A green perspective. *Science of the Total Environment, 859*, 160391.

Farooq, U., Kozinski, J. A., Khan, M. A., & Athar, M. (2010). Biosorption of heavy metal ions using wheat based biosorbents–a review of the recent literature. *Bioresource Technology, 101*(14), 5043–5053.

Favas, P. J., Pratas, J., Paul, M. S., Sarkar, S. K., & Prasad, M. N. V. (2016). Phytofiltration of metal (loid)-contaminated water: The potential of native aquatic plants. *Phytoremediation: Management of Environmental Contaminants, 3*, 305–343.

Forti, V., Baldé, C. P., Kuehr, R., & Bel, G. (2020). *The global e-waste monitor 2020*. United Nations University (UNU), International Telecommunication Union (ITU) & International Solid Waste Association (ISWA), Bonn/Geneva/Rotterdam, 120.

Fu, K., Tian, L., Hou, P., Long, M., Chen, S., & Lin, H. (2021). Stirred-tank leaching of coarse-grained waste, printed circuit boards with Acidithiobacillus ferrooxidans. *Physicochemical Problems of Mineral Processing, 57*(5), 153–163.

Gadd, G. M. (2009). Biosorption: Critical review of scientific rationale, environmental importance and significance for pollution treatment. *Journal of Chemical Technology & Biotechnology: International Research in Process, Environmental & Clean Technology, 84*(1), 13–28.

Ghassa, S., Farzanegan, A., Gharabaghi, M., & Abdollahi, H. (2020). Novel bioleaching of waste lithium ion batteries by mixed moderate thermophilic microorganisms, using iron scrap as energy source and reducing agent. *Hydrometallurgy, 197*, 105465.

Ghosh, S., & Das, A. P. (2017). Bioleaching of manganese from mining waste residues using Acinetobacter sp. *Geology, Ecology, and Landscapes, 1*(2), 77–83. https://doi.org/10.1080/24749508.2017.1332847

Ghosh, M., Paul, J., Jana, A., De, A., & Mukherjee, A. (2015). Use of the grass, Vetiveria zizanioides (L.) Nash for detoxification and phytoremediation of soils contaminated with fly ash from thermal power plants. *Ecological Engineering, 74*, 258–265.

Ghosh, S., Mohanty, S., Akcil, A., Sukla, L. B., & Das, A. P. (2016). A greener approach for resource recycling: Manganese bioleaching. *Chemosphere, 154*, 628–639. https://doi.org/10.1016/J.CHEMOSPHERE.2016.04.028

Ghosh, S., Tripathy, B., Dey, S., & Das, A. P. (2023). Comparative investigation of fungal and bacterial manganese biomining mechanisms. In *Manganese mining microorganisms* (pp. 99–115). https://doi.org/10.1016/b978-0-12-822147-1.00016-9

Gilbert, P. U., Bergmann, K. D., Boekelheide, N., Tambutté, S., Mass, T., Marin, F., et al. (2022). Biomineralization: Integrating mechanism and evolutionary history. *Science Advances, 8*(10), eabl9653.

Golzar-Ahmadi, M., & Mousavi, S. M. (2021). Extraction of valuable metals from discarded AMOLED displays in smartphones using Bacillus foraminis as an alkali-tolerant strain. *Waste Management, 131*, 226–236.

González, H., Fernández-Fuego, D., Bertrand, A., & González, A. (2019). Effect of pH and citric acid on the growth, arsenic accumulation, and phytochelatin synthesis in Eupatorium cannabinum L., a promising plant for phytostabilization. *Environmental Science and Pollution Research, 26*, 26242–26253.

Gopikrishnan, V., Vignesh, A., Radhakrishnan, M., Joseph, J., Shanmugasundaram, T., Doble, M., & Balagurunathan, R. (2020). Microbial leaching of heavy metals from e-waste: Opportunities and challenges. In *Biovalorisation of wastes to renewable chemicals and biofuels* (pp. 189–216).

Gul, I., Manzoor, M., Hashim, N., Yaqoob, K., Kallerhoff, J., & Arshad, M. (2019). Comparative effectiveness of organic and inorganic amendments on cadmium bioavailability and uptake by Pelargonium hortorum. *Journal of Soils and Sediments, 19*, 2346–2356.

Gumulya, Y., Boxall, N. J., Khaleque, H. N., Santala, V., Carlson, R. P., & Kaksonen, A. H. (2018). In a quest for engineering acidophiles for biomining applications: Challenges and opportunities. *Genes, 9*(2), 116.

Gundupalli, S. P., Hait, S., & Thakur, A. (2018). Classification of metallic and non-metallic fractions of e-waste using thermal imaging-based technique. *Process Safety and Environmental Protection, 118*, 32–39.

Gupta, N. K., Gupta, A., Ramteke, P., Sahoo, H., & Sengupta, A. (2019). Biosorption-a green method for the preconcentration of rare earth elements (REEs) from waste solutions: A review. *Journal of Molecular Liquids, 274*, 148–164.

Harikrushnan, B., Shreyass, G., Hemant, G., & Pandimadevi, M. (2016). Recovery of metals from printed circuit boards (pcbs) using a combination of hydrometallurgical and biometallurgical processes. *International Journal of Environmental Research, 10*(4), 511–518.

He, Y., Langenhoff, A. A., Sutton, N. B., Rijnaarts, H. H., Blokland, M. H., Chen, F., et al. (2017). Metabolism of ibuprofen by Phragmites australis: Uptake and phytodegradation. *Environmental Science & Technology, 51*(8), 4576–4584.

Hernández, A., Loera, N., Contreras, M., Fischer, L., & Sánchez, D. (2019). Comparison between Lactuca sativa L. and Lolium perenne: Phytoextraction capacity of Ni, Fe, and Co from galvanoplastic industry. In *Energy technology 2019: Carbon dioxide management and other technologies* (pp. 137–147). Springer International Publishing.

Hlihor, R. M., Diaconu, M., Leon, F., Curteanu, S., Tavares, T., & Gavrilescu, M. (2015). Experimental analysis and mathematical prediction of Cd (II) removal by biosorption using support vector machines and genetic algorithms. *New Biotechnology, 32*(3), 358–368.

Huang, X., Luo, D., Chen, X., Wei, L., Liu, Y., Wu, Q., et al. (2019). Insights into heavy metals leakage in chelator-induced phytoextraction of Pb-and Tl-contaminated soil. *International Journal of Environmental Research and Public Health, 16*(8), 1328.

Hubau, A., Minier, M., Chagnes, A., Joulian, C., Silvente, C., & Guezennec, A. G. (2020). Recovery of metals in a double-stage continuous bioreactor for acidic bioleaching of printed circuit boards (PCBs). *Separation and Purification Technology, 238*, 116481.

Ilyas, S., Lee, J. C., & Kim, B. S. (2014). Bioremoval of heavy metals from recycling industry electronic waste by a consortium of moderate thermophiles: Process development and optimization. *Journal of Cleaner Production, 70*, 194–202.

Iravani, S., & Varma, R. S. (2022). Genetically engineered organisms: Possibilities and challenges of heavy metal removal and nanoparticle synthesis. *Clean Technologies, 4*(2), 502–511.

Işıldar, A., van de Vossenberg, J., Rene, E. R., van Hullebusch, E. D., & Lens, P. N. (2016). Two-step bioleaching of copper and gold from discarded printed circuit boards (PCB). *Waste Management, 57*, 149–157.

Jagannath, A., Shetty, V., & Saidutta, M. B. (2017). Bioleaching of copper from electronic waste using Acinetobacter sp. Cr B2 in a pulsed plate column operated in batch and sequential batch mode. *Journal of Environmental Chemical Engineering, 5*(2), 1599–1607.

Jally, B., Laubie, B., Chour, Z., Muhr, L., Qiu, R., Morel, J. L., et al. (2021). A new method for recovering rare earth elements from the hyperaccumulating fern Dicranopteris linearis from China. *Minerals Engineering, 166*, 106879.

Jiang, J., Wang, Y., Yu, D., Hou, R., Ma, X., Liu, J., et al. (2022). Combined addition of biochar and garbage enzyme improving the humification and succession of fungal community during sewage sludge composting. *Bioresource Technology, 346*, 126344.

Johnston, C. W., Wyatt, M. A., Li, X., Ibrahim, A., Shuster, J., Southam, G., & Magarvey, N. A. (2013). Gold biomineralization by a metallophore from a gold-associated microbe. *Nature Chemical Biology, 9*(4), 241–243.

Joradon, P., Poolpak, T., Kruatrachue, M., Yang, K. M., Saengwilai, P., Upatham, S., & Pokethitiyook, P. (2024). Phytoremediation technology for recovery of Ni by Acacia plants in association with Bacillus amyloliquefaciens isolated from E-waste contaminated site. *International Journal of Phytoremediation, 26*(6), 903–912.

Juwarkar, A. A., & Yadav, S. K. (2010). Bioaccumulation and biotransformation of heavy metals. In *Bioremediation technology: Recent advances* (pp. 266–284). Springer Netherlands.

Kadivar, S., Pourhossein, F., & Mousavi, S. M. (2021). Recovery of valuable metals from spent mobile phone printed circuit boards using biochar in indirect bioleaching. *Journal of Environmental Management, 280*, 111642.

Kafle, A., Timilsina, A., Gautam, A., Adhikari, K., Bhattarai, A., & Aryal, N. (2022). Phytoremediation: Mechanisms, plant selection and enhancement by natural and synthetic agents. *Environmental Advances, 8*, 100203.

Kaksonen, A. H., Boxall, N. J., Gumulya, Y., Khaleque, H. N., Morris, C., Bohu, T., et al. (2018). Recent progress in biohydrometallurgy and microbial characterisation. *Hydrometallurgy, 180*, 7–25.

Kalve, S., Sarangi, B. K., Pandey, R. A., & Chakrabarti, T. (2011). Arsenic and chromium hyperaccumulation by an ecotype of Pteris vittata–prospective for phytoextraction from contaminated water and soil. *Current Science*, 888–894.

Kaur, P., Sharma, S., Albarakaty, F. M., Kalia, A., Hassan, M. M., & Abd-Elsalam, K. A. (2022). Biosorption and bioleaching of heavy metals from electronic waste varied with microbial genera. *Sustainability, 14*(2), 935.

Kaya, M. (2016). Recovery of metals and nonmetals from electronic waste by physical and chemical recycling processes. *Waste Management, 57*, 64–90.

Kaya, M. (2019). Electronic waste and printed circuit board recycling technologies.

Khalid, N., Noman, A., Aqeel, M., Masood, A., & Tufail, A. (2019). Phytoremediation potential of Xanthium strumarium for heavy metals contaminated soils at roadsides. *International journal of Environmental Science and Technology, 16*, 2091–2100.

Khan, S., Alam, F. S., & Borah, D. (2024). Microbial cell factories in the remediation of e-wastes: An insight. *Biotechnology for the Environment, 1*(1), 4.

Khatri, B. R., Sodha, A. B., Shah, M. B., Tipre, D. R., & Dave, S. R. (2018). Chemical and microbial leaching of base metals from obsolete cell-phone printed circuit boards. *Sustainable Environment Research, 28*(6), 333–339.

Kim, M. J., Seo, J. Y., Choi, Y. S., & Kim, G. H. (2016). Bioleaching of spent Zn–Mn or Ni–Cd batteries by Aspergillus species. *Waste Management, 51*, 168–173.

Kinoshita, T., Akita, S., Kobayashi, N., Nii, S., Kawaizumi, F., & Takahashi, K. (2003). Metal recovery from non-mounted printed wiring boards via hydrometallurgical processing. *Hydrometallurgy, 69*(1–3), 73–79.

Kiran, N. S., Yashaswini, C., Chatterjee, A., & Shah, M. P. (2024). Biotechnological approaches for metal recovery from electronic wastes. *Current Microbiology, 81*(12), 419.

Koptsik, G. N. (2014). Problems and prospects concerning the phytoremediation of heavy metal polluted soils: A review. *Eurasian Soil Science, 47*, 923–939.

Kumar, A., Saini, H. S., & Kumar, S. (2018). Bioleaching of gold and silver from waste printed circuit boards by Pseudomonas balearica SAE1 isolated from an e-waste recycling facility. *Current Microbiology, 75*, 194–201.

Kuyucak, N., & Akcil, A. (2013). Cyanide and removal options from effluents in gold mining and metallurgical processes. *Minerals Engineering, 50*, 13–29.

Lee, J. C., & Pandey, B. D. (2012). Bio-processing of solid wastes and secondary resources for metal extraction–a review. *Waste Management, 32*(1), 3–18.

Li, J., Zhang, J., Larson, S. L., Ballard, J. H., Guo, K., Arslan, Z., et al. (2019). Electrokinetic-enhanced phytoremediation of uranium-contaminated soil using sunflower and Indian mustard. *International Journal of Phytoremediation, 21*(12), 1197–1204.

Li, J., Wen, J., Guo, Y., An, N., Liang, C., & Ge, Z. (2020). Bioleaching of gold from waste printed circuit boards by alkali-tolerant Pseudomonas fluorescens. *Hydrometallurgy, 194*, 105260.

Li, X., Wang, X., Ren, C., Palansooriya, K. N., Wang, Z., & Chang, S. X. (2024). Microplastic pollution: Phytotoxicity, environmental risks, and phytoremediation strategies. *Critical Reviews in Environmental Science and Technology, 54*(6), 486–507.

Liang, S. H., Chen, S. C., Chen, C. Y., Kao, C. M., Yang, J. I., Shieh, B. S., et al. (2011). Cadmium-induced earthworm metallothionein-2 is associated with metal accumulation and counteracts oxidative stress. *Pedobiologia, 54*(5–6), 333–340.

Liao, X., Ye, M., Liang, J., Guan, Z., Li, S., Deng, Y., et al. (2022). Feasibility of reduced iron species for promoting Li and Co recovery from spent LiCoO2 batteries using a mixed-culture bioleaching process. *Science of the Total Environment, 830*, 154577.

Liapun, V., & Motola, M. (2023). Current overview and future perspective in fungal biorecovery of metals from secondary sources. *Journal of Environmental Management, 332*, 117345.

Liu, X., Hu, C., & Zhang, S. (2005). Effects of earthworm activity on fertility and heavy metal bioavailability in sewage sludge. *Environment International, 31*(6), 874–879.

Liu, R., Li, J., & Ge, Z. (2016). Review on Chromobacterium violaceum for gold bioleaching from e-waste. *Procedia Environmental Sciences, 31*, 947–953.

Liu, Q., Bai, J. F., Gu, W. H., Peng, S. J., Wang, L. C., Wang, J. W., & Li, H. X. (2020). Leaching of copper from waste printed circuit boards using Phanerochaete chrysosporium fungi. *Hydrometallurgy, 196*, 105427.

Liu, Z. W., Guo, X. Y., Tian, Q. H., & Zhang, L. (2022). A systematic review of gold extraction: Fundamentals, advancements, and challenges toward alternative lixiviants. *Journal of Hazardous Materials, 440*, 129778.

Liu, K., Tan, Q., Yu, J., & Wang, M. (2023). A global perspective on e-waste recycling. *Circular Economy, 2*(1), 100028.

Machineni, L. (2023). Waste to wealth and health: Bio-recovery and applications of chitin and its derivatives. *International Journal of Environmental Technology and Management, 26*(3–5), 341–359.

Manzoor, M., Gul, I., Manzoor, A., Kamboh, U. R., Hina, K., Kallerhoff, J., & Arshad, M. (2020). Lead availability and phytoextraction in the rhizosphere of Pelargonium species. *Environmental Science and Pollution Research, 27*, 39753–39762.

Massoud, R., Hadiani, M. R., Hamzehlou, P., & Khosravi-Darani, K. (2019). Bioremediation of heavy metals in food industry: Application of Saccharomyces cerevisiae. *Electronic Journal of Biotechnology, 37*, 56–60.

Mishra, S., Dash, D., Al-Tawaha, A. R. M. S., & Das, A. P. (2022). A review on heavy metal ion adsorption on synthetic microfiber surface in aquatic environments. *Applied Biochemistry and Biotechnology, 194*(10), 4639–4654. https://doi.org/10.1007/S12010-022-04029-W/METRICS

Mishra, S., Ghosh, S., van Hullebusch, E. D., Singh, S., & Das, A. P. (2023). A critical review on the recovery of base and critical elements from electronic waste-contaminated streams using microbial biotechnology. *Applied Biochemistry and Biotechnology, 2023*, 1–30. https://doi.org/10.1007/S12010-023-04440-X

Mohanty, S., Ghosh, S., Nayak, S., & Das, A. P. (2017). Bioleaching of manganese by Aspergillus sp. isolated from mining deposits. *Chemosphere, 172*, 302–309. https://doi.org/10.1016/J.CHEMOSPHERE.2016.12.136

Nadar, S., & Khan, T. (2024). Chapter role of microbial biofilms in bioremediation: Current perspectives. In *Microbes based approaches for the management of hazardous contaminants* (pp. 257–276).

Náhlík, V., Čížková, M., Singh, A., Mezricky, D., Rucki, M., Andresen, E., & Vítová, M. (2023). Growth of the red alga Galdieria sulphuraria in red mud-containing medium and accumulation of rare earth elements. *Waste and Biomass Valorization, 14*(7), 2179–2189.

Nannoni, F., Rossi, S., & Protano, G. (2014). Soil properties and metal accumulation by earthworms in the Siena urban area (Italy). *Applied Soil Ecology, 77*, 9–17.

Narayanasamy, M., Dhanasekaran, D., Vinothini, G., & Thajuddin, N. (2018). Extraction and recovery of precious metals from electronic waste printed circuit boards by bioleaching acidophilic fungi. *International journal of Environmental Science and Technology, 15*, 119–132.

Narayanasamy, M., Dhanasekaran, D., & Thajuddin, N. (2021). Bioremediation of noxious metals from e-waste printed circuit boards by Frankia. *Microbiological Research, 245*, 126707.

Naser, A. Z., Deiab, I., & Darras, B. M. (2021). Poly (lactic acid) (PLA) and polyhydroxyalkanoates (PHAs), green alternatives to petroleum-based plastics: A review. *RSC Advances, 11*(28), 17151–17196.

Natarajan, K. A. (2018). *Biotechnology of metals: Principles, recovery methods and environmental concerns*. Elsevier.

Natarajan, K. A. (2019). Biotechnology for environmentally benign gold production. In *Horizons in bioprocess engineering* (pp. 263–299).

Natarajan, K. A., & Kumari, A. (2014). Role of applied potentials on bioleaching of chalcopyrite concentrate and growth of Acidithiobacillus ferrooxidans. *Mining, Metallurgy & Exploration, 31*(4), 215–222.

Natarajan, G., & Ting, Y. P. (2015). Gold biorecovery from e-waste: An improved strategy through spent medium leaching with pH modification. *Chemosphere, 136*, 232–238.

Natarajan, G., Tay, S. B., Yew, W. S., & Ting, Y. P. (2015). Engineered strains enhance gold biorecovery from electronic scrap. *Minerals Engineering, 75*, 32–37.

Nguyen, T. H., Won, S., Ha, M. G., Nguyen, D. D., & Kang, H. Y. (2021). Bioleaching for environmental remediation of toxic metals and metalloids: A review on soils, sediments, and mine tailings. *Chemosphere, 282*, 131108.

Nnaji, N. D., Onyeaka, H., Miri, T., & Ugwa, C. (2023). Bioaccumulation for heavy metal removal: A review. *SN Applied Sciences, 5*(5), 125.

Pandey, V. C. (2012). Invasive species based efficient green technology for phytoremediation of fly ash deposits. *Journal of Geochemical Exploration, 123*, 13–18.

Parajuly, K., Fitzpatrick, C., Muldoon, O., & Kuehr, R. (2020). Behavioral change for the circular economy: A review with focus on electronic waste management in the EU. *Resources, Conservation & Recycling: X, 6*, 100035.

Patel, S., & Kasture, A. (2014). E (electronic) waste management using biological systems-overview. *International Journal of Current Microbiology and Applied Sciences, 3*(7), 495–504.

Peng, F. J., Ying, G. G., Pan, C. G., Selck, H., Salvito, D., & Van den Brink, P. J. (2018). Bioaccumulation and biotransformation of triclosan and galaxolide in the freshwater oligochaete Limnodrilus hoffmeisteri in a water/sediment microcosm. *Environmental Science & Technology, 52*(15), 8390–8398.

Pérez-Sanz, A., Millán, R., Sierra, M. J., Alarcón, R., García, P., Gil-Díaz, M., et al. (2012). Mercury uptake by Silene vulgaris grown on contaminated spiked soils. *Journal of Environmental Management, 95*, S233–S237.

Pinto, J., Colónia, J., Abdolvaseei, A., Vale, C., Henriques, B., & Pereira, E. (2023). Algal sorbents and prospects for their application in the sustainable recovery of rare earth elements from E-waste. *Environmental Science and Pollution Research, 30*(30), 74521–74543.

Pourhossein, F., Rezaei, O., Mousavi, S. M., & Beolchini, F. (2021). Bioleaching of critical metals from waste OLED touch screens using adapted acidophilic bacteria. *Journal of Environmental Health Science and Engineering, 19*(1), 893–906.

Prabhakar, A., Mishra, S., & Das, A. P. (2019). Isolation and identification of Lead (Pb) solubilizing bacteria from automobile waste and its potential for recovery of Lead from end of life waste batteries. *Geomicrobiological Journal, 36*(10), 894–903. https://doi.org/10.1080/01490451.2019.1654044

Pratas, J., Favas, P. J., Paulo, C., Rodrigues, N., & Prasad, M. N. V. (2012). Uranium accumulation by aquatic plants from uranium-contaminated water in Central Portugal. *International Journal of Phytoremediation, 14*(3), 221–234.

Preeti, Raj, A., Joshi, N., Rawat, J. M., Tapas, S., Pattnaik, B., & Rawat, B. (2023). Genetically modified microbes in E-waste management: A perspective. In *Microbial technology for sustainable E-waste management* (pp. 259–271). Springer International Publishing.

Priya, A., & Hait, S. (2020). Biometallurgical recovery of metals from waste printed circuit boards using pure and mixed strains of Acidithiobacillus ferrooxidans and Acidiphilium acidophilum. *Process Safety and Environmental Protection, 143*, 262–272.

Priyadarshini, B., Santhosh Kumar, M., & Das, A. P. (2024). Microbial degradation of synthetic plastics released from COVID-19 generated single use disposable face masks. *Environmental Science and Engineering, Part, F2827*, 1–16. https://doi.org/10.1007/978-3-031-50840-0_1

Priyadarsini, S., & Das, A. P. (2024). Analytical and structural characterization of waste lithium-ion batteries for their effective recycling strategy. *Environmental Chemistry and Ecotoxicology*. https://doi.org/10.1016/j.enceco.2024.12.004

Priyadarsini, S., & Das, A. P. (2025). Lithium bioleaching: A review on microbial-assisted sustainable technology for lithium bio-circularity. *Journal of Water Process Engineering, 69*, 106744. https://doi.org/10.1016/j.jwpe.2024.106744

Rana, S., Mishra, P., Ab Wahid, Z., Thakur, S., Pant, D., & Singh, L. (2020). Microbe-mediated sustainable bio-recovery of gold from low-grade precious solid waste: A microbiological overview. *Journal of Environmental Sciences, 89*, 47–64.

Rawlings, D. E., & Johnson, D. B. (2007). The microbiology of biomining: Development and optimization of mineral-oxidizing microbial consortia. *Microbiology, 153*(2), 315–324.

Reith, F., Etschmann, B., Grosse, C., Moors, H., Benotmane, M. A., Monsieurs, P., et al. (2009). Mechanisms of gold biomineralization in the bacterium Cupriavidus metallidurans. *Proceedings of the National Academy of Sciences, 106*(42), 17757–17762.

Richardson, J. B., Görres, J. H., Jackson, B. P., & Friedland, A. J. (2015). Trace metals and metalloids in forest soils and exotic earthworms in northern New England, USA. *Soil Biology and Biochemistry, 85*, 190–198.

Rienzie, R., Perera, A. T. D., & Adassooriya, N. M. (2019). Biorecovery of precious metal nanoparticles from waste electrical and electronic equipments. In *Electronic waste management and treatment technology* (pp. 133–152). Butterworth-Heinemann.

Roy, J. J., Rarotra, S., Krikstolaityte, V., Zhuoran, K. W., Cindy, Y. D. I., Tan, X. Y., et al. (2022). Green recycling methods to treat lithium-ion batteries E-waste: A circular approach to sustainability. *Advanced Materials, 34*(25), 2103346.

Şahan, M. (2016). Evaluation of recovery potential for precious and rare earth metals from e-waste.

Sahni, A., & Kumar, S. (2016). Bioleaching of Glod from obsolete mobile SIM cards.

Sakakibara, M., Ohmori, Y., Ha, N. T. H., Sano, S., & Sera, K. (2011). Phytoremediation of heavy metal-contaminated water and sediment by Eleocharis acicularis. *CLEAN–Soil, Air, Water, 39*(8), 735–741.

Şenol, Z. M., Gül, Ü. D., Gurbanov, R., & Şimşek, S. (2021). Optimization the removal of lead ions by fungi: Explanation of the mycosorption mechanism. *Journal of Environmental Chemical Engineering, 9*(2), 104760.

Shahabuddin, M., Uddin, M. N., Chowdhury, J. I., Ahmed, S. F., Uddin, M. N., Mofijur, M., & Uddin, M. A. (2023). A review of the recent development, challenges, and opportunities of electronic waste (e-waste). *International journal of Environmental Science and Technology, 20*(4), 4513–4520.

Sheel, A., & Pant, D. (2018). Recovery of gold from electronic waste using chemical-assisted microbial biosorption (hybrid) technique. *Bioresource Technology, 247*, 1189–1192.

Sheoran, V., Sheoran, A. S., & Poonia, P. (2013). Phytoremediation technologies for the reclamation of organic and inorganic polluted soils and water: A review. *Environmental Research Journal, 7*(1).

Shet, A. R., Patil, L. R., Hombalimath, V. S., Yaraguppi, D. A., & Udapudi, B. B. (2011). Enrichment of Saccharomyces cerevisiae with zinc and their impact on cell growth. *Biotechnology, Bioinformatics and Bioengineering, 1*(4), 523–527.

Singh Jadaun, J., Bansal, S., Sonthalia, A., Rai, A. K., & Singh, S. P. (2022). Biodegradation of plastics for sustainable environment. *Bioresource Technology, 347*, 126697. https://doi.org/10.1016/J.BIORTECH.2022.126697

Singh, V., Pandey, B., & Suthar, S. (2019). Phytotoxicity and degradation of antibiotic ofloxacin in duckweed (Spirodela polyrhiza) system. *Ecotoxicology and Environmental Safety, 179*, 88–95.

Sinha, R. K., Herat, S., Valani, D., & Chauhan, K. (2010). Earthworms–the environmental engineers: Review of vermiculture technologies for environmental management and resource development. *International Journal of Global Environmental Issues, 10*(3–4), 265–292.

Sivakumar, S. (2015). Effects of metals on earthworm life cycles: A review. *Environmental Monitoring and Assessment, 187*, 1–16.

Srivastava, M., Ma, L. Q., & Santos, J. A. G. (2006). Three new arsenic hyperaccumulating ferns. *Science of the Total Environment, 364*(1–3), 24–31.

Sun, Y., Lu, T., Pan, Y., Shi, M., Ding, D., Ma, Z., et al. (2022). Recovering rare earth elements via immobilized red algae from ammonium-rich wastewater. *Environmental Science and Ecotechnology, 12*, 100204.

Tabak, H. H., Lens, P., Van Hullebusch, E. D., & Dejonghe, W. (2005). Developments in bioremediation of soils and sediments polluted with metals and radionuclides–1. Microbial processes and mechanisms affecting bioremediation of metal contamination and influencing metal toxicity and transport. *Reviews in Environmental Science and Bio/Technology, 4*, 115–156.

Tabinda, A. B., Irfan, R., Yasar, A., Iqbal, A., & Mahmood, A. (2020). Phytoremediation potential of Pistia stratiotes and Eichhornia crassipes to remove chromium and copper. *Environmental Technology*.

Tay, S. B., Natarajan, G., Rahim, M. N. B. A., Tan, H. T., Chung, M. C. M., Ting, Y. P., & Yew, W. S. (2013). Enhancing gold recovery from electronic waste via lixiviant metabolic engineering in Chromobacterium violaceum. *Scientific Reports, 3*(1), 2236.

Tecchio, P., Ardente, F., Marwede, M., Clemm, C., Dimitrova, G., & Mathieux, F. (2018). Analysis of material efficiency aspects of personal computers product group. Luxembourgh. doi, 10, 89220.

Thatoi, H., Das, S., Mishra, J., Rath, B. P., & Das, N. (2014). Bacterial chromate reductase, a potential enzyme for bioremediation of hexavalent chromium: A review. *Journal of Environmental Management, 146*, 383–399.

Trivedi, A., & Hait, S. (2020). Bioleaching of selected metals from e-waste using pure and mixed cultures of Aspergillus species. In *Measurement, analysis and remediation of environmental pollutants* (pp. 271–280).

Trivedi, A., Vishwakarma, A., Saawarn, B., Mahanty, B., & Hait, S. (2022). Fungal biotechnology for urban mining of metals from waste printed circuit boards: A review. *Journal of Environmental Management, 323*, 116133.
Tsang, Y. F., Kumar, V., Samadar, P., Yang, Y., Lee, J., Ok, Y. S., et al. (2019). Production of bioplastic through food waste valorization. *Environment International, 127*, 625–644.
Ueno, D., Iwashita, T., Zhao, F. J., & Ma, J. F. (2008). Characterization of Cd translocation and identification of the Cd form in xylem sap of the Cd-hyperaccumulator Arabidopsis halleri. *Plant and Cell Physiology, 49*(4), 540–548.
Van Der Ent, A., Malaisse, F., Erskine, P. D., Mesjasz-Przybyłowicz, J., Przybyłowicz, W. J., Barnabas, A. D., et al. (2019). Abnormal concentrations of Cu–Co in Haumaniastrum katangense, Haumaniastrum robertii and Aeolanthus biformifolius: Contamination or hyperaccumulation? *Metallomics, 11*(3), 586–596.
Vardanyan, A., Vardanyan, N., Aâtach, M., Malavasi, P., & Gaydardzhiev, S. (2022). Bio-assisted leaching of non-ferrous metals from waste printed circuit boards—Importance of process parameters. *Metals, 12*(12), 2092.
Vijayaraghavan, K., & Yun, Y. S. (2008). Bacterial biosorbents and biosorption. *Biotechnology Advances, 26*(3), 266–291.
Wang, W., Tian, Y., Zhu, Q., & Zhong, Y. (2017). Barriers for household e-waste collection in China: Perspectives from formal collecting enterprises in Liaoning Province. *Journal of Cleaner Production, 153*, 299–308.
Wei, S., Xu, L., Dai, H., & Hu, Y. (2018). Ornamental hyperaccumulator Mirabilis jalapa L. phytoremediating combine contaminated soil enhanced by some chelators and surfactants. *Environmental Science and Pollution Research, 25*, 29699–29704.
Wilson, D. C., Rodic, L., Modak, P., Reka Soos, A., Carpintero, K. V., Iyer, M., & Simonett, O. (2015). *Global waste management outlook*. UNEP.
Won, S. W., Kotte, P., Wei, W., Lim, A., & Yun, Y. S. (2014). Biosorbents for recovery of precious metals. *Bioresource Technology, 160*, 203–212.
Wu, C. H., Wood, T. K., Mulchandani, A., & Chen, W. (2006). Engineering plant-microbe symbiosis for rhizoremediation of heavy metals. *Applied and Environmental Microbiology, 72*(2), 1129–1134.
Wu, W., Liu, X., Zhang, X., Zhu, M., & Tan, W. (2018). Bioleaching of copper from waste printed circuit boards by bacteria-free cultural supernatant of iron–sulfur-oxidizing bacteria. *Bioresources and Bioprocessing, 5*(1), 1–13.
Wu, Y., Chen, C., Wang, G., Xiong, B., Zhou, W., Xue, F., et al. (2020). Mechanism underlying earthworm on the remediation of cadmium-contaminated soil. *Science of the Total Environment, 728*, 138904.
Xu, H., Gong, W., Syltebo, L., Izzo, K., Lutze, W., & Pegg, I. L. (2014). Effect of blast furnace slag grades on fly ash based geopolymer waste forms. *Fuel, 133*, 332–340.
Yan, A., Wang, Y., Tan, S. N., Mohd Yusof, M. L., Ghosh, S., & Chen, Z. (2020). Phytoremediation: A promising approach for revegetation of heavy metal-polluted land. *Frontiers in Plant Science, 11*, 359.
Yu, Z., Han, H., Feng, P., Zhao, S., Zhou, T., Kakade, A., et al. (2020). Recent advances in the recovery of metals from waste through biological processes. *Bioresource Technology, 297*, 122416.
Zeng, G., Deng, X., Luo, S., Luo, X., & Zou, J. (2012). A copper-catalyzed bioleaching process for enhancement of cobalt dissolution from spent lithium-ion batteries. *Journal of Hazardous Materials, 199*, 164–169.
Zeng, Z., Huo, X., Zhang, Y., Xiao, Z., Zhang, Y., & Xu, X. (2018). Lead exposure is associated with risk of impaired coagulation in preschool children from an e-waste recycling area. *Environmental Science and Pollution Research, 25*, 20670–20679.
Zhang, L., Dong, H., Liu, Y., Bian, L., Wang, X., Zhou, Z., & Huang, Y. (2018). Bioleaching of rare earth elements from bastnaesite-bearing rock by actinobacteria. *Chemical Geology, 483*, 544–557.
Zhang, Y., Hu, X., Wang, Y., & Jiang, N. (2023). A critical review of biomineralization in environmental geotechnics: Applications, trends, and perspectives. *Biogeotechnics, 1*(1), 100003.

Zhang, X., Yang, Y., Liu, L., Sui, X., Bermudez, R. S., Wang, L., et al. (2024). Insights into the efficient degradation mechanism of extracellular proteases mediated by Purpureocillium lilacinum. *Frontiers in Microbiology, 15*, 1404439.

Zhao, F. J., Lombi, E., & McGrath, S. P. (2003). Assessing the potential for zinc and cadmium phytoremediation with the hyperaccumulator Thlaspi caerulescens. *Plant and Soil, 249*, 37–43.

Zhao, C., Li, X. H., Dong, M. Q., Li, Y. Y., & Qin, Z. F. (2021). Measurement of polychlorinated biphenyls with hand wipes and matched serum collected from Chinese E-waste dismantling workers: Exposure estimates and implications. *Science of the Total Environment, 799*, 149444.

Zhu, N., Xiang, Y., Zhang, T., Wu, P., Dang, Z., Li, P., & Wu, J. (2011). Bioleaching of metal concentrates of waste printed circuit boards by mixed culture of acidophilic bacteria. *Journal of Hazardous Materials, 192*(2), 614–619.

Zhu, N., Zhang, B., & Yu, Q. (2020). Genetic engineering-facilitated coassembly of synthetic bacterial cells and magnetic nanoparticles for efficient heavy metal removal. *ACS Applied Materials & Interfaces, 12*(20), 22948–22957.

Biotechnical Method for Recovery of Precious Minerals from E-Waste

Rajalaxmi Badra, Sailaja Priyadarsini, Koteswara Reddy, Chandra Mohan, Kiran Bala, and Alok Prasad Das

Abstract The growing number of outmoded electronic gadgets has made e-waste a major environmental and economic concern. Electronic waste contains priceless precious metals including gold, silver, and platinum, as well as dangerous compounds that require proper disposal and environmentally friendly recycling solutions. Biotechnological approaches using microbes and enzymes offer a cost-effective alternative to classic extraction methods. Microorganisms like *Acidithiobacillus ferrooxidans*, *Leptospirillum ferrooxidans*, and *Pseudomonas* can bleach metals via acidification, complexation, and redox processes. Enzymatic methods, such as hydrolases and oxidoreductases, can improve metal recovery while minimizing environmental impact. This study emphasizes the efficiency, scalability, and low ecological footprint of biotechnological methods for recovering precious minerals from e-waste. Integrating these techniques into e-waste recycling frameworks could significantly contribute to sustainable resource recovery and pollution reduction.

Keywords E-waste · Precious mineral · Biotechnological methods · Environmentally sustainable extraction · Resource recovery from waste

R. Badra · S. Priyadarsini · A. P. Das (✉)
Department of Life Sciences, Rama Devi Women's University, Bhubaneswar, Odisha, India

K. Reddy
Department of Biotechnology, Koneru Lakshmaiah Education Foundation (Deemed to be University), Guntur, Andhra Pradesh, India

C. Mohan
School of Basic and Applied Sciences, K R Mangalam University, Gurugram, India

K. Bala
Department of Zoology, Deshbandhu College, University of Delhi, New Delhi, Delhi, India

A. P. Das, S. Priyadarsini (eds.), *Electronic Waste and Environmental Pollution*, Sustainable Environmental Waste Management Strategies,
https://doi.org/10.1007/978-3-031-90287-1_16

1 Introduction

Electronic devices are becoming more important in today's world. These devices are vigorous for communiqué and accessibility. Sixty five diverse metals can be present in current electronics, and paying for any secondary education is gaining popularity due to a probable scarcity. In the technological advancement era, several pollutants like metals (Das et al., 2012b, 2015; Ghosh et al., 2020; Mohanty et al., 2016), plastics (Das et al., 2024a; Dey et al., 2024a; Jagatee et al., 2024a, 2024b; Mishra et al., 2019, 2021), biomedical (Khatun et al., 2023; Sadasivuni et al., 2019), and chemicals (Bhattacharjee et al., 2021a, 2021b; Das & Swain, 2013) are increasing in a significant rate. Electronic metals are primarily found in electronic trash (Isildar, 2016). The progression of electronic and electrical productions as well as the extensive acceptance of electronic technologies have controlled the creation of numerous new electronic devices (Al-Salem et al., 2022). When electronic items reach their end of life, they generate noteworthy quantities of e-waste. In 2020, over 55 MT of e-waste was collected, that anticipated to rise to 75 MT by 2030 (Singh, 2022). A total of 10.4 MT of e-waste was together and valorized, accounting for 17.5% of the total e-waste produced (Dutta et al., 2023). According to the United Nations' worldwide e-waste surveillance report, South Africa generates roughly 365,000 tons of e-waste annually, with Gauteng province accounting for over 55% due to its economic center status (Ichikowitz & Hattingh, 2020). Since the primary twentieth century, remarkable inventions in science, manufacturing, and technology have resulted in farfetched instruments and systems that have revolutionized our world and become vivacious in our day-to-day lives. Higher one-use incomes and expansion in industrialized countries have led to rapid technology advancements and increased adoption of electronic devices. Contemporary technology has enhanced living standards and expeditions, but it has also had undesirable effects on the atmosphere. E-waste contains a variety of metals, including expensive metals and sporadic Earth elements. The composition varies by product line and category. E-waste contains more than 1200 dangerous compounds, including lead, chromium, cadmium, and mercury, as well as nonferrous metals like copper and aluminum (Wang et al., 2021; Prabhakar et al., 2019; Raj & Das, 2023). Furthermore, to the base metallic element, e-waste, including PCBs, also comprises valuable metals like silver, gold, and platinum. Metals with good chemical stability and great conductivity are commonly employed in electrical products (Suja et al., 2018). PCBs play a decisive part in connecting electronic components in various devices. According to its configuration alignment, PCBs can be one-sided, two-sided, or multilayer-sided (Magoda & Mekuto, 2022). Globally, 1.2 MT of PCBs are manufactured each year, with an average annual increase rate of 4%. Similarly, computers' primary processor units contain significant amounts of valuable metal (Wang et al., 2021). E-waste is classified according to metallic content, kind, source, age, constructor, and constituent parts (Mourshed et al., 2024). E-waste comprises valuable metals, as well as risky metals like lead and mercury (Koti et al., 2023). E-waste is classified into three types: big household appliances, small electronics, and IT equipment, each of which requires a unique recycling technique (Wang et al., 2012). The origin of e-waste, whether from consumer electronics or industrial equipment, influences its management and recycling (Vadoudi et al., 2015). Older technologies frequently contain hazardous components, but newer ones can be built for easier recycling (Mourshed et al., 2024). The USA classified e-waste into six categories: temperature exchange equipment, major domestic equipment, tiny devices, and small IT and telecommunication devices (Ichikowitz & Hattingh, 2020). Numerous studies highlighting the high-metal content of printed circuit boards back up the assertion that they can contain up to ten times more precious metals than commercially mined materials. These boards, which account for 3.5% of electronic trash, can contain up to 40% metals, including valuable metals like gold, silver, and palladium (Blumbergs et al., 2024). Electronic devices frequently contain machinery comprised of hazardous metals and chemicals. These materials include phthalates, beryllium, antimony, PVC, and BFRs (Iglesias-González et al., 2024). These chemicals pose significant health and environmental risks, necessitating stringent regulations and effective waste management systems (Gongwei et al., 2023). Phthalates, which are extensively used as plasticizers, can leach from electronic components and cause serious health problems such as endocrine disruption (Gongwei et al., 2023).

Lithium metal is utilized in several electronic components; however, it is extremely toxic, causing lung problems and other health issues when exposed (Suja et al., 2014; Priyadarsini & Das, 2024, 2025). Antimony, which is widely present in flame retardants, can be damaging to both human health and the environment, causing air and water pollution (Castro & Bassin, 2022). The high incidence of undesirability in the electronic sector adds significantly to the rapid growth of e-waste, particularly in developing nations (Chalana et al., 2023). The electronic sector is rapidly becoming obsolete, with worldwide e-waste predicted to increase from 54.2 MT in 2020 to 75.1 MT by 2030 (Chalana et al., 2023). Short product lifespans and limited repair choices intensify this trend, resulting in a huge accumulation of e-waste (Rani & Ahirwar, 2022). Electronic products are usually replaced every year, have a quick ledge life expectancy, and are frequently abandoned as second-hand commodities in poor countries (Ogboru, 2014). The cessation and clearance of e-waste have harmed the ecosystem (Chalana et al., 2023). Over 90% of superfluous electrical and e-waste is disposed of in landfills. As a result, landfills produce more e-waste, which accelerates environmental degradation (Kumar et al., 2024). Cancer risk rises with developmental and neurological issues (Dhivya & Premalatha, 2024). Ignition hurts the environment, especially the quality of solids, soil, and air. Computer chip dirt and acid are thrown on the ground (Goyal et al., 2023). The earth turns acidic, polluting water supplies. Much research has shown that harmful metals and chemicals included in e-waste might cause health difficulties. Prolonged exposure to chemicals in e-waste has been related to significant health consequences, as well as respiratory diseases, renal damage, and various malignancies (Nyeko et al., 2023). Metals like Cd, As, Li, Hg, Se, Cr, and Pb have been shown to have hazardous effects. Americium on the human body (Morais et al., 2012). Heavy metals can enter the body from a variety of sources, the most common of which are contaminated food and water (Mishra et al., 2022a). Industrial discharges (Das & Ghosh,

2022; Mishra & Das, 2023), mining (Dey et al., 2023b, 2024b, Pradhan et al., 2023), plastics (Priyadarshini et al., 2024a, 2024b; Jadaun et al., 2022), agricultural practices, and urban runoff all contribute significantly to metal contamination in the environment. For example, lead and cadmium exposure is closely related to occupational dangers in areas such as battery manufacture and mining (Viroonudomphol et al., 2016). In recent years, microorganisms, specifically cyanogenic bacteria, have been painstakingly studied as an alternate approach for the gold cyanidation process (Ghosh et al., 2016a; Ghosh & Das, 2017; Mohanty et al., 2017). Microorganisms such as *C. violaceum*, *A. ferrooxidans,* and *Aspergillus* sp. improved gold and copper retrieval from e-waste (Natarajan & Ting, 2014). This bacterium has a high probability for environmental gold recapture (Batista et al., 2025). *Chromobacterium violaceum* is a mesophilic, mobile Gram-negative bacteria that prosper in a variety of temperate habitats. As a facultative anaerobe, it can use oxygen when it is accessible and switch to anaerobic metabolism when it is not. This versatility allows it to thrive in an extensive range of ecological niches with changing oxygen levels (Batista et al., 2025). Several functional and nutritious characteristics including glycine concentration, pH, temperature, pulp density, and medium composition may disturb the result of the technique. It assists researchers in identifying crucial aspects that influence outcomes, allowing them to efficiently optimize operations (Morshedi & Akbarian, 2014). The regaining of Au and Ag metals from e-waste is extensively investigated. E-waste mobile PCBs were processed into lumps and leached in a batch-stirred tank reactor method (Vuppaladadiyam et al., 2024). Electro-electronic trash makes up most of the overall garbage and grows at a pace of 3–5% per year, resulting in around 20–50 million tons of abandoned materials (Zhao et al., 2023).

2 Sources of E-Waste

Even though it is referred to as “waste,” electronic garbage is quite valuable. Electronic components often contain precious metals such as palladium, gold, silver, and copper. Gold is the most economically useful asset, accounting for around 50% of potential resale revenue (Al Sultan & Benli, 2023). E-waste is primarily disposed of through three sources: mineral beneficiation technologies, such as chemical and metallurgical procedures, are used for secondary recovery of PCBs, spent batteries, and LCD/LED panels (Kudpeng et al., 2020). E-waste contains a wide range of perilous and harmless compounds, such as polymers, glass fiber, flame retardants, ferrous and nonferrous metals (Mishra et al., 2023). Inappropriate disposal can be harmful to both persons and the environment (Dutta et al., 2022). These metals have excellent electrical conductivity and chemical resistance, making them useful in a variety of industries including electronics, computers, automotive, jewelry, dental, and aviation. According to some, Numbers show that there is between 10 and 1000 g of gold for every ton of e waste, which is 17 times more than mineral ore (Van Yken et al., 2021). The source of these materials is concentrated in a few lands. For example, South Africa supplies the great substance (70–90%) of Platinum Group Elements (PGE). China accounts for 58% of the global primary mining of rare earth elements, presenting a grave scenario. China develops 80% of the world's Rare Earth Elements (REE) from other countries (Pawar & Ewing, 2022). Electronic garbage is commonly known as e-waste, a growing industrial danger due to the rising use of circuit boards and screens. With the right equipment, waste may be a sustainable source of precious metals. Selectivity for valuable metals is critical for this solution, surpassing capacity. A porphyrin-based polymer with selective binding and a reductive method may recycle valuable metals, including gold, from e-waste at an unprecedented pace (Hong et al., 2020). Urban mining, which collects and purifies rich metals from waste streams, is a crucial resource security strategy in response to increased global demand and limited geological resources. Recycling e-waste offers a cost-effective alternative to mining and contains higher metal concentrations than ores (Zeng et al., 2018; Bal et al., 2018; Sahu et al., 2024). Main PGM is in high demand globally due to its various applications. Catalytic converters are widely used in automobiles to reduce dangerous exhaust pollution. Catalytic converters are often "thrifted" by changing their composition based on PGM market prices. However, stringent automobile emission restrictions are projected to limit further decreases in PGM loadings. Fuel cells will power the next group of "clean" hydrogen-powered cars. PGM electrocatalysts are utilized in both anodic and cathodic processes in low-temperature hydrogen fuel cells. Due to limited resources, there may be competing needs for carbon-neutral transportation now and in the future.

E-waste contains harmful compounds such as lead and mercury that represent substantial environmental dangers if not properly managed (Wang et al., 2023). E-waste contains abandoned items such as computers, televisions, and mobile phones that contain dangerous elements like lead and cadmium (Kawasaki & Barati, 2024). Copper occurs naturally as minerals such as covellite, chalcocite, azurite, chalcopyrite, and malachite. Furthermore, because copper occurs in the form of minerals, metallurgical processing is required to separate metallic copper from these minerals. Copper is commonly utilized in the manufacturing of wires, cables, and metal alloys due to its ductility, malleability, and strong electrical conductivity. Critical raw materials like REE can be obtained by drawing out and dispensation natural ores or recovering them from spent goods like e-waste (Parajuly et al., 2020). Common sources of e-waste include domestic appliances, workplace equipment, and consumer electronics such as computers and mobile phones, which have short lifespans and are constantly replaced (Chatterjee & Abraham, 2017).

E-waste is off the record into various categories, including domestic appliances, office machines, and consumer items. The increase of the problem is caused by a lack of appropriate disposal facilities for e-waste. Examples of consumer electronics include mobile phones, Tablets, Personal computers and laptops, Televisions (CRT, LED, LCD), Printers and scanners, cameras, and audio equipment (radios, headphones, speakers). The quick update cycle in consumer electronics, driven by innovation and shorter product lifespans, creates a substantial volume of e-waste. There are two sorts of home appliances. Large appliances with examples: Refrigerators, washers, air conditioners, and dishwashers. Small appliances with examples: Vacuum cleaners, toasters, coffee machines, irons, and electrical fans. Obsolete appliances abandoned due to deterioration, inefficiency, or replacement with energy-efficient ones contribute to mass e-waste. IT and telecom equipment examples include routers and modems, servers, hard drives and memory devices, networking cables, and monitors. The continual desire for quicker and more secure IT infrastructure results in frequent replacement, causing enormous e-waste. Lighting devices include LED and fluorescent bulbs, halogen lamps, and compact fluorescent lamps (CFLs). Discarded lighting devices are a source of dangerous compounds, such as mercury, requiring careful disposal. Medical equipment like x-ray machines, dialysis machines, MRI scanners, ECG devices, and laboratory analyzers. Hospitals and clinics trash old or nonfunctional equipment, contributing to the e-waste problem. Industrial equipment examples include machinery control systems, power tools, assembly line robotics, and electric motors. Industries often modernize machinery to maintain efficiency and decrease downtime resulting in enormous e-waste. Batteries like lithium-ion batteries lead-aciate batteries, and nickel-cadmium batteries are discarded gadgets and electric cars, are toxic and contribute to e-waste through inappropriate disposal, and are toxic and contribute to e-waste through inappropriate disposal (Priyadarsini & Das, 2024, 2025). Office equipment examples include photocopiers, shredders, fax machines, projectors, and desktop PCs. Offices and organizations routinely replace outmoded technology, resulting in enormous e-waste generation. Wearabl devices and smart tools include smartwatches, fitness trackers, and cybernetic reality headsets.

The growing popularity of wearables with short lifespans contributes to increased e-waste in this category. Cables & Accessories include USB cables, chargers, earphones, adaptors, and power strips. Nonfunctional or mismatched cables and accessories frequently wind up in landfills. Miscellaneous devices include toys with embedded electronics, electronic musical instruments, gaming consoles, and digital clocks. E-waste is created when lifestyle-related electronic goods are abandoned due to upgrades or malfunctions. As of 2021, global e-waste production stands at around 53.6 million metric tons, with forecasts to reach 74 million by 2030.

3 Effects of E-Waste

Inappropriate e-waste disposal in landfills and public health hazards (Adetunji & Olaniran, 2021). E-waste contains toxic organic and inorganic impurities, which are exposed to ecosystems, vegetation, and animals. These compounds can be discharged as heavy metals, acids, or deadly chemicals (Li et al., 2020, Das et al., 2011; Ghosh et al., 2016b, 2023a, 2023b). The gasp of tiny and rough particles, skin contact with unsafe materials or swallowing of contaminated dust all result in direct exposure. Recycling electronic waste reduces reliance on primary metal sources, promoting a circular economy and reducing environmental impact (Malhotra & Jain, 2023).

Exposure to e-waste increases its concentration in the food chain, posing health risks including DNA damage, birth deformities, heart failure, cancer, and skin diseases (Das & Mishra, 2008, 2009; Paul et al., 2023a, 2023b). Improper e-waste disposal discharges dust particles and chemicals into the environment, causing air pollution and lung damage (Li et al., 2020). E-waste incineration emits toxic gases that affect the environment. Hazardous substances in e-waste can pollute surface and subterranean water, making it corrosive and poisonous (Mishra et al., 2022b, 2024a, 2024b; Mishra & Das, 2021). This has a deleterious influence on marine and freshwater species, biodiversity, ecosystems, and soil ecology (Rautela et al., 2021). Developing countries frequently lack the infrastructure for effective e-waste disposal, with over 90% of e-waste being treated informally (Bhat et al., 2023). Toxic ingredients in e-waste, such as lead and cadmium, pose serious health concerns to workers and local communities. Improper disposal methods, such as incineration and landfilling, have major environmental consequences (Dhivya & Premalatha, 2024). Improper dumping of e-waste poses significant health risks because dangerous substances may escape into the environment (Jain et al., 2023). Formal recycling techniques, which are popular in poorer countries, exacerbate these risks, resulting in substantial environmental degradation (Ogboru, 2014). Heavy metal emissions can damage plants and animals across the food chain, making them a significant environmental pollutant with detrimental consequences for human health and the ecosystem (Giese et al., 2021). Cadmium is known to cause kidney damage and act as a carcinogen, cadmium exposure is especially harmful to the renal system, potentially leading to chronic kidney disease. Lead exposure, especially in children can cause developmental abnormalities, reduced IQ, and various cognitive deficiencies. Mercury (Hg) is heavy metal, neurotoxic, and can negatively impair brain function, causing symptoms such as quakes and intellectual deterioration. Long-term arsenic exposure has been associated to an amplified hazard of cancer, skin lesions, and developmental problems in children (Morais et al., 2012).

4 Recovery Techniques

E-waste reprocessing strategies have been documented in America (Li et al., 2015). Microbial approaches have been used in biotechnology since prehistoric times, in addition to traditional methods (Bala et al., 2024; Das et al., 2023; Dey et al., 2023a; Kanungo et al., 2023; Mishra et al., 2020). Microbial leaching, particularly copper leaching, has been used for mine water since the Greek and Roman periods, maybe over 2000 years ago. Bacterial leaching occurs when a bacterial cell makes physical contact with a mineral surface, resulting in enzymatic activity. Bacteria favor crystal defective areas, however, the process of adhesion and metal solubility is not entirely understood. Electrochemical interactions occur throughout the mineral surface, and metal solubility is the outcome (Jones et al., 2024). Figure 1 depicts different recycling techniques for the recovery of metal from e-waste. Bacteria and fungi aid the bioleaching process by oxidizing metal sulfides, resulting in metal solubilization

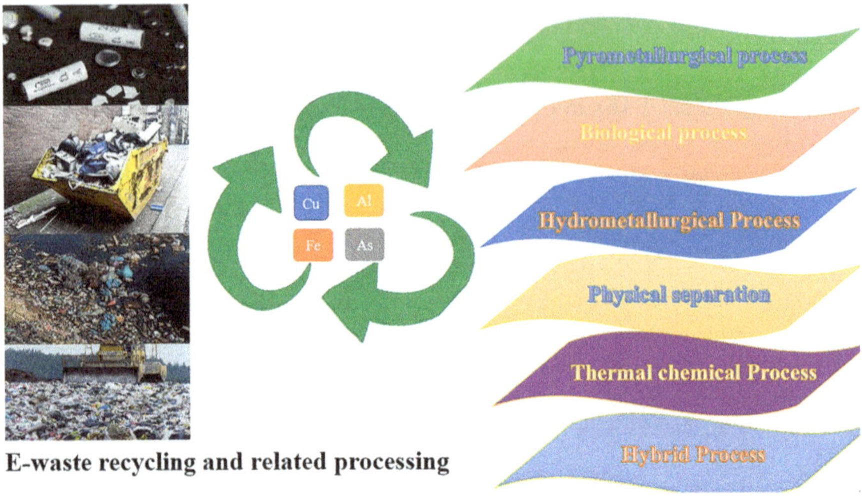

Fig. 1 Depicts different recycling techniques for the recovery of metal from e-waste

(Das & Ghosh, 2018, 2023c, 2023d, 2023g). Certain microbial consortia have been demonstrated to effectively mobilize heavy metals, including copper, from mining tailings. Enzymatic bio-nanoparticle production is limited for aggressive PGM wastes, such as those resulting from acid dissolution. In comparison to traditional mining technologies, microbial leaching is less expensive and has a lower environmental impact (Das & Ghosh, 2023a, 2023b, 2023f, 2023h). It can clean up contaminated sites by converting hazardous materials into less dangerous forms, facilitating ecosystem restoration (Bala et al., 2024). Indirect bioleaching happens when bacteria produce a lixiviate, which oxidizes the mineral. The interaction between *A. ferrooxidans* cells and 500–1000 μm PCB particles was thought to be positive if the van der Waals attraction force was stronger than the electrostatic repulsive force, which occurs in high ionic strength solutions. *A. ferrooxidans* used waste PCBs to bioleach copper by directly oxidizing elemental copper after bio-oxidizing ferrous ions. *A. ferrooxidans* converts ferrous ions (Fe^{2+}) to ferric ions (Fe^{3+}), which are necessary for extracting copper from sulfide minerals in PCBs (Tonietti et al., 2024). Table 1 describes the bioleaching efficiency of various microorganisms for metal recovery from e-waste.

The ferric ions can react with copper sulfides, leading to copper solubilization (Vardanyan et al., 2023). *A. ferrooxidans* can achieve copper recovery rates of over 95% under ideal conditions, including regulated pH and temperature. The bacteria can adapt to high copper concentrations and improve bioleaching performance through immobilization techniques that limit metal toxicity.

The complexolysis mechanism includes the way in which fungus cyanogenic microorganisms create cyanide during the latter stationary phase (Mishra et al., 2021). Metal recapture from e-waste saves primary resources, protects the environment, and contributes to the spherical economy. E-waste has been designated as a secondary resource in municipal mining due to the presence of valuable metals at

Table 1 Describes the bioleaching efficiency of various microorganisms for metal recovery from e-waste

S.L. no.	Microorganism	Metal source	Recovery efficiency	Bioleaching condition	Reference
1	*Pseudomonas biofilm*	Spent computer PCBs	Ag: 14.7%	Pulp density: 2% pH: 7.0 Days: 7	Hu et al. (2021)
2	*Pseudomonas putida*	Spent PCBs	Au: 44%	Pulp density: 1% pH: 7.3–8.6 Days: 2	Isildar (2016)
3	*Bacillus foraminis*	Smartphones	Ag:100% Ag: 56.8% Mo: 41.4%	Days: 12 Rpm:160 C: 40	Golzar-Ahmadi and Mousavi (2021)
4	*Pseudomonas fluorescens*	Spent PCBs	Au: 54%	pH: 9.0 Pulp density: 0.33%	Li et al. (2020)
5	*Pseudomonas balearica* SAE1	Spent PCBs	Au: 68.5% Ag: 33.8%	pH:9.0 Pulp density: 10 g/L	Kumar, et al. (2018)
6	*Aspergillus fumigatus* A2DS	Spent mobile phone PCB	Ni: 42.37%,32.29%,27.07% Cu: 62%,58%,61.8%	Pulp density: 0.5% pH: 6.0	Patel and Lakshmi (2021)
7	*Penicillium simplicissimum*	Spent mobile PCBs	Cu: 90% Ni: 89%	Cu: pH: 7.0 Ni: pH: 2.0	Arshadi et al. (2019)
8	*Aspergillus tubingensis*	Spent computer PCBs	Cu: 34% Zn: 54% Ni: 8%	pH: 5.0 Pulp density: 0.25–1% Days: 33	Trivedi and Hait (2020)
9	*Phanerochaete chrysosporium*	Spent PCBs	Cu: 60.96%	pH: 5.0 Pulp density: 2% Days:14	Liu et al. (2020)
10	*Candida orthopsilosis*	Spent cellphone PCBs	Cu: 1%	pH: 4.4; 0.5 g Days: 45	Díaz-Martínez et al. (2019)
11	*Aspergillus niger*	Spent mobile phone batteries	Li: 100% Mn: 72% Mn: 45% Ni: 38% Co: 94% Cu: 62%	Days: 9 pH: 4.6	Bahaloo-Horeh et al. (2018)
12	*Aspergillus niger*	Spent PCBs	Zn:100% Ni: 80.39% Cu: 85.88%	Pulp density: 0.5–20 g/L	Faraji et al. (2018)
13	*Trichoderma atroviride*	Spent computer PCB	Pd: 1% Au: 13%	pH: 5.0 Days: 30	Díaz-Martínez et al. (2019)
14	*Saccharomyces cerevisiae*	PCBs crap	Au: 98%	5.0×10^{14} cells; 10 min	Saitoh et al. (2018)

higher absorptions than those in primary ore. The cyanogenic characteristics of *C. violaceum* kind it ideal for metal recovery procedures such as gold extraction from ores. The aptitude of the bacterium to convert gold nanoparticles to soluble forms may improve metal dissolution rates (McGivney et al., 2018). The bacterium has the possibility to recover metals from e-waste, which is widely produced globally (Das et al., 2024b; Das & Ghosh, 2023e, 2023i; Nag et al., 2023). Microorganisms struggle to efficiently extract various metals from complicated e-waste substrates. Mixed cultures and microbial consortia are becoming increasingly popular. Different microbial strains are needed for bioleaching base metals compared to precious metals. Using mixed cultures can promote synergy among microorganisms, improving nutrient availability and functional robustness. Fungi, which produce biogenic cyanide or organic acids, play a critical role in recovering valuable metals from e-waste due to their ability to act as ordinary chelating and complexing agents (Beiki et al., 2023). Organic acids are also known for being biodegradable and less corrosive than inorganic acids. Dilute carboxylic acid solutions, depending on their pKa values, are moderately acidic in the pH 3–5 range. Fungi can withstand hazardous metals and grow in a wide pH range. Bio-cyanidation using *A. niger* is a sustainable method for extracting Au and Ag. Bioleaching can occur under both aerobic and anaerobic conditions. Temperature, pulp density, and pH significantly influence the efficiency of this process.

Despite having over 90% recyclable compounds, Brazil lacks reusing plants and a designed management system, despite the passage of the Solid Waste National Policy requiring e-waste reverse logistics. Some e-waste is sold in informal markets, while others are disposed of in landfills, despite environmental concerns. A small part is collected, crushed, and exported for metal recovery (Zeng et al., 2012). The Brazilian Agency for Manufacturing Development's Technical Feasibility Study found that recycling electronic waste poses a difficulty due to the high cost of reverse logistics in an intricate continental country. Recycling must be cost-effective and profitable to compensate for transportation distances. Lack of venture in recycling plants prevents gold recovery, which might fund the process. E-waste assembly businesses can only export after eliminating PCBs. While precious metals are the primary focus of e-waste recycling, bulky polymers are also valuable and should be considered (Priyadarsini & Das, 2025). Plastics make up nearly 37% of WEEE. However, recovering e-waste plastics is challenging due to their mixed material configuration and synthetic polymer properties. E-waste plastics typically contain ten different polymer types, including high-impact polystyrene (HIPS, 27%), polycarbonate (PC, PC-ABS, 7%), acrylonitrile butadiene styrene (ABS, 24%), polyethylene (PE, 2%), and polypropylene (PP, 5%), as well as polymer blends and other thermoplastics (Baldé et al., 2017). Various methods are employed for metal recovery, including mechanical pretreatment and electrochemical leaching (Blumbergs et al., 2024). Biohydrometallurgy offers an environmentally friendly approach to copper recovery (Iglesias-González et al., 2024). Chemical leaching using ionic liquids and other solvents, achieving over 90% recovery efficiency (Oke & Potgieter, 2024; Das et al., 2012a). Hydrometallurgy uses chemical leaching to recover metals from e-waste, which is especially successful for valuable metals like gold and silver. Pyrometallurgy is a high-temperature process that extracts metals from materials such as printed circuit boards (Tipre et al., 2021).

While the potential for high-metal recovery from PCBs is great, there are still concerns about the recycling methods' economic feasibility and environmental impact. Balancing these aspects is critical for sustainable e-waste supervision procedures. Despite limits, improper e-waste disposal remains a problem, endangering the environment and human health (Castro & Bassin, 2022). Effective e-waste management systems are vital for recycling and reducing the environmental impact of hazardous materials. Current techniques usually prioritize precious metals while ignoring the recovery of critical basic elements. While standards like RoHS strive to decrease the dangers associated with hazardous substances in electronics, the ongoing difficulties of e-waste management and recycling highlight the need (Oke & Potgieter, 2024).

5 Conclusion

The recovery of valuable metals from e-waste using biotechnological technologies provides a long-term and ecologically responsible answer to the rising problem of electronic waste management. Microorganisms, with bacteria and fungi, remove valuable metals like gold, silver, platinum, and palladium from unrestricted electronics through metabolic processes. Biotechnological methods for recovering precious metals from e-waste are a worthwhile and ecological alternative to traditional mining and chemical extraction.

These approaches are environmentally benign since they eliminate the need for harsh chemicals such as cyanide and aqua regia, which are commonly used in conventional processes. Precious metals can be successfully recovered with minimal environmental impact by utilizing microorganisms capable of bioleaching and biosorption. Furthermore, this technology promotes the circular economy by transforming trash into resources, so addressing both resource scarcity and environmental degradation. Despite their potential, biotechnological techniques face challenges such as process efficiency, metal selectivity, and e-waste optimization. Microbial systems often operate at slower rates than chemical processes, necessitating advances in genetic modification, bioreactor design, and process optimization. Furthermore, implementing these solutions into existing e-waste recycling systems requires a large investment in R&D. The biotechnological salvage of precious metals from e-waste is a unique and ecologically friendly approach with the potential to revolutionize waste management and resource recovery. While challenges remain, future advances in biotechnology, together with supportive legislation and public awareness, have the potential to play a significant role in achieving a justifiable and resource-efficient future. Investing in this field can help society address the dual issues of e-waste management and precious metal scarcity.

References

Adetunji, A. I., & Olaniran, A. O. (2021). Treatment of industrial oily wastewater by advanced technologies: A review. *Applied Water Science, 11*(6), 98. https://doi.org/10.1007/s13201-021-01430-4

Al Sultan, M. S., & Benli, B. (2023). Recent sustainable trends for e-waste bioleaching. *Physicochemical Problems of Mineral Processing, 59*(5). https://doi.org/10.37190/ppmp/167375

Al-Salem, S. M., Leeke, G. A., El-Eskandarany, M. S., Van Haute, M., Constantinou, A., Dewil, R., & Baeyens, J. (2022). On the implementation of the circular economy route for E-waste management: A critical review and an analysis for the case of the State of Kuwait. *Journal of Environmental Management, 323*, 116181. https://doi.org/10.1016/j.jenvman.2022.116181

Arshadi, M., Yaghmaei, S., & Mousavi, S. M. (2019). Study of plastics elimination in bioleaching of electronic waste using Acidithiobacillus ferrooxidans. *International journal of Environmental Science and Technology, 16*, 7113–7126. https://doi.org/10.1007/s13762-018-2120-1

Bahaloo-Horeh, N., Mousavi, S. M., & Baniasadi, M. (2018). Use of adapted metal tolerant Aspergillus niger to enhance bioleaching efficiency of valuable metals from spent lithium-ion mobile phone batteries. *Journal of Cleaner Production, 197*, 1546–1557. di.

Bal, B., Ghosh, S., & Das, A. P. (2018). Microbial recovery and recycling of manganese waste and their future application: A review. *Geomicrobiological Journal, 36*(1), 85–96. https://doi.org/10.1080/01490451.2018.1497731

Bala, K., Dogra, A., Lata, P., & Das, A. P. (2024). Discussing sources and biological-chemical remediation approaches for microplastic pollution. In *Sustainable microbial technology for synthetic and cellulosic microfiber bioremediation* (pp. 37–59). https://doi.org/10.1007/978-3-031-62600-5_3

Baldé, C. P., Forti, V., Gray, V., Kuehr, R., & Stegmann, P. (2017). *The global e-waste monitor 2017: Quantities, flows and resources*. United Nations University, International Telecommunication Union, and International Solid Waste Association. https://doi.org/10.59117/20.500.11822/44316

Batista, B. B., de Lima, V. M., Will, W. R., Fang, F. C., & da Silva Neto, J. F. (2025). A cytochrome bd repressed by a MarR family regulator confers resistance to metals, nitric oxide, sulfide, and cyanide in Chromobacterium violaceum. *Applied and Environmental Microbiology*, e02360-24. https://doi.org/10.1101/2024.08.06.606881

Beiki, V., Naseri, T., & Mousavi, S. M. (2023). An efficient approach for enhancement of gold and silver bioleaching from spent telecommunication printed circuit boards using cyanogenic bacteria: Prevention of biofilm formation. *Waste Management, 171*, 590–598. https://doi.org/10.1016/j.wasman.2023.10.007

Bhat, A. P., Dhoble, S. J., Vengadaesvaran, B., & Rewatkar, K. G. (2023). E-waste: Application ethics towards reutilizations. In *360-Degree waste management* (Vol. 2, pp. 169–201). Elsevier. https://doi.org/10.1016/b978-0-323-90909-9.00003-4

Bhattacharjee, I., Hayashi, K., & Hirata, S. (2021a). Key of suppressed triplet nonradiative transition-dependent chemical backbone for spatial self-tunable afterglow. *JACS Au, 1*(7), 945–954. https://doi.org/10.1021/jacsau.1c00132

Bhattacharjee, J., Mishra, S., & Das, A. P. (2021b). Recent advances in sensor-based detection of toxic dyes for bioremediation application: A review. *Applied Biochemistry and Biotechnology, 2021*, 1–20. https://doi.org/10.1007/S12010-021-03767-7

Blumbergs, E., Shishkin, A., Markus, K., Serga, V., Goljandin, D., Klauson, A., et al. (2024). Economic aspects of mechanical pre-treatment's role in precious metals recovery from electronic waste. *Metals, 14*(1), 95. https://doi.org/10.3390/met14010095

Castro, F. D., & Bassin, J. P. (2022). Electronic waste: Environmental risks and opportunities. In *Hazardous waste management* (pp. 421–458). Elsevier. https://doi.org/10.1016/b978-0-12-824344-2.00002-1

Chalana, A., Singh, K., Sharma, S., Bhardwaj, V., & Rai, R. K. (2023). E-waste management: Prospects and strategies. In *Microbial technology for sustainable e-waste management* (pp. 303–318). Springer. https://doi.org/10.1007/978-3-031-25678-3_19

Chatterjee, A., & Abraham, J. (2017). Efficient management of e-wastes. *International journal of Environmental Science and Technology, 14*, 211–222. https://doi.org/10.1007/s13762-016-1072-6

Das, A. P., & Ghosh, S. (2018). Bioleaching of manganese from mining waste materials. *Materials Today: Proceedings, 5*(1), 2381–2390. https://doi.org/10.1016/J.MATPR.2017.11.459

Das, A. P., & Ghosh, S. (2022). Role of microorganisms in extenuation of mining and industrial wastes. *Geomicrobiology Journal, 39*(3–5), 173–175. https://doi.org/10.1080/01490451.2022.2038953

Das, A. P., & Ghosh, S. (2023a). Cost-effectiveness of the biomining process. In *Manganese mining microorganisms* (pp. 143–151). https://doi.org/10.1016/B978-0-12-822147-1.00008-X

Das, A. P., & Ghosh, S. (2023b). Evolution of biomining technology. In *Manganese mining microorganisms* (pp. 5–19). https://doi.org/10.1016/B978-0-12-822147-1.00012-1

Das, A. P., & Ghosh, S. (2023c). Factors affecting bioleaching. In *Manganese mining microorganisms* (pp. 77–86). https://doi.org/10.1016/B978-0-12-822147-1.00004-2

Das, A. P., & Ghosh, S. (2023d). Mechanism of manganese bioleaching. In *Manganese mining microorganisms* (pp. 87–98). https://doi.org/10.1016/B978-0-12-822147-1.00014-5

Das, A. P., & Ghosh, S. (2023e). Metagenomic insights into unculturable microbial diversity. In *Manganese mining microorganisms* (pp. 63–76). https://doi.org/10.1016/B978-0-12-822147-1.00002-9

Das, A. P., & Ghosh, S. (2023f). Microorganisms in manganese biomining processes. In *Manganese mining microorganisms* (pp. 55–61). https://doi.org/10.1016/B978-0-12-822147-1.00013-3

Das, A. P., & Ghosh, S. (2023g). Mixed culture bioleaching: An insight into manganese biomining process and efficacy. In *Manganese mining microorganisms* (pp. 129–141). https://doi.org/10.1016/B978-0-12-822147-1.00001-7

Das, A. P., & Ghosh, S. (2023h). Role of metabolites and metabolomics for understanding of manganese bioleaching. In *Manganese mining microorganisms* (pp. 117–127). https://doi.org/10.1016/B978-0-12-822147-1.00011-X

Das, A. P., & Ghosh, S. (2023i). *Manganese mining microorganisms*.

Das, A. P., & Mishra, S. (2008). Hexavalent chromium (VI): Health hazards and environmental pollutant. *Journal of Environmental Research and Development, 2*(3), 386–392.

Das, A. P., & Mishra, S. (2009). Hexavalent chromium [Cr (VI)]: Yellow water pollution and its remediation. *Sarovar Saurabh ENVIS Newsletter on Wetland Ecosystems, 5*(2).

Das, A. P., & Swain, S. (2013). Algal biosorption of toxic dye methylene blue. In *A potential source of food, feed, biochemicals, biofuels and biofertilizers. International Conference on Algal Biorefinery* (Vol. 13). Indian Institute of Technology. https://doi.org/10.7726/ajbb.2013.1002

Das, A. P., Sukla, L. B., Pradhan, N., & Nayak, S. (2011). Manganese biomining: A review. *Bioresource Technology, 102*(16), 7381–7387. https://doi.org/10.1016/J.BIORTECH.2011.05.018

Das, A. P., Swain, S., Panda, S., Pradhan, N., & Sukla, L. B. (2012a). Reductive acid leaching of low grade manganese ores. *Geomaterials, 2012*(04), 70–72. https://doi.org/10.4236/GM.2012.24011

Das, A. P., Sukla, L. B., & Pradhan, N. (2012b). Microbial recovery of manganese using *Staphylococcus epidermidis*. *International Journal of Nonferrous Metallurgy, 2012*(02), 9–12. https://doi.org/10.4236/IJNM.2012.12002

Das, A. P., Ghosh, S., Mohanty, S., & Sukla, L. B. (2015). Consequences of manganese compounds: A review. *Toxicological & Environmental Chemistry, 96*(7), 981–997. https://doi.org/10.1080/02772248.2015.1005428

Das, A. P., Mallavarapu, M., & Ghosh, S. (2023). Ecotoxicity of mining pollutants on the environment and their remediation. *Environmental Chemistry and Ecotoxicology, 5*, 165–167. https://doi.org/10.1016/J.ENCECO.2023.08.002

Das, A., Dey, S., & Das, A. P. (2024a). Ecotoxicological impacts of synthetic microfiber pollutants and development of sustainable mitigation strategies. *Environmental Chemistry and Ecotoxicology, 7*, 201. https://doi.org/10.1016/J.ENCECO.2024.12.003

Das, A., Dey, S., Yadav, R. N., Dutta, P., Dhiman, S., Boruah, P. J., et al. (2024b). Unleashing naphthopyranopyrimidine's anticancer potential: A deep eutectic solvent (DES) study. *New Journal of Chemistry, 48*(17), 7566–7578. https://doi.org/10.1039/d4nj00796d

Dey, S., Das, A., Mallick, K., Sahu, A., & Das, A. P. (2023a). Environmental petroleum waste: Pollution, toxicity, sustainable remediation (pp. 159–175). https://doi.org/10.1007/978-3-031-48220-5_7

Dey, S., Tripathy, B., Kumar, M. S., & Das, A. P. (2023b). Ecotoxicological consequences of manganese mining pollutants and their biological remediation. *Environmental Chemistry and Ecotoxicology, 5*, 55. https://doi.org/10.1016/J.ENCECO.2023.01.001

Dey, S., Bissoyi, A., Koteswara Reddy, G., & Das, A. P. (2024a). Synthetic microfibers: An urban emerging pollutant demanding for sustainable environmental solutions. In *Sustainable microbial technology for synthetic and cellulosic microfiber bioremediation* (pp. 107–121). https://doi.org/10.1007/978-3-031-62600-5_7

Dey, S., Dash, A., & Das, A. P. (2024b). Bioremediation as a sustainable and circular solution for mining waste management and bioenergy generation. In *Sustainable management of mining waste and tailings* (pp. 107–119). CRC Press. https://doi.org/10.1201/9781003442455-6

Dhivya, K., & Premalatha, G. (2024). E-waste challenges & solutions. In *Sustainable management of electronic waste* (pp. 255–275). https://doi.org/10.1002/9781394166923.ch13

Díaz-Martínez, M. E., Argumedo-Delira, R., Sánchez-Viveros, G., Alarcón, A., & Mendoza-López, M. R. (2019). Microbial bioleaching of Ag, Au and Cu from printed circuit boards of mobile phones. *Current Microbiology, 76*, 536–544. https://doi.org/10.1007/s00284-019-01646-3

Dutta, D., Goel, S., & Kumar, S. (2022). Health risk assessment for exposure to heavy metals in soils in and around E-waste dumping site. *Journal of Environmental Chemical Engineering, 10*(2), 107269. https://doi.org/10.1016/j.jece.2022.107269

Dutta, D., Rautela, R., Gujjala, L. K. S., Kundu, D., Sharma, P., Tembhare, M., & Kumar, S. (2023). A review on recovery processes of metals from E-waste: A green perspective. *Science of the Total Environment, 859*, 160391. https://doi.org/10.1016/j.scitotenv.2022.160391

Faraji, F., Golmohammadzadeh, R., Rashchi, F., & Alimardani, N. (2018). Fungal bioleaching of WPCBs using Aspergillus niger: Observation, optimization and kinetics. *Journal of Environmental Management, 217*, 775–787. https://doi.org/10.1016/j.jenvman.2018.04.043

Ghosh, S., & Das, A. P. (2017). Bioleaching of manganese from mining waste residues using Acinetobacter sp. *Geology, Ecology, and Landscapes, 1*(2), 77–83. https://doi.org/10.1080/24749508.2017.1332847

Ghosh, S., Mohanty, S., Akcil, A., Sukla, L. B., & Das, A. P. (2016a). A greener approach for resource recycling: Manganese bioleaching. *Chemosphere, 154*, 628–639. https://doi.org/10.1016/j.chemosphere.2016.04.028

Ghosh, S., Mohanty, S., Nayak, S., Sukla, L. B., & Das, A. P. (2016b). Molecular identification of indigenous manganese solubilising bacterial biodiversity from manganese mining deposits. *Journal of Basic Microbiology, 56*(3), 254–262. https://doi.org/10.1002/jobm.201500477

Ghosh, S., Gandhi, M., van Hullebusch, E. D., & Das, A. P. (2020). Proteomic insights into Lysinibacillus sp.-mediated biosolubilization of manganese. *Environmental Science and Pollution Research, 28*(30), 40249–40263. https://doi.org/10.1007/S11356-020-10863-4

Ghosh, B., Pham, L., Gubareva, M., & Teplova, T. (2023a). Energy transition metals and global sentiment: Evidence from extreme quantiles. *Resources Policy, 86*, 104170. https://doi.org/10.1016/j.resourpol.2023.104170

Ghosh, S., Tripathy, B., Dey, S., & Das, A. P. (2023b). Comparative investigation of fungal and bacterial manganese biomining mechanisms. In *Manganese mining microorganisms* (pp. 99–115). https://doi.org/10.1016/b978-0-12-822147-1.00016-9

Giese, G., Nagy, Z., & Lee, L. E. (2021). Deconstructing ESG ratings performance: Risk and return for E, S, and G by time horizon, sector, and weighting. *Journal of Portfolio Management, 47*(3), 94–111. https://doi.org/10.3905/jpm.2020.1.198

Golzar-Ahmadi, M., & Mousavi, S. M. (2021). Extraction of valuable metals from discarded AMOLED displays in smartphones using Bacillus foraminis as an alkali-tolerant strain. *Waste Management, 131*, 226–236. https://doi.org/10.1016/j.wasman.2021.06.006

Gongwei, Q., Yueqiao, L., Yong, C., Yuedong, L., Huijuan, Y., Xiaoqin, S., et al. (2023). Determination of four certain phthalates in the electronic and electrical product by high performance liquid chromatography. *Academic Journal of Materials & Chemistry, 4*(2), 59–63. https://doi.org/10.25236/ajmc.2023.040210

Goyal, S., Sharma, S., Kharayat, P. S., & Gupta, S. (2023, December). E-waste: Promoting the need for green electronics. In *2023 International Conference on Power Energy, Environment & Intelligent Control (PEEIC)* (pp. 326–329). IEEE. https://doi.org/10.1109/peeic59336.2023.10451991

Hong, Y., Thirion, D., Subramanian, S., Yoo, M., Choi, H., Kim, H. Y., et al. (2020). Precious metal recovery from electronic waste by a porous porphyrin polymer. *Proceedings of the National Academy of Sciences, 117*(28), 16174–16180. https://doi.org/10.1073/pnas.2000606117

Hu, J., Tang, Y., Ai, F., Lin, M., & Ruan, J. (2021). Biofilm for leaching precious metals from waste printed circuit boards using biocyanidation technology. *Journal of Hazardous Materials, 403*, 123586. https://doi.org/10.1016/j.jhazmat.2020.123586

Ichikowitz, R., & Hattingh, T. S. (2020). Consumer e-waste recycling in South Africa. *South African Journal of Industrial Engineering, 31*(3), 44–57. https://doi.org/10.7166/31-3-2416

Iglesias-González, N., Dorado, A. D., Ramírez, P., & Mazuelos, A. (2024). A high productivity bioprocess for obtaining metallic copper from printed circuit boards (PCBs). *Minerals Engineering, 205*, 108459. https://doi.org/10.1016/j.mineng.2023.108459

Isildar, A. (2016). *Biological versus chemical leaching of electronic waste for copper and gold recovery* (Doctoral dissertation, Université Paris-Est; Università degli studi (Cassino, Italie)). https://doi.org/10.1201/9780429023903

Jadaun, J. S., Bansal, S., Sonthalia, A., Rai, A. K., & Singh, S. P. (2022). Biodegradation of plastics for sustainable environment. *Bioresource Technology, 347*, 126697. https://doi.org/10.1016/j.biortech.2022.126697

Jagatee, S., Priyadarshini, S., Rath, C. C., & Das, A. P. (2024a). Synthetic microfiber: An enduring environmental problem linked to sustainable development. In *Renewable energy generation and value addition from environmental microfiber pollution through advanced greener solution* (pp. 93–112). Springer. https://doi.org/10.1007/978-3-031-51792-1_6

Jagatee, S., Priyadarshini, S., Rath, C. C., & Das, A. P. (2024b). Synthetic microfiber: An enduring environmental problem linked to sustainable development. *Environmental Science and Engineering, Part, F2333*, 93–112. https://doi.org/10.1007/978-3-031-51792-1_6

Jain, M., Kumar, D., Chaudhary, J., Kumar, S., Sharma, S., & Verma, A. S. (2023). Review on E-waste management and its impact on the environment and society. *Waste Management Bulletin, 1*(3), 34–44. https://doi.org/10.1016/j.wmb.2023.06.004

Jones, T. R., Poitras, J., Levett, A., da Silva, G., Gunathunga, S., Ryan, B., et al. (2024). Microbe-mineral interactions within kimberlitic fine residue deposits: Impacts on mineral carbonation. *Frontiers in Climate, 6*, 1345085. https://doi.org/10.3389/fclim.2024.1345085

Kanungo, J., Sahoo, T., Swain, L. P., & Behera, I. D. (2023). *Toxicity of persistent hydrocarbon pollutants, sources and sustainable remediation process* (pp. 39–65). https://doi.org/10.1007/978-3-031-48220-5_2.

Kawasaki, T., & Barati, M. (2024, August). Sulfidation reaction of metallic copper in flash smelting furnace conditions. In *Conference of Metallurgists* (pp. 841–849). Springer. https://doi.org/10.1007/978-3-031-67398-6_143

Khatun, R., Singh, S., Dubey, N. K., & Prasad Das, A. (2023). A review on marine-based phytochemicals and their application in biomedical research. In *Recent frontiers of phytochemicals: Applications in food, pharmacy, cosmetics, and biotechnology* (pp. 383–395). Elsevier. https://doi.org/10.1016/B978-0-443-19143-5.00013-X

Koti, A., Kolluri, K., Khare, A., Khare, P., & Meduri, A. (2023, March). Computational framework for the sustainable management and optimization of E-waste. In *2023 10th International*

Conference on Computing for Sustainable Global Development (INDIACom) (pp. 1263–1266). IEEE. https://doi.org/10.1007/978-3-031-67398-6_143

Kudpeng, K., Bohu, T., Morris, C., Thiravetyan, P., & Kaksonen, A. H. (2020). Bioleaching of gold from sulfidic gold ore concentrate and electronic waste by Roseovarius tolerans and Roseovarius mucosus. *Microorganisms, 8*(11), 1783. https://doi.org/10.1007/978-3-031-67398-6_143

Kumar, S., Stecher, G., Li, M., Knyaz, C., & Tamura, K. (2018). MEGA X: molecular evolutionary genetics analysis across computing platforms. Molecular biology and evolution, 35(6), 1547–1549. https://doi.org/10.1093/molbev/msy096. PMID: 29722887; PMCID: PMC5967553.

Kumar, R., Saxena, A., Chawla, J., & Kumar, V. (2024). Environmental impacts of e-waste: Pollution and resource depletion. In *Sustainable solutions for e-waste and development* (pp. 12–32). IGI Global Scientific Publishing. https://doi.org/10.4018/979-8-3693-1018-2.ch002

Li, J., Liang, C., & Ma, C. (2015). Bioleaching of gold from waste printed circuit boards by Chromobacterium violaceum. *Journal of Material Cycles and Waste Management, 17*, 529–539. https://doi.org/10.1007/s10163-014-0276-4

Li, J., Wen, J., Guo, Y., An, N., Liang, C., & Ge, Z. (2020). Bioleaching of gold from waste printed circuit boards by alkali-tolerant Pseudomonas fluorescens. *Hydrometallurgy, 194*, 105260. https://doi.org/10.1016/j.hydromet.2020.105260

Liu, Z., Ciais, P., Deng, Z., Lei, R., Davis, S. J., Feng, S., ... & Schellnhuber, H. J. (2020). Near-real-time monitoring of global CO2 emissions reveals the effects of the COVID-19 pandemic. Nature communications, 11(1), 5172. https://doi.org/10.1038/S41467-020-18922-7;TECHMETA

Magoda, K., & Mekuto, L. (2022). Biohydrometallurgical recovery of metals from waste electronic equipment: Current status and proposed process. *Recycling, 7*, 67. https://doi.org/10.3390/recycling7050067

Malhotra, P., & Jain, A. (2023). Chemical methods for the treatment of e-waste. In *Waste management and resource recycling in the developing world* (pp. 181–204). Elsevier. https://doi.org/10.1016/b978-0-323-90463-6.00019-1

McGivney, E., Gao, X., Liu, Y., Lowry, G. V., Casman, E., Gregory, K. B., et al. (2018). Biogenic cyanide production promotes dissolution of gold nanoparticles in soil. *Environmental Science & Technology, 53*(3), 1287–1295. https://doi.org/10.1021/acs.est.8b05884

Mishra, S., & Das, A. P. (2021). Current treatment technologies for removal of microplastic and microfiber pollutants from wastewater. In *Wastewater treatment: Cutting-edge molecular tools, techniques and applied aspects* (pp. 237–251). https://doi.org/10.1016/B978-0-12-821881-5.00011-8

Mishra, S., & Das, A. P. (2023). Microbial remediation of synthetic microfiber contaminated wastewater. In M. P. Shah (Ed.), *Microbial technologies in industrial wastewater treatment* (pp. 337–350). Springer. https://doi.org/10.1007/978-981-99-2435-6_17

Mishra, S., Rath, C., & Das, A. P. (2019). Marine microfiber pollution: A review on present status and future challenges. *Marine Pollution Bulletin, 140*, 188–197. https://doi.org/10.1016/J.MARPOLBUL.2019.01.039

Mishra, S., Singh, R. P., Rath, C. C., & Das, A. P. (2020). Synthetic microfibers: Source, transport and their remediation. *Journal of Water Process Engineering, 38*, 101612. https://doi.org/10.1016/J.JWPE.2020.101612

Mishra, S., Rout, P. K., & Das, A. P. (2021). Emerging microfiber pollution and its remediation. In R. Prasad (Ed.), *Environmental and microbial biotechnology* (pp. 247–266). Springer. https://doi.org/10.1007/978-981-15-5499-5_9

Mishra, S., Dash, D., Al-Tawaha, A. R. M. S., & Das, A. P. (2022a). A review on heavy metal ion adsorption on synthetic microfiber surface in aquatic environments. *Applied Biochemistry and Biotechnology, 194*(10), 4639–4654. https://doi.org/10.1007/S12010-022-04029-W/METRICS

Mishra, S., Dash, D., & Das, A. P. (2022b). Detection, characterization and possible biofragmentation of synthetic microfibers released from domestic laundering wastewater as an emerging source of marine pollution. *Marine Pollution Bulletin, 185*, 114254. https://doi.org/10.1016/J.MARPOLBUL.2022.114254

Mishra, S., Ghosh, S., van Hullebusch, E. D., Singh, S., & Das, A. P. (2023). A critical review on the recovery of base and critical elements from electronic waste-contaminated streams using microbial biotechnology. *Applied Biochemistry and Biotechnology, 2023*, 1–30. https://doi.org/10.1007/S12010-023-04440-X

Mishra, P., Oster, N., & Henriksen, D. (2024a). Generative AI, teacher knowledge and educational research: Bridging short-and long-term perspectives. *TechTrends, 68*(2), 205–210. https://doi.org/10.1007/s11528-024-00938-1

Mishra, S., Dash, D., & Das, A. P. (2024b). Isolation and molecular characterization of polyester-degrading bacterial strains from municipal wastewater and their application in microplastic degradation in contaminated wastewater. *Geomicrobiology Journal, 41*, 732. https://doi.org/10.1080/01490451.2024.2376101

Mohanty, S., Ghosh, S., Nayak, S., & Das, A. P. (2016). Isolation, identification and screening of manganese solubilizing fungi from low-grade manganese ore deposits. *Geomicrobiological Journal, 34*(4), 309–316. https://doi.org/10.1080/01490451.2016.1189016

Mohanty, S., Ghosh, S., Nayak, S., & Das, A. P. (2017). Bioleaching of manganese by Aspergillus sp. isolated from mining deposits. *Chemosphere, 172*, 302–309. https://doi.org/10.1016/J.CHEMOSPHERE.2016.12.136

Morais, S., Costa, F. G., & Pereira, M. D. L. (2012). Heavy metals and human health. *Environmental Health–Emerging Issues and Practice, 10*(1), 227–245. https://doi.org/10.5772/29869

Morshedi, A., & Akbarian, M. (2014). Application of response surface methodology: Design of experiments and optimization: A mini review. *Indian Journal of Fundamental and Applied Life Sciences, 4*(4), 2434–2439. https://doi.org/10.3390/app131810428

Mourshed, M., Khatun, S., Islam, K., Masuk, N. I., & Masud, M. H. (2024). Generation, composition, collection, and treatment of e-waste. In *Management of electronic waste: Resource recovery, technology and regulation* (pp. 39–72). https://doi.org/10.1002/9781119894360.ch3

Nag, M., Lahiri, D., Ghosh, S., Sarkar, T., Pati, S., Das, A. P., Ram, D. K., Bhattacharya, D., & Ray, R. R. (2023). Application of microorganisms in biotransformation and bioremediation of environmental contaminant: A review. *Geomicrobiology Journal, 41*, 374. https://doi.org/10.1080/01490451.2023.2261443

Natarajan, G., & Ting, Y. P. (2014). Pretreatment of e-waste and mutation of alkali-tolerant cyanogenic bacteria promote gold biorecovery. *Bioresource Technology, 152*, 80–85. https://doi.org/10.1016/j.biortech.2013.10.108

Nyeko, S. J., Mlay, S. V., Amerit, B., Abima, B., Among, J., Nyero, A. I., et al. (2023). The impact of electronic-electrical waste on human health and environment: A systematic literature review. *Journal of Engineering and Technology Research, 15*(1), 1–16. https://doi.org/10.56279/bmrj.v26i2.1

Ogboru, I. (2014). *Episiotomy of Nigeria's economic malady: Its depth and the way out.* https://doi.org/10.15640/jeds.v5n2a11

Oke, E. A., & Potgieter, H. (2024). Recent chemical methods for metals recovery from printed circuit boards: A review. *Journal of Material Cycles and Waste Management, 26*(3), 1349–1368. https://doi.org/10.1007/s10163-024-01944-4

Parajuly, K., Fitzpatrick, C., Muldoon, O., & Kuehr, R. (2020). Behavioral change for the circular economy: A review with focus on electronic waste management in the EU. *Resources, Conservation & Recycling: X, 6*, 100035. https://doi.org/10.1016/j.rcrx.2020.100035

Patel, F., & Lakshmi, B. (2021). Bioleaching of copper and nickel from mobile phone printed circuit board using Aspergillus fumigatus A2DS. *Brazilian Journal of Microbiology, 52*(3), 1475–1487. https://doi.org/10.1007/s42770-021-00526-y

Paul, A., Dey, S., Ram, D. K., & Das, A. P. (2023a). Hexavalent chromium pollution and its sustainable management through bioremediation. *Geomicrobiology Journal, 41*, 1–11. https://doi.org/10.1080/01490451.2023.2218377

Paul, J., Ueno, A., & Dennis, C. (2023b). ChatGPT and consumers: Benefits, pitfalls and future research agenda. *International Journal of Consumer Studies, 47*(4), 1213–1225. https://doi.org/10.1111/ijcs.12928

Pawar, G., & Ewing, R. C. (2022). Recent advances in the global rare-earth supply chain. *MRS Bulletin, 47*(3), 244–249. https://doi.org/10.1557/s43577-022-00305-6

Prabhakar, A., Mishra, S., & Das, A. P. (2019). Isolation and identification of lead (Pb) solubilizing bacteria from automobile waste and its potential for recovery of lead from end-of-life waste batteries. *Geomicrobiological Journal, 36*(10), 894–903. https://doi.org/10.1080/01490451.2019.1654044

Pradhan, G., Tripathy, B., Ram, D. K., Digal, A. K., & Das, A. P. (2023). Bauxite mining waste pollution and its sustainable management through bioremediation. *Geomicrobiology Journal, 41*, 1–10. https://doi.org/10.1080/01490451.2023.2235353

Priyadarshini, B., Santhosh Kumar, M., & Das, A. P. (2024a). Microbial degradation of synthetic plastics released from COVID-19 generated single use disposable face masks. In *Impact of COVID-19 waste on environmental pollution and its sustainable management: COVID-19 waste and its management* (pp. 1–16). Springer. https://doi.org/10.1007/978-3-031-50840-0_1

Priyadarshini, B., Santhosh Kumar, M., & Das, A. P. (2024b). Microbial degradation of synthetic plastics released from COVID-19 generated single use disposable face masks. *Environmental Science and Engineering, Part, F2827*, 1–16. https://doi.org/10.1007/978-3-031-50840-0_1

Priyadarsini, S., & Das, A. P. (2024). Analytical and structural characterization of waste lithium-ion batteries for their effective recycling strategy. *Environmental Chemistry and Ecotoxicology*, 7, 182–191.

Priyadarsini, S., & Das, A. P. (2025). Lithium bioleaching: A review on microbial-assisted sustainable technology for lithium bio-circularity. *Journal of Water Process Engineering, 69*, 106744. https://doi.org/10.1016/j.jwpe.2024.106744

Priyadarsini, A., Mohanty, C., Nanda, S., Mishra, A., Das, N., Swain, N., et al. (2024). Synergistic cobalt oxide/reduced graphene oxide/biochar nano-composite catalyst: Harnessing the power of the catalyst for sustainable remediation of organic dyes and chromium (VI). *RSC Advances, 14*(14), 10089–10103. https://doi.org/10.1039/d4ra01031k

Raj, K., & Das, A. P. (2023). Lead pollution: Impact on environment and human health and approach for a sustainable solution. *Environmental Chemistry and Ecotoxicology, 5*, 79–85. https://doi.org/10.1016/J.ENCECO.2023.02.001

Rani, R., & Ahirwar, R. (2022). Management of E-waste. In *Waste technology for emerging economies* (pp. 49–78). CRC Press. https://doi.org/10.1201/9781003132349-4

Rautela, R., Arya, S., Vishwakarma, S., Lee, J., Kim, K. H., & Kumar, S. (2021). E-waste management and its effects on the environment and human health. *Science of the Total Environment, 773*, 145623. https://doi.org/10.1016/j.scitotenv.2021.145623

Sadasivuni, K. K., Rattan, S., Waseem, S., Brahme, S. K., Kondawar, S. B., Ghosh, S., Das, A. P., Chakraborty, P. K., Adhikari, J., Saha, P., & Mazumdar, P. (2019). Silver nanoparticles and its polymer nanocomposites—Synthesis, optimization, biomedical usage, and its various applications. In *Lecture notes in bioengineering* (pp. 331–373). https://doi.org/10.1007/978-3-030-04741-2_11

Sahu, A., Mallick, K., & Das, A. P. (2024). *Environmental impact of single-use synthetic face mask and its recycling: A sustainable approach* (pp. 197–212). https://doi.org/10.1007/978-3-031-50840-0_11

Saitoh, N., Fujimori, R., Nakatani, M., Yoshihara, D., Nomura, T., & Konishi, Y. (2018). Microbial recovery of gold from neutral and acidic solutions by the baker's yeast Saccharomyces cerevisiae. *Hydrometallurgy, 181*, 29–34. https://doi.org/10.1016/j.hydromet.2018.08.011

Singh, O. (2022). Forecasting trends in the generation and management of hazardous waste. In *Hazardous waste management* (pp. 465–489). Elsevier. https://doi.org/10.1016/b978-0-12-824344-2.00015-x

Suja, F., Abdul Rahman, R., Yusof, A., & Masdar, M. S. (2014). E-waste management scenarios in Malaysia. *Journal of Waste Management, 2014*(1), 609169. https://doi.org/10.1155/2014/609169

Suja, M., Debnath, B., Bashar, S. B., Su, L., Lake, R., & Liu, J. (2018). Electrically driven plasmon-exciton coupled random lasing in ZnO metal-semiconductor-metal devices. *Applied Surface Science, 439*, 525–532. https://doi.org/10.1016/j.apsusc.2018.01.075

Tipre, D. R., Khatri, B. R., Thacker, S. C., & Dave, S. R. (2021). The brighter side of e-waste—A rich secondary source of metal. *Environmental Science and Pollution Research, 28*(9), 10503–10518. https://doi.org/10.1007/s11356-020-12022-1

Tonietti, L., Esposito, M., Cascone, M., Barosa, B., Fiscale, S., Tomajoli, M. T. M., et al. (2024). Unveiling the bioleaching versatility of Acidithiobacillus ferrooxidans. *Microorganisms, 12*(12), 2407. https://doi.org/10.3390/microorganisms12122407

Trivedi, A., & Hait, S. (2020). Efficacy of metal extraction from discarded printed circuit board using Aspergillus tubingensis. In *Bioresource utilization and bioprocess* (pp. 167–175). https://doi.org/10.1007/978-981-15-1607-8_16

Vadoudi, K., Kim, J., Laratte, B., Lee, S. J., & Troussier, N. (2015). E-waste management and resources recovery in France. Waste management & research: the journal of the International Solid Wastes and Public Cleansing Association, ISWA. 33. https://doi.org/10.1177/0734242X15597775.

Van Yken, J., Boxall, N. J., Cheng, K. Y., Nikoloski, A. N., Moheimani, N. R., & Kaksonen, A. H. (2021). E-waste recycling and resource recovery: A review on technologies, barriers and enablers with a focus on Oceania. *Metals, 11*(8), 1313. https://doi.org/10.3390/met11081313

Vardanyan, A., Vardanyan, N., & Gaydardzhiev, S. (2023). Biological extraction of Cu and Ni from printed circuit boards via redoxolysis with concomitant material characterization. *Hydrometallurgy, 221*, 106145. https://doi.org/10.1016/j.hydromet.2023.106145

Viroonudomphol, D., Suwanton, L., Pinyosirikul, U., Satsue, S., & Harnroongroj, T. (2016). Effect of active and passive smoking on heavy metals toxic and antioxidant trace elements. *Journal of Medical and Bioengineering, 5*, 58. https://doi.org/10.12720/jomb.5.1.58-62

Vuppaladadiyam, S. S. V., Thomas, B. S., Kundu, C., Vuppaladadiyam, A. K., Duan, H., & Bhattacharya, S. (2024). Can e-waste recycling provide a solution to the scarcity of rare earth metals? An overview of e-waste recycling methods. *Science of the Total Environment, 924*, 171453. https://doi.org/10.1016/j.scitotenv.2024.171453

Wang, F., Huisman, J., Meskers, C. E., Schluep, M., Stevels, A., & Hagelüken, C. (2012). The Best-of-2-Worlds philosophy: Developing local dismantling and global infrastructure network for sustainable e-waste treatment in emerging economies. *Waste Management, 32*(11), 2134–2146.

Wang, K., Qian, J., & He, S. (2021). Contested worldings of e-waste environmental justice: Nonhuman agency and e-waste scalvaging in Guiyu, China. *Annals of the American Association of Geographers, 111*(7), 2165–2184. https://doi.org/10.1080/24694452.2021.1889353

Wang, Q., Duan, H., Miao, Q., Li, H., Liu, J., Wang, N., & Xu, Q. (2023). Environmental and economic impact assessment of synergistic organic-waste treatment strategies in eco-industrial parks: A pilot-scale case study in Shenzhen, China. Environmental Impact Assessment Review, 103, 107250. https://doi.org/10.1016/J.EIAR.2023.107250

Zeng, X., Zheng, L., Xie, H., Lu, B., Xia, K., Chao, K., et al. (2012). Current status and future perspective of waste printed circuit boards recycling. *Procedia Environmental Sciences, 16*, 590–597. https://doi.org/10.1016/j.proenv.2012.10.081

Zeng, X., Mathews, J. A., & Li, J. (2018). Urban mining of e-waste is becoming more cost-effective than virgin mining. *Environmental Science & Technology, 52*(8), 4835–4841. https://doi.org/10.1021/acs.est.7b04909

Zhao, W., Xu, J., Fei, W., Liu, Z., He, W., & Li, G. (2023). The reuse of electronic components from waste printed circuit boards: A critical review. *Environmental Science: Advances, 2*(2), 196–214. https://doi.org/10.1039/d2va00266c

Integrating Circular Economy Principles with Fuzzy Multi-criteria Decision-Making for Sustainable E-Waste Management: A Theoretical Framework

Nandini Gupta, Ajoy Kanti Das, Suman Patra, Suman Das, Rakhal Das, Anjan Mukherjee, and Abhijit Bhattacharya

Abstract The rapid increase in electronic waste (e-waste) poses significant environmental and public health challenges, particularly in urban areas of developing countries like Agartala, India. This chapter presents a case study on e-waste management (EWM) in Agartala, employing a fuzzy multicriteria decision-making (FMCDM) model to evaluate and identify the most sustainable EWM strategy. The study focuses on three key criteria—environmental impact, cost of infrastructure, and public acceptance—and compares three alternatives: formal recycling facilities, informal sector involvement, and public-private partnerships (PPP). We apply fuzzified evaluations to handle uncertainties and apply weighting factors to prioritize each criterion. Results indicate that a public-private partnership provides the most balanced approach to reducing environmental risks and addressing the economic and social factors influencing public participation. Our findings suggest fuzzy decision-making models' capability to provide a basis for a comprehensive and sustainable EWM strategy coherent with urban EWM for environmental sustainability and circular economy objectives.

N. Gupta (✉)
Department of Environmental Science, Bir Bikram Memorial College, Agartala, Tripura, India

A. K. Das · S. Patra · S. Das
Department of Mathematics, Tripura University, Agartala, Tripura, India

R. Das
Department of Education (ITEP), NIT Agartala, Jirania, Tripura, India

A. Mukherjee
Department of Mathematics, The ICFAI University Tripura, Mohanpur, Tripura, India

A. Bhattacharya
Department of Chemistry, Bir Bikram Memorial College, Agartala, Tripura, India

A. P. Das, S. Priyadarsini (eds.), *Electronic Waste and Environmental Pollution*, Sustainable Environmental Waste Management Strategies,
https://doi.org/10.1007/978-3-031-90287-1_17

Keywords E-waste management · Fuzzy multicriteria decision-making · Public-private partnerships · Environmental impact · Urban waste management · Sustainable development · Circular economy · Informal sector · Recycling infrastructure

1 Introduction

Rapid technological advances and rapid increase in the use of electronics have led to exponential growth in the generation of electronic waste, which is now becoming a significant global concern (Priyadarsini & Das, 2025). Contained in e-waste are both valuable and hazardous materials, and the proper disposal and recycling of which it should undergo is both an environmental and public health concern. The complications of e-waste have found traditional waste management practices insufficient and require us to develop sustainable, data-driven models. In the past few decades, however, electrical and electronic equipment (EEE) production has been dramatically accelerated, and with it has come an inevitable increase in electrical and electronic waste. E with EEE (electronic equipment and electronic parts) discarded e-waste was stated as e-waste (Gregory et al., 2009) and is very harmful to the environment and health if not well managed (Babu et al., 2007). EWM is still at its nascent stage in developing countries, and it has growing concerns over e-waste, which have become critical in the last few years (Balabanic et al., 2011; Breivik et al., 2014). With the exponential growth in electronic devices, e-waste has become a big problem because of the rapid proliferation of e-waste (Mishra et al., 2023). E-waste is scrapped electronic products including computers, smartphones, televisions, and household appliances (Priyadarsini & Das, 2024). As with all electronic components, these contain valuable metals like gold, silver, and copper, but these also pose risks from hazardous materials including lead, mercury, and cadmium. Emerging as a rapidly escalating issue, e-waste disposal can lead to severe environmental contamination that damages soil, air, and water, and, if not adequately controlled, results in rapidly growing quantities of hazardous substances requiring disposal (Das et al., 2015; Mishra et al., 2022b). The soil can become heavily contaminated, leaching into the groundwater and agricultural products (Baldé et al., 2014). Aquatic ecosystems are at risk when toxic elements from e-waste infiltrate water sources, posing serious risks to human health (Barba-Gutiérrez et al., 2008; Mishra & Das, 2021, 2023). Moreover, incinerating e-waste releases hazardous chemicals into the atmosphere to worsen air pollution and cause health problems (Duan et al., 2009).

A new transformative framework and a sustainable approach to tackle these challenges is being presented by the circular economy (CE), a closed-loop system which shifts norms from the traditional "take-make-dispose" approach, to a resource efficient, product life extension one in which waste is a resource (Dey et al., 2024; Sahu et al., 2024). The recycling, reuse, and remanufacturing techniques recommended within CE principles seek to maximize the value of resources. This gives a material-efficient use by fully using the materials and not prematurely discarding them (Chang, 1996). CE promotes the design of durable, repairable, and upgradable products, extending the lifetime of electronic devices and reducing the amount of e-waste generated (Hasan, 2004). Closely incorporating CE principles into EWM reduces the use of virgin resources and minimizes the generation of waste, as well as the reduction of emissions of toxic material into the environment. This approach leads to better resource efficiency and sustainable practices. CE is mainly about resource recovery: e-waste is perceived as a potential source of essential raw materials, that is, not a burden (Karapetrovic & Rosenbloom, 1999). E-waste materials lose energy through the process of improper disposal and distribution, and therefore e-wastes contain high quantities of materials, which means that they can be used to construct new items. Advanced recycling technologies enable the efficient recovery of the valuable metals and components from the e-waste and hence reduce the new materials demand and conserve natural resources (Khetriwal et al., 2009). Furthermore, CE can be stimulated by policies and regulations that support the design for product and recycling technologies, which will result in an innovation in the product design as well as recycling technology, a more sustainable approach for EWM (Damghani et al., 2008). CE has recently become a trend as a sustainable model to resolve EWM challenges.

According to the CE, products and materials should be recycled and kept in use as long as possible (Kiddee et al., 2013). However, despite recognizing the value of these resources in e-waste, interests in CE principles within waste management strategies appear to be minimal (Chagnes et al., 2016). Many of these materials are recycled and recovered which can lessen the requirement for virgin raw materials, abatement of the environment, and promote economic development (Cassidy & Kreitner, 2011; Choi et al., 2006). CE is known to entail a number of benefits, although many countries, in particular those from low- and middle-income brackets, have difficulty following this path with effective EWM systems because of infrastructural and regulatory constraints. For example, there is an alarming growing amount of e-waste being produced due to population increase, urbanization, and tremendous technological advances (in the case of Iran and other developing nations) (Balabanic et al., 2011; Babu et al., 2007). This consequently led to the continuation of use of suboptimal EWM practices, for instance, informal recycling, incineration, and landfilling, which worsen environmental pollution and human health risk (Kiddee et al., 2013). E-waste management involves a multidimensional approach from both environmental, social, and economic, and technical perspectives. Extant literature proposes several strategies to handle EWM, such as Extended Producer Responsibility (EPR), Life Cycle Assessment (LCA), and Material Flow Analysis (MFA) (Ahluwalia & Nema, 2007; Apisitpuvakul et al., 2008). The contributions of these approaches include making e-waste management more sustainable by placing the responsibility for product end of life with manufacturers, evaluating the environmental impact of a product through its entire life cycle, and tracking material flow in order to achieve higher recycling rates with less product generation. Such strategies have an important impact on designing a more sustainable and efficient EWM system. Nevertheless, due to their complexity as well as the fact that the criteria themselves may contradict one another, recent years have seen a growing application of Multicriteria Decision-Making (MCDM) methods (Das and Granados, 2021; 2022a; 2022b; 2023).

MCDM approaches provide an essential framework for systematically using these complexities in the process of EWM. These methods facilitate decisions-makers to compare many of at times, contradictory criteria such as environmental impact, economic cost, social acceptance, and technical feasibility when searching for both economic and sustainable solutions (Das et al., 2023; 2025a; 2025b). Consequently, traditional methods like the Analytic Hierarchy Process (AHP) and its fuzzy counterpart (FAHP) are extensively employed in EWM for their capacity of combining quantitative and qualitative factors in a structured way and of comparing alternatives. In particular, FAHP efficiently deals with the inherent uncertainties and imprecise information which often characterize e-waste situations, and delivers more specific and precise results (Kahraman, 2008). Other than further amplifying the effectiveness of the CE principles in promoting sustainable EWM, the integration of CE principles with fuzzy MCDM techniques is introduced. In this sense, CE strategies like resource recovery, reuse, recycling, and product life extension are adopted to satisfy the general objectives of decreasing the environmental footprint and the economic viability. This study facilitates decision-making on these strategies as a whole, taking into account the environmental soundness of the selected alternatives, as well as their economic feasibility and social acceptability (De Souza et al., 2016; Das et al., 2024). For instance, FAHP can rank initiatives according to criteria like the recoverable potential of valuable materials, the reduction of environmental hazards, operational costs, scalable, and social acceptability. Fuzzy MCDM approaches have been applied in practical applications for designing good e-waste collection systems, identifying best options for recycling technology, and those for waste treatment. By tackling the critical issues of selection of eco-friendly processes, minimization of toxic byproducts, and efficient resource allocation, these methods help us to achieve quality of life for the coming generations. An example of that is to rank recycling technologies using such material recovery capacity, energy efficiency, operational safety, and compliance with environmental regulations (Kahraman, et al., 2004). It can also help in the design of the collection system in terms of accessibility, cost efficiency, and community participation. Fuzzy MCDM also handles dynamic decision-making, that is, with time, they can accommodate technological advancements, policy changes, societal needs, etc (Das, 2018). These methods incorporate real-time data and stakeholder inputs such that the EWM strategy becomes flexible and future-ready. This paper shows that the complexity of the CE-oriented e-waste solutions is well captured by the fuzzy MCDM as an indispensable tool for achieving long-term sustainability in waste management practices.

The objectives of this study are to develop a theoretical framework for integrating CE principles into FMCDM to increase EWM system effectiveness. It uses a systematic, flexible, and adaptable approach to address the complex and multifaceted nature of EWM by giving freedom to the user to specify problem attributes and to incorporate any of the methods for the evaluation of remote sensing archive data included in both CE and FMCDM. It places emphasis not only on environmental sustainability, but also on socioeconomic issues of waste management, from which a balanced view of waste management for the policymakers and practitioners is availed. The significance of resource recovery and efficient recycling processes that convert waste into useful resources and that minimize the environmental footprint is highlighted by the framework. It solves precisely the uncertainties encountered in EWM decision-making by integrating fuzzified evaluations. This makes it possible to evaluate alternatives fully, taking into account a wide array of criteria ranging from environmental impact to economic viability, public acceptance, and technological capabilities. One important feature of this framework is its ability to operate in multiple geographical and cultural as well as regulatory contexts. It can be adapted to urban and rural settings and differing degrees of infrastructure and community awareness. Moreover, integration of CE principles promotes the transition from linear waste disposal and disposal systems to the closed-loop models for long product life cycle extension, material recovery, and least waste generation. It also points out that governments, private sectors, and communities are essential partners in identifying and developing EWM strategies that succeed. Through fostering such partnerships and the use of a creative decision-making tool, the framework provides the potential road to sustainable development in the fight against e-waste on a global level. All the different sections of the chapter are described in the Fig. 1.

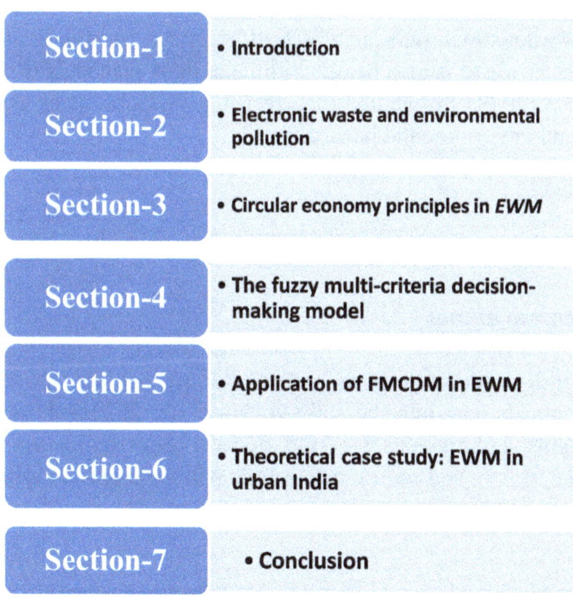

Fig. 1 Displays the following parts that comprise this chapter

2 Electronic Waste and Environmental Pollution

2.1 E-Waste

E-waste, which is a discarded electronic device comprises of waste from computers, smartphones, televisions, and home appliances, is considered as a major environmental challenge. Despite the fact that these devices have very valuable resources, they also shelter a great diversity of toxic substances that can be very harmful if they are not handled correctly.

2.2 Composition of E-Waste

2.2.1 Valuable Metals

Valuable precious metals like silver, gold, and copper are found in many electronic devices and can be recovered and recycled in environmentally friendly ways, lowering you demand for new resources and helping toward sustainable material use. The latter are economically valuable and also critical for technological advancement.

2.2.2 Toxic Substances

Along with hazardous materials such as lead, mercury, cadmium, and brominated flame retardants, e-waste is also heavy with hazardous materials (Prabhakar et al., 2019; Raj & Das, 2023). If handled incorrectly, these substances can be a serious health hazard and environmental hazard.

2.3 Environmental Impacts

2.3.1 Soil Contamination

If e-waste is not disposed of properly and you dump machines into landfills, heavy metals and chemicals leak into the soil (Mishra et al., 2022a). It can contaminate soil, destroy plants, and make its way into the food chain. One example is that lead can inhibit plant growth, and cadmium can end up in the crops (on which consumers may feed).

2.3.2 Water Pollution

Leachate from landfills and informal recycling sites, and runoff from informal recycling sites can also leak toxic elements of e-waste into groundwater and surface water. Particularly worrying are heavy metals, such as mercury and lead, which can contaminate drinking water sources and aquatic ecosystems. For example, fish can contain mercury that can accumulate and harm aquatic life, and thereby put humans at risk from fish consumption.

2.3.3 Air Pollution

Hazardous chemicals can be released into the air when e-wastes are disposed of improperly, for instance, when people take e-wastes to burn them, or when crude methods are used in recycling those e-wastes. E-waste can be incinerated, emitting dioxins, furans, and other toxic compounds. The airborne pollutants can contribute to respiratory problems and cardiovascular problems, and other health problems for nearby populations. Furthermore, the improper recycling process can release poisonous gases and particulate matter, worsening air quality conditions more.

2.4 Resource Depletion and Environmental Degradation

2.4.1 Loss of Recoverable Materials

Lost precious materials would otherwise be recovered and reused due to inefficient EWM. For instance, gold and copper can be extracted and recycled from e-waste, saving the mining process and conserving natural resources. Not only do these materials represent a depletion of finite resources used up in greenhouse gas-intensive extraction and processing, they also add to the environmental impact of extracting fresh materials.

2.4.2 Exacerbation of Resource Depletion

Without proper recycling and recovery, the generation of e-waste continues indefinitely depleting natural resources. It also adds to pressure on the environment to find new resources to compensate for those lost in e-waste.

2.5 *Inadequate EWM Practices*

2.5.1 Informal Recycling

Many other places that e-waste is sometimes processed in an informal way; minimal safety and environment regulations apply. Environmental and health impacts of these practices can be severe, and prevent effective recycling efforts.

2.5.2 Regulatory Challenges

Improper handling of e-waste is possible due to inadequate policies and enforcement mechanisms. Mitigation of the adverse impacts of e-waste requires effective regulation and well-designed management strategies.

Therefore, there is a need to have different effective EWM strategies such as the right recycling methods, public awareness campaigns as well as strong regulation. EWM ensures that Ecosystems and Human health are protected from releasing pollution into the ecosystem, also promoting resource optimization. It recovers valuable materials and minimizes waste and thus decreasing the ecological footprint related to the production, use, and disposal of electronic devices. This approach gives us the possibility to conserve finite resources, decrease consumption of energy, and support a more circular economy that contributes to long-term environmental sustainability and reducing pressure on the natural systems.

3 Circular Economy Principles in *EWM*

The CE, therefore, represents a paradigm shift away from the linear product paradigm of "take-make-dispose" to a closed-loop system. The purpose of this model is to lengthen the life cycle of products, reduce waste, and enhance the capacity for the reutilization of resources. Furthermore, when applied to EWM, CE principles can improve environmental sustainability and resource efficiency. Here's a detailed look at how these principles are integrated into EWM:

3.1 *Resource Efficiency*

Material Utilization: CE focuses on a material-efficient design of electronic products during the product life cycle. This means designing products so that they would make use of available resources, and would also reduce waste. For instance, selecting materials easier to recycle or to lengthen the product life span will result in the production of reduced e-waste.

Minimization of Resource Consumption: CE focuses on resource efficiency in order to limit the consumption of raw materials. So, manufacturing process optimization to reduce waste, product design which makes less use of resources, and improvement of resource recovery system efficiency.

3.2 *Product Life Extension*

Design for Durability: According to CE, durable products with a longer operational life should be designed. That is using the best and most durable materials and components, keeping replacements to a minimum and reducing the e-waste generated in this process.

Repairability and Upgradability: Products should be easily repairable and upgradable. That means including features that will facilitate easy dismantling of the keyboard, replacement of its parts, and performing software updates. One of the biggest ways of increasing lifespan is enhancing the repairability of the electronic devices, so the consumer would wait with its replacement for a longer period. This limits demand for new devices which in turn decreases e-waste production and the use of raw materials. Repairability makes repair easier and cheaper, reducing the need for repeated new electronics manufacture, which cuts resource extraction and energy use and greenhouse gas emissions. Repairability is good for the environment, for environmental sustainability, and for the principles of a circular economy.

Refurbishment and Remanufacturing: CE advocates for the remanufacturing and refurbishment of electronic products. Refurbishment restores worn-out, used products to as good as new condition, whereas remanufacturing takes products apart and puts them back together in accordance with original specifications. Both practices extend the life cycle and reduce e-waste of electronic devices.

3.3 *Waste as a Resource*

Resource Recovery: In the CE framework, e-waste is not seen as waste since it is considered as a valuable resource through the lens. E-waste is a rich source of valuable materials including precious metals like silver, gold, and copper, plus other recoverable materials such as plastics and rare earth elements. These materials are then reclaimed by advanced recycling technologies, extracted, and reintroduced into the production cycle, relieving the reliance on virgin resource mining and extraction. It not only preserves natural reserves, but also saves energy and does harm to the environment by reducing resource extraction and the processes that accompany them. Additionally, rehabilitation of valuable materials from e-waste promotes a circular economy and helps us support sustainability while decreasing the amount of waste produced.

Recycling and Reuse: The company encourages electronic component and material recycling and reuse. Valuable metals and components can be recovered through effective recycling processes to be further used to produce new products. Reuse is the use of used electronic devices or components and therefore reduces the need for new materials, mitigating environmental harm.

3.4 Sustainability and Environmental Impact

Reduction of Virgin Resource Use: CE focuses on the recovery and extension of the products' life, and therefore minimizes the demand of new raw materials. Besides conserving vital natural resources, this approach manages to dramatically minimize the environmental damage resulting from extracting and processing these resources, including habitat destruction, energy usage, and pollution.

Minimization of Hazardous Substances: CE principles can be used to reduce the release of hazardous substances. The recycling and prudent management of e-waste disposal are critically important to prevent dangerous chemicals, such as cadmium, mercury, and lead from getting into the environment. If not managed properly, these hazardous substances seep into the soil and bad into water systems; contaminating them and causing the whole of it including ecosystems, human health, and biodiversity to be at risk. Importantly, by adopting responsible recycling practices, we can make a huge difference in stemming these threats to the environment and bring about cleaner soil, safer water sources, and a healthier environment for our communities.

3.5 Policy and Economic Incentives

Regulatory Support: Generally, CE principles have to be implemented under supportive policies and regulations. Incentives—such as recycling—mandates—such as eco-design—and research and development support—for sustainable technologies can all help governments encourage circular practices.

Economic Benefits: Involving CE principles into EWM might result in economic benefits including the creation of jobs in the recycling and refurbishing sectors, decreased costs related to waste disposal, and offer for innovation in product design and manufacturing.

By application of CE principles to EWM, a more rational and sustainable approach is realized by emphasizing the continuous usage of resources, minimizing environmental effects, and increasing the recovery of valuable materials. Beyond that, it helps in better management of e-waste and contributes to the overall objectives of environmental sustainability and resource conservation.

4 The FMCDM Model

FMCDM model is a powerful decision-making tool to evaluate the alternatives and rank them in terms of multiple criteria, in particular, when one or more criteria describing decision problem are uncertain, imprecise or incomplete. It integrates fuzzy logic to represent the inherent "vagueness" that is prevalent in real-world data with a view to give more "flexibility" to decision-makers to operate under uncertainty.

4.1 Decision-Making with Fuzzy Logic

Originally developed by Lotfi Zadeh in 1965, fuzzy logic is an extension of existing binary logic (true/false) that does not necessarily represent reality. Fuzzy logic assigns degrees of truth, not true or false values, but values between 0 and 1. More specifically, this proves to be useful for making decisions in the case of EWM with multiple criteria, where parameters (e.g., impact on environment, cost, feasibility) may not have a precise dimension.

4.2 Components of the FMCDM Model

Identify what the key factors are that influence EWM decisions such as economic feasibility, environmental impact, social acceptability, and technological capability. This is about converting crisp data (e.g., "high," "medium," "low") of information to fuzzy sets to handle future (desired level of) uncertainties, particularly when we are dealing with subjective or incomplete information. A fuzzy matrix of alternatives (for example, recycling technologies, policy strategies) versus criteria (for example, strict, trade-off), with fuzzy values to each criterion based upon its importance to the decision-maker.

The aggregation of fuzzy scores into a final ranking of alternatives considers key factors that influence environmental impact, economic feasibility, social acceptability, and technological capability.

Fuzzification: Converting crisp data into fuzzy sets using linguistic variables (e.g., "high," "medium," "low") to accommodate uncertainties in information, especially when it is subjective or incomplete.

Decision-Making Matrix: Constructing a matrix where alternatives (e.g., recycling technologies, policy strategies) are evaluated against selected criteria, with fuzzy values assigned to each criterion based on its relative importance.

Aggregation and Weighting: Aggregating fuzzy scores to generate a final ranking of alternatives. The importance of different criteria can be changed according to their importance for making the decision under consideration (for example, environmental criteria may be more important when the pollution is severe).

5 Application of FMCDM in EWM

FMCDM is applied to EWM by evaluating the potential solutions in many dimensions. This framework can guide the development of sustainable strategies by addressing key aspects such as:

Environmental Impact: Factors influencing management practices (e.g., recycling, reuse, landfilling) that influence pollution of soil, air, and water.

Economic Feasibility: The cost of using CE-driven practices for collection, transportation, and recycling technologies.

Social Acceptability: EWM programs require public awareness and cooperation.

Technological Capability: How existing technologies can recover valuable materials efficiently, despite often being able to handle only simple e-waste streams.

5.1 Model Implementation Process

5.1.1 Step 1: Alternatives Identification

It is possible that other waste management technologies (e.g., formal recycling centers, informal sector involvement) or other policies (e.g., extended producer responsibility, government subsidies for recycling) may be an alternative. Pollution reduction, resource recovery of Environmental protection. Hydraulic costs (investment in infrastructure, operational costs) and economic costs (investment in infrastructure, operational costs). Job creation, public awareness, and social impact of Alternatives.

Alternatives could include different waste management technologies (e.g., formal recycling centers, informal sector involvement) or policies (e.g., extended producer responsibility, government subsidies for recycling).

5.1.2 Step 2: Defining Criteria

Environmental protection (pollution reduction, resource recovery)
Economic costs (investment in infrastructure, operational costs)
Social impact (job creation, public awareness)

5.1.3 Step 3: Fuzzification of Input Data

Linguistic terms (e.g., “low environmental impact,” “moderate cost”) are used to fuzzify input data that will be collected and recycled through e-waste collection and recycling activities.

5.1.4 Step 4: Evaluation and Ranking

It uses the FMCDM model to rank alternatives based on the aggregated fuzzy scores, so decision-makers are free to choose the most sustainable EWM.

6 Theoretical Case Study: EWM in Urban India

A hypothetical case study based on EWM in urban India is used to illustrate the FMCDM technique. It presents the case where fuzzy logic is employed to evaluate for different waste management strategies for the city of Agartala with the comparison of environmental benefits, social impact, and economic feasibility.

6.1 Case Study Overview: EWM in Agartala, India

This case study shows how the FMCDM model can be used to select the most sustainable EWM strategy for Agartala, India. It uses three key criteria—environmental impact, cost of infrastructure, and public acceptance—along with three potential alternatives for managing e-waste: The secondary involved informal sector participation or formal recycling facilities, or public-private partnerships (PPP). The model evaluates each alternative under conditions of uncertainty using fuzzy logic to provide a broad, adaptable solution.

6.1.1 Criteria for Evaluation

Three main factors that have a direct impact on the sustainability of EWM practices form the basis of the decision-making process:

1. **Environmental Impact**

This criterion assesses each alternative's contribution to resource recovery, pollution reduction, and environmental hazard mitigation.

- A decrease in hazardous waste (such as lead and mercury) is one indicator used to evaluate the influence on the environment.
- The recovery of valuable resources, such as gold and copper metals.
- The possibility of soil, water, and air pollution.

2. **Cost of Infrastructure**

This criterion evaluates each alternative's feasibility from a financial standpoint.

- The initial outlay for the establishment of collection centers and recycling facilities is one of the key indications.

- Maintenance and operating expenses.
- Incentives or subsidies from the government are needed to make the model sustainable.

3. **Public Acceptance**

This criterion assesses public collaboration, awareness, and participation to determine the solution's social viability. It assesses:

- The degree of knowledge and readiness to participate in official recycling initiatives.
- Households and businesses are working together to manage and separate the disposal of electronic trash.
- Whether policy changes (such as extended producer responsibility) are acceptable.

6.1.2 Alternatives for EWM

The following three options are assessed by the FMCDM model:

1. **Conventional Recycling Establishments**

Formally establishing e-waste recycling facilities with state-of-the-art machinery and safety procedures guarantees that hazardous trash is handled correctly and optimizes material recovery. Environmental authorities oversee formal facilities, making sure that safety and pollution control regulations are followed.

Benefits include decreased pollution hazards, efficient resource recovery, and high environmental protection.

Difficulties include lengthy lead times for establishment, high setup expenses, and low public engagement without sufficient incentives.

2. **Involvement of the Informal Sector**

The informal sector is crucial to EWM in many Indian cities. Informal recyclers manually gather and disassemble electronic garbage, removing valuable parts for future sale. Although this approach is economical, it usually lacks protections for the environment and human health.

Benefits: Existing labor, high adaptability, and low infrastructure expenses.

Difficulties: Significant harm to the environment, hazards to employees' health, a lack of regulations, and a lack of public confidence.

3. **Public-Private Partnerships**

In this option, government and private organizations work together to oversee the collection, recycling, and disposal of e-waste. PPPs make use of the resources and experience of the private sector, while the government handles public awareness initiatives and regulatory supervision.

Benefits include the possibility of innovation, quicker implementation, and shared expenditures.

Difficulties include the need for robust regulatory structures, complicated discussions, and possible public distrust.

6.1.3 Fuzzification of Criteria

A key step in FMCDM that enables the management of ambiguities and imprecise data is fuzzification. Fuzzification helps overcome the difficulties of assessing criteria that are intrinsically ambiguous or challenging to accurately quantify by transforming qualitative assessments into fuzzy values between 0 and 1. A more thorough explanation of the fuzzification of the criteria applied in the context of EWM may be found here:

Environmental Impact

Formal Recycling Facilities: High (0.8)

E-waste is handled and processed in an environmentally responsible manner because to formal recycling facilities' frequent adherence to strict environmental norms and standards. Because of controlled activities, appropriate management of hazardous materials, and efficient pollution control techniques, there is a lower chance of environmental impact, as shown by this high fuzzy value.

Informal Sector Involvement: Low (0.3)

The informal industry typically has a lower environmental impact score because of its unregulated and frequently primitive recycling procedures. This low number explains the improper handling of dangerous materials and the absence of official oversight, which can have serious negative effects on the environment and human health.

Public-Private Partnerships (PPP): Moderate (0.7)

Governmental-commercial partnerships are a well-rounded strategy in which commercial and governmental organizations work together to control e-waste. The efficacy of these collaborations in leveraging the advantages of both industries to reduce environmental hazards is indicated by the moderate fuzzy value. Although the effect may differ depending on the performance and dedication of the parties concerned, the existence of both public scrutiny and private efficiency aids in minimizing environmental harm.

Cost of Infrastructure

Formal Recycling Facilities: High (0.2)

Because formal recycling facilities require specialized staff, sophisticated technology, and adherence to regulations, they are expensive to build and maintain. The large amount of money needed is represented by this low fuzzy value, which may act as a deterrent to the growth of official recycling systems.

Informal Sector Involvement: Low (0.8)

The informal sector, on the other hand, has very low infrastructural expenses. Since informal activities frequently make use of already-existing resources and are not burdened by the costs of formal facilities, the low value illustrates how cost-effective they are. However, there are risks to the environment and human health as a result of this cost-effectiveness.

Public-Private Partnerships (PPP): Moderate (0.5)

Because the public and private sectors share duties, infrastructure costs in a PPP scenario are minimal. This balanced value shows that even while the costs are shared, which increases the system's financial viability, both partners must still make sizable expenditures to guarantee efficient EWM.

Public Acceptance

Formal Recycling Facilities: Moderate (0.5)

Because it depends on things like accessibility, public awareness, and the facilities' perceived efficacy, public acceptability of formal recycling facilities is only moderate. The intermediate number suggests that although the public may embrace formal facilities, this acceptance is not always assured and may differ depending on the experience and involvement of the community.

Informal Sector Involvement: Moderate (0.6)

Because of its established methods and current existence, the informal sector is also somewhat accepted by the general public. In contrast to recently established formal systems, the informal sector may be more widely accepted because it is well-known to the general public, even though it may not have official status.

Public-Private Partnerships (PPP): High (0.7)

Because PPP efforts incorporate government control, which can increase trust and involvement, they typically receive high public acceptance. Public institutions' participation in e-waste management promotes more accountability and transparency, which boosts public trust and support for recycling initiatives.

To sum up, the criteria's fuzzification offers a sophisticated comprehension of the several facets of EWM. A more flexible and thorough examination is made possible by the fuzzy values given to each criterion, which represent the inherent uncertainties and qualitative character of the evaluation process. Decision-makers can more effectively evaluate the trade-offs of various EWM approaches and make better-informed choices that balance social, economic, and environmental considerations by taking these fuzzy values into account.

Table 1 The fuzzified values for each option under the three criteria are compiled in the decision matrix

Criteria	Formal recycling	Informal sector	Public-private partnership
Environmental impact	0.8	0.3	0.7
Cost of infrastructure	0.2	0.8	0.5
Public acceptance	0.5	0.6	0.7

Decision Matrix: Evaluating Alternatives

The fuzzified values for each option under the three criteria are compiled with the decision matrix and shown in Table. 1.

Aggregation and Ranking

The fuzzy values for each possibility are combined according to their relative relevance using FMCDM. Depending on the priorities of the policy, weighting factors might be used to give some criterion more weight. For instance, environmental effect might be given the most weight during the decision-making process if it is thought to be the most important factor.

Assumed Weights for the Criteria

- Environmental Impact: 0.5
- Cost of Infrastructure: 0.3
- Public Acceptance: 0.2

Step-by-Step Calculation (Example)

1. **Formal Recycling Facilities**
   - Environmental Impact: 0.8 × 0.5 = 0.4
   - Cost of Infrastructure: 0.2 × 0.3 = 0.06
   - Public Acceptance: 0.5 × 0.2 = 0.1
   - **Total Score**: 0.4 + 0.06 + 0.1 = 0.56
2. **Informal Sector Involvement**
   - Environmental Impact: 0.3 × 0.5 = 0.15
   - Cost of Infrastructure: 0.8 × 0.3 = 0.24
   - Public Acceptance: 0.6 × 0.2 = 0.12
   - **Total Score**: 0.15 + 0.24 + 0.12 = 0.51
3. **Public-Private Partnerships (PPP)**
   - Environmental Impact: 0.7 × 0.5 = 0.35
   - Cost of Infrastructure: 0.5 × 0.3 = 0.15
   - Public Acceptance: 0.7 × 0.2 = 0.14
   - **Total Score**: 0.35 + 0.15 + 0.14 = 0.64

Results and Interpretation

With a score of 0.64, the fuzzy scores show that the Public-Private Partnership (PPP) model is the most viable choice for EWM in Agartala. Compared to official recycling or depending on the unorganized sector, this option more successfully strikes a balance between environmental impact, cost, and public acceptance. The PPP offers a workable compromise that addresses the high infrastructure costs of formal recycling while enhancing the health and environmental risks associated with informal sector participation. It does this by utilizing both private sector efficiency and government monitoring.

Public-Private Partnerships are the most promising avenue for sustainable EWM in Agartala, according to the FMCDM strategy. Long-term sustainability is ensured by this strategy, which also balances cost and public acceptance while reducing environmental threats. The fuzzy decision-making framework is a flexible tool for environmental managers and policymakers since it can be enhanced and improved with real-world data.

7 Conclusion

In order to handle the intricacies of sustainable EWM, this chapter has examined the integration of CE principles with FMCDM models. This work gives a fresh viewpoint on decision-making in this crucial area and advances sustainable EWM practices by proposing a theoretical framework that integrates these two methods. This chapter's main contribution is its creative fusion of FMCDM and CE principles to create a strong theoretical foundation for EWM. This integrated approach highlights the significance of sustainable resource consumption, effective recycling procedures, and the avoidance of environmental consequences, in contrast to traditional systems that frequently concentrate just on recycling or waste reduction. This framework offers a thorough solution that takes into account the economic, environmental, and social facets of EWM.

7.1 Key Findings and Relative Importance

The chapter emphasizes a number of important findings:

1. Improved Decision-Making: By combining a wider variety of factors, the integration of CE principles with FMCDM improves decision-making processes and raises the sustainability results of EWM plans.
2. Increased Resource Efficiency: The framework addresses the urgent problem of resource depletion and encourages the best possible use of resources through recycling and reuse, which is consistent with the CE's tenets.

3. Holistic Approach: In contrast to conventional techniques, the suggested model provides a more comprehensive approach to EWM by taking into account a number of factors, including social acceptability, economic viability, and environmental impact.

These results, which represent a substantial breakthrough in the literature on EWM, highlight the significance of implementing a holistic framework that tackles several aspects of sustainability.

Limitations This study has certain limitations that should be acknowledged, notwithstanding its contributions:

1. Data Restrictions: The theoretical framework may not adequately represent the complexity of actual EWM situations because it is predicated on extant literature and speculative scenarios.
2. Methodological Limitations: Because FMCDM is theoretical in nature, additional empirical validation may be necessary to evaluate its efficacy in real-world applications.

The following areas should be the focus of future research:

1. Empirical Validation: Carrying out empirical investigations to verify the theoretical framework and evaluate its usefulness in diverse geographic and industrial situations.
2. Criteria Improvement: To incorporate new developments in EWM trends and technologies, the FMCDM model's criteria were expanded and improved.
3. Integration with Other Models: To improve the suggested framework's resilience and application, it is investigated how it may be integrated with other sustainability and decision-making models.

Implementation in Similar Environmental Problems The findings and strategies discussed in this chapter have wider ramifications for related environmental issues. Hazardous materials and plastic garbage are two examples of alternative waste management techniques that can benefit from the application of CE principles to decision-making models. Policymakers and practitioners may create more sustainable and successful plans for handling different environmental problems by taking a comprehensive and multicriteria approach, which will ultimately help achieve the goals of global sustainability.FundingNo funding was obtained.

Ethical Approval Any of the authors did not conducted any human or animal experiments for this publication.

References

Ahluwalia, P. K., & Nema, A. K. (2007). A life cycle based multi-objective optimization model for the management of computer waste. *Resources, Conservation and Recycling, 51*(4), 792–826.

Apisitpuvakul, W., Piumsomboon, P., Watts, D., & Koetsinchai, W. (2008). LCA of spent fluorescent lamps in Thailand at various rates of recycling. *Journal of Cleaner Production, 16*(10), 1046–1061.

Babu, R., Parande, A. K., & Basha, A. C. (2007). Electrical and electronic waste: A global environmental problem. *Waste Management & Research, 25*(4), 307–318.
Balabanic, D., Rupnik, M., & Klemencic, A. K. (2011). Negative impact of endocrine disrupting compounds on human reproductive health. *Reproduction, Fertility and Development, 23*(3), 403–416.
Baldé, C. P., Wang, F., Kuehr, R., & Huisman, J. (2014). *The global e-waste monitor*. United Nations University Institute for Advanced Study of Sustainability.
Barba-Gutiérrez, Y., Adenso-Díaz, B., & Hopp, M. (2008). An analysis of some environmental consequences of European electrical and electronic waste regulation. *Resources, Conservation and Recycling, 52*(4), 481–495.
Breivik, K., Armitage, J. M., Wania, F., & Jones, K. C. (2014). Tracking the global generation and exports of e-waste. Do existing estimates add up? *Environmental Science & Technology, 48*(15), 8735–8743.
Cassidy, C. M., & Kreitner, R. (2011). *Principles of management*. South-Western/Cengage Learning.
Chagnes, A., Cote, G., Ekberg, C., Nilsson, M., & Retegan, T. (2016). *WEEE recycling: Research, development, and policies*. Elsevier Science.
Chang, D. (1996). Applications of the extent analysis method on fuzzy AHP. *European Journal of Operational Research, 95*(3), 649–655.
Choi, B. C., Shin, H. S., Lee, S. Y., & Hur, T. (2006). Life cycle assessment of a personal computer and its effective recycling rate. *International Journal of Life Cycle Assessment, 11*(2), 122–128.
Damghani, A. M., Savarypour, G., Zand, E., & Deihimfard, R. (2008). Municipal solid waste management in Tehran: Current practices, opportunities, and challenges. *Waste Management, 28*(5), 929–934.
Das, A. K. (2018). Weighted fuzzy soft multiset and decision-making, *International Journal of Machine Learning and Cybernetics, 9*: 787–794. https://doi.org/10.1007/s13042-016-0607-y
Das, A. K., Granados, C. (2021). An advanced approach to fuzzy soft group decision-making using weighted average ratings, *SN Computer Science, 2*: 471. https://doi.org/10.1007/s42979-021-00873-5
Das, A. K., Granados, C. (2022a). A new fuzzy parameterized intuitionistic fuzzy soft multiset theory and group decision-making. *Journal of Current Science and Technology, 12*(3), 547–567. https://doi.org/10.14456/jcst.2022.42
Das, A. K., Granados, C. (2022b). FP-intuitionistic multi fuzzy N-soft set and its induced FP-Hesitant N soft set in decision-making. *Decision Making: Applications in Management and Engineering, 5*(1): 67–89. https://doi.org/10.31181/dmame181221045d
Das, A. K., Granados, C. (2023). IFP-intuitionistic multi fuzzy N-soft set and its induced IFP-hesitant N-soft set in decision-making, *Journal of Ambient Intelligence and Humanized Computing, 14*: 10143–10152. https://doi.org/10.1007/s12652-021-03677-w
Das, A. K., Gupta, N., Granados, C. (2023). Weighted hesitant bipolar-valued fuzzy soft set in decision-making. *Songklanakarin Journal of Science and Technology, 45*(6): 681–690, https://sjst.psu.ac.th/journal/45-6/10.pdf
Das, A. K., Gupta, N., Mahmood, T., Tripathy, B. C., Das, R., Das, S. (2024). Assessing anthropogenic influences on the water quality of Gomati River using an innovative weighted fuzzy soft set based water pollution rating system. *Discover Water 4*, 73. https://doi.org/10.1007/s43832-024-00136-3
Das, A. K., Gupta, N., Mahmood, T., Tripathy, B. C., Das, R., Das, S. (2025a). An efficient water quality evaluation model using weighted hesitant fuzzy soft sets for water pollution rating. In A. Kumar, P. Kumar, S. Rathee, & B. Kumar (Eds.), Mechatronics: Concepts, Tools, Applications, and New Trends (1st ed., p. 17). CRC Press. https://doi.org/10.1201/9781003494478-10
Das, A. K., Gupta, N., Mahmood, T., Tripathy, B. C., Das, R., Das, S. (2025b). An innovative fuzzy multi-criteria decision making model for analyzing anthropogenic influences on urban river water quality. *Iran Journal of Computer Science, 8*, 103–124. https://doi.org/10.1007/s42044-024-00211-x

Das, A. P., Ghosh, S., Mohanty, S., & Sukla, L. B. (2015). Advances in manganese pollution and its Bioremediation. In L. B. Sukla, N. Pradhan, S. Panda, & B. K. Mishra (Eds.), *Environmental microbe biotechnology* (1st ed., pp. 313–328). Springer. https://doi.org/10.1007/978-3-319-19018-1_16

De Souza, R. G., Clímaco, J. C. N., Sant'Anna, A. P., Rocha, T. B., do Valle, R. D. A. B., & Quelhas, O. L. G. (2016). Sustainability assessment and prioritisation of e-waste management options in Brazil. *Waste Management, 57*, 46–56.

Dey, S., Dash, A., & Das, A. P. (2024). Bioremediation as a sustainable and circular solution for mining waste management and bioenergy generation. In *Sustainable management of mining waste and tailings: A circular economy approach* (pp. 107–119). https://doi.org/10.1201/9781003442455-6/BIOREMEDIATION-SUSTAINABLE-CIRCULAR-SOLUTION-MINING-WASTE-MANAGEMENT-BIOENERGY-GENERATION-SUDESHNA-DEY-AKANSHYA-DASH-ALOK-PRASAD-DAS

Duan, H., Eugster, M., Hischier, R., Streicher-Porte, M., & Li, J. (2009). Life cycle assessment study of a Chinese desktop personal computer. *Science of the Total Environment, 407*(5), 1755–1764.

Gregory, J., Magalini, F., Kuehr, R., & Huisman, J. (2009). E-waste take-back system design and policy approaches. In *Solving the e-waste problem (StEP), White Paper.*

Hasan, S. E. (2004). Public awareness is key to successful waste management. *Journal of Environmental Science and Health, Part A, 39*(2), 483–492.

Kahraman, C. (Ed.). (2008). *Fuzzy multi-criteria decision making: Theory and applications with recent developments*. Springer.

Kahraman, C., Cebeci, U., & Ruan, D. (2004). Multi-attribute comparison of catering service companies using fuzzy AHP: The case of Turkey. *International Journal of Production Economics, 87*(2), 171–184.

Karapetrovic, S., & Rosenbloom, E. S. (1999). A quality control approach to consistency paradoxes in AHP. *European Journal of Operational Research, 119*(3), 704–718.

Khetriwal, D. S., Kraeuchi, P., & Widmer, R. (2009). Producer responsibility for e-waste management: Key issues for consideration—Learning from the Swiss experience. *Journal of Environmental Management, 90*(1), 153–165.

Kiddee, P., Naidu, R., & Wong, M. H. (2013). Electronic waste management approaches: An overview. *Waste Management, 33*(5), 1237–1250.

Mishra, S., & Das, A. P. (2021). Current treatment technologies for removal of microplastic and microfiber pollutants from wastewater. In *Wastewater treatment: Cutting-edge molecular tools, techniques and applied aspects* (pp. 237–251). https://doi.org/10.1016/B978-0-12-821881-5.00011-8

Mishra, S., & Das, A. P. (2023). Microbial remediation of synthetic microfiber contaminated wastewater. In M. P. Shah (Ed.), *Microbial technologies in industrial wastewater treatment* (pp. 337–350). Springer. https://doi.org/10.1007/978-981-99-2435-6_17

Mishra, S., Dash, D., Al-Tawaha, A. R. M. S., & Das, A. P. (2022a). A review on heavy metal ion adsorption on synthetic microfiber surface in aquatic environments. *Applied Biochemistry and Biotechnology, 194*(10), 4639–4654. https://doi.org/10.1007/S12010-022-04029-W/METRICS

Mishra, S., Dash, D., & Das, A. P. (2022b). Detection, characterization and possible biofragmentation of synthetic microfibers released from domestic laundering wastewater as an emerging source of marine pollution. *Marine Pollution Bulletin, 185*, 114254. https://doi.org/10.1016/J.MARPOLBUL.2022.114254

Mishra, S., Ghosh, S., van Hullebusch, E. D., Singh, S., & Das, A. P. (2023). A critical review on the recovery of base and critical elements from electronic waste-contaminated streams using microbial biotechnology. *Applied Biochemistry and Biotechnology, 2023*, 1–30. https://doi.org/10.1007/S12010-023-04440-X

Prabhakar, A., Mishra, S., & Das, A. P. (2019). Isolation and identification of lead (Pb) solubilizing bacteria from automobile waste and its potential for recovery of lead from end of life waste batteries. *Geomicrobiological Journal, 36*(10), 894–903. https://doi.org/10.1080/01490451.2019.1654044

Priyadarsini, S., & Das, A. P. (2024). Analytical and structural characterization of waste lithium-ion batteries for their effective recycling strategy. *Environmental Chemistry and Ecotoxicology, 7*, 182. https://doi.org/10.1016/j.enceco.2024.12.004

Priyadarsini, S., & Das, A. P. (2025). Lithium bioleaching: A review on microbial-assisted sustainable technology for lithium bio-circularity. *Journal of Water Process Engineering, 69*, 106744. https://doi.org/10.1016/j.jwpe.2024.106744

Raj, K., & Das, A. P. (2023). Lead pollution: Impact on environment and human health and approach for a sustainable solution. *Environmental Chemistry and Ecotoxicology, 5*, 79–85. https://doi.org/10.1016/J.ENCECO.2023.02.001

Sahu, A., Dey, S., Mallick, K., Priyadarsini, S., Behera, I. D., & Das, A. P. (2024). A review on advanced technology for sustainable management of synthetic microplastic waste. In *Sustainable management of urban plastic waste through circular economic approaches* (pp. 262–279). https://doi.org/10.1201/9781003465447-15/REVIEW-ADVANCED-TECHNOLOGY-SUSTAINABLE-MANAGEMENT-SYNTHETIC-MICROPLASTIC-WASTE-AISHWARYA-SAHU-SUDESHNA-DEY-KRISHNAMAYEE-MALLICK-SAI-LAJA-PRIYADARSINI-IPSITA-DIPAMITRA-BEHERA-ALOK-PRASAD-DAS

Impact of Electronic Waste on Environmental Pollution and Its Sustainable Management Through Circular Economy

Shalini, Sailaja Priyadarsini, Alok Prasad Das, and V. P. Sharma

Abstract The improper handling of electronic garbage, or "e-waste," contributes to contamination of the air, water, and land, as well as negative health effects, making it a major global environmental concern. We have attempted to investigate the implications of e-waste on environmental degradation and human health while emphasizing sustainable management through circular economy strategies. Current trends in e-waste generation, coupled with global policies and innovative technologies, are discussed to provide insights into effective solutions. This chapter also presents case studies and examines how international collaborations can facilitate the transition from a linear to a circular economy. By integrating environmental sustainability with economic resilience, the proposed solutions aim to address e-waste challenges comprehensively.

Keywords Electronic waste · Circular economy · Environmental sustainability · E-waste management · Pollution · Recycling · Innovation

1 Introduction

Electronic waste, or "e-waste," is a major global environmental concern because inappropriate treatment of it can lead to negative health impacts and contamination of the air, water, and land (Mishra et al., 2023).

In certain nations, improper or insufficient scientific processing can have a negative impact on human health and pollute the environment. A global e-waste problem and possible environmental concern with harmful toxic effects have resulted from

Shalini · A. P. Das
CSIR-IITR, Lucknow, Uttar Pradesh, India

S. Priyadarsini · V. P. Sharma (✉)
Department of Life Sciences, Rama Devi Women's University, Bhubaneswar, Odisha, India

A. P. Das, S. Priyadarsini (eds.), *Electronic Waste and Environmental Pollution*, Sustainable Environmental Waste Management Strategies,
https://doi.org/10.1007/978-3-031-90287-1_18

the increasing use of electronic goods brought about by digitalization and advancements in research and development.

An electronic product generates electronic rubbish, or e-waste when its useful life is over. The consumer-centric society, the quick advancements in technology, and the fascination of young people with various instruments for wanting fast moving consumer goods (FMCG) or products are the main causes of these, which are typically produced in big quantities.

A decrease in the average life duration of computers and accessories, short innovation cycles, low recycling rates, and the frequent introduction of new models and needless purchases of electrical and electronic equipment (EEE) are the main causes of the exponentially growing amount of e-waste.

Environmental sustainability is seriously threatened by the commensurate rise in electronic trash, or "e-waste," brought on by the quick speed of technological advancement. Discarded electrical and electronic equipment (EEE) that contains hazardous compounds like lead and mercury as well as valuable materials like palladium, silver, and gold is referred to as e-waste.

Electronic garbage, or "e-waste," is being generated at previously unheard-of amounts due to the twenty-first century's rapid technological advancements. One of the waste streams with the fastest rate of growth in the world is e-waste, which includes abandoned electronic items including computers, smartphones, and home appliances. More than 50 million metric tons of e-waste are produced each year, according to the Global E-waste Monitor 2023 and other sources. If the right actions are not done, the amount is expected to increase by 2075 or the next century. Because of the migration and leaching of heavy metals like lead, mercury, cadmium, chromium, and brominated flame retardants, inappropriate handling and disposal of e-waste poses serious threats to the environment and human health.

Air quality deterioration, water pollution, and soil contamination are all caused by inappropriate disposal techniques including landfilling and incineration (Bala et al., 2024; Paul et al., 2023; Pradhan et al., 2023). The management of these hazardous materials presents major environmental and socioeconomic issues for developing nations, which are frequently overburdened by imported e-waste. Employees in the unregulated recycling industry are regularly exposed to hazardous substances, which can have a serious negative impact on their health. The impact of e-waste on the environment is described in Fig. 1.

E-waste offers a substantial chance for resource recovery despite these obstacles. Gold, silver, palladium, and other precious metals are among the valuable materials found in many modern products. By treating e-waste as a resource rather than waste, a circular economy might reduce environmental harm, preserve natural resources, and spur economic expansion (Dey et al., 2024a; Priyadarsini & Das, 2025). In order to reduce waste and prolong product lifecycles, the circular economy framework places a strong emphasis on the ideas of reducing, reusing, and recycling resources.

Global initiatives to manage e-waste have accelerated in recent years. The goal of regional programs like the European Union's Waste Electrical and Electronic Equipment (WEEE) Directive and international accords like the Basel Convention is to control the processing of e-waste and encourage environmentally friendly

Fig. 1 Impact of electronic waste on the environment

practices. These frameworks emphasize that in order to effectively address the e-waste challenge, governments, businesses, and consumers must act together.

E-waste is a complex topic, with research being done on its effects on the environment, human health, and sustainable treatment techniques. It draws attention to how the circular economy may help turn the problem of e-waste into a chance to achieve both economic and environmental sustainability. Mahatma Gandhi believed that the damage we are causing to the world's forests is merely a mirror image of the harm we are causing to ourselves and each other.

2 Global E-Waste Generation: Trends and Statistics

2.1 Global Trends

Global e-waste creation has reached about 53.6 million metric tons per year, according to recent data from the Global E-Waste Monitor (2024) and is expected to rise to 74 million metric tons by 2030 (GEM_2020_Executive_summery, 2021).

Globally, 75 million tons of e-waste are expected to be produced by 2030 and 111 million tons by 2050 (Parajuly et al., n.d.).

This increase is largely driven by the accelerating pace of digital transformation, shorter lifecycles of electronic devices, and consumer demand for new technologies. Asia remains the largest contributor, accounting for nearly 24.9 million metric tons annually, driven by rapid urbanization, industrialization, and technological consumption. Europe and the Americas follow, contributing 12 and 11.9 million metric tons, respectively. Africa and Oceania, though generating comparatively lower amounts, face significant challenges in sustainable waste management due to limited recycling infrastructure and regulatory frameworks (Adrian et al., n.d.).

The environmental implications of these trends are profound. E-waste contributes to approximately 70% of the toxic waste found in landfills globally. Furthermore, when electronic devices are disposed of improperly, toxic materials such as heavy metals like lead, cadmium, and mercury, as well as persistent organic pollutants (POPs), are released into the air, water, and soil, which have long-term effects on the environment and human health.

2.2 Regional Contributions

- Asia-Pacific: The region houses major e-waste producers, such as China and India, driven by their large populations, expanding middle class, and rapid economic growth. China alone generates over ten million metric tons annually, with significant contributions from the manufacturing and technology sectors. India, though producing less per capita, faces critical challenges with informal recycling sectors, where workers often dismantle electronic devices without adequate safety measures.
- Europe: Europe leads in regulatory compliance and e-waste recycling efforts, driven by stringent policies such as the Waste Electrical and Electronic Equipment (WEEE) Directive and the Circular Economy Action Plan. Countries like Germany and Switzerland boast advanced recycling facilities, achieving a formal collection rate of 55%, the highest globally. However, the region still struggles with the illegal export of e-waste to developing countries.
- Americas: In North America, the United States contributes significantly to global e-waste, generating approximately seven million metric tons annually. Despite high consumer electronics turnover, formal recycling rates remain low at around 15%, highlighting a critical gap in public awareness and infrastructure. Latin America faces unique challenges, including informal sector dominance and inadequate regulatory frameworks, leading to unregulated dumping and health risks (Baldé et al., n.d.).
- Africa: Despite having the lowest per capita e-waste production, Africa suffers greatly from the illicit dumping of e-waste from developed nations. Countries like Ghana and Nigeria are hotspots for informal recycling activities, where hazardous methods such as burning cables for metal recovery result in environmental

pollution and health risks. Local initiatives like the Agbogbloshie Makerspace Platform (AMP) are emerging to promote sustainable practices.

- Oceania: Despite high per capita e-waste generation, Oceania has limited recycling infrastructure, with Australia and New Zealand generating approximately 716,000 metric tons annually. Although they encounter obstacles in rural and isolated locations, initiatives such as Australia's National Television and Computer Recycling Scheme seek to address these issues (Van Yken et al., 2021).

Efforts to address e-waste issues globally must prioritize collaborative action, including strengthening international agreements like the Basel Convention and encouraging innovation in recycling technologies. Only through collective responsibility and sustainable practices can the adverse impacts of e-waste be mitigated.

3 Environmental Impacts of E-Waste on a Global Scale

3.1 Air Pollution

The open-air burning of electronic waste in informal dumpsites is a major contributor to air pollution globally. This practice releases a range of hazardous chemicals, including dioxins, furans, and other persistent organic pollutants (POPs), into the atmosphere. In addition to degrading local air quality, these pollutants release greenhouse gases into the atmosphere, which fuels global warming. These emissions are concentrated in areas like Ghana's Agbogbloshie, a well-known centre for unofficial e-waste recycling. The environment and public health are negatively impacted both immediately and over time by the poisonous clouds that are released when plastics, circuit boards, and wires are burned. Moreover, these pollutants can travel long distances, affecting air quality in neighbouring regions (Adrian et al., n.d.).

3.2 Soil and Groundwater Contamination

The improper disposal and processing of e-waste lead to severe contamination of soil and groundwater. The soil gradually becomes infertile due to the leaching of toxic heavy metals from abandoned electronic gadgets, including lead, cadmium, mercury, and arsenic. Additionally, these contaminants infiltrate groundwater systems, polluting drinking water supplies and compromising public health. The bioaccumulation of these substances in plants and animals enters the human food chain, causing chronic health issues. Urban areas in developing countries, where landfilling is a common practice, are particularly vulnerable. For instance, regions in South Asia and Africa, lacking stringent waste management regulations, experience widespread contamination from e-waste dumping.

3.3 Marine Pollution

Improper disposal of electronic waste has significant implications for marine environments. Plastics (Dey et al., 2024b; Singh Jadaun et al., 2022), heavy metals (Mishra et al., 2022), and toxic chemicals from e-waste often find their way into rivers and oceans, contributing to marine pollution. This type of pollution not only harms marine biodiversity but also poses risks to the food chain. Marine organisms ingest microplastics and toxic chemicals, which then accumulate in larger predators and eventually reach humans who consume seafood. Studies indicate that the presence of lead, cadmium, and flame retardants in marine organisms has escalated due to e-waste pollution. Coastal areas in regions like Southeast Asia, where informal e-waste processing is prevalent, are especially vulnerable to such contamination.

3.4 Human Health Hazards

The health effects of e-waste on people are concerning, especially for those who engage in unregulated recycling industries (Das & Mishra, 2008, 2009). Unprotected workers are exposed to hazardous substances like lead, cadmium, and brominated flame retardants. Numerous health issues, such as respiratory ailments, brain damage, kidney abnormalities, and some types of cancer, can result from long-term exposure to these compounds. Children working in these hazardous environments are at an even greater risk, often suffering from developmental issues and cognitive impairments due to their higher vulnerability to toxins. Furthermore, communities living near informal e-waste recycling hubs face elevated risks of disease due to contaminated air, water, and soil (Ankit et al., 2021).

Burning e-waste in open dumpsites releases toxic gases such as dioxins, contributing to air pollution and climate change. This issue is particularly prominent in regions like Agbogbloshie, Ghana (Njoku et al., 2024).

4 Global Regulatory Frameworks and Initiatives

4.1 International Agreements

- **Basel Convention**: One important international agreement addressing the regulation of transboundary transfers of hazardous waste, including e-waste, was the Basel Convention, which was established in 1989. Its main goals are to reduce the amount of hazardous waste produced, ensure that it is managed sustainably, and stop illicit dumping in underdeveloped nations. The Ban Amendment is one of the recent amendments to the convention that attempts to forbid the export of hazardous waste, including e-waste, from industrialized to underdeveloped countries. To increase compliance, nations are urged to create regional capacity-building initiatives and bolster enforcement systems.

- **Sustainable Development Goals (SDGs)**: The sustainable development goals (SDGs), which were adopted by the UN in 2015, provide an international framework for tackling environmental issues.
  - Goal 12: Recycling, reusing, and increasing resource efficiency are ways to reduce waste, according to responsible consumption and production. It explicitly promotes sustainable e-waste management practices to minimize environmental harm.
  - Goal 13: Climate action emphasizes the importance of reducing greenhouse gas emissions associated with e-waste generation and promoting sustainable technology design. Integrating these goals into national policies is vital to align e-waste management with broader sustainability objectives.
- **Minamata Convention on Mercury**: This treaty addresses mercury pollution and specifically targets e-waste, as many electronics contain mercury. The convention seeks to phase out products containing mercury and encourage the adoption of alternative materials, reducing the environmental and health risks associated with improper disposal of mercury-containing e-waste.

4.2 Regional Policies

- **European Union (EU):**
  - To encourage the collection, recycling, and recovery of e-waste, the Waste Electrical and Electronic Equipment Directive (WEEE Directive) was first established in 2003 and updated in 2012 (EU on Waste Electrical and Electronic Equipment (WEEE) (Text with EEA Relevance), 2012). It establishes required collection and recycling goals for member states and mandates that manufacturers pay for the end-of-life management of their products.
  - The WEEE Directive is supplemented by the Restriction of Hazardous Substances (RoHS) Directive, which limits the use of some hazardous materials in electronic equipment, including lead, cadmium, and brominated flame retardants. This ensures that products entering the EU market are less toxic and easier to recycle.
  - Initiatives like the Green Deal aim to transition the EU toward a circular economy by emphasizing resource efficiency and stricter recycling standards, with a focus on reducing the environmental impact of e-waste.
- **China**:
  - The Circular Economy Promotion Law, enacted in 2008, focuses on sustainable production and resource conservation. This legislation encourages

industries to adopt green manufacturing practices and improve e-waste recycling rates.
  - The establishment of collection and recycling systems is required of producers under the Regulation for the Administration of the Recovery and Disposal of Waste Electrical and Electronic Products (China WEEE). It also requires companies to provide detailed information about their recycling methods and compliance with environmental standards.
  - To lessen reliance on imported resources, China's "Urban Mining" programs encourage the recovery of valuable metals from e-waste. Government-funded programs and public-private partnerships have significantly expanded the capacity of formal recycling facilities.

- **United States**:
  - In the absence of a nationwide e-waste management law, certain states have put in place different rules. In order to finance recycling initiatives, California's Electronic Waste Recycling Act, for instance, mandates that consumers pay an upfront recycling fee when buying electronics.
  - The R2 (responsible recycling) Standard and e-Stewards Certification are voluntary industry standards that promote responsible e-waste recycling practices among recyclers.

- **India**:
  - India's E-Waste (Management) Rules, introduced in 2011 and revised in 2016, mandate extended producer responsibility (EPR) (Guidelines on Implementation of E-Waste (Management) Rules, 2016 Implementation Guidelines for E-Waste (Management) Rules, 2016 Central Pollution Control Board, Delhi, n.d.). Producers are required to establish collection mechanisms, such as take-back programs and collection centres.
  - The government has also partnered with NGOs and private organizations to raise awareness about e-waste disposal and encourage formal recycling practices.

- **Africa**:
  - Managing e-waste is a major problem for many African countries since they are frequently the destinations of illegal exports of discarded technology. In response, countries like Nigeria have developed national e-waste management policies to formalize the recycling sector and combat informal practices.
  - The Bamako Convention highlights the importance of environmentally responsible waste management and forbids the importation of hazardous garbage, including e-waste, into African countries.

5 Circular Economy Models for E-Waste Management

5.1 *Principles of Circular Economy*

The circular economy aims to rethink economic progress by emphasizing advantages for society and the environment. Its tenets in relation to e-waste include:

- **Design for Longevity and Repairability**: To increase their lifespan, products are made to be easily repaired, last longer, and have interchangeable parts.
- **Closing Resource Loops**: By creating closed-loop systems, materials recovered from e-waste are reintroduced into the production cycle, reducing dependence on virgin resources.
- **Shared Economy Practices**: Encouraging shared ownership models (e.g., leasing electronic devices instead of owning them) to maximize product utilization.
- **Regenerating Natural Systems**: Reducing waste and pollution through sustainable manufacturing practices, and ensuring that end-of-life products are processed in ways that regenerate natural ecosystems (Longhurst et al., 2023).
- **Digital Tools and Traceability**: Tracking the life cycle of electronic devices and increasing recycling and reuse transparency by utilizing technologies like blockchain.

5.2 *Global Success Stories*

- **Japan**:
  - In order to recover valuable metals like palladium, silver, and gold from abandoned electronics, Japan has implemented sophisticated urban mining procedures. Initiatives such as the Tokyo 2020 Olympics' "Medals Project" showcased this approach, where 5000 medals were produced using recycled metals from old devices.
  - Japan also emphasizes the Eco-Town Program, which promotes the integration of e-waste recycling facilities within urban centres, reducing logistical costs and encouraging community participation.
- **European Union**:
  - The EU's WEEE Directive mandates high recycling and recovery targets. Member states like Germany and Sweden have achieved compliance rates above 85% due to efficient collection systems and state-of-the-art recycling plants.
  - Circular Business Models: Companies like Fair Phone in the EU have pioneered modular smartphones that are easy to repair, upgrade, and recycle, reducing electronic waste generation.

- Eco-Design Directive: This initiative complements the WEEE Directive by requiring manufacturers to design products with improved energy efficiency and recyclability.

- **South Korea**:
  - South Korea employs smart collection systems using IoT-enabled bins that track e-waste deposits and provide incentives to users. The country's recycling facilities use automated technologies to sort and process e-waste, achieving high recovery rates for critical materials like cobalt and lithium.
- **India**:
  - India's extended producer responsibility (EPR) framework has formalized e-waste recycling, leading to the growth of eco-parks dedicated to e-waste management. Startups like Karo Sambhav partner with corporations to manage their e-waste streams, focusing on end-to-end solutions from collection to recycling.

5.3 *Role of Corporations*

The role of the corporation in e-waste management is described in Fig. 2.

- **Apple**:
  - Apple has implemented several initiatives to promote a circular economy:

    Trade-in Programs: Consumers are encouraged to trade in old devices, which Apple refurbishes or recycles.

Fig. 2 Role of corporation

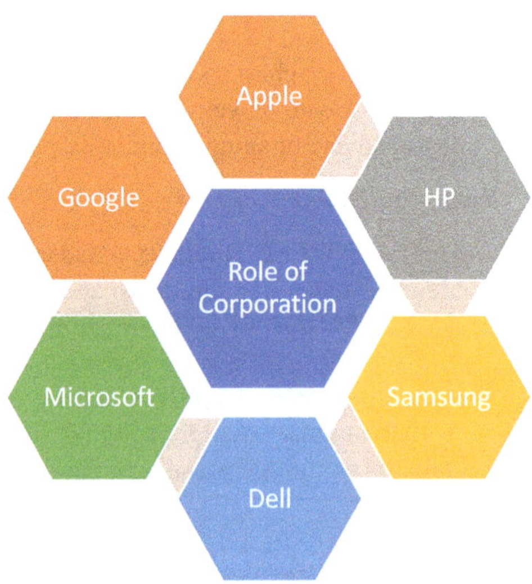

Liam and Daisy: Apple's disassembly robots recover valuable materials like cobalt, tungsten, and rare earth elements from old iPhones, reducing reliance on mining.
Apple has committed to using 100% recycled rare earth elements in its devices.

- **Dell**:
  - Dell runs a closed-loop recycling program, using plastics from old devices to manufacture new ones. The company also uses recycled carbon fibre in its laptops and has set ambitious goals for material recovery.
  - Their circular design framework emphasizes repairable and upgradable product designs to extend the lifespan of devices.
- **Samsung**:
  - Samsung's Galaxy Upcycling program transforms old smartphones into IoT devices, reducing e-waste generation.
  - The company has pledged to use sustainable packaging and has integrated recycled materials into its newer product lines.
- **HP**:
  - In 1991, HP Planet Partners launched a program for product returns and recycling. At the moment, HP gathers goods for recycling and resale in 76 nations and territories across the globe (HP Statement on E-Waste & Used Electronic Equipment, 2023).
  - The company uses ocean-bound plastics and post-consumer recycled content in manufacturing, showcasing its commitment to sustainability.
- **Microsoft**:
  - Microsoft has embraced a carbon-negative and waste-free strategy, focusing on refurbishing servers and electronic equipment from its data centres. These refurbished products are then resold or reused, aligning with circular economy principles.
- **Google**:
  - Google integrates circularity into its hardware division by designing products with recycled aluminium, plastics, and other materials. Its sustainability bond finances projects that include e-waste recycling and sustainable technology development.

6 Technological and Innovative Approaches

6.1 Artificial Intelligence (AI)

AI has revolutionized e-waste management by introducing automation and precision in various stages of recycling:

- **AI-Driven Sorting**: Robotics and advanced AI algorithms allow for the remarkably accurate identification and separation of e-waste components such as circuit boards, metals, and plastics. These systems use image recognition, infrared sensors, and machine learning to categorize materials quickly.
- **Predictive Maintenance**: AI tools are being used to predict the lifespan of electronic devices, enabling proactive refurbishment and repair, which extends product life and reduces waste (Olawade et al., 2024).
- **Data-Driven Decision-Making**: AI models analyse large datasets on e-waste generation patterns, enabling policymakers to optimize collection points, recycling centres, and transportation networks.
- **Circular Supply Chains**: Companies use AI to trace materials across the supply chain, ensuring compliance with circular economy principles and preventing illegal e-waste exports.

6.2 Advanced Recycling Techniques

Innovative recycling technologies have emerged to recover valuable materials while minimizing environmental harm:

- Hydrometallurgical Processes:
  - These involve the use of aqueous solutions to extract valuable metals, such as gold, silver, cobalt, and rare earth elements, from e-waste with high recovery efficiency (Ghulam & Abushammala, 2023).
  - Compared to traditional pyrometallurgical methods, hydrometallurgy produces less air pollution and requires lower energy consumption.
- **Pyrometallurgical Techniques**: While energy-intensive, these methods are used to smelt and recover metals like copper and aluminium. Advanced furnaces equipped with emission control systems have made this process more sustainable.
- **Cryogenic Recycling**: Cooling e-waste to extremely low temperatures makes it brittle, enabling easier separation of components such as metals and plastics. This technique reduces the need for chemical processing.
- **Electrochemical Recovery**: This technique recovers high-purity metals from streams of e-waste by electrolyzing them. Palladium and platinum are two rare metals that can be successfully removed from circuit boards using this method.
- **Laser Separation Technology:** In order to salvage reusable materials from printed circuit boards, lasers are used to de-solder and disassemble components.

6.3 Bioremediation

Bioremediation offers an eco-friendly approach to managing hazardous e-waste components:

- **Microbial Recovery of Metals**:
  - Some microbes may extract metals like copper, zinc, and gold from e-waste, including *Acidithiobacillus ferrooxidans* and *Leptospirillum ferrooxidans*.
  - In comparison to chemical leaching, this process, called bioleaching, is less expensive and has a smaller environmental impact.
- **Fungal Bioremediation**: Hazardous substances like brominated flame retardants are detoxified, and valuable metals are recovered from electronic waste using fungi like *Aspergillus niger*.
- **Phytoremediation**: Certain plants, referred to as hyperaccumulators, have the ability to absorb heavy metals from contaminated soils where e-waste has been inappropriately disposed of, including lead and cadmium.
- **Enzymatic Treatment**: Enzymes produced by microbes are being tested for their ability to break down hazardous materials like polybrominated diphenyl ethers (PBDEs) found in plastics.

6.4 Internet of Things (IoT)

IoT technologies facilitate effective resource use and real-time monitoring, which improves e-waste management:

- **Smart Bins:** Sensor-equipped IoT-enabled trash cans monitor the quantity of e-waste dumped, streamlining collection routes and cutting expenses.
- **Device Tracking**: IoT sensors embedded in electronics help monitor their condition and location, facilitating efficient take-back programs and recycling.
- **Reverse Logistics**: IoT systems optimize the collection, transportation, and processing of discarded electronics, ensuring timely recycling.

6.5 Blockchain Technology

Blockchain enhances transparency and traceability in e-waste management:

- **Material Traceability**: Blockchain tracks the journey of raw materials from mining to product disposal, ensuring compliance with ethical sourcing and recycling practices.

- **Preventing Counterfeit Recycling**: The technology helps verify that recycled materials are processed in certified facilities, preventing the use of informal recycling networks.
- **Smart Contracts**: Blockchain-based contracts automate payments for returned e-waste, incentivizing consumers to participate in recycling programs.

6.6 Robotics and Automation

- **Robotic Disassembly**: Robots equipped with vision systems and AI can disassemble complex electronic devices, such as smartphones and laptops, more efficiently than manual labour.
- Automation in Shredding: Automated shredding systems equipped with magnetic separators and eddy current separators enhance the recovery of metals from shredded e-waste.

6.7 3D Printing with Recycled Materials

- **Circular Manufacturing**: Recycled plastics from e-waste are used as feedstock for 3D printing, reducing the demand for virgin materials. This technology has applications in manufacturing spare parts and prototype components.

7 Public Awareness and Community Involvement

7.1 Role of NGOs

Non-governmental organizations (NGOs) play a critical role in bridging the gap between policies and grassroots implementation by engaging in education, advocacy, and action:

- **Basel Action Network (BAN)**: BAN monitors and campaigns against the illegal dumping of hazardous e-waste in developing countries. It advocates for transparent and ethical recycling practices to prevent environmental and health hazards.
- **Greenpeace**: Greenpeace campaigns for corporate accountability in e-waste management, urging companies to adopt environmentally friendly production and recycling practices. It also conducts awareness drives about the ecological and social impact of e-waste.
- **Toxics Link (India)**: This NGO promotes the formalization of the recycling industry and works to increase public awareness of the risks associated with inappropriate e-waste disposal. It collaborates with nearby communities to set up programs for recycling and collection.

- **Step Initiative**: This global initiative supports sustainable e-waste management by promoting research, capacity building, and public-private partnerships to tackle e-waste challenges.
- **E-Waste Coalition**: Supported by the United Nations, this coalition brings together multiple agencies to address the global e-waste crisis through education, capacity-building programs, and policy advocacy.

NGOs also engage in community-based programs that empower local stakeholders to adopt ethical recycling practices and reduce informal handling of e-waste.

7.2 Consumer Awareness

Consumer involvement is crucial for effective e-waste management. Awareness campaigns and education initiatives aim to inform individuals about the consequences of e-waste and their role in mitigating its impact:

- **Right to Repair Campaigns:**
  - These campaigns advocate for legislation that gives consumers the right to repair their devices rather than replace them. This movement reduces e-waste by encouraging repairability and extending the lifespan of electronics.
  - By promoting repair-friendly design and providing access to repair manuals and spare parts, these campaigns empower consumers to make sustainable choices.
- **E-Waste Collection Drives**:
  - Community programs organized by local governments, NGOs, and corporations encourage individuals to dispose of old electronics responsibly. Events like "E-Waste Recycling Week" provide accessible drop-off points for e-waste collection.
- **Public Education Initiatives**:
  - To inform the public about the dangers incorrect e-waste disposal poses to the environment and human health, schools, colleges, and community centres host lectures and seminars.
  - Interactive campaigns use social media platforms to share information on safe recycling practices, collection centres, and the value of circular economy principles.
- **Eco-Labeling:**
  - Labels like "Energy Star" or "EPEAT" help consumers identify eco-friendly and recyclable products. This encourages the purchase of sustainable electronics and fosters demand for green technologies.

- **Digital Platforms for Recycling**:
  - Apps and websites such as Recycle Nation and ZonTail provide information on the nearest certified recycling centres, making it easier for consumers to recycle responsibly.

7.3 Role of Youth and Educational Institutions

- **Student-Led Campaigns**: Students often take the lead in organizing awareness drives and conducting surveys to map e-waste disposal habits. They act as catalysts for change within their communities.
- **E-Waste Hackathons**: Universities and innovation hubs host hackathons to encourage young innovators to develop creative solutions for e-waste recycling and management.
- **Incorporating E-Waste Management in Curriculum**: Educational institutions integrate topics on sustainability and e-waste management into science and environmental studies, fostering eco-conscious behaviour from an early age.

7.4 Community Involvement

Engaging communities in e-waste management ensures a bottom-up approach to tackling the crisis:

- **E-Waste Banks**: Communities establish "e-waste banks" where individuals can deposit unused electronics in exchange for incentives such as discounts or vouchers for new devices.
- **Incentive Programs**: Municipalities and NGOs collaborate to offer incentives like tax credits, discounts, or free services to households and businesses that participate in e-waste recycling initiatives.
- **Volunteer Programs**: Volunteer networks help disseminate information about e-waste collection points, assist in organizing recycling events, and guide citizens on how to responsibly handle e-waste.

7.5 Media and Social Influence

Media campaigns play a vital role in shaping public attitudes and behaviours regarding e-waste:

- **Television and Radio Campaigns**: Informative advertisements and public service announcements (PSAs) broadcast on traditional media platforms reach wider audiences, including those in remote areas.

- **Social Media Outreach**: Platforms like Instagram, Facebook, and Twitter are used to run creative campaigns, challenges, and videos highlighting the importance of e-waste management. Hashtags like #RecycleElectronics and #StopEwaste have gained traction in global conversations.
- **Influencers and Celebrities**: Prominent figures endorse eco-friendly behaviour by promoting responsible consumption and disposal of electronics, thereby influencing their followers to adopt similar practices.

Organizations such as Basel Action Network advocate for ethical recycling practices and raise global awareness.

8 Case Studies

8.1 Europe

8.1.1 Norway

Norway's e-waste management system has emerged as a global benchmark, with over 68% e-waste recovery rates. The country has implemented strict regulations under the EU's Waste Electrical and Electronic Equipment (WEEE) Directive, focusing on producer responsibility and advanced recycling technologies. State-of-the-art recycling facilities in Norway extract valuable materials such as gold, silver, and rare earth elements from discarded electronics with high precision and minimal environmental impact. Additionally, public awareness campaigns encourage households and businesses to participate in formal recycling programs.

- **Key Drivers of Success**:
  - Robust infrastructure for e-waste collection and processing
  - Financial incentives for consumers to return obsolete electronics
  - Strict enforcement of recycling targets and penalties for non-compliance

8.1.2 Sweden

Sweden has adopted innovative practices, including "e-waste malls" like **ReTuna**, where discarded electronics are repaired and resold. This approach not only reduces waste but also promotes a culture of reuse and repair. The Swedish government supports these initiatives through subsidies and public-private partnerships, enhancing their scalability.

8.1.3 Germany

Germany has leveraged digital tools, such as the "E-Schrott App," to streamline e-waste disposal by providing users with real-time information on collection centres. High compliance with the WEEE Directive has resulted in Germany achieving recycling rates above 65%, significantly contributing to circular economy goals.

8.2 Asia

8.2.1 India

India's extended producer responsibility (EPR) framework has shown promising results in formalizing e-waste collection and recycling. Key highlights include:

- **Increased Compliance**: There were 4868 manufacturers registered under the framework in the fiscal year 2023–2024; 2209 of them had no EPR targets assigned to them, while 2659 had specified criteria to meet. The compliance returns of 3084 registered producers were successfully filed. By December 10, 2024, a remarkable 92% of the designated EPR targets had been met, demonstrating outstanding compliance progress (Report of CPCB in EA No. 04 of 2024 IN OA No. 512 of 2018 (Shailesh Singh Vs. State of U.P and Ors.), n.d.).
- **Eco-Parks**: The establishment of e-waste eco-parks in cities like Delhi has provided centralized hubs for recycling and refurbishing electronic goods.
- **Community Involvement**: NGOs and private entities work with local communities to raise awareness about proper e-waste disposal practices, helping divert waste from informal sectors.

8.2.2 China

China's Circular Economy Promotion Law and its "Home Appliance Old for New" program have significantly boosted e-waste recycling rates. Modern recycling facilities guarantee the recovery of precious and rare earth metals, and consumers are encouraged to exchange their old gadgets for savings on new purchases. Moreover, artificial intelligence (AI) technologies are deployed in sorting centres to enhance recycling efficiency.

8.2.3 Japan

Japan's "Home Appliance Recycling Law" mandates manufacturers to recycle four key appliances—televisions, air conditioners, refrigerators, and washing machines. The nation has achieved recycling rates exceeding 90% for some categories, with innovative technologies enabling the recovery of high-value metals. Initiatives such as the Tokyo 2020 Olympic Games sourcing recycled metals for medals from e-waste showcased Japan's leadership in sustainable practices.

8.2.4 South Korea

South Korea's pay-as-you-throw (PAYT) waste management system includes e-waste collection, encouraging citizens to dispose of electronics responsibly. Public-private partnerships have led to the development of efficient e-waste processing facilities, contributing to the country's high recycling rates.

8.3 Africa

8.3.1 Ghana

Pilot projects in Ghana have focused on formalizing the e-waste recycling sector, particularly in regions like Agbogbloshie, infamous for informal recycling practices. The introduction of certified collection centres and safe dismantling facilities has significantly reduced hazardous practices. Training programs for local workers provide education on safe recycling methods, minimizing health and environmental risks.

8.3.2 Nigeria

Nigeria has launched initiatives to combat the influx of illegal e-waste imports and encourage local recycling efforts. Through collaboration with international organizations, Nigeria has established state-of-the-art recycling facilities in Lagos and Abuja, improving resource recovery and reducing environmental pollution.

8.3.3 South Africa

The National Waste Management Strategy of South Africa emphasizes the integration of informal workers into the formal sector and sets particular goals for recycling e-waste. Public-private partnerships have driven the establishment of recycling facilities and provided training for workers, promoting a safer and more sustainable approach to e-waste management.

8.4 North America

8.4.1 United States

Several states, including California and New York, have implemented e-waste recycling laws requiring manufacturers to fund recycling programs. Initiatives like the "e-Stewards Certification" ensure ethical recycling practices, while innovative start-ups focus on recovering rare earth elements from electronic waste.

8.4.2 Canada

Producer responsibility is a key component of Canada's e-waste management initiatives, and governments like British Columbia have set high criteria for their recycling and collection practices. Public awareness campaigns and partnerships with retailers facilitate convenient drop-off points for e-waste recycling.

8.5 Australia

Australia's **National Television and Computer Recycling Scheme (NTCRS)** has achieved notable success in diverting millions of tonnes of e-waste from landfills. Public campaigns emphasize responsible consumption, while advancements in recycling technologies enable efficient recovery of valuable materials.

9 Challenges in Global E-Waste Management

9.1 Informal Recycling Sector

The informal recycling sector remains a significant challenge in developing nations, where the lack of regulatory enforcement and infrastructure enables unsafe practices. To extract valuable metals, workers in informal recycling frequently employ crude techniques like open burning and acid baths, which pose serious health risks and contaminate the environment.

- Health Risks: Toxic chemicals like lead, mercury, and cadmium are exposed to workers, causing chronic illnesses, neurological damage, and respiratory problems.
- Environmental Impacts: Improper disposal releases harmful chemicals into soil and water systems, degrading ecosystems and contaminating food chains.

- Barriers to Formalization: Efforts to transition workers from the informal to the formal sector face challenges such as lack of awareness, inadequate funding, and resistance due to loss of livelihood.

9.2 *Financial Constraints*

Advanced recycling technologies, such as hydrometallurgical and pyrometallurgical processes, require significant capital investment, making them inaccessible to low-income regions.

- Infrastructure Costs: Developing collection, transportation, and processing facilities requires substantial investment, which many countries lack.
- Limited Government Support: Many governments in low- and middle-income nations do not allocate sufficient budgets to e-waste management programs.
- Dependence on External Funding: International aid and grants often serve as primary sources of funding, which may not be sustainable in the long-term.

9.3 *Lack of Consumer Awareness*

In addition to a lack of knowledge about appropriate e-waste disposal procedures, many customers accumulate gadgets or dispose of them in landfills.

- Cultural Attitudes: In some regions, e-waste is not perceived as hazardous, resulting in low participation in recycling programs.
- Absence of Incentives: The lack of financial or convenience-based incentives discourages consumers from engaging in responsible disposal.
- Insufficient Education Campaigns: Public awareness campaigns are often limited in scope and fail to address the root causes of improper e-waste handling.

9.4 *Regulatory Gaps*

The absence of robust regulatory frameworks in many countries hampers effective e-waste management.

- Inconsistent Policies: Regulations vary widely between countries, with some nations having comprehensive laws and others having none.
- Weak Enforcement: Even in countries with existing e-waste laws, enforcement is often weak, allowing illegal dumping and informal recycling to persist.

- Global Trade of E-Waste: The transboundary movement of e-waste to countries with lax regulations exacerbates the problem, creating environmental justice issues.

9.5 Technological Limitations

While advancements in recycling technologies hold promise, their adoption remains limited.

- Low Recovery Rates: Current technologies may not efficiently recover all valuable materials, leading to resource loss.
- Complexity of Modern Electronics: The increasing miniaturization and complexity of electronic devices make dismantling and material separation more challenging.
- Dependence on Manual Labour: In many recycling facilities, manual labour is still required for sorting and dismantling, increasing costs and risks.

9.6 Evolving Nature of E-Waste

The fast pace of technological advancement results in shorter product lifecycles, which raises the amount of e-waste produced.

- Planned Obsolescence: Manufacturers design products with limited lifespans, encouraging frequent replacements.
- Emergence of New Waste Streams: Devices such as wearables and smart home technologies add complexity to waste management.
- Data Security Concerns: The presence of sensitive information in discarded electronics deters some consumers from recycling.

9.7 International Disparities

The global nature of e-waste management reveals stark disparities between developed and developing nations.

- Export of E-Waste: High-income countries often export e-waste to low-income nations, where it is processed under hazardous conditions.
- Technological Divide: Developed nations have access to advanced recycling technologies while developing countries rely on basic methods.
- Inadequate Support for Developing Nations: International initiatives often fall short of providing the necessary resources and training to build local capacity.

10 Policy Recommendations

10.1 Strengthen International Collaborations to Harmonize Regulations

- Unified Standards: Develop globally accepted standards for e-waste classification, handling, and recycling to reduce regulatory discrepancies.
- Cross-Border Enforcement: To stop illicit transboundary e-waste movements, nations should work together more closely while upholding accords like the Basel Convention.
- Global Fund for E-Waste Management: To help low- and middle-income countries build e-waste infrastructure, create an international fund backed by developed countries, multinational firms, and international organizations.
- Knowledge Sharing: Facilitate the exchange of technological advancements, best practices, and research through platforms like UN bodies or regional alliances.

10.2 Enhance Extended Producer Responsibility (EPR) Frameworks Globally

- Mandatory Compliance: Implement stronger EPR laws that mandate manufacturers assume operational and financial accountability for the safe disposal, recycling, and recovery of their goods.
- Incentivize Eco-Design: Designing items that are simpler to upgrade, repair, and recycle will encourage manufacturers to embrace the concepts of the circular economy.
- Third-Party Auditing: Implement third-party audits to monitor EPR compliance and penalize non-adherence.
- Inclusion of Informal Sector: Develop mechanisms to integrate informal recyclers into formal EPR systems, providing training and financial support to improve their practices.

10.3 Invest in Public Awareness and Education to Promote Recycling Behaviours

- Comprehensive Awareness Campaigns: Launch widespread educational programs highlighting the environmental, health, and economic impacts of improper e-waste disposal.
- School and University Curriculums: Integrate e-waste management topics into environmental science and sustainability courses at various educational levels.

- Behavioural Incentives: Introduce incentives such as discounts, vouchers, or tax credits for consumers who return their old electronics for recycling.
- Targeted Outreach: Develop campaigns tailored to specific demographics, such as youth, rural communities, and urban professionals, using relatable messages and local languages.

10.4 Foster Innovation in Recycling Technologies

- Government Grants: Provide funding and tax benefits for research and development in advanced recycling techniques, such as AI-driven sorting systems and bioremediation.
- Collaborative Innovation Hubs: Establish public-private partnerships to create innovation hubs focused on sustainable e-waste management solutions.
- Scaling Proven Technologies: Facilitate the global deployment of successful technologies from countries like Japan and the European Union in developing regions.

10.5 Strengthen Local and Regional Policies

- Decentralized Waste Management: Empower local governments to develop region-specific strategies for e-waste collection, recycling, and disposal.
- Mandatory Recycling Targets: Set legally binding recycling targets for both corporations and municipalities, with transparent reporting mechanisms.
- Formalizing Informal Sector: Allocate funds to formalize and regulate informal e-waste handlers, ensuring safer working conditions and compliance with environmental laws.

10.6 Encourage Corporate Responsibility Beyond EPR

- Transparent Supply Chains: Require companies to disclose their sourcing and recycling processes to promote accountability.
- Voluntary Sustainability Pledges: Encourage companies to exceed regulatory requirements by adopting voluntary sustainability goals.
- Product Longevity Programs: Promote programs like trade-ins, buybacks, and refurbishment to extend product lifecycles.

10.7 Address Economic Barriers to Recycling

- Subsidize Infrastructure: Provide subsidies or low-interest loans to set up recycling and waste-processing facilities, especially in underdeveloped regions.
- Reduce Taxes on Recycled Products: Lower taxes and duties on products made from recycled materials to encourage market competitiveness.
- Job Creation: Promote recycling as a sector that creates jobs by investing in training programs and establishing recycling hubs.

10.8 Expand Research and Data Collection

- Global E-Waste Database: Create a centralized database to track e-waste generation, movement, and recycling rates worldwide.
- Health Impact Studies: Fund longitudinal studies on the health impacts of improper e-waste recycling, particularly in developing countries.
- Technology Assessment: Regularly evaluate emerging recycling technologies for scalability, cost-effectiveness, and environmental impact.

10.9 Promote Consumer Responsibility

- Right-to-Repair Legislation: Pass laws mandating manufacturers to provide consumers with tools, parts, and information necessary for repairing their devices.
- Deposit-Refund Schemes: Introduce programs that allow customers to pay a deposit for electronics purchases and receive their money back when they return the device for recycling.
- E-Waste Collection Drives: Organize community collection events to make recycling convenient and accessible for all.
- Strengthen international collaborations to harmonize regulations.
- Enhance extended producer responsibility (EPR) frameworks globally.
- Invest in public awareness and education to promote recycling behaviours.

11 Conclusion

E-waste management requires a unified global effort that integrates circular economy principles. Countries can reduce the negative effects of e-waste on the environment and human health while opening up new economic opportunities by coordinating their regulatory policies, technical advancements, and community involvement. Illegal recycling, inappropriate landfill disposal, clandestine exports, and other problems are some of the ways that e-waste tends to impact the world. The environment is seriously harmed by discarded electronic waste because it releases toxicants such as heavy metals, non-ferrous materials, glass, microplastics, and others that have an adverse effect on the land, water, and air, raising concerns about ecosystem health. The need for trash management and monitoring has been emphasized by both national and international organizations.

Innovative e-waste policies from many nations, such as extended producer responsibility (EPR), WEEE directives, and regional regulatory standards with effective tactics, including recycling incentives and more stringent laws on hazardous materials, must be understood.

References

Adrian, S., Drisse, M. B., Cheng, Y., Devia, L., Deubzer, O., Goldizen, F., Gorman, J., Herat, S., Honda, S., Iattoni, G., Jingwei, W., Jinhui, L., Khetriwal, D. S., Linnell, J., Magalini, F., Nnororm, I. C., Onianwa, P., Ott, D., Ramola, A., … Zeng, X. (n.d.). *Quantities, flows, and the circular economy potential The Global E-waste Monitor 2020.*

Ankit, Saha, L., Kumar, V., Tiwari, J., Sweta, Rawat, S., Singh, J., & Bauddh, K. (2021). Electronic waste and their leachates impact on human health and environment: Global ecological threat and management. *Environmental Technology and Innovation, 24*. https://doi.org/10.1016/j.eti.2021.102049

Bala, K., Dogra, A., Lata, P., & Das, A. P. (2024). Discussing sources and biological-chemical remediation approaches for microplastic pollution. In *Sustainable microbial technology for synthetic and cellulosic microfiber bioremediation* (pp. 37–59). https://doi.org/10.1007/978-3-031-62600-5_3

Baldé, C. P., Kuehr, R., Yamamoto, T., McDonald, R., Althaf, S., Bel, G., Deubzer, O., Fernandez-Cubillo, E., Forti, V., Gray, V., Herat, S., Honda, S., Iattoni, G., Khetriwal, D. S., & Luda di Cortemiglia, V. (n.d.). *Required Citation*. https://www.itu.int/itu-d/sites/environment

Das, A. P., & Mishra, S. (2008). Hexavalent chromium (VI): Health hazards and environmental pollutant. *Journal of Environmental Research and Development, 2*(3), 386–392.

Das, A. P., & Mishra, S. (2009). Hexavalent chromium [Cr (VI)]: Yellow water pollution and its remediation. *Sarovar Saurabh ENVIS Newsletter on Wetland Ecosystems, 5*(2).

Dey, S., Dash, A., & Das, A. P. (2024a). Bioremediation as a sustainable and circular solution for mining waste management and bioenergy generation. In *Sustainable management of mining waste and tailings: A circular economy approach* (pp. 107–119). https://doi.org/10.1201/9781003442455-6/BIOREMEDIATION-SUSTAINABLE-CIRCULAR-SOLUTION-MINING-WASTE-MANAGEMENT-BIOENERGY-GENERATION-SUDESHNA-DEY-AKANSHYA-DASH-ALOK-PRASAD-DAS

Dey, S., Dash, N., Ray, S. K., Priyadarsini, S., Chhotaray, C., & Das, A. P. (2024b). Occurrence and quantification of tire wear plastics pollutants from highway city dust of Bhubaneswar city of Odisha. In *Sustainable microbial technology for synthetic and cellulosic microfiber bioremediation* (pp. 77–89). https://doi.org/10.1007/978-3-031-62600-5_5

EU on waste electrical and electronic equipment (WEEE) (Text with EEA relevance). (2012). http://data.europa.eu/eli/dir/2024/884/oj

Global Entrepreneurship Monitor (GEM) 2020/2021 Global Report; uploaded on 03-05-2021, 2021

Ghulam, S. T., & Abushammala, H. (2023). Challenges and opportunities in the management of electronic waste and its impact on human health and environment. *Sustainability (Switzerland), 15*(3), 1837. https://doi.org/10.3390/su15031837

Global E waste Monitor, (2024). e report published by United Nations Institute for Training and Research (UNITAR) and International Telecommunication Union, (4th Edn; 2024, pages 1–140)

Guidelines on Implementation of E-Waste (Management) Rules, 2016 Implementation Guidelines for E-Waste (Management) Rules, 2016 Central Pollution Control Board, Delhi. (n.d.).

HP Statement on E-Waste & Used Electronic Equipment. (2023). http://hp.com/recycle

Longhurst, M., Arribas, M., Blériot, J., Boullenger, G., Buckley, M., Cantu, A., Chen, X., Defruyt, S., Gravis, L., Miserocchi, A., Opsomer, R., Vojvodic, T., De Vries, J., & Wachholz, C. (2023). *Initial considerations for global rules in the international legally binding instrument to end plastic pollution*.

Mishra, S., Dash, D., Al-Tawaha, A. R. M. S., & Das, A. P. (2022). A review on heavy metal ion adsorption on synthetic microfiber surface in aquatic environments. *Applied Biochemistry and Biotechnology, 194*(10), 4639–4654. https://doi.org/10.1007/S12010-022-04029-W/METRICS

Mishra, S., Ghosh, S., van Hullebusch, E. D., Singh, S., & Das, A. P. (2023). A critical review on the recovery of base and critical elements from electronic waste-contaminated streams using microbial biotechnology. *Applied Biochemistry and Biotechnology, 2023*, 1–30. https://doi.org/10.1007/S12010-023-04440-X

Njoku, A., Agbalenyo, M., Laude, J., Ajibola, T. F., Attah, M. A., & Sarko, S. B. (2024). Environmental injustice and electronic waste in Ghana: Challenges and recommendations. *International Journal of Environmental Research and Public Health, 21*(1). https://doi.org/10.3390/ijerph21010025

Olawade, D. B., Wada, O. Z., David-Olawade, A. C., Fapohunda, O., Ige, A. O., & Ling, J. (2024). Artificial intelligence potential for net zero sustainability: Current evidence and prospects. *Next Sustainability, 4*, 100041. https://doi.org/10.1016/j.nxsust.2024.100041

Parajuly, K., Kuehr, R., Awasthi, A. K., Fitzpatrick, C., Lepawsky, J., Smith, E., Widmer, R., & Zeng, X. (n.d.). *Future e-waste scenarios*. www.unu.edu

Paul, A., Dey, S., Ram, D. K., & Das, A. P. (2023). Hexavalent chromium pollution and its sustainable management through bioremediation. *Geomicrobiology Journal, 41*, 1–11. https://doi.org/10.1080/01490451.2023.2218377

Pradhan, G., Tripathy, B., Ram, D. K., Digal, A. K., & Das, A. P. (2023). Bauxite mining waste pollution and its sustainable management through bioremediation. *Geomicrobiology Journal, 41*, 1–10. https://doi.org/10.1080/01490451.2023.2235353

Priyadarsini, S., & Das, A. P. (2025). Lithium bioleaching: A review on microbial-assisted sustainable technology for lithium bio-circularity. *Journal of Water Process Engineering, 69*, 106744. https://doi.org/10.1016/j.jwpe.2024.106744

Report of CPCB in EA No. 04 of 2024 IN OA No. 512 of 2018 (Shailesh Singh Vs. State of U.P & Ors.). (n.d.).

Singh Jadaun, J., Bansal, S., Sonthalia, A., Rai, A. K., & Singh, S. P. (2022). Biodegradation of plastics for sustainable environment. *Bioresource Technology, 347*, 126697. https://doi.org/10.1016/J.BIORTECH.2022.126697

Van Yken, J., Boxall, N. J., Cheng, K. Y., Nikoloski, A. N., Moheimani, N. R., & Kaksonen, A. H. (2021). E-waste recycling and resource recovery: A review on technologies, barriers and enablers with a focus on Oceania. *Metals, 11*(8), 1313. https://doi.org/10.3390/met11081313

Challenges and Opportunities for Effective E-Waste Management in Odisha: A Household-Level Study

Diptimayee Mishra and Pranaya Patra

Abstract The twenty-first-century human beings are experiencing rampant growth of electronic devices. The increasing necessity of human society has been challenged by the explosive e-waste generation. In this connection, this research work focuses on e-waste awareness among households in Odisha. The research is based on primary data. Data has been gathered through an extensive household survey. Using a structured questionnaire, information is gathered from a heterogeneous group of households to assess their knowledge, attitudes, and behaviours concerning e-waste. Our findings reveal that mobile phones, televisions, cables, and chargers are the most prevalent e-waste items in households. The primary cause of e-waste is the upgradation in technology combined with a lack of proper repair services. The analysis reveals that most people are unaware of the hazardous substances in e-waste, with only some sections recognizing the potential risks posed by these substances. The study results also reveal that the majority of people either discard their obsolete devices in regular trash or sell them to scrap dealers. The challenges backed by demographic and socio-economic variables inhibiting the proper disposal practices have also been highlighted in the study. The findings of the study suggest an urgent need for heightened information and awareness campaigns, along with the establishment of additional collection points, recycling centres, and flexible policies and regulations for effective e-waste management.

Keywords Electronic devices · E-waste · Odisha

D. Mishra (✉)
PG Department of Economics, Berhampur University, Brahmapur, Odisha, India

P. Patra
Department of Integrated B.ED-M.ED, Maharaja Sri Ramachandra Bhanja Deo University, Baripada, Odisha, India

A. P. Das, S. Priyadarsini (eds.), *Electronic Waste and Environmental Pollution*, Sustainable Environmental Waste Management Strategies,
https://doi.org/10.1007/978-3-031-90287-1_19

1 Introduction

The adoption of electronic technology has surged dramatically in a few decades, leading to the alarming rise of waste electronics or e-waste globally (Priyadarsini & Das, 2025). This toxic e-waste represents a significant burden on our planet. Also known as waste electrical and electronic equipment (WEEE) or end-of-life electronics, e-waste includes discarded devices ranging from large and small household appliances to telecommunication equipment, automatic dispensers, lighting fixtures, and various related components deemed obsolete or unwanted by users (European Commission, 2013; Qu et al., 2013). Global e-waste generation currently escalates by 2.6 million tonnes yearly, projecting a staggering 82 million by 2030 (The global e-waste Monitor, 2024). E-waste poses a critical environmental threat and can be categorized into seven major types: temperature exchange equipment, lamps, screens and monitors, large equipment, and small IT and telecommunication devices (Kumar et al., 2017). The quick expansion in e-waste generation is a pressing challenge that demands immediate attention from developed and underdeveloped nations (Mishra et al., 2023). Exposure to e-waste is linked to serious health issues, such as berylliosis, hard metal lung diseases, Itai Itai disease, and Minamata disease (Gauba et al., 2016, (Das & Mishra, 2008; Das & Singh, 2011; Raj & Das, 2023). The health risks associated with e-waste exposure are profound, leading to significant changes in cellular function, behavioural and emotional issues, and severe neonatal complications including increased rates of spontaneous abortions, premature births, stillbirths, and diminished birth weights (Grant et al., 2013; Behera et al., 2020; Ghosh et al., 2018). WEEE creates an undeniable global threat. Its toxic emissions infiltrate soil and air, wreaking havoc on biodiversity (Das et al., 2015b; Ghosh & Das, 2015). The direct consequences include releasing hazardous compounds, acids, heavy metals, and carcinogens, while indirect effects involve the dangerous biomagnification of heavy metals (Sivaramanan, 2013). The condition and density of e-waste directly determine its harmful impact. E-waste including discarded televisions, computers, stereos, and batteries causes soil acidification and groundwater contamination (Sankhla et al., 2016; Priyadarsini & Das, 2024). Alarmingly, the cadmium in just one mobile phone battery can pollute 600,000 L of water, rendering it unsafe for human consumption and devastating to aquatic ecosystems (Polak & Drapalova, 2012). Therefore, there is a need to address and prioritize the handling of e-waste to mitigate its dire repercussions.

E-waste is undeniably a significant source of noxious elements that threaten human health and environmental sustainability, at the same time it also serves as a vital reservoir of valuable materials (Paul et al., 2023; Pradhan et al., 2023). The recycling of e-waste is crucial for recovering valuable metals such as platinum, gold, manganese, and silver, along with rare earth elements (Das et al., 2011, 2015a; Ghosh et al., 2016). This presents substantial opportunities for the recycling sector. Electronic devices consist of various precious metals, including gold, silver, and platinum, as well as rare earth minerals like cerium, neodymium, and yttrium, alongside base metals such as copper, zinc, and nickel (Rene et al., 2021). For instance, recycling one million smartphones yields an impressive 24 kg of gold, 900 kg of copper, 250 kg of silver, and 9 kg of palladium. Notably, it was found that the global e-waste inventory contained 300 tonnes of gold, valued at 10.4 billion euros in 2014 (Ramanayaka et al., 2019). Proper e-waste recycling generates economic value and converts recovered materials into secondary resources for new electronic devices, significantly reducing our dependency on primary raw materials (Tanskanen, 2013). The Global E-waste Generation Report reveals that e-waste generation impacted 53.6 million tonnes (Mt) in 2019, with a mere 17.4% officially collected and responsibly recycled. This amount has surged from 44.7 Mt in 2016 and is projected to escalate further to 74.7 Mt by 2030, with an annual growth rate nearing 2 Mt (Forti et al., 2020; Ahirwar & Tripathi, 2021). The e-waste generation is shaped by critical elements like time, geographical destination, lifestyle choices, population dynamics, and socio-economic conditions. In 2019, Asia emerged as the leading region in e-waste generation, producing 24.9 Mt, followed by America with 13.1 Mt, Europe at 12 Mt, Africa generating 2.9 Mt, and Oceania contributing 0.7 Mt (Rautela et al., 2021).

India stands as one of the top global consumers of tech gadgets. The rapid progress in social and economic development driven by significant advancements in information technology (IT) has spotlighted a critical global challenge. Developing nations, including India, must confront the urgent social and environmental impacts of electronic waste (e-waste) as they navigate the complexities of sustainable collection, recycling, and treatment of discarded electronics (Chancerel et al., 2009). As of 2016, India ranks as the fifth-largest producer of e-waste globally, generating 1.85 million tonnes—approximately 12% of total global e-waste production. Medical and electrical equipment are responsible for 8% and 7% of this output, respectively (Garg & Adhana, 2019). An overwhelming 75% of e-waste originates from computer devices, with the telecom sector contributing an additional 12%. The Indian e-waste recycling industry is predominated by the informal sector, where workers are subjected to low-tech skills that pose significant health and environmental threats. Conditions such as open-air incineration and wet chemical leaching of metals occur without adequate solutions for addressing hazardous components in e-waste (United Nations Environment Programme and United Nations University, 2009). The relentless pace of technological breakthroughs and the widespread adoption of electronic appliances have led to a staggering expansion in e-waste generation in major metropolitan areas, including Delhi, Bangalore, Pune, and Kolkata. In 2011 alone, annual e-waste generation reached 470,000 tonnes. Alarmingly, a mere 6% of this e-waste was recycled, with the informal sector managing a startling 95% (Singla & Bansal, 2012). Among the leading states and Union Territories in e-waste generation, Maharashtra tops the list with 19.8%, followed closely by Tamil Nadu at 13.1%, Andhra Pradesh at 12.5%, and Uttar Pradesh at 10.1%. About city-wise contributions, Mumbai leads at 24.0%, followed by Delhi at 21.2% and Bangalore at 10.1% (Chatterjee, 2011). The average lifespan of electronic devices in India is diminishing rapidly. Currently, the average life span of a cell phone is 2 years, while personal computers and electric kettles typically last around 3 years. Other appliances, such as food mixers, televisions, DVDs, and toasters, usually last about

5 years. Items like vacuum cleaners, irons, freezers, hair dryers, and radios have an average lifespan of 10 years, whereas air conditioners can last up to 12 years (Gupta & Kumar, 2014). Wath et al. (2010) in their study highlighted that the Indian e-waste system demands urgent reforms that embrace scientific procedural standards and environmentally sound technologies for more effective handling.

Odisha is one of India's fastest-growing states, with a robust economy valued at $100 billion and an impressive average growth rate of 7.25% since 2015–2016, surpassing the national average by 1%. The state demonstrates its commitment to progress by allocating 7.61% of its budget to Electronics and Information Technology (People Guide to Budget, Govt of Odisha, 2024–2025). As technology becomes an essential aspect of everyday life, the boom in using electronic gadgets like personal computers, smartphones, and home appliances in Odisha is undeniable. Waste accumulation in the state has dramatically increased. In 1999–2000, daily waste generation was 646 tonnes, but that number soared to 2239 tonnes by 2011–2012 and continues to rise annually. Out of the waste produced, an impressive 1837 tonnes are collected per day; however, only 33 tonnes are treated (Das et al., 2017). Bhubaneswar, the capital of Odisha, epitomizes a smart city. It is the second-fastest growing urban area in Eastern India, following Kolkata and boasts a densely populated environment. Presently, Bhubaneswar's solid waste generation reaches approximately 520 tonnes per day. Of this solid waste, 61.81% is biodegradable, 27.15% is an inert material, 7.8% consists of plastics and leather, and 1.25% is metals and glass (Mohanty et al., 2022). A recent report on e-waste reveals India has 312 authorized recycling facilities that can collectively process 782,080 metric tonnes of waste. Whereas Odisha contributes a processing capacity of 3680 metric tonnes of e-waste each year (Biswas & Singh, 2020). Against this backdrop, the present study tries to analyse the major e-waste items in households and to examine the e-waste disposal practices, attitudes, and opinions of individuals at the household level. The study also attempts to unravel the level of awareness among respondents.

2 Review of Literature

Over the past few years, the number of published studies on waste electronic (e-waste) has surged, reflecting the urgent concerns surrounding this issue. The literature comprehensively addresses a wide array of topics related to e-waste. We have systematically categorized these studies into two groups: those focused on developed nations and those examining developing nations.

2.1 Evidence from Developed Countries

A study by Sthiannopkao and Wong (2013) on developed countries prioritizes the regulation of e-waste disposal through extended producer responsibility for e-waste disposal; a significant amount of e-waste is being shipped to developing countries such as China, India, Pakistan, and Nigeria. Kiddee et al. (2013) conclusively found that e-waste contains toxic substances that severely threaten the environment and human health if not managed properly. Various effective tools such as life cycle assessment (LCA), material flow analysis (MFA), multi-criteria analysis (MCA), and extended producer responsibility (EPR) have been designed to manage electronics waste, particularly in developed nations. Another important observation by Kumar et al. (2017) is the lifespan of electronic devices, particularly smaller gadgets like cell phones, lightweight laptops, and tablets, is shrinking. As a result, within 4–5 years, nearly 1 billion devices will be discarded. According to Robinson (2009), global e-waste production is estimated to be 20–25 million tonnes per year, with the majority produced in Europe, the US, and Australasia. It is a foregone conclusion that Eastern Europe, Latin America, and China will emerge as a leading generator of electronic waste over the next decades. A study by Tanskanen (2013) reveals that two-thirds of e-waste generated in European countries is still being either landfilled or exported to developing countries with proper recycling methods. A study by Mmereki et al. (2016) compellingly reveals that the governments of many developed countries have implemented environmentally sound management systems to minimize the reuse, recovery, and recycling of valuable materials, setting a high priority compared to developing countries. A recent study conducted in UAE by Attia et al. (2021) shows that only 21% of people have a high awareness of the concept of e-waste and e-waste recycling, whereas 33.3% are completely unaware of e-waste. Shockingly, the majority (73.3%) of young respondents (age groups of about 18–24 and 25–29) do not know about e-waste, and only 50.9% of respondents are familiar with e-waste recycling. Besides a major portion of young people, about 77.2% have no idea about e-waste recycling centres. A study by Alghazo et al. (2019) estimated the e-waste in the Gulf Cooperation Council (GCC) region to be 857 Kt (Kilotonnes), with 3–5% annual growth of e-waste. A forecasting model from 2018 to 2040 estimated e-waste generation in GCC to grow significantly to 1.094 Mt. by 2040. Therefore, for the capable handling of e-waste in the GCC region and more efficient recycling facilities, strong legislation with more advanced technology is required.

2.2 *Evidence from Developing Countries*

The paper by Wath et al. (2010) presents a comprehensive review and assessment of e-waste management systems in developed and developing countries. It highlights the challenges faced by rapidly developing countries like India in applying the e-waste disposal framework of developed countries. Those challenges include adverse socio-economic conditions, poor infrastructure, and inappropriate legislation for electronic waste among others. Another study by Bhatia et al. (2019) on e-waste management awareness unequivocally found that awareness increases with income. Many electronic device users discard their devices much earlier than their recommended lifespan. Often unaware of the proper disposal methods they keep the e-waste at home or sell it to informal recyclers instead of handing it over to authorized e-waste collectors. A study in South Africa by Uhunamure et al. (2021) revealed alarming environmental consequences and health risks linked with e-waste. The study unequivocally found that most respondents believe e-waste has negatively affected the environment and that most households are willing to contribute towards proper e-waste practices. A study by Arya and Kumar (2020) emphasized that while e-waste is a valuable source of precious metals, it poses significant challenges due to toxic chemicals, especially in developing countries like India. Essential approaches to sustainable e-waste regulation in India include environment-friendly product design, circular resource utilization, extended producer responsibility, and formalization of the unregulated recycling sector. A review study by Osibanjo and Nnorom (2007) conducted in Nigeria highlighted the rapid growth in ICT as the main reason for the decreasing lifespan of electronic equipment. It emphasized the absence of infrastructure for appropriate waste management in developing countries. The authors strongly suggested the creation of product reuse through remanufacturing and establishing efficient recycling facilities as essential measures for addressing e-waste in different countries. According to Gangwar et al. (2019), e-waste burning by the informal sector causes high levels of air particulate matter (PM10) concentration, which is a reason for cardiovascular morbidity and a series of health risks to humans and animals. In a comprehensive study by Sivaramanan (2013), approximately 75% of e-waste lacks clear alternatives, such as remanufacturing, refurbishment repairing their part for reuse, etc. Most are junk occupying useable space at apartments, firms, industries, and houses. Most e-waste recyclers were exporting toxic materials like circuit boards, mercury lamps, and leaded glass usually to India, China, and Africa. A report by The United Nations Environment Programme (UNEP) and United Nations University (United Nations Environment Programme and United Nations University, 2009) issued a bold and startling statement for developing nations that by 2020 e-waste from old computers would increase by 500% in India, and e-waste from thrown-away mobile phones would be about seven times higher than the 2007 level in China and about 18 times high in India as compared to 2007 baseline. E-waste generation and management is a current concern all over the world. Many unexplored regions need immediate attention, particularly in the case of developing countries. Furthermore, there are very few studies in the case of Odisha. Therefore, the study of effective e-waste management in Odisha is an emerging domain based on gaps in extant literature. Therefore, this study stands as a pivotal foundation with good inception for next-generation thinkers and groundbreaking improvement in the realm of e-waste management in Odisha.

3 Methodology

3.1 Data and Methods

The study is based on primary data. The information on e-waste generation and awareness analysis done in this study has focused on the state of Odisha. The study measuredly concentrates on the household's behaviour because households have a wide contribution to electrical and electronic waste products due to their growing consumption demand. The present study focuses on their choice, habits, behaviours, awareness, and knowledge of hazardous e-waste management and its potential health risks. The survey questions were first drafted for a pilot survey in the context of e-waste awareness in Odisha. The pilot survey was conducted via the Internet using emails and social media apps and was distributed among 20 unspecified participants. The participants belong to different age groups and genders to get more insightful feedback. The pilot survey was conducted to check the gaps in the survey questions or to know any other drawbacks. The final survey was executed via household survey as well as via the Internet through Google Forms for a limited duration. The survey questionnaire was sent randomly to different individuals residing in Odisha through the author's contacts, social media apps, and emails. This survey includes all population categories and different age groups. The questions were written in English and targeted to Odisha's permanent residents. As a result, 239 authentic responses were gathered and thoroughly analysed in this study. The accuracy of the survey sample is based on a margin of error with a 95% level of confidence interval using the 2011 population census of Odisha. Equation 1 illustrates the formula for the calculation of margin of error (Cochran, 1977).

$$\text{Margin Error} = Z \times \sqrt{\frac{p(1-p)}{n} \times \frac{N-n}{N-1}} \quad (1)$$

Z is the Z-score corresponding to the desired level of confidence which is taken as 95% for the proposed study. Thus, the Z-score is 1.96 taken from the Z table. P is an estimate of the proportion of the population with the attribute or the maximum variability of the population, i.e., 0.5. n is the sample size, and N is the population size. The estimated result of the marginal error is 6.31%, which is quite acceptable between 3% and 10% and reasonable for survey-based research-like attitudes, opinions, or behaviour (Attia et al., 2021). The findings are based on descriptive analysis and are depicted through graphical representation (Table 1).

Table 1 General description of the questionnaire

Section	Description
Section 1	Includes questions 1–9 which are sociodemographic questions about the respondent's name, age group, gender, level of education, occupation, area belongingness, size of households, social category, and household annual income
Section 2	Includes questions 10–15 about the respondent's awareness of the type of e-waste at home, the duration of e-waste stored at home, the main reason behind the disposal of e-waste, awareness of hazardous substances found in e-waste, and how to handle electronic device which contains hazardous substances
Section 3	Includes questions 16–21, specific to disposal practices, knowledge about local e-waste collection and recycling programs in the area, usually disposal of non-working electronic devices, knowledge of the importance of disposal of e-waste, measures that could improve the e-waste management, neighbour's knowledge about e-waste management, and any additional comments and suggestions

3.2 Study Area

The study is conducted in the state of Odisha, India. Its prime location is in the northeast part of the Indian Peninsula with 17.7800°N and 22.7260°N latitudes and 81.0376°E and 87.0987°E longitudes. The state is divided into 30 districts. The population of the state is more than 40 million, making it the 11th largest state in population (2011 census). Furthermore, it has cities like Bhubaneswar, Cuttack, Rourkela, Berhampur, Sambalpur, and Puri acting as hubs of technological development. Odisha stands as one of the fastest-growing economies with a growth rate of 7.25% (People Guide to Budget, Govt of Odisha, 2024–2025). As a growing economy, it has increased the adoption of technology which leads to a rise in the usage of electronic gadgets and subsequent e-waste generation. The study covers both rural and urban areas with all categories of population in a specific period.

4 Results and Discussion

Based on socio-demographic insights from the study samples, 54.8% are found to be male, and 45.2% are female. Similarly, 43.5% are rural participants, and 56.5% are from urban areas. As can be viewed in Table 2, the highest percentage (25.9%) of respondents is in the 18–24 age group category. Additionally, 25.9% and 25.1% of respondents have post-graduate and undergraduate degrees, respectively (the two highest percentages in the level of education). Furthermore, 45.6% of respondents have 3–4 members in their households. About 25.1% of respondents are students followed by private sector employees (16.7%). The highest percentage of respondents (46%) fall in the category of annual household income of less than ₹ 60,000. An extensive overview of the sociodemographic scenario is given in Table 2.

4.1 Types of E-Waste Found in the Households

Figure 1 represents the presence of different types of e-waste in the studied households. Since there can be a combination of items under e-waste, we have taken it as the percentage of response for each item. It reveals mobile phones have the highest share in e-waste with the response of 85.4% (highest) of respondents, driven by time-to-time upgrades and shorter life spans. It has become a necessary good for people. Cable and chargers are the associated goods with mobile phones with 67.5% of responses for the same. This is followed by lighting devices like LED bulbs and CFLs with 52.7% response, reflecting issues like a routine replacement. Television at 47.3% and kitchen appliances at 40.6% had a notable share due to the transition to newer models. Devices like printers/scanners and microwave ovens shared the lowest percentage at 12.6% and 13.8% respectively, likely due to their long

Table 2 Socio-demographic overview

Sociodemographic variable	Options	Percentage
Gender	Female	45.2
	Male	54.8
Age group	Under 18	5.8
	18–25 years	25.9
	26–35 years	22.6
	36–45 years	15.9
	46–60 years	22.2
	Over 60	7.5
Education level	No formal education	7.1
	Primary education	7.9
	Secondary education	18.4
	Higher secondary education	10.9
	Undergraduate degree	25.1
	Postgraduate degree	25.9
	Doctorate	4.6
Size of households	1–2 members	5
	3–4 members	45.6
	5–6 members	29.7
	More than 6 members	19.7
Occupation	Student	25.1
	Farmer	12.6
	Homemaker	15.5
	Government employee	13.8
	Private sector employee	16.7
	Business owner	11.7
	Retired	2.5
	Any other (please specify)	2.1
Area	Rural	43.5
	Urban	56.5
Household annual income	Less than ₹ 60,000	46
	₹ 60,000–₹ 120,000	29.7
	₹120,001–₹ 300,000	12.6
	₹300,001–₹ 600,000	8.4
	More than ₹ 600,000	3.3

durability and extended period of lifespan. Air conditioners, refrigerators, washing machines, and tablets are categorized under the medium range of e-waste.

Figure 2 reveals that the highest percentage (41.4%) of people have stored electronics waste at home for over 2 years. Mobile phones, cable and chargers, television, lighting devices, and audio video equipment fall under this category. This is followed by 20.9% of people who have had electronic devices for 6 months–1 year. Moreover, very few people about 10.5% had their e-waste less than 6 months, and

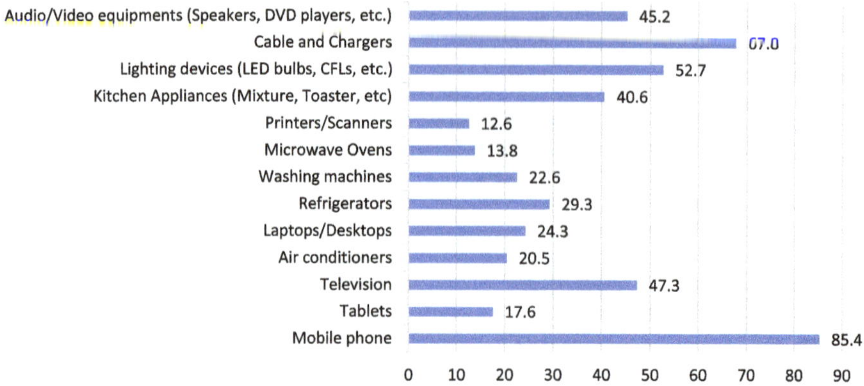

Fig. 1 Distribution of e-waste (as a percentage of respondents)

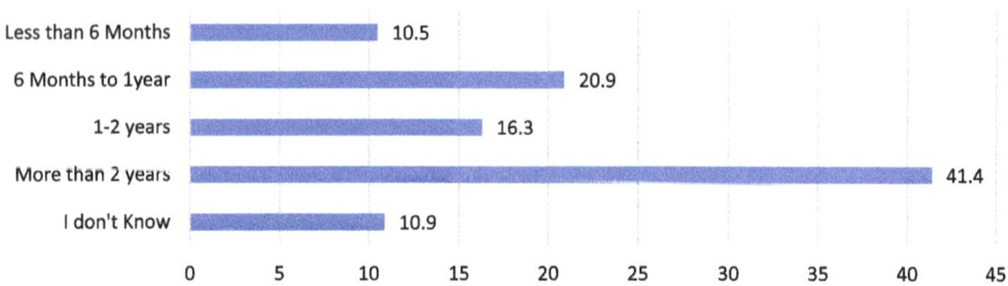

Fig. 2 Duration of the stored obsolete electronic devices in households (in percentage)

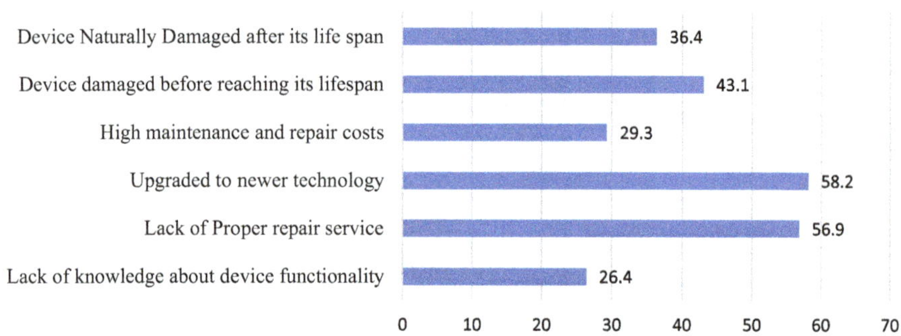

Fig. 3 Reasons for accumulation of e-waste (as a percentage of responses)

10.9% don't know the duration of e-waste at their homes. In addition, it is revealed that mobile phones, cable and chargers, television, lighting devices, and refrigerators are the common devices almost everyone has as e-waste. Respondents in the category of income level of less than ₹60,000 and family size of 3–4 members responded more in the category of having e-waste for more than 2 years.

Figure 3 illustrates the respondent's opinion about the reasons for the accumulation of e-waste in the study area. Each reason provides insights into understanding consumer behaviour and awareness regarding the main causes of the accumulation of e-waste. Upgradation to new technology (58.2%) has become the predominant factor reflecting a quick technological transaction with the latest and advanced features. Closely following this is the lack of proper repair service (56.9%), indicating a lack of sufficient service centres in the locality. Additionally, 43.1% of respondents believe electronic devices are damaged before their lifespan, possibly due to poor quality subassembly parts, manufacturing defects, or inappropriate usage of electronic gadgets. About 36.4% of responses came in favour of devices being naturally damaged after their lifespan. Finally, 29.3% of respondents believed a high maintenance cost, and 26.4% reported a lack of knowledge about device functionality, emphasizing the importance of user education.

4.2 *Awareness of the Presence of Hazardous Substances in E-Waste*

Figure 4 shows the awareness of the presence of harmful substances in e-waste among households in Odisha. Since here the respondents are allowed to choose multiple options, the result can be interpreted in terms of the percentage of respondents opting for the particular option. The results unfold that lead is the most recognized substance, with 39.3% of responses. Lead is found in a denser amount in mechanical breakings and the removal of solder from microchips. Chronically, it affects the reproductive system, kidneys, and mental development of children (Vats & Singh, 2014). A significant portion, i.e., 37.7% of respondents are found to be unaware of any of the hazardous substances listed above. This shows a lack of realization or education about discarded electronics and their toxic components. The level of unawareness was high among age group between below 18 and above 60. Farmers and homemakers have a lower awareness rate. Mercury at 26.8% is the second most identified substance. It often finds a place in tilt switches, solid-state devices, flat-screen monitors, and fluorescent tubes which are responsible for health hazards like sensory impairment, dermatitis, weakness in muscles, and memory loss (Kumar & Singh, 2014). Substances like nickel at 18.4%, fluorescent lamp phosphors at 15.9%, and cadmium at 15.5% demonstrate moderate understanding. Nickel furnishes the largest health risk for carcinogens in mobile phones. However, the contribution of beryllium is also having a notable impact (Singh et al., 2019).

Figures 4, 5, 6, 7, and 8 illustrate the awareness of health risks associated with e-waste. Health risks further categorized into three sections are awareness of specific risks like cancer and respiratory issues, awareness of general risks like it is only harmful to health, and the third category is not aware of the risk. A significant portion of respondents (38.5%) reveal they are unaware of the associated risks. Furthermore, respondents who come under the income category of less than ₹60,000 have a low percentage of awareness rate. Results show respondents who come under the income category of ₹120,001–300,000 are highly aware of the risks associated

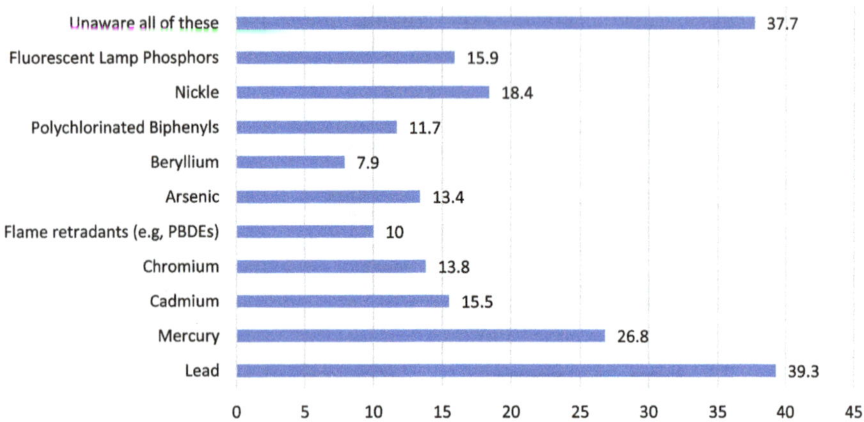

Fig. 4 Awareness of the presence of hazardous substances in e-waste (as a percentage of respondents)

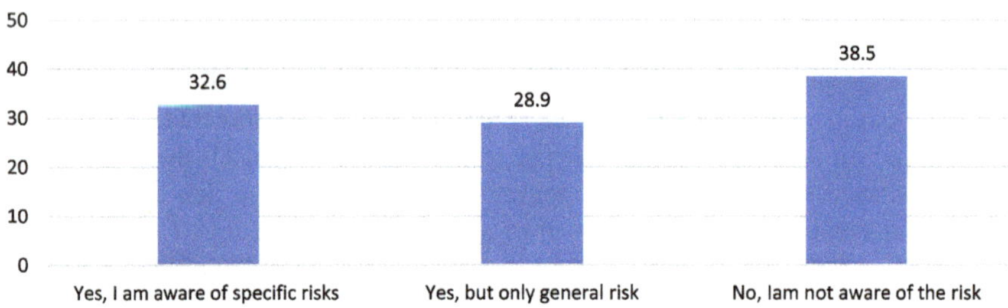

Fig. 5 Awareness of health risks associated with e-waste (in percentage)

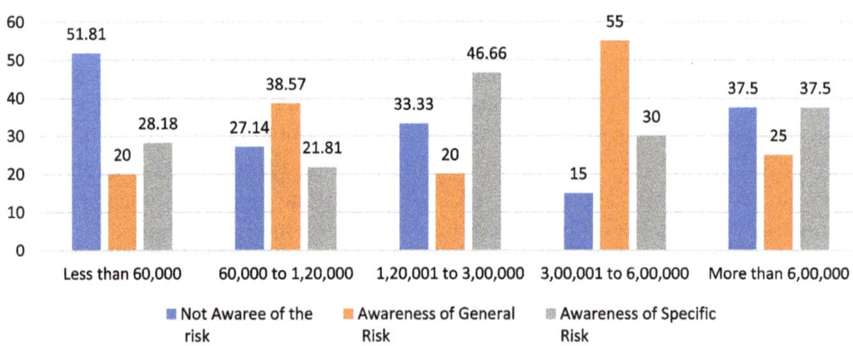

Fig. 6 Income distribution and awareness of health risks (in percentage)

with e-waste. Income categories ₹300,001–600,000 show a very high level of general health awareness. Taking age groups into account the results unfold that the age group above 60, followed by the age group 46–60 and 36–45 indicating a low awareness of health risks associated with e-waste. The age group 18–25 exhibits a

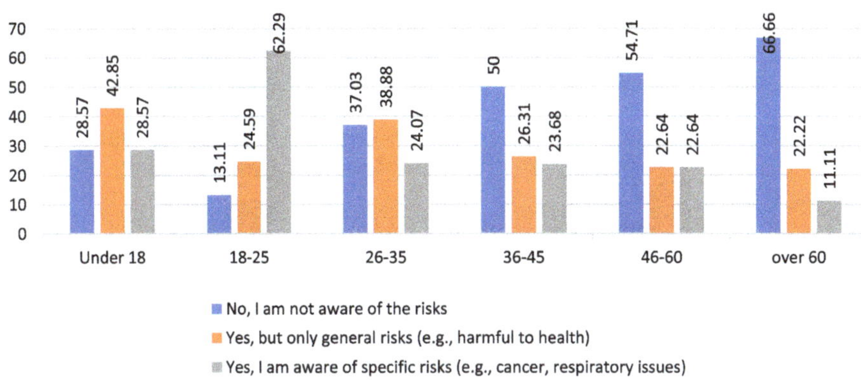

Fig. 7 Awareness of health risks within different age groups (in percentage)

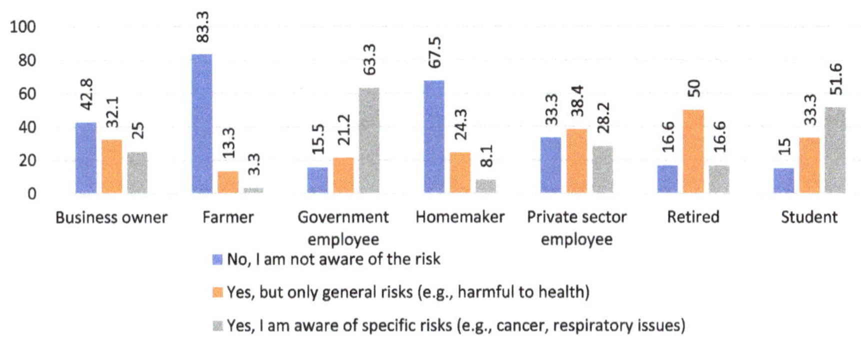

Fig. 8 Awareness of health risks within different occupation categories (in percentage)

high awareness of specific risks, and respondents under the age groups of 18 and 26–35 have shown the highest response on awareness of the associated general risks. Figure 8 highlights that the farmers and homemakers are unaware of the risks related to hazardous e-waste. Respondents coming under the category of government employees and students have more knowledge about the specific risk. Retired persons are highly aware of general risks; furthermore, private sector employees, students, and business owners also have good knowledge about associated general risks.

4.3 E-Waste Disposal Practices, Attitudes, and Opinions

Figure 9 reveals that the majority of respondents (71.5%) are unaware of any recycling programs or local e-waste collection centres when asked if there are any in their area. It is worth noticing that only 28.5% of people are aware of the recycling program available in their nearby locality. Additionally, when asked how they often

Fig. 9 Respondent's awareness of e-waste collection and recycling programs (in percentage)

Fig. 10 Disposal behaviour of respondents of the unused/ discarded electronic items

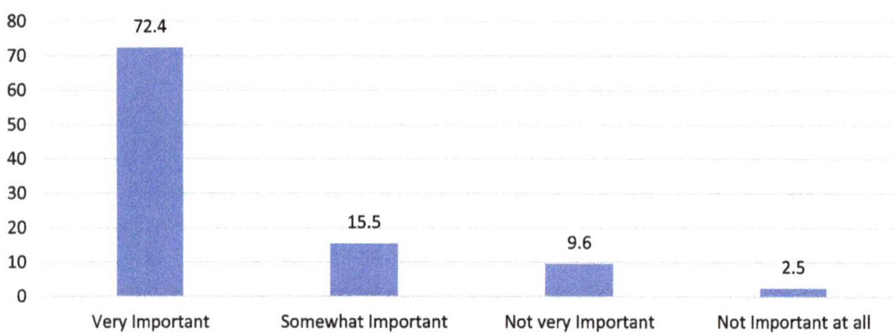

Fig. 11 Opinion of respondents on how important to dispose of e-waste

dispose of their discarded electronic devices, 30.5% of respondents said they sell them to scrap dealers. A noticeable amount of people store them at home (23%), while 20.5% of people throw them in general trash. Only 6.7% of people are concerned about returning them to the retailers or manufacturers for recycling. On a positive note, 72.4% of respondents think it is crucial to properly use discarded e-waste, and a few of them, about 2.5% believe that it is not important to dispose of the e-waste. These people are farmers, and few of them are homemakers who have no direct connection with the use of electronic gadgets. Figures 10 and 11 describe the disposal behaviour of respondents of the e-waste items and the opinion of the respondents about the importance of disposal of e-waste, respectively.

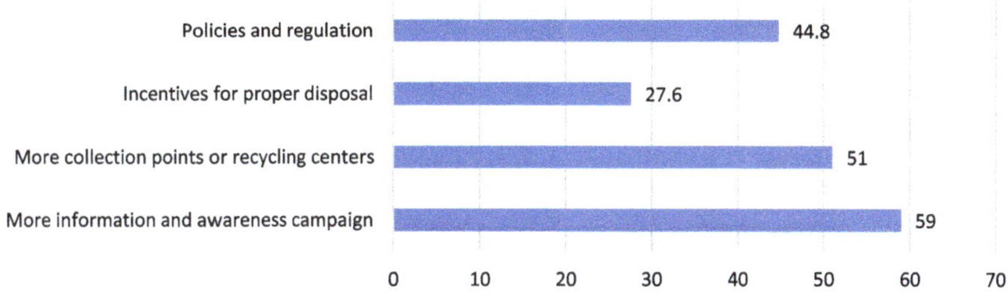

Fig. 12 Measures to improve e-waste management (as a percentage of responses)

Figure 12 illustrates the survey findings regarding opinions on refining e-waste management. An overwhelming 59% of respondents emphasize the critical need for an awareness campaign. It explains how people tend to ignore or overlook the guidelines and instructions the manufacturer provides when they buy a product. Furthermore, it emphasizes the need for public education, large-scale awareness campaigns, and targeted advertisement across the current trending social media platforms. Moreover, 51% of respondents convey the importance of more collection points and recycling centres in their area. Whereas 44.8% expressed the required appropriate policies and regulations, 27.6% of respondents opted for incentives for proper disposal.

5 Conclusion

Rapid e-waste growth is an ongoing process that can never be denied in the twenty-first century. E-waste has become the biggest threat not only because it has hazardous substances but also because of data safety protocols to ensure the protection of all sensitive information. Therefore, e-waste is a global concern and a menace. However, analysis of e-waste management in the context of Odisha remains relatively unexplored. This study concentrates on household-level practice and examines their awareness level, illuminating the challenges regarding the e-waste management framework. A comprehensively organized questionnaire survey was administered among households across Odisha revealing critical insights into the current scenario regarding e-waste awareness, management, and disposal behaviour. The findings signify age, occupation, gender, and income-based disparity in e-waste awareness and disposal behaviour, specifying the priority for targeted interventions.

Mobile phones, cable, and chargers rank as the highest share in e-waste generation, driven by the latest advancements and short product lifespans. Other categories, such as lighting devices and some kitchen appliances, also contribute a notable amount to e-waste generation, reflecting maintenance swaps, and transition to newer models. However, microwave ovens and printers represent a smaller fraction of e-waste generation due to their extended usage period. Again, the findings show that

41.4% of households have stored e-waste for over 2 years, mostly households coming under the low-income brackets. A lack of centres for repair services and pricey maintenance needs emerged as crucial aspects driving the accumulation of e-waste, alongside device damage before the expected lifespan is part of the concern. Awareness of electrical hazardous substances and associated health risks is alarmingly low, with a significant share of participants lacking knowledge about the toxic materials in e-waste. While substances like lead and mercury are slightly identified, others like cadmium and nickel are poorly comprehended. When asked about the awareness of health risks, particular health risks like cancer and respiratory issues are bitterly acknowledged by the younger generation and affluent participants, suggesting an unequal spread of knowledge. Apart from this, most households lack awareness of recycling programs and local e-waste collection centres, with 28.5% of respondents informed about such initiatives. Moreover, many unaware participants are mostly rural people and farmers. Disposal practice overwhelmingly consists of the restoration of e-waste in their home and a dominant number of participants sell this toxic waste to scrap dealers (30.5%), while small fractions of households engage in recycling practices (6.7%). Most people who are involved in the recycling practice belong to the urban area. The study highlights the gaps in e-waste management and urges policymakers to adopt inclusive practices that benefit all sections of the population. Technology and social media can play a major role in raising awareness among people. This study is limited to the state of Odisha for a relatively small sample representing its population and mostly focusing on households' awareness of e-waste. Despite this, this study's findings are a solid foundation for future research. Accordingly, decision-makers may use this study's result to gain deeper insight into household's perceptions and behaviour toward e-waste disposal and recycling practices. Future studies could deepen their analysis on a large sample and widen it to other electronics and electrical waste categories. This study will help identify deficiencies in Odisha's e-waste management and recycling system, thereby creating a comprehensive e-waste tracking system for better management practices.

References

Ahirwar, R., & Tripathi, A. K. (2021). E-waste management: A review of recycling process, environmental and occupational health hazards, and potential solutions. *Environmental Nanotechnology, Monitoring & Management, 15*, 100409.

Alghazo, J., Ouda, O., Alanezi, F., Asam, Z.-u.-Z., Rehan, M., Salameh, M. H., & Nizami, A.-S. (2019). Potential of electronic waste recycling in Gulf Cooperation Council states: An environmental and economic analysis. *Environment Science and Pollution Research, 26*, 35610. https://doi.org/10.1007/s11356-019-04956-y

Arya, S., & Kumar, S. (2020). E-waste in India at a glance: Current trends, regulations, challenges, and management strategies. *Journal of Cleaner Production, 271*, 122707. https://doi.org/10.1016/j.jclepro.2020.122707

Attia, Y., Soori, P. K., & Ghaith, F. (2021). Analysis of households e-waste awareness, disposal behaviour, and estimation of potential waste mobile phones towards an effective e-waste management system in Dubai. *Toxic, 9*, 236. https://doi.org/10.3390/toxics9100236

Behera, I. D., Basak, G., Kumar, R. R., Sen, R., & Meikap, B. C. (2020). Treatment of petroleum refinery sludge by petroleum degrading bacterium Stenotrophomonas pavanii IRB19 as an efficient novel technology. *Journal of Environmental Science and Health, Part A, 56*(2), 226–239. https://doi.org/10.1080/10934529.2020.1866924

Bhatia, A., Kiran, C., & Kapoor, M. (2019). A study on awareness of consumers towards e-waste management in the city of Jaipur. *Journal of Management, 6*, 200–208.

Biswas, A., & Singh, G. S. (2020). *E-waste management in India: Challenges and agenda.* Centre for Science and Environment.

Chancerel, P., Meskers, C. E. M., Hageluken, C., & Rotter, S. (2009). Assessment of precious metal flows during preprocessing of waste electrical and electronic equipment. *Journal of Industrial Ecology, 13*, 791–810.

Chatterjee, S. (2011). *Electronic waste and India.* Department of Information Technology.

Cochran, W. G. (1977). *Sampling techniques* (3rd ed.). Wiley.

Das, A. P., & Mishra, S. (2008). Hexavalent chromium (VI): Health hazards and environmental pollutant. *Journal of Environmental Research and Development, 2*(3), 386–392.

Das, A. P., & Singh, S. (2011). Occupational health assessment of chromite toxicity among Indian miners. *Indian Journal Of Occupational And Environmental Medicine, 15*(1), 6. https://doi.org/10.4103/0019-5278.82998

Das, A. P., Sukla, L. B., Pradhan, N., & Nayak, S. (2011). Manganese biomining: A review. *Bioresource Technology, 102*(16), 7381–7387. https://doi.org/10.1016/J.BIORTECH.2011.05.018

Das, A. P., Ghosh, S., Mohanty, S., & Sukla, L. B. (2015a). Advances in manganese pollution and its bioremediation. In L. B. Sukla, N. Pradhan, S. Panda, & B. K. Mishra (Eds.), *Environmental microbe biotechnology* (1st ed., pp. 313–328). Springer. https://doi.org/10.1007/978-3-319-19018-1_16

Das, A. P., Ghosh, S., Mohanty, S., & Sukla, L. B. (2015b). Consequences of manganese compounds: A review. *Toxicological & Environmental Chemistry, 96*(7), 981–997. https://doi.org/10.1080/02772248.2015.1005428

Das, D., Panda, S., & Kisku, T. (2017). A survey on waste to energy potential in Odisha. *International Journal of Innovative Research in Science, Engineering and Technology, 6*, 16110–16115.

European Commission. (2013). *Official page on waste electrical and electronic equipment.* http://ec.europa.eu/environment/waste/weee/index_en.htm

Forti, V., Balde, C. P., Kuehr, R., & Bel, G. (2020). *The Global E-waste Monitor 2020: Quantities, flows and the circular economy potential.* United Nations University (UNU)/United Nations Institute for Training and Research (UNITAR)—Co-hosted SCYCLE Programme/International Telecommunication Union (ITU)/International Solid Waste Association (ISWA).

Gangwar, C., Choudhari, R., Chauhan, A., Kumar, A., Singh, A., & Tripathi, A. (2019). Assessment of air pollution caused by illegal e-waste burning to evaluate the human health risk. *Environment International, 125*, 191–199.

Garg, N., & Adhana, D. K. (2019). E-waste management in India: A study of current scenario. *International Journal of Management, Technology, and Engineering, 9*, 2791–2803.

Gauba, P., Sharma, A., & Singh, G. (2016). E-waste and its impacts. *Asian Journal of Multidisciplinary Studies, 4*, 6–8.

Ghosh, S., & Das, A. P. (2015). Modified titanium oxide (TiO_2) nanocomposites and its array of applications: A review. *Toxicological & Environmental Chemistry, 97*(5), 491–514. https://doi.org/10.1080/02772248.2015.1052204

Ghosh, S., Mohanty, S., Akcil, A., Sukla, L. B., & Das, A. P. (2016). A greener approach for resource recycling: Manganese bioleaching. *Chemosphere, 154*, 628–639. https://doi.org/10.1016/J.CHEMOSPHERE.2016.04.028

Ghosh, S., Kumar, M. S., Bal, B., & Das, A. P. (2018). Application of bioengineering in revamping human health. In *Synthetic biology: Omics tools and their Applications* (pp. 21–37). https://doi.org/10.1007/978-981-10-8693-9_2

Grant, K., Goldizen, F. C., Sly, P. D., Brune, M.-N., Berg, M. V. D., & Norman, R. (2013). Health consequences of exposure to e-waste: A systematic review. *The Lancet Global Health, 1*, 350–361.

Gupta, V., & Kumar, A. (2014). E-waste status and management in India. *Journal of Information Engineering and Applications, 4*, 41–48.

Kiddee, P., Naidu, R., Wong, H. M. (2013). *Electronics waste management approaches an overview*. www.Elsevier.com/locate/wasman

Kumar, U., & Singh, D. N. (2014). Electronics waste: Concerns & hazardous threats. *International Journal of Current Engineering and Technology, 4*(2), 802–811.

Kumar, A., Holuszko, M., & Espinosa, D. C. R. (2017). E-waste: An overview on generation, collection, legislation and recycling practices. *Resources, Conservation and Recycling, 122*, 32–42.

Mishra, S., Ghosh, S., van Hullebusch, E. D., Singh, S., & Das, A. P. (2023). A critical review on the recovery of base and critical elements from electronic waste-contaminated streams using microbial biotechnology. *Applied Biochemistry and Biotechnology, 2023*, 1–30. https://doi.org/10.1007/S12010-023-04440-X

Mmereki, D., Baizhan, L., Baldwin, A., Hong, L. (2016). *The generation, composition, treatment and disposal system, and impact of E-waste*. https://www.intechopen.com

Mohanty, S., Mishra, S., & Mohanty, A. (2022). Municipality solid waste management. A case study of Smart City Bhubaneswar, Odisha. *Journal of Environment Management and Tourism, 13*, 1361–1337.

Osibanjo, O., & Nnorom, I. C. (2007). The challenge of electronic waste (e-waste) management in developing countries. *Waste Management Research, 25*, 489. http://wmr.sagepub.com/cgi/

Paul, A., Dey, S., Ram, D. K., & Das, A. P. (2023). Hexavalent chromium pollution and its sustainable management through bioremediation. *Geomicrobiology Journal, 41*, 1–11. https://doi.org/10.1080/01490451.2023.2218377

People Guide to Budget, Govt of Odisha. (2024–2025). https://finance.odisha.gov.in

Polak, M., & Drapalova, L. (2012). Estimation of end-of-life mobile phones generation: The case study of The Czech Republic. *Waste Management, 32*, 1583–1591.

Pradhan, G., Tripathy, B., Ram, D. K., Digal, A. K., & Das, A. P. (2023). Bauxite mining waste pollution and its sustainable management through bioremediation. *Geomicrobiology Journal, 41*, 1–10. https://doi.org/10.1080/01490451.2023.2235353

Priyadarsini, S., & Das, A. P. (2024). Analytical and structural characterization of waste lithium-ion batteries for their effective recycling strategy. *Environmental Chemistry and Ecotoxicology., 7*, 182. https://doi.org/10.1016/j.enceco.2024.12.004

Priyadarsini, S., & Das, A. P. (2025). Lithium bioleaching: A review on microbial-assisted sustainable technology for lithium bio-circularity. *Journal of Water Process Engineering, 69*, 106744. https://doi.org/10.1016/j.jwpe.2024.106744

Qu, Y., Zhu, Q. H., Sarkis, J., Geng, Y., & Zhong, Y. (2013). A review of developing an e-waste collection system in Dalian, China. *Journal of Cleaner Production, 52*(8), 176–184.

Raj, K., & Das, A. P. (2023). Lead pollution: Impact on environment and human health and approach for a sustainable solution. *Environmental Chemistry and Ecotoxicology, 5*, 79–85. https://doi.org/10.1016/J.ENCECO.2023.02.001

Ramanayaka, S., Keerthanan, S., & Vithanage, M. (2019). Urban mining of e-waste: Treasure hunting for precious nanometals. In *Handbook of electronic waste management* (pp. 19–54).

Rautela, R., Arya, S., Vishwakarma, S., Lee, J., Kim, K.-H., & Kumar, S. (2021). E-waste management and its effects on the environment and human health. *Science of the Total Environment, 773*, 145623.

Rene, E. R., Sethurajan, M., Ponnusamy, V. K., Kumar, G., Dung, T. N. B., Brindhadevi, K., & Pugazhendhi, A. (2021). Electronic waste generation, recycling, and resource recovery: Technological perspectives and trends. *Journal of Hazardous Materials, 416*, 125664.

Robinson, H. B. (2009). E-waste: An assessment of global production and environmental impacts. *Science of the Total Environment, 408*, 183–191.

Sankhla, M. S., Kumari, M., Nanda, M., Mohril, S., Singh, G. P., Chaturvedi, B., & Kumar, R. (2016). Effect of electronic waste on environmental & human health—A review. *IOSR Journal of Environmental Science, Toxicology and Technology, 10*, 98–104.

Singh, N., Duan, H., Ogunseitan, A. O., Li, J., & Tang, Y. (2019). Toxicity trends in e-waste: A comparative analysis of metal in discarded mobile phones. *Journal of Hazardous Materials, 380*, 1–9.

Singla, S., & Bansal, M. (2012). Management of electronic waste: An integrated approach. *International Journal of Engineering, Science, and Metallurgy, 2*(3), 781–786.

Sivaramanan, S. (2013). E-waste management, disposal and its impacts on the environment. *Universal Journal of Environmental Research and Technology, 3*, 531–537.

Sthiannopkao, S., & Wong, H. M. (2013). Handling e-waste in developed and developing countries: Initiatives, practices, and consequences. *Science of the Total Environment, 436–464*, 1147–1153.

Tanskanen, P. (2013). Management and recycling of electronic waste. *Acta Materialia, 61*, 1001–1011.

The Global E-Waste Monitor. (2024). https://ewastemonitor.info/the-global-e-waste-monitor-2024/

Uhunamure, E. S., Nethengwe, S. N., Shale, K., Mudau, V., Mokgoebo, M. (2021). *Appraisal of households' knowledge and perception towards e-waste management in Limpopo province, South Africa*. https://doi.org/10.3390/recycling6020039.

United Nations Environment Programme and United Nations University. (2009, July). *Recycling-from E-waste to resources*. https://www.researchgate.net/publication/278849195

Vats, M. C., & Singh, S. K. (2014). E-waste characteristics and its disposal. *International Journal of Ecological Science and Environmental Engineering, 1*(2), 49–61.

Wath, S. B., Vaidya, A. N., Dutt, P. S., & Chakrabarti, T. (2010). A roadmap for development of sustainable E-waste management system in India. *Science of the Total Environment, 409*(1), 19–32.

Bio-leaching as a Bio-circular Economy Approach for Recovery of Precious Minerals from E-Waste

Ankita Priyadarshini, Annapurna Maharana, Sailaja Priyadarsini, Vinod Pravin Sharma, Chandra Mohan, and Alok Prasad Das

Abstract The sharp rise in electronic waste (e-waste) creates major environmental concerns, stemming from the accumulation of hazardous substances and the depletion of valuable materials. It encompasses discarded outdated electronic devices, and improper management of this waste leads to significant environmental pollution and poses potential risks to human health. Many electronic devices are designed with limited durability, making them outdated quickly. The increasing prevalence of bacterial infections, and toxicological and environmental impacts, alongside the anticipated growth of e-waste trends, highlights the urgent need to address this issue with effective policies and sustainable waste management practices. Bioleaching represents a transformative method for e-waste organization within a bio-circular economy context, highlighting sustainability and resource recovery. It can be adapted to various types of e-waste, assembling it as a multipurpose solution for labelling the growing waste crisis. This process exploits the metabolic activity of bacteria and fungi, which can solubilize metals like lead, nickel, aluminium, and lithium from complex matrices seen in electronic components. Bioleaching presents a promising alternative that is both eco-friendly and cost-effective for the recovery of precious minerals. It ensures that valuable materials are reclaimed and reused rather than discarded. This is a key solution to the challenges modelled by e-waste in our increasing society. This chapter offers a detailed summary of the various sources of e-waste and explores the bio-recovery of precious minerals through bioleaching, highlighting its economic significance.

Keywords Bioleaching · Environmental sustainability · Bio-recovery · Bio-circular economy

---

A. Priyadarshini · A. Maharana · S. Priyadarsini · A. P. Das (✉)
Department of Life Sciences, Rama Devi Women's University, Bhubaneswar, Odisha, India

V. P. Sharma
CSIR-IITR, Lucknow, Uttar Pradesh, India

C. Mohan
School of Basic and Applied Sciences, K R Mangalam University, Gurugram, India

A. P. Das, S. Priyadarsini (eds.), *Electronic Waste and Environmental Pollution*, Sustainable Environmental Waste Management Strategies,
https://doi.org/10.1007/978-3-031-90287-1_20

1 Introduction

E-waste is termed as any outdated electronics items that have touched the final stage of their suitable life, such as discarded computers, computer accessories, mobile phones, televisions and others that encompass harmful substances and can pose a threat to human and animal well-being and the atmosphere (Ahirwar & Tripathi, 2021). According to another definition, e-waste includes an extensive and increasing series of electronic devices extending from large domestic appliances and small gadgets to computers and devices that have been rejected by their users. Simply described, it is garbage generated by discarded, abandoned, or otherwise undesirable electronic and electrical goods (Rajesh et al., 2022). It includes not only locally created garbage but also e-waste from other areas of the nation and beyond, which is then discarded after recycling. With the massive import of e-garbage that is coming as a charity to developing nations like India, the challenge is where to dump this load and how to handle this waste, which has become a serious environmental threat nowadays (Arya & Kumar, 2020).

E-waste is an intricate combination of metallic element and heavy metallic element (Ni, Li, Mg, Cu, and Co) (Mishra et al., 2022a; Prabhakar et al., 2019; Raj & Das, 2023), plastics (Jagatee et al., 2024; Mishra et al., 2020; Paul et al., 2024; Singh Jadaun et al., 2022; Singh et al., 2020), biomedical waste (Khatun et al., 2023; Sadasivuni et al., 2019), petroleum (Behera et al., 2020, 2021; Dey et al., 2023a; Lahiri et al., 2021; Mohanta et al., 2024), and chemicals (Bhattacharjee et al., 2021; Das & Swain, 2013) that are toxic and pose risks towards the atmosphere and environment. The basic content of e-waste is often defined through the electronic device's manufacturer, typical and kind, production date, and lifespan. IT and digital statement chips have more valuable metals than common household goods (Bagwan, 2024; Priyadarsini & Das, 2024, 2025). A smartphone, contains more than 40 fundamentals, together with valuable metallic elements such as palladium, silver, and gold as well as base metallic elements like copper and tin. Electrical and electronic devices contain a mixture of metals like lead, mercury, cadmium, nickel, copper, zinc, and other metal compounds, many of which pose environmental risks (Kumar et al., 2017). Electrical and electronic devices contain a blend of metals that include mercury, lead, zinc cadmium, manganese, etc., most of which are environmental hazards (Bal et al., 2018; Das & Mishra, 2008; Dey et al., 2023b; Mohanty et al., 2018; Sanket et al., 2016). E-waste also gives a source of plastics that include esters of polyvinyl chloride, polyethylene terephthalate (PET), polypropylene, and polystyrene which have all proven to cause environmental effects upon disposal (Dey et al., 2024a; Mishra et al., 2021, 2022b, 2024). Toxic organic compounds found in e-waste include PBDDs, DP, PAHs, PCDDs, and PCBs (Das & Singh, 2011; Kanungo et al., 2023; Mallick et al., 2024).

The rapid growth can also be seen in the aspect of e-waste which hit 54 million metric tonnes in 2018, where Asia-Pacific region has its share of about 24.9 million tonnes. E-waste has generated at a rate three times the population growth, and predictions are that it will have to triple by 2030 to 74 million tonnes per year. Recycling only covers about 13% of e-waste (Widmer et al., 2005). According to global e-waste data, of the 52 million metric tonnes of e-waste formed worldwide in 2020, only 18% was properly composed and used. The Global Telecommunication Union (ITU) announced a plan to raise the worldwide e-waste reusing percentage to 32% by 2023 at its 2018 Plenipotentiary Conference (Ghimire & Ariya, 2020; Rai et al., 2021).

In India, electronics manufacturing engineering has arisen as a rapidly expanding part of the Indian economy, both in terms of goods and services and exportation. India is now one of the top four countries in terms of e-waste creation as well as the world's second-biggest population (Arya & Kumar, 2020). Over 60% of India's e-waste is produced by 65 cities, while ten states account for 62% of the total e-waste produced in India. Maharashtra (Mumbai) is the top e-waste-producing state in India, followed by Tamil Nadu, Punjab, Uttar Pradesh, West Bengal, Delhi, Karnataka, Gujarat, Madhya Pradesh, West Bengal, and Delhi. Mumbai leads India's top ten cities in e-waste generation, followed by Hyderabad, Pune, Surat, Delhi, Bangalore, Chennai, Pune, Surat, Kolkata, Ahmadabad, and Nagpur (Arya & Kumar, 2020; Kumar Induria Pt Ravishankar et al., n.d.). Increasing revenues, urbanization, and the affordability of low-cost electronics have collectively contributed to an enormous rise in the use of electric apparatus such as mobiles, laptop computers, and domestic uses in India. Macroeconomic factors have contributed to India's status as a potentially rising economy, with increased ambitions for a lavish lifestyle, fast urbanization, and substantial international commercial trade growth. The recent increase in electronic usage has spread to a variety of industries, including education, healthcare, and entertainment, indicating its widespread impact (Borthakur & Govind, 2017).

The mainstream of e-waste recycling takes place in the casual sector, utilizing unpolished and dangerous processes. To solve the issue, adequate legislative measures as well as cost-effective, sustainable technological remedies must be implemented (Ismail & Hanafiah, 2020). One of the peak worries about e-waste, especially in rising nations like China and India, is the discarding of e-waste from industrialized countries. Large quantities of outdated electronic equipment are frequently traded to different nations in the Asia Pacific, including China and India (Lucier & Gareau, 2020; Halim & Suharyanti, 2019; Kaushik & Herat, 2020). These devices have high repairability and raw solid demand. This can lead to massive growths of residue in underdeveloped places without dense eco-friendly regulations. The primary causes for these transfers are inexpensive labour and a lack of ecological and professional values in Asia (Lucier & Gareau, 2020; Halim & Suharyanti, 2019).

Every year, millions of electric and electronic apparatuses are rejected when they disrupt or become outmoded. These abandoned devices are categorized as e-waste and might cause a threat to both human well-being and the atmosphere if improperly discarded and recycled. Computers, mobile phones, big household electronics, and medical equipment are common things found in e-waste sources. Loads of electronic waste are inappropriately recycled, kept in homes and warehouses, discarded, and illegally exported (Das et al., 2011; Das & Ghosh, 2018, 2023f; Ghosh et al., 2016, 2023b). When e-waste is cast using flawed practices, it can discharge up to

100 distinct chemical compounds into the atmosphere, with recognized neurotoxins like lead. Globally, e-waste is growing at a quicker rate than any other trash category, with yearly amounts rising sharply. The amount of electronic garbage created worldwide is millions of metric tonnes annually, and estimates suggest that this amount will rise by 4–5% per year. Urbanization, industrialization, and the worldwide necessity of technological devices are major causes of this increase (Akram et al., 2019; Shahabuddin et al., 2023). The electronics sector in India is one of the rapidly growing industries, which makes the problem of e-waste even worse. The primary concern with e-waste is the scarcity of dismantlement and reprocessing facilities that can recover usable components as well as precious metals from discarded e-items (Sahu et al., 2024). Recycling e-waste is an alarming risk in the world (Behera et al., 2020; Das et al., 2015b; Ghosh & Das, 2015). E-waste has increased overall trash volume, and when dismantled, it emits a variety of harmful and carcinogenic gases and metals, endangering human well-being and contaminating the atmosphere (Ahirwar & Tripathi, 2021; López et al., 2022; Man et al., 2013; Nandan et al., 2023).

Additionally, most of these activities are performed by workers in unsafe environments. Additionally, much of the e-waste created is disposed of in landfills, a standard approach for the organization of electronic waste. Hence, recycling of e-waste has become crucial. It helps to reduce atmospheric strains. Every form of waste material has its own chemical and physical properties; however, e-waste comprises both useful materials that can be improved and toxic materials that must be handled with care to prevent pollution and hazards to health (Lucier & Gareau, 2020; Nandan et al., 2023). Recovery of reusable parts and basic materials is possible through recycling, in particular Cu and precious metals. Nevertheless, wealthy countries are generally not willing to recycle e-waste due to the absence of such reusing facilities, expensive labour, and restrictive environmental regulations. Rather, it is either disposed of in a landfill or sent from affluent countries to poor countries where it is likely to be recycled with undeveloped methods and without regard to any health (Cobbing et al., 2008; Debnath et al., 2018).

Export of e-waste especially where such activities generate high income has become a common practice even when it is prohibited under the Basel Convention. Due to the increase in the amount of electronic waste, both human beings and the atmosphere are facing various negative consequences. However, due to the inappropriate handling and progressions in technology, this condition has finally turned into a threat to humanity. The main essentials that are severely affected by electronic waste are the environment and population health (López et al., 2022; Nandan et al., 2023). Bioleaching is an innovative method of improving precious and hazardous metallic elements from electronic waste by employing various microorganisms. An environmentally beneficial method of extracting valuable metals from the earth, such as gold, copper, and silver is called bioleaching. It makes use of microbes to simplify mineral matrices in order to recover valuable metals (Rai et al., 2021; Sen et al., 2024; Sinha et al., 2021).

2 E-Waste Generation and Its Different Categories

2.1 Major Sources and the Components

Electronic waste is generated by the dumping of electronic equipment, including televisions, computers, and mobile phones. The e-waste configuration is diverse and consists of a wide group of materials that can be systematically categorized into organic, ceramic, and inorganic categories. A deep understanding of the sources and structure of e-waste is critical to the design of active management and recycling strategies. One of the crucial difficulties is the international production of e-waste, primarily due to the heavy stipulation of electronic devices in human society (Ahirwar & Tripathi, 2021). The tremendous progress of calculating and statement apparatus is mainly accountable for the comprehensive bloom of e-waste manufacture among the various electrical consumables. Other reasons behind the massive generation of e-waste include: (i) quick progress in data and message industries, (ii) flexibility in electronic appliances, (iii) quick technical progress and current innovations, and (iv) the deteriorating movement of prices of electronic devices (Althaf et al., 2019; Singh et al., 2016).

E-waste production will probably increase by 70% in 2030. Major components of e-waste are metals 60%, plastics 15%, and a mix of metaplastics. As and when the more toxic amount of the metal goes above the permissible limit, contained in the environment and with chromium, cadmium, lead, selenium, mercury, arsenic, and chromium may cause risks and threats to the human and the environment directly or indirectly (Bhattarai et al., 2024; Mishra et al., 2023; Murthy & Ramakrishna, 2022; Robinson, 2009). The most major type contributors are waste household equipment at 45% of the total, followed by data and communications technology equipment at 33.9%, and consumer electronics at 13.7%. The e-waste is very heterogeneous and composite with respect to the type of apparatuses and materials. Yet, however, copper and expensive metals account for over 80% of the value of most of the e-waste samples. The eight metals, Pb, Cu, Cr, As, Zn, Cd, Hg, and Ni, are the most significant and harmful to the atmosphere (Cayumil et al., 2016; Jaiswal & Srivastava, 2024; Pant et al., 2012; Shaikh et al., 2020). The main constituents of e-waste are PCBs, which provide a diverse and composite mixture of elements that classically consist of ~30% polymers, ~30% refractories, and ~40% metals. Aluminium oxide, alkaline, silicon oxide, and alkaline-earth oxides are typical refractory components of PCBs. Other phases, which include potassium, magnesium barium titanate, and aluminium silicates, might exist in some boards. However, the contents of copper, between 10% and 20%, are so much more than ores, which range from ~0.5% to 1%, that such a form must be treated differently, for metal to be recovered (Arshadi et al., 2018; Hanafi et al., 2012). Figure 1 presents an overview of e-waste pollution, highlighting its sources, associated impacts, and strategies for sustainable remediation and management.

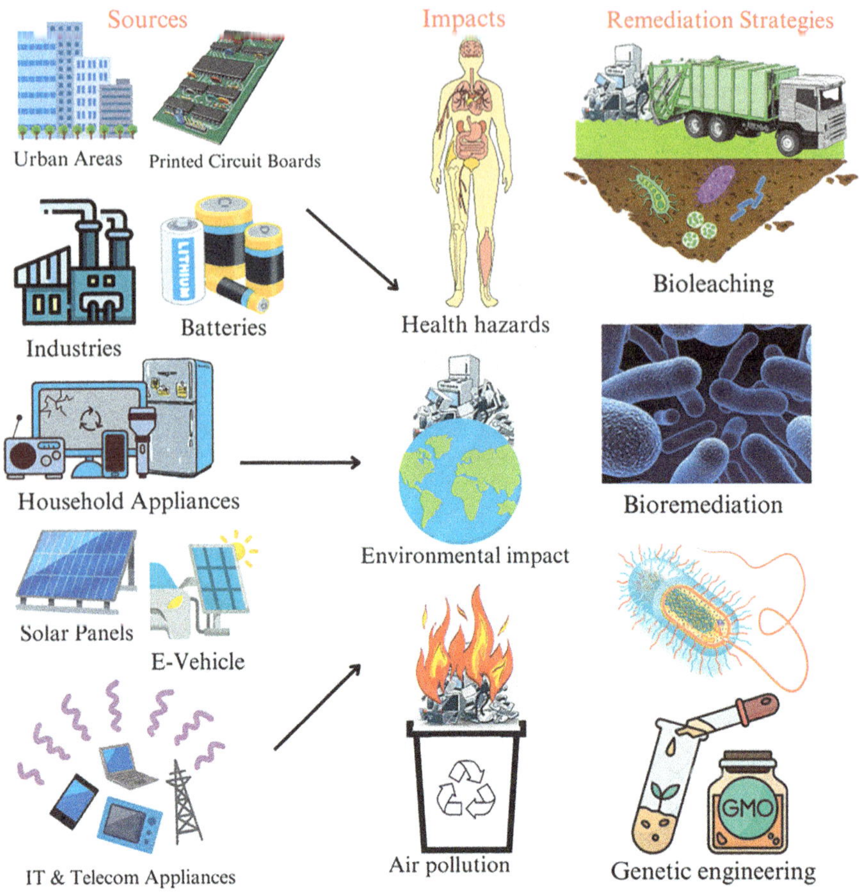

Fig. 1 Sources of e-waste pollution, its environmental and societal impacts, and the remediation strategies for sustainable management

2.2 *Status of E-Waste in the World*

Electric and electronic items form a hallmark of development worldwide. It encompasses a wide range of products that are basically utilized in daily life activities. They are found in homes and organizations throughout the globe. However, the ownership varies with income level per capita. About 19 billion kg of e-waste were recovered as secondary resources mainly from metals like iron, which are found in ample quantities and recycled in almost every route of e-waste management. Platinum-group metals and precious metals were some of the most valuable but found in much lower quantities. Share patent applications on e-waste management increased from 148 per million in 2010 to 787 in 2022 (Alblooshi et al., 2022; Hossain & Sahajwalla, 2024; Ismail & Hanafiah, 2020). The majority of those being applied relate to technology for recycling cables with little indication of an increase in patents filed for related technologies in the recovery of critical raw materials. Although the rare earth elements are extremely useful and have specific characteristics that will prove vital to future technologies, including the generation of renewable energy or e-mobility, the world remains breathtakingly dependent on the production chain by a few countries. Recycling such elements even though involved with devices that show relatively higher content proves challenging in economic terms (Cucchiella et al., 2015; Mishra et al., 2023; Seif et al., 2024). As such, recycling activities can recover only roughly 1% of the current demand for recycling of rare-earth elements. The market value for the recovery of rare earth elements remains too low to sustain bigger-scale commercial recycling programs (Bhattarai et al., 2024; Ramprasad et al., 2022; *Rare Earth Metals in Solar Panels - Victory Metals*, n.d.). An astonishing 63 million tonnes (Mt) of e-waste was generated in 2022, up 83% from 2010; it is set to increase further by 33%, to 83 million tonnes, in 2031; billions of dollars of strategically valuable possessions are being wasted and dumped in the ground; just 2% of rare earth element demand can be covered through recycling from e-waste. About 63 million tonnes of e-waste a year in 2021—that is 1.55 million 41-ton trucks or approximately the number of trucks bumper to bumper that could go all the way around the equator. Meanwhile, only a quarter, 22.4%, of the mass of e-waste generated for the year reported collection and recycling in 2022 (Bagwan, 2024; Kumar et al., 2017; Robinson, 2009). The overall generation in the world is 2.6 million tonnes. It will rise with this rate to 82 million by 2030 which is almost 33% more than the number anticipated to be witnessed during 2022. On a different regional basis, Asia, the biggest at 24.9 Mt during the year 2019 has been followed by Americas roughly at 13.1 Mt and then about 12 Mt from Europe. As if that is not particularly telling, the largest volume of production of e-waste has been registered in the USA, China, and India, among others. In 2019, for example, the US generated an estimated 6.9 Mt of e-waste, and one would assume that only around 15% of this was recycled (Awasthi & Li, 2017; Garlapati, 2016; Osibanjo & Nnorom, 2007; Perkins et al., 2014; Priyashantha et al., 2022).

3 Sustainable E-Waste Management Practices

Sustainable e-waste control is known as handling electronic trash in a way that harms the environment and its constituents, people, as little as possible. With further technological development, more electronics are thrown away, creating critical environmental problems. Steps involved: this is a methodological strategy for managing electronic waste in a fully maintainable manner (Jawadand & Randive, 2021; Murthy & Ramakrishna, 2022; Poonia et al., 2024).

(a) E-Waste Generation Reduction: The very first step in the effective supervision of e-waste is its reduction. The culture of buying less and use for an extended time has to be developed. Consumerism is challenged to buy sturdier and repairable goods. The manufacturers need to adopt the principles of eco-design by designing products that would easily get disassembled, repaired, and upgraded to extend the life series of the products (Ismail & Hanafiah, 2020).

(b) Electronic Material Recycling: Another important aspect of environmentally friendly e-waste management is that electronic material must be recycled for an extended period of time. Companies may establish plans to renovate and recycle old devices for new clients or organizations. Besides waste minimization, recycling or donation of used but useful devices can open doors of technology to societies (Jiang et al., 2012).

(c) E-Waste Recycling: Recycling is the most fundamental element of responsible and effective waste management because it permits the recovery of valuable materials from retired electronics while reducing the harm that the extraction of raw materials for use can cause. A well-designed recycling system can ensure that hazardous materials are handled properly and safely. In addition, your licensed e-waste recycler will make sure they are adhering to some of the industries' most stringent data security and environmental impacts (Tansel, 2017).

(d) Environmental Impact Issues: All these parties—the government, manufacturer, merchant, and customer—should know that improperly used e-waste hurts the environment. Some dangerous materials that e-waste contains are cadmium, lead, and mercury. If not disposed of well, they will poison soil and water. The sustainable processes ensure that such materials will be disposed of in proper ways (Kumar et al., 2024; Robinson, 2009).

(e) Extended Producer Responsibility and Industry Involvement: To make the e-waste management policy a success, there needs to be an involvement of all the parties involved. EPR ensures that producers are responsible for their products from the stage of manufacturing to the end of their life cycle; hence it encourages them to take responsibility for collecting and recycling the e-waste (Faibil et al., 2023; Khetriwal et al., 2009).

(f) Education and Awareness Campaign: Public education should be applied to still responsible attitudes toward the suitable dumping of e-waste both in communal and isolated sectors. Education and awareness can be used as means of sensitizing customers about recycling, teaching clients appropriate methods of disposal, and making them aware of negative impacts on the atmosphere as an outcome of e-waste (Chauhan et al., 2020).

(g) Technological Developments: Technological advancements enhance the e-waste management process greatly. Through blockchain technology, all electronic waste can be traced from the producers to disposal and, hence, the right procedures can be followed in recycling and processing (Borthakur & Govind, 2017).

(h) International Cooperation: This is a global problem since e-waste affects everybody worldwide. Organizations, countries, and business leaders should cooperate in solving the issue. It demands national and international networks to support stronger e-waste management regulation (Chen & Wu, 2010; Ni & Zeng, 2009).

4 Bioleaching of E-Waste

4.1 Historical Perspective of Bioleaching

Bioleaching has its roots in the sixteenth century when natural mineral leaching was first observed. More systematic studies were initiated back in the 1960s. Scientists isolated specific microbes that could improve metal solubilization. Therefore, initial efforts did indeed exist, some of them carried out by Dr. H. J. W. (Hans) S. L. McCarthy in the 1960s, on identifying how much bacteria helped in recovering metals (White et al., 1997). Meanwhile, it became evident that naturally occurring bacteria were used in the mines of Rio Tinto, Spain, to dissolve the iron and copper from the ores, which became the first major commercial application of bioleaching. It verified that these biological processes could be applied to a biomining context and therefore opened routes to further study microbial metal extraction (Amils, 2016; Das & Ghosh, 2023a; Xavier et al., 2019).

It refers to the extraction of metallic elements from rocks using natural materials found in the atmosphere, such as microorganisms, air, and water. In brief, it can be said that bioleaching corresponds to the exploitation of the size of certain archaea and bacteria, naturally occurring, for the catalysis of mineral oxidation for marketable purposes, or also for mining waste handling or remediation, where microorganisms are used for the removal of toxic metallic element and metalloids (Bala et al., 2024; Das & Ghosh, 2023b, e; Ghosh & Das, 2017; Mohanty et al., 2017).

There are two methods via which bioleaching may be carried out. They are direct and indirect. Direct bioleaching involves the use of easily oxidizable minerals with the microorganisms in which there is a reaction involving enzymes that efficiently separates iron from the minerals. With the indirect bioleaching method, there is no straight connection between microorganisms and the minerals; instead, it is the production of representatives by the microorganisms used for mineral oxidation (Tributsch, 2001; Vera et al., 2022). Such are some of the important aspects that bioleaching deals with, which also include, attaining the extraction of metals from ores in an efficient and effective way, without harmful activities in the environment, using a natural and efficient process, recovery of the remaining lower quality ores that would not be economical to mine otherwise, generation of lower wastes than the traditional process, obtaining high-purity grades for numerous manufacturing applications, or just the reduction of consumable energy and other resources (Ghosh et al., 2018; Watling et al., 2014). Bacterial escape of metal sulphide technology has advanced very speedily over the last few years. Microorganisms have entered a well-established biotechnological practice of metal retrieval; most frequently, this is used to recover nickel, zinc, copper, cobalt, and uranium metals, which are usually removed from unsolvable sulphides or, in the case of uranium, from rusted minerals (Cui & Zhang, 2008; Islam et al., 2021; Pokorna & Zabranska, 2015; Vera et al., 2022; Watling et al., 2014).

4.2 Mechanism of Microbial Metal Leaching

Microbiological bioleaching is of high potential and innovation, which could be materialized in metal extraction from complex ores. Optimized biochemical and microbiological operations in bioleaching decide whether the technology applied will better fit with resource recovery or environment remediation operations (Ghosh et al., 2016). Depending upon the mineralytic effect of bacteria and fungi against minerals is determined through three major principles: acidolysis, complexolysis, and redoxolysis (Dutta et al., 2023). The mechanisms through which microorganisms assemble alloys include the production of biological or mineral acids (protons), the processes of corrosion and depletion, and the elimination of complexing agents. The most prevalent lifeless acid found in discharge environments is sulphuric acid, which results from the sulphur-oxidizing microorganism known as thiobacilli. Bacterial (and also fungal) metabolism produces a series of organic acids, leading to complex and chelate formation and organic acidolysis (Brandl & Faramarzi, 2006; BRANDL Zürich, n.d.). The authors established a dynamic model that has been applied in explaining coordination chemistry in mineral solubilization by focusing on the suspension of oxides via protonation of the mineral peripheral and coupled with the absorption of appropriate complex-forming ligands. Ligands may co-exist with the proton-leading and ligand-leading mineral fluidification in an acidic solution (BRANDL Zürich, n.d.; Kaya, 2016; White et al., 1997).

There are two types of operations that include the bacteriological utilization of metals.

4.2.1 Direct Bacterial Leaching

There is a straight connection sandwiched between the cell of bacteria and the mineral sulphide external area in direct bacterial leaching, where oxidation to sulphate occurs through several steps catalysed enzymatically. According to Torma, the subsequent crease-resistant metal sulphides' can be oxidized through direct contact by *T. ferrooxidans*: chalcocite, molybdenite, stibnite, cobaltite, millerite, chalcocite, sphalerite, and galena (Bosecker, 1997; Sand et al., 2001; Tributsch, 2001).

4.2.2 Indirect Bacterial Leaching

In the process of indirect bioleaching, microbes are not required to physically interact with the mineral's external area because they produce a sap that chemically oxidizes sulphide minerals. Such bacteria, however, facilitate ferrous iron's normally slow reoxidation as they can act as catalysts. Chemical oxidation alone can achieve ferrous iron oxidation in the pH levels of 2–3 but introduces bacterial assistance of over 10^5–10^6 to enhance oxidation efficiency. *T. thiooxidans* assists bioleaching processes in that it helps create an acidic pH where iron-oxidizing microbes such as *T. ferrooxidans* and *L. ferrooxidans* can thrive (Bosecker, 1997; Dutta et al., 2023; Sand et al., 2001) (Bosecker, 1997).

Overall, bioleaching is constructed on the contact of organic and biochemical corrosion processes. Primarily, it mainly emphasizes the ferric and ferrous iron cycle.

4.3 Diversity of Microorganisms Used in Bioleaching

4.3.1 *Thiobacillus*

In the process of bioleaching, the most active bacteria involved belong to the genus *Thiobacillus*. They are gram-negative, non-spore-forming rod-shaped microbes that thrive under aerobic conditions and are chemolithoautotrophic (Das & Ghosh, 2023a). They require CO_2 as a carbon source to produce new cell material and gain liveliness through the oxidation of sulphur compounds with sulphate as the final corrosion product. Bacterial leaching takes place in acidic conditions at a pH level between 1.5 and 3, which keeps most metal ions dissolved. The acidophilic species plays a crucial role in the bioleaching process. Other thiobacilli that oxidize sulphur and sulphides thrive only at advanced pH values, where metal ions can't stay dissolved (Bosecker, 1997; Yaashikaa et al., 2022).

4.3.2 *Leptospirillum*

Leptospirillum ferrooxidans is a chemolithotrophic ferrous iron-oxidizing bacteria that survives at lower pH levels and upper levels of molybdenum, uranium, and silver than *T. ferrooxidans*, but it is more delicate to copper and thus is incapable of oxidizing sulphur or sulphur compounds. Thus, *L. ferrooxidans* requires the presence of *T. ferrooxidans*. These mesophilic bacteria can thrive at temperatures between 25 and 35 °C (Gopikrishnan et al., 2020).

4.3.3 Heterotrophic Microorganisms

Heterotrophic bacteria and fungi require organic additions for growth and energy and are also involved in metal bioleaching. In manganese bioleaching, metal solubilization takes place by the production of organic acids by Mn solubilizing bacteria and fungus or through enzymatic reduction of extremely rusty metallic compounds (Das & Ghosh, 2023c, d; Ghosh et al., 2023b). These organisms excrete compounds with two hydrophilic responsive groups, which dissolve heavy metallic elements by displacing metallic ions from the ore matrix with hydrogen ions and thereby form resolvable metal complexes (Bosecker, 1997; Ghosh et al., 2016; Ghosh & Das, 2018; Ilyas et al., 2007). Tables 1 and 2 outline various bacterial and fungal species utilized in metal bioleaching and their corresponding leaching efficiency percentages.

Table 1 Different bacterial species involved in metal bioleaching along with the leaching efficiency percentage

Sl no	Bacteria involved	Metal	% of leaching	References
1.	*Acidithiobacillus ferrooxidans*	Copper (Cu)	85	Adetunji et al. (2023)
2.	*Acidithiobacillus thiooxidans*	Iron (Fe)	78	Ghassa et al. (2021)
3.	*Bacillus megaterium*	Lithium (Li)	66	Biswal and Balasubramanian (2023)
4.	*Bacillus subtilis*	Lead (Pb)	60	Liu et al. (2022)
5.	*Desulfovibrio desulfuricans*	Uranium (U)	73	Abhilash et al. (2011)
6.	*Desulfovibrio vulgaris*	Germanium (Ge)	62	Blayda (2017)
7.	*Leptospirillum ferriphilum*	Nickel (Ni)	90	Seif et al. (2024)
8.	*Leptospirillum ferris*	Selenium (Se)	69	Rao et al. (2021)
9.	*Leptospirillum ferrooxidans*	Chromium (Cr)	67	Mishra et al. (2005)
10.	*Pseudomonas fluorescens*	Gold (Au)	70	Dutta et al. (2023)
11.	*Pseudomonas putida*	Manganese (Mn)	82	Bal et al. (2019)
12.	*Rhodococcus opacus*	Cobalt (Co)	92	Zhang et al. (2020)
13.	*Saccharomyces cerevisiae*	Platinum (Pt)	55	Karim and Ting (2022)
14.	*Sulfolobus acidocaldarius*	Vanadium (V)	64	Li et al. (2021)
15.	*Sulfolobus metallicus*	Rhodium (Rh)	80	Karim and Ting (2022)
16.	*Thiobacillus cuprinus*	Gallium (Ga)	81	Maneesuwannarat et al. (2016)
17.	*Thiobacillus ferrooxidans*	Palladium (Pd)	88	Adetunji et al. (2023)
18.	*Thiobacillus thiooxidans*	Silver (Ag)	75	Liu et al. (2022)

Table 2 Different fungal species involved in metal bioleaching along with their leaching efficiency percentages

Sl no.	Fungus involved	Metal	% of leaching	References
1.	*Aspergillus niger*	Zinc (Zn)	80	Kompalitch and van Hullebusch (2024)
2.	*Aspergillus oryzae*	Aluminium (Al)	74	Kremser et al. (2022)
3.	*Aspergillus tereus*	Antimony (Sb)	61	Aghazadeh et al. (2023)
4.	*Penicillium chrysogenum*	Indium (In)	77	Jowkar et al. (2018)
5.	*Penicillium janthinellum*	Osmium (Os)	68	Feruza et al. (2022)
6.	*Penicillium simplicissimum*	Tin (Sn)	65	Igiri et al. (2018)
7.	*Saccharomyces cerevisiae*	Platinum (Pt)	55	Karim & Ting (2022)

4.4 Role of Microbial Enzymes in E-Waste Bioleaching

Microorganisms like bacteria (*T. thiooxidans, A. ferrooxidans, Pseudomonas putida, E. asburiae,* and *B. cereus*), fungi (*A. niger, P. simplicissimum,* and *F. oxysporum*), and archae (*F. acidarmanus* and *S. acidocaldarius*) have potential to leach precious minerals (Khan et al., 2024; Das et al., 2012; Dey et al., 2024b).

A. ferrooxidans is a gram-negative bacterium that thrives in extremely acidic environments and has been acknowledged as a prototypical organism in bioleaching because of its capacity to mobilize a variety of elements, including Fe, Ni, Cu, Zn, Ga, As, Mo, Li, P, V, Cr, W, Pb, and U, as well as its major contribution to the oxidation of ferrous iron. *A. ferrooxidans* promotes metal extraction through the formation of iron (III) ions in aerobic conditions, which can subsequently interact with sulphide minerals (Tonietti et al., 2024). Certain species of *Pseudomonas* spp. use sulphate reductases to carry out the bio-reduction of sulphate ions, which aid in metal solubilization (Kaur et al., 2022).

S. thermosulfidooxidans, a thermophilic bacterium is known for its efficiency in bioleaching at elevated temperatures, which can enhance the recovery of precious minerals from e-waste. Its enzymatic activity is vital for breaking down complex metal sulphides. *S. thermosulfidooxidans* is used in the bioleaching of chalcopyrite and pyrite (Zeng et al., 2020).

Iron-oxidizing bacteria (*A. ferrooxidans, L. ferriphilum,* and *Sulfobacillus* spp.) and acidophilic sulphur oxidizing bacteria (*A. thiooxidans, S. thermosulfidooxidans,* and *Alicyclobacillus* spp.) are extensively used for valuable metals recovery from batteries (Barrie Johnson & Hallberg, 2008; Biswal & Balasubramanian, 2023).

Microorganisms belong to *Pseudomonas* spp. and *Chromobacterium* spp. and produce cyanide, a catalyst for metal recovery from e-waste (Natarajan & Ting, 2015). The bacterium *Chromobacterium violaceum* is notable for its ability to produce cyanide, which aids in gold recovery from electronic scrap materials, increasing gold dissolution efficiency (Dangton & Leepowpanth, 2014). *Frankia* spp. produce organic acids that significantly enhance the leaching of metals from e-waste, achieving up to 70% gold recovery (Marappa et al., 2020).

Fungal species such as *A. niger, A. tubingensis,* and *Penicillium* spp. are intended for e-wastes recycling and mineral recovery. In metal extraction from lithium-ion batteries, *Aspergillus* spp. is extensively used (Biswal & Balasubramanian, 2023).

Laccase, which is produced by fungi like *Pleurotus florida*, is essential for the oxidation of phenolic compounds and aids in the solubilization of metals such as iron and copper from e-waste (Kaur et al., 2022). These enzymes contribute significantly to the bioleaching process by increasing metal solubilization and recovery. Collectively, these microbes and their enzymatic activities constitute a long-term and efficient approach to recycling electronic waste that is effective in mitigating both sustainability challenges and resource recovery (Bal et al., 2019).

4.5 *Factors Affecting Bioleaching*

4.5.1 pH

Modification of the accurate pH rate is significantly required for microbial activity and the solubilization of metallic elements. Bacterial oxidation of ferrous iron and sulphide requires an optimum pH range of 2.0–2.5 (Bosecker, 1997; Yaashikaa et al., 2022).

4.5.2 Temperature

The temperature range for ferric iron and sulphide corrosion by *Thiobacillus* spp. is optimized between 28 and 30 °C. The thermophilic microorganisms can be employed at the advanced temperature of 50 °C–80 °C for the leaching process (Bosecker, 1997).

4.5.3 Heavy Metals

Associated with metal sulphide leaching is the rise in absorption of metals in the leachate. Generally, leaching organisms are highly tolerant to heavy iron sensitivity. Fully dissimilar sensitivities may be shown to diverse strains of the same types of heavyweight metals. The technique often allows strain to be adapted to ever-higher absorptions of metals or to precise substrates by adding the metallic element or substrates in increasing concentrations (Gopikrishnan et al., 2020; Ilyas et al., 2007).

4.5.4 Atmospheric Gases

Many leaching microbes need abundant oxygen for the best growth and activity. Methods such as ventilating, moving, or trembling can complete this in the laboratory atmosphere. However, providing adequate oxygen becomes more difficult in larger operations like dump or stack leaching. It attains high vintage where only carbon dioxide carbon sources are needed without any additional CO2 accumulation (Adetunji et al., 2023; Bosecker, 1997).

4.6 *Impact of Metal Toxicity on Microbial Activity in Bioleaching*

Metal toxicity refers to the deteriorating effect of heavy metals on organisms, especially on microbes. Metal toxicity in bioleaching can significantly influence the growth rate of microbes and their metabolic activity, thereby reducing the efficiency of the metal recovery process.

Heavy metal toxification encompasses various mechanisms like redox catalysts in reactive oxygen species (ROS) production, ion regulation annihilation, DNA and protein formation, and breakdown of enzymatic functions (Igiri et al., 2018). The presence of heavy metals can alter the biochemical and physiological properties and can lead to oxidative stress, microbial denaturation, and impaired bioremediation capacity (Das et al., 2015a, 2024; Kanungo et al., 2024; Nag et al., 2023; Paul et al., 2023; Pradhan et al., 2023). Aluminium can stabilize reactive oxygen species through metal ion displacement or ligand interactions that can cause DNA damage (Olaniran et al., 2013; Booth et al., 2015).

Nickel, copper, and aluminium have been shown to inactivate microorganisms, particularly when exposure time is considered, even at concentrations previously deemed non-toxic. Iron concentrations are crucial for microbial metabolism but excessive iron can also inhibit growth, demonstrating a complex relationship between nutrient and toxic metal levels (Benzal et al., 2020). Microbial ecotoxicity due to high metal concentration results in environmental health hazards and mutagenicity and carcinogenicity, cell destruction that leads to microbial community degradation (Bhakta, 2017; Bhattacharjee et al., 2022; Das et al., 2015a, 2023; Dey et al., 2023b).

5 Innovative GMO Approaches for Revolutionary E-Waste Recycling and Metal Recovery

Genetic engineering serves as a promising opportunity for the extraction of metals and other components and provides beneficiary approaches for the remediation of precious minerals due to their environmentally friendly strategies and less adverse health effects than any other methods that have potential dangers to life and are toxic to the environment (Iravani, 2011). Microorganisms are genetically modified by different techniques such as transfection, conjugation, protoplast fusion, and mutagenesis that increase microbial adaptation, microbial resistance mechanisms, and metabolic performance in high-metal environments (Liu et al., 2022).

For example, *E. coli* is genetically modified for greater absorption of Cd^{2+} because of the expression of the metallothionein and cadmium transport system (Deng et al., 2007). *Bacillus subtilis* is genetically modified with the gene CmarsM, found in heat-resistant algae *Cyanidioschyzon merolae*, which is functional in methylating and volatilizing arsenic (As) (Huang et al., 2015). *Shewanella oneidensis* is genetically modified to enhance the electron transfer capacity that is effective in heavy metal extraction (Li et al., 2020).

The application of genetically modified organisms is a sustainable method for extensive metal recovery operations (Abashina & Vainshtein, 2023). For example, genetic engineering has led to improved arsenic bioleaching by *A. ferrooxidans* (Yachkula et al., 2022). Certain enzymes are reportedly known to detoxify heavy metals, e.g., mercury resistance merA gene from *D. radiodurans* in *E. coli* can tolerate high Hg(II) levels and convert into less toxic elemental state Hg(0) (Brim et al., 2000). Heavy metal binding is facilitated by certain peptides such as zinc (Zn)(II), pbrT and pbrD, cadB, copM for lead (Pb)(II), cadmium (Cd)(II), and copper (Cu) (II), respectively (Xavier et al., 2019).

6 Integration of Bioleaching with Phytomining for Enhanced Metal Recovery

Integration of bioleaching with phytomining represents an exciting approach toward mineral recovery from e-waste. E-waste minimization is a serious problem nowadays due to increasing global electronic consumption that leads to environmental metal accumulation, posing significant ecological and health threats. Green technologies, for example, phytomining and bioleaching focus on improving the e-waste management and thereby making it more sustainable and cost-effective (Zulkernain et al., 2023).

Phytomining involves cultivating hyperaccumulating plants that absorb metals from contaminated substrates, effectively concentrating valuable metals for recovery (Sinha et al., 2021). In this integrated approach, the bioleaching process first releases metals from e-waste into a bioavailable form and once metals are solubilized by microorganisms, they are taken up by hyperaccumulator plants, thus enhancing the overall recovery rate of metals from e-waste.

The term "hyperaccumulator" defines plants from various families that can thrive, tolerate, and accumulate heavy metals in aerial tissues (Rascio & Navari-Izzo, 2011). For example, plants like *Brassica juncea* (Indian mustard) and *Alyssum murale* can absorb metals like copper, nickel, and cadmium from soil. After the harvest, such metal-rich biomass can be burnt and its metals extracted from the ash (Rani et al., 2023; Rathore et al., 2019; Srivastava, 2020).

The synergy in this scenario lies in that the bioleaching microbes can aid in the mobilization of metals so that their access to uptake by the plant is made easier. Where phytomining has its limitation in metal accessibility in soil, bioleaching could release the metal into the environment and thus not present a route to recover it. There exists an efficient pathway toward sustainable metal extraction from e-waste between two methods (Debnath et al., 2018; Islam et al., 2020).

The crushed e-waste into fine particles increases the surface area and makes the metals available for the microorganisms. In controlled conditions, the e-waste material is inoculated with microorganisms that cause the release of metals through the oxidation of metal sulphides or the production of organic acids. The metal changes from its solid state to an ionic form that can be dissolved in soil solution through microbial intervention. The treated soil is planted with hyperaccumulator plants, which absorb the metal ions released during the process of bioleaching. These metals will accumulate in the tissue of the plant over time. When the plants have matured, they are harvested, and the biomass is processed to extract and purify the metals contained in the plant tissue (Khaliq et al., 2014; Ng, 2023; Zulkernain et al., 2023).

Compared to traditional mining, bioleaching, and phytomining produce extremely low levels of toxic effluents that have a very low impact on the environment during metal extraction. Bioleaching and phytomining are relatively cheap technologies that require low energy input and rather simple infrastructure. Combining bioleaching and phytomining can enhance recovery rates by utilizing the microbial action to pre-treat substrates, making metals more bioavailable for plant uptake (Kompalitch & van Hullebusch, 2024).

Here are some examples and case studies illustrating the integration of bioleaching with phytomining for enhanced recovery of minerals from electronic trash:

(a) **Restoration of Gold and Copper from Electronic Equipment**

A. ferrooxidans is an iron as well as sulphur oxidizer used in bioleaching for the release of copper ions from crushed PCB materials. Weeks later, *Brassica juncea* (Indian mustard plants) is cultivated in a bioleached solution, which contains dissolved copper ions. The combination of the two processes, bioleaching and phytomining, increased copper recovery by 30% relative to that obtained through only bioleaching. It has been observed that *Brassica juncea* can accumulate a considerable amount of copper, which may be potentially used to accumulate metals within plant tissues (Hanafi et al., 2012; Jayabarath et al., 2022; Rani et al., 2023; Sinha et al., 2021).

In fungi species, such as *Aspergillus niger*, gold ions are released from electric equipment through bioleaching. Following the solubilization of the gold by bioleaching, *Brassica juncea* or other hyperaccumulator crops are grown in the fertilizer-enriched soil that absorbs the solubilized gold (Argumedo-Delira et al., 2019; Brandl et al., 2001; Rascio & Navari-Izzo, 2011).

(b) **Nickel Recovery from Electronic Waste Residues Using Hyperaccumulators**

The nickel residues of e-waste are supplemented with *Leptospirillum ferrooxidans,* having metal-oxidizing characteristics. This process helps facilitate the leaching out of ions from the soil. Nickel becomes thus more readily absorbed by a hyperaccumulator like *Alyssum murale.* Eventually, the plants can then be harvested and processed such that the nickel content of the plant material is made available. The amount of residual waste will then correspondingly be decreased as resource recovery occurs (Islam et al., 2021; Rascio & Navari-Izzo, 2011; Tognacchini et al., 2020).

(c) **Experimental Combination for Multi-Metal Recovery from Smartphone Batteries**

Some smartphone batteries hold a lot of lithium, cobalt, and manganese metals. These are increasing in value but remain very difficult to extract since the extraction requires the destruction of this complicated matrix of materials. Cobalt and manganese are then released by applying a bioleaching solution containing *A. thiooxidans* on the crushed residues. Then the soil treated with hyperaccumulator plants, like *Phyllanthus amarus*, which has the potential for metal uptake are grown in the treated soil (Biswal & Balasubramanian, 2023; Noruzi et al., 2022).

7 Incorporation of Minerals Recovered from E-Waste in Green Technologies for the Advancement of Renewable Energy

Some of the minerals obtained from e-waste include copper, silver, gold, and rare earth elements. They are highly utilized in most green technologies, particularly renewable energy source technologies, including solar panels, wind turbine generators, electric vehicle batteries, or energy storage systems because these minerals possess outstanding electrical conductivity and catalytic properties that will facilitate a much more sustainable energy cycle (Gopikrishnan et al., 2020; Mallick et al., 2023).

Green technologies are important to recover valuable minerals from electronic trash to promote sustainability and prevent climate change (Das & Ghosh, 2018). E-waste is an abundant source of rare earth elements (REEs) which is essential for modern technologies such as renewable energy applications. Neodymium and dysprosium are some of the REEs found in electronic waste. These are highly critical components in the manufacture of extremely efficient magnets with wide-ranging applications in renewable energy technologies like wind turbines and electric vehicles (Bhattarai et al., 2024; Ghosh et al., 2016; Hossain & Sahajwalla, 2024; Lukowiak et al., 2020). Recycling the materials reduces the demand from mining, and thus demand for sustainability in green technologies increases. The idea is therefore important since it is a prominent aspect linked to electronic waste and is becoming a source of critical material for renewable energy applications (Bhattarai et al., 2024).

Lithium, a component of lithium-ion batteries (LIBs), can be recovered from e-waste to produce new batteries intended for energy storage systems and electric vehicles (Lithium-Ion Battery Recycling | US EPA, n.d.). Platinum group metals (PGMs), for example, palladium and platinum, are essential catalysts in fuel cells. These metals can be retrieved from e-waste to reduce their reliance on mining.

Copper is known for its excellent conductivity which makes it vital for solar panels, and copper recovered from electronics trash is a great method by considerably reducing such environmental impacts from manufacturing on its way toward supporting the circular economy since it focuses more on the reuse of materials to give new, renewable energy solutions. Increased use of electric vehicles and renewable energy requires the increased utilization of rechargeable batteries, which use lithium (Aghazadeh et al., 2023; Dey et al., 2024a; Ma et al., 2022; *Rare Earth Metals in Solar Panels - Victory Metals*, n.d.). This lithium contains a significant amount of e-waste, which mitigates the environmental damage caused by mining lithium at the site of extraction. This understanding gives one an insight into the ways through which e-waste can be included as a viable component in the supply chain of the batteries of green technologies (Bhattarai et al., 2024).

8 Regulatory Challenges in E-Waste Handling and Bioleaching Processes

Despite being an efficient and promising process, metal bioleaching by microbes faces limitations that hamper its commercialization. Production of metabolites by microorganisms is confined because of the different pH conditions needed for microbial growth and activity. Optimization of pH conditions is crucial to maximize metabolite production while maintaining microbial growth. The efficiency of microbial bioleaching is low in comparison to chemical leaching due to the production of leachate by microbes in lower concentrations. Enhanced metal dissolution can be achieved through process optimization (Arya & Kumar, 2020; Gopikrishnan et al., 2020; Ilankoon et al., 2018).

The slow kinetics of microbial leaching limit the large-scale application for e-waste metal recovery. The development of innovative and cost-effective techniques for recovering metals from bioleaching processes is a significant challenge. Using native autochthonous organisms enhances metal bioleaching by reducing the toxicity of metal present in the e-waste (Arab et al., 2020; Fashola et al., 2020; Ghosh et al., 2018; Wang et al., 2008; Wen et al., 2013).

E-waste and bioleaching processes, therefore, have multifaceted regulatory concerns, being majorly on environmental safety and public health. Although bioleaching is an eco-friendly method for metal extraction, it carries the threat of leachates that are hazardous when not kept in a closed environment such as mine tailings. International trade further complicates these issues where weak regulation in some jurisdictions leads to illegal dumping that causes degradation of the environment. All these regulatory aspects must be tackled for sustainable e-waste management practices while mitigating environmental implications (Ahirwar & Tripathi, 2021; Jawadand & Randive, 2021; Mallick et al., 2023; Tsydenova & Bengtsson, 2011).

9 Bio-Circular Economy Approach in Recovery of Minerals form E-Waste

Rapid advancement in technology has led to a surge in e-waste generation consequently leading to global e-waste pollution. E-waste primarily contains complex mixture of plastics and microplastics from terrestrial and marine sources, microfibres from different sources, refractory oxides, heavy metals, halogens, dyes, and combustible substances resulting in various health hazard and environmental challenges (Bala et al., 2024; Bhandari et al., 2023; Das et al., 2023, 2024; Ghosh et al., 2023a; Mallick et al., 2023). Such immense and continuous consumer demand for electrical and electronic devices coupled with its rapidly aging products due to rapidly accelerating progress in technologies, more electrical and electronic equipment are becoming waste, throughout the whole world.

Circular bioeconomy offers a promising strategy towards resolving the global challenges related to renewable energy sources as well as aims to maintain the added value of products for as long as possible. Assimilation of waste through bioprocesses to manufacture beneficial products and metabolites supports sustainable circular bioeconomy (Leong et al., 2021).

Moreover, the perspective of the circular economy approach encourages the implementation of an effective collection of e-waste to prevent improper dumping in landfills or incineration (Ezeudu et al., 2021). The bio-circular economy emphasizes the significance of the prolonged lifespan of electronic devices.

10 Conclusion

Currently, millions of tonnes of e-waste need more efficient and worthwhile recycling techniques. The recovery of major elements from electronic scraps not only reduces the reliance on natural resource mining but also decreases the pollution from hazardous chemicals (mostly organic micropollutants released from the improperly disposed of e-waste). Bioleaching is an environmental strategy for recovering valuable minerals from spent electronic goods. The microbial biodiversity and molecular pathways of bioleaching microbes play a vital function in the extraction of valuable minerals from electronic scraps.

References

Abashina, T., & Vainshtein, M. (2023). Current trends in metal biomining with a focus on genomics aspects and attention to arsenopyrite leaching-a review. *Microorganisms, 11*(1). https://doi.org/10.3390/MICROORGANISMS11010186

Abhilash, Mehta, K. D., Kumar, V., Pandey, B. D., & Tamrakar, P. K. (2011). Bioleaching—An alternate uranium ore processing technology for India. *Energy Procedia, 7*, 158–162. https://doi.org/10.1016/J.EGYPRO.2011.06.021

Adetunji, A. I., Oberholster, P. J., & Erasmus, M. (2023). Bioleaching of metals from e-waste using microorganisms: A review. *Minerals, 13*(6). https://doi.org/10.3390/MIN13060828

Aghazadeh, S., Abdollahi, H., Gharabaghi, M., & Mirmohammadi, M. (2023). Bioleaching of zinc, copper and antimony from a tetrahedrite concentrate using acidophilic microorganisms. *Hydrometallurgy, 219*, 106075. https://doi.org/10.1016/J.HYDROMET.2023.106075

Ahirwar, R., & Tripathi, A. K. (2021). E-waste management: A review of recycling process, environmental and occupational health hazards, and potential solutions. *Environmental Nanotechnology, Monitoring & Management, 15*, 100409. https://doi.org/10.1016/J.ENMM.2020.100409

Akram, R., Natasha, Fahad, S., Hashmi, M. Z., Wahid, A., Adnan, M., Mubeen, M., Khan, N., Rehmani, M. I. A., Awais, M., Abbas, M., Shahzad, K., Ahmad, S., Hammad, H. M., & Nasim, W. (2019). RETRACTED ARTICLE: Trends of electronic waste pollution and its impact on the global environment and ecosystem. *Environmental Science and Pollution Research, 26*(17), 16923–16938. https://doi.org/10.1007/S11356-019-04998-2

Alblooshi, B. G. K. M., Ahmad, S. Z., Hussain, M., & Singh, S. K. (2022). Sustainable management of electronic waste: Empirical evidences from a stakeholders' perspective. *Business Strategy and the Environment, 31*(4), 1856–1874. https://doi.org/10.1002/BSE.2987

Althaf, S., Babbitt, C. W., & Chen, R. (2019). Forecasting electronic waste flows for effective circular economy planning. *Resources, Conservation and Recycling, 151*, 104362. https://doi.org/10.1016/J.RESCONREC.2019.05.038

Amils, R. (2016). Lessons learned from thirty years of geomicrobiological studies of Río Tinto. *Research in Microbiology, 167*(7), 539–545. https://doi.org/10.1016/J.RESMIC.2016.06.001

Arab, B., Hassanpour, F., Arshadi, M., Yaghmaei, S., & Hamedi, J. (2020). Optimized bioleaching of copper by indigenous cyanogenic bacteria isolated from the landfill of e-waste. *Journal of Environmental Management, 261*, 110124. https://doi.org/10.1016/J.JENVMAN.2020.110124

Argumedo-Delira, R., Gómez-Martínez, M. J., & Soto, B. J. (2019). Gold bioleaching from printed circuit boards of mobile phones by Aspergillus niger in a culture without agitation and with glucose as a carbon source. *Metals, 9*(5), 521. https://doi.org/10.3390/MET9050521

Arshadi, M., Yaghmaei, S., & Mousavi, S. M. (2018). Content evaluation of different waste PCBs to enhance basic metals recycling. *Resources, Conservation and Recycling, 139*, 298–306. https://doi.org/10.1016/J.RESCONREC.2018.08.013

Arya, S., & Kumar, S. (2020). E-waste in India at a glance: Current trends, regulations, challenges and management strategies. *Journal of Cleaner Production, 271*, 122707. https://doi.org/10.1016/J.JCLEPRO.2020.122707

Awasthi, A. K., & Li, J. (2017). Management of electrical and electronic waste: A comparative evaluation of China and India. *Renewable and Sustainable Energy Reviews, 76*, 434–447. https://doi.org/10.1016/J.RSER.2017.02.067

Bagwan, W. A. (2024). Electronic waste (E-waste) generation and management scenario of India, and ARIMA forecasting of E-waste processing capacity of Maharashtra state till 2030. *Waste Management Bulletin, 1*(4), 41–51. https://doi.org/10.1016/J.WMB.2023.08.002

Bal, B., Ghosh, S., & Das, A. P. (2018). Microbial recovery and recycling of manganese waste and their future application: A review. *Geomicrobiological Journal, 36*(1), 85–96. https://doi.org/10.1080/01490451.2018.1497731

Bal, V. H., Kim, S. H., Fok, M., & Lord, C. (2019). Autism spectrum disorder symptoms from ages 2 to 19 years: Implications for diagnosing adolescents and young adults. *Autism Research, 12*(1), 89– 99. https://doi.org/10.1002/aur.2004

Bala, K., Dogra, A., Lata, P., & Das, A. P. (2024). Discussing sources and biological-chemical remediation approaches for microplastic pollution. In *Sustainable microbial technology for synthetic and cellulosic microfiber bioremediation* (pp. 37–59). https://doi.org/10.1007/978-3-031-62600-5_3

Barrie Johnson, D., & Hallberg, K. B. (2008). Carbon, iron and sulfur metabolism in acidophilic micro-organisms. *Advances in Microbial Physiology, 54*, 201–255. https://doi.org/10.1016/S0065-2911(08)00003-9

Behera, I. D., Basak, G., Kumar, R. R., Sen, R., & Meikap, B. C. (2020). Treatment of petroleum refinery sludge by petroleum degrading bacterium Stenotrophomonas pavanii IRB19 as an efficient novel technology. *Journal of Environmental Science and Health, Part A, 56*(2), 226–239. https://doi.org/10.1080/10934529.2020.1866924

Behera, I. D., Nayak, M., Biswas, S., Meikap, B. C., & Sen, R. (2021). Enhanced biodegradation of total petroleum hydrocarbons by implementing a novel two-step bioaugmentation strategy using indigenous bacterial consortium. *Journal of Environmental Management, 292*, 112746. https://doi.org/10.1016/J.JENVMAN.2021.112746

Benzal, E., Cano, A., Solé, M., Lao-Luque, C., Gamisans, X., & Dorado, A. D. (2020). Copper recovery from PCBs by Acidithiobacillus ferrooxidans: Toxicity of bioleached metals on biological activity. *Waste and Biomass Valorization, 11*(10), 5483–5492. https://doi.org/10.1007/S12649-020-01036-Y

Bhakta, J. N. (2017). Metal toxicity in microorganism. In *Handbook of research on inventive bioremediation techniques* (pp. 1–23). https://doi.org/10.4018/978-1-5225-2325-3.CH001

Bhandari, G., Gupta, S., Chaudhary, P., Chaudhary, S., & Gangola, S. (2023). Bioleaching: A sustainable resource recovery strategy for urban mining of e-waste. In *Microbial technology for sustainable e-waste management* (pp. 157–175). https://doi.org/10.1007/978-3-031-25678-3_10

Bhattacharjee, J., Mishra, S., & Das, A. P. (2021). Recent advances in sensor-based detection of toxic dyes for bioremediation application: A review. *Applied Biochemistry and Biotechnology, 2021*, 1–20. https://doi.org/10.1007/S12010-021-03767-7

Bhattacharjee, R., Ghosh, S., Nath, A., Basu, A., Biswas, O., Patil, C. R., & Kundu, C. N. (2022). Theragnostic strategies harnessing the self-renewal pathways of stem-like cells in the acute myeloid leukemia. *Critical Reviews in Oncology/Hematology, 177*, 103753. https://doi.org/10.1016/J.CRITREVONC.2022.103753

Bhattarai, A., Gardas, R. L., Yadav, S. N., Chatterjee, S. K., & Dev, R. K. (2024). *Recovery of Rare Earth Elements (REEs) from different sources of e-waste and their potential applications: A critical review*. https://doi.org/10.20944/PREPRINTS202404.1659.V1

Biswal, B. K., & Balasubramanian, R. (2023). Recovery of valuable metals from spent lithium-ion batteries using microbial agents for bioleaching: A review. *Frontiers in Microbiology, 14*, 1197081. https://doi.org/10.3389/FMICB.2023.1197081/BIBTEX

Blayda, I. A. (2017). Variance analysis for optimization of the germanium bioleaching process from coal beneficiation dumps. *Biotechnologia Acta, 10*(4), 44–52. https://doi.org/10.15407/BIOTECH10.04.044

Booth, S. C., Weljie, A. M., & Turner, R. J. (2015). Metabolomics reveals differences of metal toxicity in cultures of Pseudomonas pseudoalcaligenes KF707 grown on different carbon sources. *Frontiers in Microbiology, 6*(JUL). https://doi.org/10.3389/FMICB.2015.00827

Borthakur, A., & Govind, M. (2017). Emerging trends in consumers' E-waste disposal behaviour and awareness: A worldwide overview with special focus on India. *Resources, Conservation and Recycling, 117*, 102–113. https://doi.org/10.1016/J.RESCONREC.2016.11.011

Bosecker, K. (1997). Bioleaching: Metal solubilization by microorganisms. *FEMS Microbiology Reviews, 20*(3–4), 591–604. https://doi.org/10.1111/j.1574-6976.1997.tb00340.x

Brandl, H., & Faramarzi, M. A. (2006). Microbe-metal-interactions for the biotechnological treatment of metal-containing solid waste. *China Particuology, 4*(2), 93–97. https://doi.org/10.1016/S1672-2515(07)60244-9

BRANDL Zürich, H. (n.d.). *Bacterial Attachment on Mineral Surfaces 200 5 Microbial Diversity in Bioleaching Environments 200 6 Case Studies of Bioleaching Applications 208 6.1 Commercial-Scale Copper Ore Bioleaching 208 6.2 Reactor Bioleaching of Fly Ash 210 6.3 Shake Flask Bioleaching of Electronic Scrap 212 7 Economics of Metal Bioleaching 213 8 Perspectives of Bioleaching Technology 214 8.1 Heterotrophic Leaching 214 8.2 Leaching Under Thermophilic Conditions 215 8.*

Brandl, H., Bosshard, R., & Wegmann, M. (2001). Computer-munching microbes: Metal leaching from electronic scrap by bacteria and fungi. *Hydrometallurgy, 59*(2–3), 319–326. https://doi.org/10.1016/S0304-386X(00)00188-2

Brim, H., McFarlan, S. C., Fredrickson, J. K., Minton, K. W., Zhai, M., Wackett, L. P., & Daly, M. J. (2000). Engineering Deinococcus radiodurans for metal remediation in radioactive mixed waste environments. *Nature Biotechnology, 18*(1), 85–90. https://doi.org/10.1038/71986

Cayumil, R., Khanna, R., Rajarao, R., Mukherjee, P. S., & Sahajwalla, V. (2016). Concentration of precious metals during their recovery from electronic waste. *Waste Management, 57*, 121–130. https://doi.org/10.1016/J.WASMAN.2015.12.004

Chauhan, J., Gautam, V. K., & Mishra, P. N. (2020). To study Information, Education, and Communication (IEC) campaign as a tool to spread awareness in understanding the support of 3R concept for waste minimization in Indore city as the cleanest city of India. In *Solid waste policies and strategies: Issues, challenges and case studies* (pp. 7–25). https://doi.org/10.1007/978-981-15-1543-9_2

Chen, Y., & Wu, T.-H. (2010). Effective e-waste management-the role of international cooperation and fragementation.

Cobbing, J. E., Hobbs, P. J., Meyer, R., & Davies, J. (2008). A critical overview of transboundary aquifers shared by South Africa. *Hydrogeology Journal, 16*(6), 1207–1214. https://doi.org/10.1007/s10040-008-0285-2

Cucchiella, F., D'Adamo, I., Lenny Koh, S. C., & Rosa, P. (2015). Recycling of WEEEs: An economic assessment of present and future e-waste streams. *Renewable and Sustainable Energy Reviews, 51*, 263–272. https://doi.org/10.1016/j.rser.2015.06.010

Cui, J., & Zhang, L. (2008). Metallurgical recovery of metals from electronic waste: A review. *Journal of Hazardous Materials, 158*(2–3), 228–256. https://doi.org/10.1016/J.JHAZMAT.2008.02.001

Dangton, J., & Leepowpanth, Q. (2014). A study of gold recovery from e-waste by bioleaching using Chromobacterium violaceum. *Applied Mechanics and Materials, 548–549*, 280–283. https://doi.org/10.4028/WWW.SCIENTIFIC.NET/AMM.548-549.280

Das, A. P., & Ghosh, S. (2018). Bioleaching of manganese from mining waste materials. *Materials Today: Proceedings, 5*(1), 2381–2390. https://doi.org/10.1016/J.MATPR.2017.11.459

Das, A. P., & Ghosh, S. (2023a). Evolution of biomining technology. In *Manganese mining microorganisms* (pp. 5–19). https://doi.org/10.1016/B978-0-12-822147-1.00012-1

Das, A. P., & Ghosh, S. (2023b). Factors affecting bioleaching. In *Manganese mining microorganisms* (pp. 77–86). https://doi.org/10.1016/B978-0-12-822147-1.00004-2

Das, A. P., & Ghosh, S. (2023c). Mechanism of manganese bioleaching. In *Manganese mining microorganisms* (pp. 87–98). https://doi.org/10.1016/B978-0-12-822147-1.00014-5

Das, A. P., & Ghosh, S. (2023d). Microorganisms in manganese biomining processes. In *Manganese mining microorganisms* (pp. 55–61). https://doi.org/10.1016/B978-0-12-822147-1.00013-3

Das, A. P., & Ghosh, S. (2023e). Mixed culture bioleaching: An insight into manganese biomining process and efficacy. In *Manganese mining microorganisms* (pp. 129–141). https://doi.org/10.1016/B978-0-12-822147-1.00001-7

Das, A. P., & Ghosh, S. (2023f). Role of metabolites and metabolomics for understanding of manganese bioleaching. In *Manganese mining microorganisms* (pp. 117–127). https://doi.org/10.1016/B978-0-12-822147-1.00011-X

Das, A. P., & Mishra, S. (2008). Hexavalent chromium (VI): Health hazards and environmental pollutant. *J Environ Res Dev, 2*(3), 386–392.

Das, A. P., & Singh, S. (2011). Occupational health assessment of chromite toxicity among Indian miners. *Indian Journal Of Occupational And Environmental Medicine, 15*(1), 6. https://doi.org/10.4103/0019-5278.82998

Das, A. P., & Swain, S. (2013). Algal biosorption of toxic dye methylene blue. In *A potential source of food, feed, biochemicals, biofuels and biofertilizers, international conference on algal biorefinery, Indian Institute of Technology* (p. 13).

Das, A. P., Sukla, L. B., Pradhan, N., & Nayak, S. (2011). Manganese biomining: A review. *Bioresource Technology, 102*(16), 7381–7387. https://doi.org/10.1016/J.BIORTECH.2011.05.018

Das, A. P., Swain, S., Panda, S., Pradhan, N., & Sukla, L. B. (2012). Reductive acid leaching of low grade manganese ores. *Geomaterials, 2012*(04), 70–72. https://doi.org/10.4236/GM.2012.24011

Das, A. P., Ghosh, S., Mohanty, S., & Sukla, L. B. (2015a). Advances in manganese pollution and its bioremediation. In L. B. Sukla, N. Pradhan, S. Panda, & B. K. Mishra (Eds.), *Environmental microbe biotechnology* (1st ed., pp. 313–328). Springer. https://doi.org/10.1007/978-3-319-19018-1_16

Das, A. P., Ghosh, S., Mohanty, S., & Sukla, L. B. (2015b). Consequences of manganese compounds: A review. *Toxicological & Environmental Chemistry, 96*(7), 981–997. https://doi.org/10.1080/02772248.2015.1005428

Das, A. P., Mallavarapu, M., & Ghosh, S. (2023). Ecotoxicity of mining pollutants on the environment and their remediation. *Environmental Chemistry and Ecotoxicology, 5*, 165–167. https://doi.org/10.1016/J.ENCECO.2023.08.002

Das, A., Dey, S., Bala, K., Gnanasekaran, R., & Das, A. P. (2024). Microbial bioremediation technology for sustainable treatment and management of synthetic microfiber waste. In *Sustainable microbial technology for synthetic and cellulosic microfiber bioremediation* (pp. 221–238). https://doi.org/10.1007/978-3-031-62600-5_12

Debnath, B., Chowdhury, R., & Ghosh, S. K. (2018). Sustainability of metal recovery from E-waste. *Frontiers of Environmental Science and Engineering, 12*(6), 1–12. https://doi.org/10.1007/S11783-018-1044-9/METRICS

Deng, X., Yi, X. E., & Liu, G. (2007). Cadmium removal from aqueous solution by gene-modified Escherichia coli JM109. *Journal of Hazardous Materials, 139*(2), 340–344. https://doi.org/10.1016/J.JHAZMAT.2006.06.043

Dey, S., Das, A., Mallick, K., Sahu, A., & Das, A. P. (2023a). *Environmental petroleum waste: Pollution, toxicity, sustainable remediation* (pp. 159–175). https://doi.org/10.1007/978-3-031-48220-5_7

Dey, S., Tripathy, B., Kumar, M. S., & Das, A. P. (2023b). Ecotoxicological consequences of manganese mining pollutants and their biological remediation. *Environmental Chemistry and Ecotoxicology*. https://doi.org/10.1016/J.ENCECO.2023.01.001

Dey, S., Bissoyi, A., Koteswara Reddy, G., & Das, A. P. (2024a). Synthetic microfibers: An urban emerging pollutant demanding for sustainable environmental solutions. In *Sustainable microbial technology for synthetic and cellulosic microfiber bioremediation* (pp. 107–121). https://doi.org/10.1007/978-3-031-62600-5_7

Dey, S., Dash, A., & Das, A. P. (2024b). Bioremediation as a sustainable and circular solution for mining waste management and bioenergy generation. In *Sustainable management of mining waste and tailings: A circular economy approach* (pp. 107–119). https://doi.org/10.1201/9781003442455-6

Dutta, D., Rautela, R., Gujjala, L. K. S., Kundu, D., Sharma, P., Tembhare, M., & Kumar, S. (2023). A review on recovery processes of metals from E-waste: A green perspective. *Science of the Total Environment, 859*, 160391. https://doi.org/10.1016/J.SCITOTENV.2022.160391

Ezeudu, O. B., Oraelosi, T. C., Agunwamba, J. C., & Ugochukwu, U. C. (2021). Co-production in solid waste management: Analyses of emerging cases and implications for circular economy in Nigeria. *Environmental Science and Pollution Research, 28*(37), 52392–52404. https://doi.org/10.1007/S11356-021-14471-8/METRICS

Faibil, D., Asante, R., Agyemang, M., Addaney, M., & Baah, C. (2023). Extended producer responsibility in developing economies: Assessment of promoting factors through retail electronic firms for sustainable e-waste management. *Waste Management and Research, 41*(1), 117–142. https://doi.org/10.1177/0734242X221105433

Fashola, M. O., Ngole-Jeme, V. M., & Babalola, O. O. (2020). Heavy metal immobilization potential of indigenous bacteria isolated from gold mine tailings. *International Journal of Environmental Research, 14*(1), 71–86. https://doi.org/10.1007/S41742-019-00240-6/METRICS

Feruza, B., Abdurassul, Z., Alma, T., Alimgazy, S., & Akmaral, S. (2022). Extraction of rhenium and osmium from lead technogenic raw materials of copper production. *Materials, 15*(12), 4071. https://doi.org/10.3390/MA15124071

Garlapati, V. K. (2016). E-waste in India and developed countries: Management, recycling, business and biotechnological initiatives. *Renewable and Sustainable Energy Reviews, 54*, 874–881. https://doi.org/10.1016/J.RSER.2015.10.106

Ghassa, S., Farzanegan, A., Gharabaghi, M., & Abdollahi, H. (2021). Iron scrap, a sustainable reducing agent for waste lithium ions batteries leaching: An environmentally friendly method to treating waste with waste. *Resources, Conservation and Recycling, 166*, 105348. https://doi.org/10.1016/J.RESCONREC.2020.105348

Ghimire, H., & Ariya, P. A. (2020). E-wastes: Bridging the knowledge gaps in global production budgets, composition, recycling and sustainability implications. *Sustainable Chemistry, 1*(2), 154–182. https://doi.org/10.3390/SUSCHEM1020012

Ghosh, S., & Das, A. P. (2015). Modified titanium oxide (TiO2) nanocomposites and its array of applications: A review. *Toxicological & Environmental Chemistry, 97*(5), 491–514. https://doi.org/10.1080/02772248.2015.1052204

Ghosh, S., & Das, A. P. (2017). Bioleaching of manganese from mining waste residues using Acinetobacter sp. *Geology, Ecology, and Landscapes, 1*(2), 77–83. https://doi.org/10.1080/24749508.2017.1332847

Ghosh, S., & Das, A. P. (2018). Metagenomic insights into the microbial diversity in manganese-contaminated mine tailings and their role in biogeochemical cycling of manganese. *Scientific Reports, 8*(1), 1–12. https://doi.org/10.1038/s41598-018-26311-w

Ghosh, S., Mohanty, S., Akcil, A., Sukla, L. B., & Das, A. P. (2016). A greener approach for resource recycling: Manganese bioleaching. *Chemosphere, 154*, 628–639. https://doi.org/10.1016/J.CHEMOSPHERE.2016.04.028

Ghosh, S., Bal, B., & Das, A. P. (2018). Enhancing manganese recovery from low-grade ores by using mixed culture of indigenously isolated bacterial strains. *Geomicrobiology Journal, 35*(3), 242–246. https://doi.org/10.1080/01490451.2017.1362080

Ghosh, S., Sinha, J. K., Ghosh, S., Vashisth, K., Han, S., & Bhaskar, R. (2023a). Microplastics as an emerging threat to the global environment and human health. *Sustainability, 15*(14), 10821. https://doi.org/10.3390/SU151410821

Ghosh, S., Tripathy, B., Dey, S., & Das, A. P. (2023b). Comparative investigation of fungal and bacterial manganese biomining mechanisms. In *Manganese mining microorganisms* (pp. 99–115). https://doi.org/10.1016/b978-0-12-822147-1.00016-9

Gopikrishnan, V., Vignesh, A., Radhakrishnan, M., Joseph, J., Shanmugasundaram, T., Doble, M., & Balagurunathan, R. (2020). Microbial leaching of heavy metals from e-waste: Opportunities and challenges. In *Biovalorisation of wastes to renewable chemicals and biofuels* (pp. 189–216). https://doi.org/10.1016/B978-0-12-817951-2.00010-9

Halim, L., & Suharyanti, Y. (2019). E-waste: Current research and future perspective on developing countries. *International Journal of Industrial Engineering and Engineering Management, 1*(2), 25–42. https://doi.org/10.24002/IJIEEM.V1I2.3214

Hanafi, J., Jobiliong, E., Christiani, A., Soenarta, D. C., Kurniawan, J., & Irawan, J. (2012). Material recovery and characterization of PCB from electronic waste. *Procedia - Social and Behavioral Sciences, 57*, 331–338. https://doi.org/10.1016/J.SBSPRO.2012.09.1194

Hossain, R., & Sahajwalla, V. (2024). Current recycling innovations to utilize e-waste in sustainable green metal manufacturing. *Philosophical Transactions A, 382*(2284), 20230239. https://doi.org/10.1098/RSTA.2023.0239

Huang, K., Chen, C., Shen, Q., Rosen, B. P., & Zhao, F. J. (2015). Genetically engineering Bacillus subtilis with a heat-resistant arsenite methyltransferase for bioremediation of arsenic-contaminated organic waste. *Applied and Environmental Microbiology, 81*(19), 6718–6724. https://doi.org/10.1128/AEM.01535-15/SUPPL_FILE/ZAM999116599SO1.PDF

Igiri, B. E., Okoduwa, S. I. R., Idoko, G. O., Akabuogu, E. P., Adeyi, A. O., & Ejiogu, I. K. (2018). Toxicity and bioremediation of heavy metals contaminated ecosystem from tannery wastewater: A review. *Journal of Toxicology, 2018*(1), 2568038. https://doi.org/10.1155/2018/2568038

Ilankoon, I. M. S. K., Ghorbani, Y., Chong, M. N., Herath, G., Moyo, T., & Petersen, J. (2018). E-waste in the international context—A review of trade flows, regulations, hazards, waste management strategies and technologies for value recovery. *Waste Management, 82*, 258–275. https://doi.org/10.1016/J.WASMAN.2018.10.018

Ilyas, S., Anwar, M. A., Niazi, S. B., & Afzal Ghauri, M. (2007). Bioleaching of metals from electronic scrap by moderately thermophilic acidophilic bacteria. *Hydrometallurgy, 88*(1–4), 180–188. https://doi.org/10.1016/J.HYDROMET.2007.04.007

Iravani, S. (2011). Green synthesis of metal nanoparticles using plants. *Green Chemistry, 13*(10), 2638–2650. https://doi.org/10.1039/C1GC15386B

Islam, A., Ahmed, T., Awual, M. R., Rahman, A., Sultana, M., Aziz, A. A., Monir, M. U., Teo, S. H., & Hasan, M. (2020). Advances in sustainable approaches to recover metals from e-waste-A review. *Journal of Cleaner Production, 244*, 118815. https://doi.org/10.1016/J.JCLEPRO.2019.118815

Islam, A., Swaraz, A. M., Teo, S. H., Taufiq-Yap, Y. H., Vo, D. V. N., Ibrahim, M. L., Abdulkreem-Alsultan, G., Rashid, U., & Awual, M. R. (2021). Advances in physiochemical and biotechnological approaches for sustainable metal recovery from e-waste: A critical review. *Journal of Cleaner Production, 323*, 129015. https://doi.org/10.1016/J.JCLEPRO.2021.129015

Ismail, H., & Hanafiah, M. M. (2020). A review of sustainable e-waste generation and management: Present and future perspectives. *Journal of Environmental Management, 264*, 110495. https://doi.org/10.1016/J.JENVMAN.2020.110495

Jagatee, S., Priyadarshini, S., Rath, C. C., & Das, A. P. (2024). Synthetic microfiber: An enduring environmental problem linked to sustainable development. *Environmental Science and Engineering, Part, F2333*, 93–112. https://doi.org/10.1007/978-3-031-51792-1_6

Jaiswal, M., & Srivastava, S. (2024). A review on sustainable approach of bioleaching of precious metals from electronic wastes. *Journal of Hazardous Materials Advances, 14*, 100435. https://doi.org/10.1016/J.HAZADV.2024.100435

Jawadand, S., & Randive, K. (2021). *A Sustainable approach to transforming mining waste into value-added products* (pp. 1–20). https://doi.org/10.1007/978-3-030-73796-2_1

Jayabarath, J., Kumar, R. S., Pandi, P. K., & Aswinth Elango, K. (2022). Mining of heavy metals from low grade ore using in situ techniques: Bioleaching and phytomining. *World Journal of Pharmaceutical Research, 11*(5). https://doi.org/10.20959/wjpr20225-23912

Jiang, P., Harney, M., Chen, B., Song, Y., Chen, Q., Chen, T., Lazarus, G., Dubois, L. H., & Korzenski, M. B. (2012). Improving the end-of-life for electronic materials via sustainable recycling methods. *Procedia Environmental Sciences, 16*, 485–490. https://doi.org/10.1016/J.PROENV.2012.10.066

Jowkar, M. J., Bahaloo-Horeh, N., Mousavi, S. M., & Pourhossein, F. (2018). Bioleaching of indium from discarded liquid crystal displays. *Journal of Cleaner Production, 180*, 417–429. https://doi.org/10.1016/J.JCLEPRO.2018.01.136

Kanungo, J., Sahoo, T., Swain, L. P., & Behera, I. D. (2023). Toxicity of persistent hydrocarbon pollutants, sources and sustainable remediation process. 39–65. https://doi.org/10.1007/978-3-031-48220-5_2

Kanungo, J., Sahoo, T., Bal, M., & Behera, I. D. (2024). Performance of bioremediation strategy in waste lubricating oil pollutants: A review. *Geomicrobiology Journal, 41*(4), 360–373. https://doi.org/10.1080/01490451.2023.2245395

Karim, S., & Ting, Y. P. (2022). Bioleaching of platinum, palladium, and rhodium from spent automotive catalyst using bacterial cyanogenesis. *Bioresource Technology Reports, 18*, 101069. https://doi.org/10.1016/J.BITEB.2022.101069

Kaur, P., Sharma, S., Albarakaty, F. M., Kalia, A., Hassan, M. M., & Abd-Elsalam, K. A. (2022). Biosorption and bioleaching of heavy metals from electronic waste varied with microbial genera. *Sustainability, 14*(2), 935–935. https://doi.org/10.3390/SU14020935

Kaushik, P. R., & Herat, S. (2020). Current state of e-waste management in India. *International Journal of Environment and Waste Management, 25*(3), 322. https://doi.org/10.1504/ijewm.2020.10027003

Kaya, M. (2016). Recovery of metals and nonmetals from electronic waste by physical and chemical recycling processes. *Waste Management, 57*, 64–90. https://doi.org/10.1016/J.WASMAN.2016.08.004

Khaliq, A., Rhamdhani, M. A., Brooks, G., & Masood, S. (2014). Metal extraction processes for electronic waste and existing industrial routes: A review and Australian perspective. *Resources, 3*(1), 152–179. https://doi.org/10.3390/RESOURCES3010152

Khan, S., Alam, F. S., & Borah, D. (2024). Microbial cell factories in the remediation of e-wastes: An insight. *Biotechnology for the Environment, 1*(1), 1–20. https://doi.org/10.1186/S44314-024-00003-4

Khatun, R., Singh, S., Dubey, N. K., & Prasad Das, A. (2023). A review on marine-based phytochemicals and their application in biomedical research. In *Recent frontiers of phytochemicals: Applications in food, pharmacy, cosmetics, and biotechnology* (pp. 383–395). https://doi.org/10.1016/B978-0-443-19143-5.00013-X

Khetriwal, D. S., Kraeuchi, P., & Widmer, R. (2009). Producer responsibility for e-waste management: Key issues for consideration—Learning from the Swiss experience. *Journal of Environmental Management, 90*(1), 153–165. https://doi.org/10.1016/J.JENVMAN.2007.08.019

Kompalitch, A. H., & van Hullebusch, E. D. (2024). Integration of bioleaching and biorefinery technologies for the recovery of base and critical elements from electronic waste. *Advances in Science, Technology and Innovation, Part, F2284*, 179–206. https://doi.org/10.1007/978-3-031-43625-3_11

Kremser, K., Gerl, P., Borrás, A. B., Espinosa, D. R., Martínez, B. M., Guebitz, G. M., & Pellis, A. (2022). Bioleaching/enzyme-based recycling of aluminium and polyethylene from beverage cartons packaging waste. *Resources, Conservation and Recycling, 185*, 106444. https://doi.org/10.1016/J.RESCONREC.2022.106444

Kumar Induria Pt Ravishankar, S., Sahu, S., Kumar Induria, S., & Sahu, T. (n.d.). *E-waste management in India Chapter 11 E-waste management in India Gestión de residuos electrónicos en India.* https://www.researchgate.net/publication/370627123_Chapter_11_E-waste_management_in_India_Gestion_de_residuos_electronicos_en_India

Kumar, A., Holuszko, M., & Espinosa, D. C. R. (2017). E-waste: An overview on generation, collection, legislation and recycling practices. *Resources, Conservation and Recycling, 122*, 32–42. https://doi.org/10.1016/J.RESCONREC.2017.01.018

Kumar, R., Saxena, A., Chawla, J., & Kumar, V. (2024) Environmental impacts of e-waste: Pollution and resource depletion. https://Services.Igi-Global.Com/Resolvedoi/Resolve.Aspx?Doi=10.4018/979-8-3693-1018-2.Ch002, 12–32. https://doi.org/10.4018/979-8-3693-1018-2.CH002

Lahiri, D., Nag, M., Dey, A., Sarkar, T., Joshi, S., Pandit, S., Das, A. P., Pati, S., Pattanaik, S., Tilak, V. K., & Ray, R. R. (2021). Biofilm mediated degradation of petroleum products. *Geomicrobiology Journal*, 1–10. https://doi.org/10.1080/01490451.2021.1968979

Leong, H. Y., Chang, C. K., Khoo, K. S., Chew, K. W., Chia, S. R., Lim, J. W., Chang, J. S., & Show, P. L. (2021). Waste biorefinery towards a sustainable circular bioeconomy: A solution to global issues. *Biotechnology for Biofuels, 14*(1), 1–15. https://doi.org/10.1186/S13068-021-01939-5/TABLES/3

Li, J., Tang, Q., Li, Y., Fan, Y. Y., Li, F. H., Wu, J. H., Min, D., Li, W. W., Lam, P. K. S., & Yu, H. Q. (2020). Rediverting electron flux with an engineered CRISPR-ddAsCpf1 system to enhance the pollutant degradation capacity of Shewanella oneidensis. *Environmental Science and Technology, 54*(6), 3599–3608. https://doi.org/10.1021/ACS.EST.9B06378/SUPPL_FILE/ES9B06378_SI_001.PDF

Li, J., Zhang, B., Yang, M., & Lin, H. (2021). Bioleaching of vanadium by Acidithiobacillus ferrooxidans from vanadium-bearing resources: Performance and mechanisms. *Journal of Hazardous Materials, 416*, 125843. https://doi.org/10.1016/J.JHAZMAT.2021.125843

Lithium-Ion Battery Recycling | US EPA. (n.d.). Retrieved October 29, 2024, from https://www.epa.gov/hw/lithium-ion-battery-recycling

Liu, Y., Feng, J., Pan, H., Zhang, X., & Zhang, Y. (2022). Genetically engineered bacterium: Principles, practices, and prospects. *Frontiers in Microbiology, 13*, 997587. https://doi.org/10.3389/FMICB.2022.997587/BIBTEX

López, M., Reche, C., Pérez-Albaladejo, E., Porte, C., Balasch, A., Monfort, E., Eljarrat, E., & Viana, M. (2022). E-waste dismantling as a source of personal exposure and environmental release of fine and ultrafine particles. *Science of the Total Environment, 833*, 154871. https://doi.org/10.1016/J.SCITOTENV.2022.154871

Lucier, C. A., & Gareau, B. J. (2020). Electronic waste recycling and disposal: An overview. In *Assessment and management of radioactive and electronic wastes*. https://doi.org/10.5772/INTECHOPEN.85983

Lukowiak, A., Zur, L., Tomala, R., LamTran, T. N., Bouajaj, A., Strek, W., Righini, G. C., Wickleder, M., & Ferrari, M. (2020). Rare earth elements and urban mines: Critical strategies for sustainable development. *Ceramics International, 46*(16), 26247–26250. https://doi.org/10.1016/J.CERAMINT.2020.03.067

Ma, Y., Liu, X., Zhou, X., Tang, J., Gan, H., & Yang, J. (2022). Reductive transformation and synergistic action mechanism in the process of treating spent lithium-ion batteries with antibiotic bacteria residues. *Journal of Cleaner Production, 331*, 129902. https://doi.org/10.1016/J.JCLEPRO.2021.129902

Mallick, K., Sahu, A., Dubey, N. K., & Das, A. P. (2023). Harvesting marine plastic pollutants-derived renewable energy: A comprehensive review on applied energy and sustainable approach. *Journal of Environmental Management, 348*, 119371. https://doi.org/10.1016/J.JENVMAN.2023.119371

Mallick, K., Sahu, S., Sahu, A., Dey, S., & Das, A. P. (2024). *A critical review of marine microfiber pollution routes, toxicity, and its sustainable remediation* (pp. 189–211). https://doi.org/10.1007/978-3-031-51792-1_10

Man, M., Naidu, R., & Wong, M. H. (2013). Persistent toxic substances released from uncontrolled e-waste recycling and actions for the future. *Science of the Total Environment, 463–464*, 1133–1137. https://doi.org/10.1016/J.SCITOTENV.2012.07.017

Maneesuwannarat, S., Vangnai, A. S., Yamashita, M., & Thiravetyan, P. (2016). Bioleaching of gallium from gallium arsenide by Cellulosimicrobium funkei and its application to semiconductor/electronic wastes. *Process Safety and Environmental Protection, 99*, 80–87. https://doi.org/10.1016/J.PSEP.2015.10.008

Marappa, N., Ramachandran, L., Dharumadurai, D., & Nooruddin, T. (2020). Recovery of gold and other precious metal resources from environmental polluted e-waste printed circuit board by bioleaching Frankia. *International Journal of Environmental Research, 14*(2), 165–176. https://doi.org/10.1007/S41742-020-00254-5

Mishra, D., Kim, D. J., Ahn, J. G., & Rhee, Y. H. (2005). Bioleaching: A microbial process of metal recovery; A review. In *Metals and materials international* (Vol. 11, Issue 3, pp. 249–256). Korean Institute of Metals and Materials. https://doi.org/10.1007/BF03027450

Mishra, S., Singh, R. P., Rath, C. C., & Das, A. P. (2020). Synthetic microfibers: Source, transport and their remediation. *Journal of Water Process Engineering, 38*, 101612. https://doi.org/10.1016/J.JWPE.2020.101612

Mishra, S., Rout, P. K., & Das, A. P. (2021). Emerging microfiber pollution and its remediation. In R. Prasad (Ed.), *Environmental and microbial biotechnology* (pp. 247–266). Springer. https://doi.org/10.1007/978-981-15-5499-5_9

Mishra, S., Dash, D., Al-Tawaha, A. R. M. S., & Das, A. P. (2022a). A review on heavy metal ion adsorption on synthetic microfiber surface in aquatic environments. *Applied Biochemistry and Biotechnology, 194*(10), 4639–4654. https://doi.org/10.1007/S12010-022-04029-W/METRICS

Mishra, S., Dash, D., & Das, A. P. (2022b). Detection, characterization and possible biofragmentation of synthetic microfibers released from domestic laundering wastewater as an emerging source of marine pollution. *Marine Pollution Bulletin, 185*, 114254. https://doi.org/10.1016/J.MARPOLBUL.2022.114254

Mishra, S., Ghosh, S., van Hullebusch, E. D., Singh, S., & Das, A. P. (2023). A critical review on the recovery of base and critical elements from electronic waste-contaminated streams using microbial biotechnology. *Applied Biochemistry and Biotechnology, 195*(12), 7859–7888. https://doi.org/10.1007/S12010-023-04440-X/METRICS

Mishra, S., Dash, D., & Das, A. P. (2024). Microbial degradation of polyester microfibers using indigenously isolated bacterial strain Exiguobacterium Sp. *CLEAN—Soil, Air, Water, 52*(12), e202300343. https://doi.org/10.1002/CLEN.202300343

Mohanta, S., Pradhan, B., & Behera, I. D. (2024). Impact and remediation of petroleum hydrocarbon pollutants on agricultural land: A review. *Geomicrobiology Journal, 41*(4), 345–359. https://doi.org/10.1080/01490451.2023.2243925

Mohanty, S., Ghosh, S., Nayak, S., & Das, A. P. (2017). Bioleaching of manganese by Aspergillus sp. isolated from mining deposits. *Chemosphere, 172*, 302–309. https://doi.org/10.1016/J.CHEMOSPHERE.2016.12.136

Mohanty, S., Ghosh, S., Bal, B., & Das, A. P. (2018). A review of biotechnology processes applied for manganese recovery from wastes. *Reviews in Environmental Science and Bio/Technology, 17*(4), 791–811. https://doi.org/10.1007/S11157-018-9482-1

Murthy, V., & Ramakrishna, S. (2022). A review on global E-waste management: Urban mining towards a sustainable future and circular economy. *Sustainability, 14*(2), 647. https://doi.org/10.3390/SU14020647

Nag, M., Lahiri, D., Ghosh, S., Sarkar, T., Pati, S., Das, A. P., Ram, D. K., Bhattacharya, D., & Ray, R. R. (2023). Application of microorganisms in biotransformation and bioremediation of environmental contaminant: A review. *Geomicrobiology Journal*. https://doi.org/10.1080/01490451.2023.2261443

Nandan, A., Suresh, A. C., Saole, P., Jeevanasai, S. A., Chandrasekaran, R., Meili, L., Wan Azelee, N. I., & Selvasembian, R. (2023). An integrated approach for electronic waste management—Overview of sources of generation, toxicological effects, assessment, governance, and mitigation approaches. *Sustainability, 15*(24), 16946. https://doi.org/10.3390/SU152416946

Natarajan, G., & Ting, Y. P. (2015). Gold biorecovery from e-waste: An improved strategy through spent medium leaching with pH modification. *Chemosphere, 136*(136), 232–238. https://doi.org/10.1016/j.chemosphere.2015.05.046

Ng, C. C. (2023). *Recovery of precious metals from e-wastes through conventional and phytoremediation treatment methods: A review and prediction*. https://philpapers.org/rec/NGROPE

Ni, H. G., & Zeng, E. Y. (2009). Law enforcement and global collaboration are the keys to containing e-waste tsunami in China. *Environmental Science and Technology, 43*(11), 3991–3994. https://doi.org/10.1021/ES802725M/SUPPL_FILE/ES802725M_SI_001.PDF

Noruzi, F., Nasirpour, N., Vakilchap, F., & Mousavi, S. M. (2022). Complete bioleaching of Co and Ni from spent batteries by a novel silver ion catalyzed process. *Applied Microbiology and Biotechnology, 106*(13–16), 5301–5316. https://doi.org/10.1007/S00253-022-12056-0/METRICS

Olaniran, A. O., Balgobind, A., & Pillay, B. (2013). Bioavailability of heavy metals in soil: Impact on microbial biodegradation of organic compounds and possible improvement strategies. *International Journal of Molecular Sciences, 14*(5), 10197–10228. https://doi.org/10.3390/IJMS140510197

Osibanjo, O., & Nnorom, I. C. (2007). The challenge of electronic waste (e-waste) management in developing countries. *Waste Management and Research, 25*(6), 489–501. https://doi.org/10.1177/0734242X07082028

Pant, D., Joshi, D., Upreti, M. K., & Kotnala, R. K. (2012). Chemical and biological extraction of metals present in E waste: A hybrid technology. *Waste Management, 32*(5), 979–990. https://doi.org/10.1016/J.WASMAN.2011.12.002

Paul, A., Dey, S., Ram, D. K., & Das, A. P. (2023). Hexavalent chromium pollution and its sustainable management through bioremediation. *Geomicrobiology Journal*, 1–11. https://doi.org/10.1080/01490451.2023.2218377

Paul, A., Priyadarsini, S., Karakuş, S., Behera, J. K., Gnanasekaran, R., & Das, A. P. (2024). Fate and accumulation of synthetic microfibers in the environment. In *Sustainable microbial technology for synthetic and cellulosic microfiber bioremediation* (pp. 61–76). https://doi.org/10.1007/978-3-031-62600-5_4

Perkins, D. N., Brune Drisse, M. N., Nxele, T., & Sly, P. D. (2014). E-waste: A global hazard. *Annals of Global Health, 80*(4), 286–295. https://doi.org/10.1016/J.AOGH.2014.10.001

Pokorna, D., & Zabranska, J. (2015). Sulfur-oxidizing bacteria in environmental technology. *Biotechnology Advances, 33*(6), 1246–1259. https://doi.org/10.1016/J.BIOTECHADV.2015.02.007

Poonia, K., Singh, P., Ahamad, T., Van Le, Q., Phan Quang, H. H., Thakur, S., Mishra, A. K., Selvasembian, R., Hussain, C. M., Nguyen, V. H., & Raizada, P. (2024). Sustainability, performance, and production perspectives of waste-derived functional carbon nanomaterials towards a sustainable environment: A review. *Chemosphere, 352*, 141419. https://doi.org/10.1016/J.CHEMOSPHERE.2024.141419

Prabhakar, A., Mishra, S., & Das, A. P. (2019). Isolation and identification of Lead (Pb) solubilizing bacteria from automobile waste and its potential for recovery of Lead from end of life waste batteries. *Geomicrobiological Journal, 36*(10), 894–903. https://doi.org/10.1080/01490451.2019.1654044

Pradhan, G., Tripathy, B., Ram, D. K., Digal, A. K., & Das, A. P. (2023). Bauxite mining waste pollution and its sustainable management through bioremediation. *Geomicrobiology Journal*, 1–10. https://doi.org/10.1080/01490451.2023.2235353

Priyadarsini, S., & Das, A. P. (2024). Analytical and structural characterization of waste lithium-ion batteries for their effective recycling strategy. *Environmental Chemistry and Ecotoxicology*. https://doi.org/10.1016/j.enceco.2024.12.004

Priyadarsini, S., & Das, A. P. (2025). Lithium bioleaching: A review on microbial-assisted sustainable technology for lithium bio-circularity. *Journal of Water Process Engineering, 69*, 106744. https://doi.org/10.1016/j.jwpe.2024.106744

Priyashantha, A. K. H., Pratheesh, N., & Pretheeba, P. (2022). E-waste scenario in South-Asia: An emerging risk to environment and public health. *Environmental Analysis, Health and Toxicology, 37*(3), e2022022. https://doi.org/10.5620/EAHT.2022022

Rai, V., Liu, D., Xia, D., Jayaraman, Y., & Gabriel, J. C. P. (2021). Electrochemical approaches for the recovery of metals from electronic waste: A critical review. *Recycling, 6*(3), 53. https://doi.org/10.3390/RECYCLING6030053

Raj, K., & Das, A. P. (2023). Lead pollution: Impact on environment and human health and approach for a sustainable solution. *Environmental Chemistry and Ecotoxicology, 5*, 79–85. https://doi.org/10.1016/J.ENCECO.2023.02.001

Rajesh, R., Kanakadhurga, D., & Prabaharan, N. (2022). Electronic waste: A critical assessment on the unimaginable growing pollutant, legislations and environmental impacts. *Environmental Challenges, 7*, 100507. https://doi.org/10.1016/J.ENVC.2022.100507

Ramprasad, C., Gwenzi, W., Chaukura, N., Izyan Wan Azelee, N., Upamali Rajapaksha, A., Naushad, M., & Rangabhashiyam, S. (2022). Strategies and options for the sustainable recovery of rare earth elements from electrical and electronic waste. *Chemical Engineering Journal, 442*, 135992. https://doi.org/10.1016/J.CEJ.2022.135992

Rani, P., Rose, P. K., Kidwai, M. K., & Meenakshi. (2023). Brassica Juncea L.: A potential crop for phytoremediation of various heavy metals. In *Heavy metal toxicity: Environmental concerns, remediation and opportunities* (pp. 285–311). https://doi.org/10.1007/978-981-99-0397-9_14

Rao, S., Liu, Y., Wang, D., Cao, H., Zhu, W., Yang, R., Duan, L., & Liu, Z. (2021). Pressure leaching of selenium and tellurium from scrap copper anode slimes in sulfuric acid-oxygen media. *Journal of Cleaner Production, 278*, 123989. https://doi.org/10.1016/J.JCLEPRO.2020.123989

Rare Earth Metals in Solar Panels - Victory Metals. (n.d.). Retrieved October 29, 2024, from https://www.victorymetalsaustralia.com/investor-centre/victory-metals-knowledgebase/rare-earth-metals-solar-panels/

Rascio, N., & Navari-Izzo, F. (2011). Heavy metal hyperaccumulating plants: How and why do they do it? And what makes them so interesting? *Plant Science, 180*(2), 169–181. https://doi.org/10.1016/J.PLANTSCI.2010.08.016

Rathore, S. S., Shekhawat, K., Dass, A., Kandpal, B. K., & Singh, V. K. (2019). Phytoremediation mechanism in Indian mustard (Brassica juncea) and its enhancement through agronomic interventions. *Proceedings of the National Academy of Sciences India Section B - Biological Sciences, 89*(2), 419–427. https://doi.org/10.1007/S40011-017-0885-5/METRICS

Robinson, B. H. (2009). E-waste: An assessment of global production and environmental impacts. *Science of the Total Environment, 408*(2), 183–191. https://doi.org/10.1016/J.SCITOTENV.2009.09.044

Sadasivuni, K. K., Rattan, S., Waseem, S., Brahme, S. K., Kondawar, S. B., Ghosh, S., Das, A. P., Chakraborty, P. K., Adhikari, J., Saha, P., & Mazumdar, P. (2019). Silver nanoparticles and its polymer nanocomposites—Synthesis, optimization, biomedical usage, and its various applications. In *Lecture notes in bioengineering* (pp. 331–373). https://doi.org/10.1007/978-3-030-04741-2_11

Sahu, A., Mallick, K., & Das, A. P. (2024). *Environmental impact of single-use synthetic face mask and its recycling: A sustainable approach* (pp. 197–212). https://doi.org/10.1007/978-3-031-50840-0_11

Sand, W., Gehrke, T., Jozsa, P. G., & Schippers, A. (2001). (Bio)chemistry of bacterial leaching—Direct vs. indirect bioleaching. *Hydrometallurgy, 59*(2–3), 159–175. https://doi.org/10.1016/S0304-386X(00)00180-8

Sanket, A. S., Ghosh, S., Sahoo, R., Nayak, S., & Das, A. P. (2016). Molecular identification of acidophilic manganese (Mn)-solubilizing bacteria from mining effluents and their application in mineral beneficiation. *Geomicrobiological Journal, 34*(1), 71–80. https://doi.org/10.1080/01490451.2016.1141340

Seif, R., Salem, F. Z., & Allam, N. K. (2024). E-waste recycled materials as efficient catalysts for renewable energy technologies and better environmental sustainability. *Environment, Development and Sustainability, 26*(3), 5473–5508. https://doi.org/10.1007/S10668-023-02925-7

Sen, M., Sinha, N., & Ghosh, D. (2024). Resource recovery from the e-wastes through bioleaching. In *Role of green chemistry in ecosystem restoration to achieve environmental sustainability* (pp. 271–280). https://doi.org/10.1016/B978-0-443-15291-7.00027-4

Shahabuddin, M., Uddin, M. N., Chowdhury, J. I., Ahmed, S. F., Uddin, M. N., Mofijur, M., & Uddin, M. A. (2023). A review of the recent development, challenges, and opportunities of electronic waste (e-waste). *International journal of Environmental Science and Technology, 20*(4), 4513–4520. https://doi.org/10.1007/S13762-022-04274-W/TABLES/1

Shaikh, S., Thomas, K., & Zuhair, S. (2020). An exploratory study of e-waste creation and disposal: Upstream considerations. *Resources, Conservation and Recycling, 155*, 104662. https://doi.org/10.1016/J.RESCONREC.2019.104662

Singh Jadaun, J., Bansal, S., Sonthalia, A., Rai, A. K., & Singh, S. P. (2022). Biodegradation of plastics for sustainable environment. *Bioresource Technology, 347*, 126697. https://doi.org/10.1016/J.BIORTECH.2022.126697

Singh, N., Li, J., & Zeng, X. (2016). Global responses for recycling waste CRTs in e-waste. *Waste Management, 57*, 187–197. https://doi.org/10.1016/J.WASMAN.2016.03.013

Singh, R. P., Mishra, S., & Das, A. P. (2020). Synthetic microfibers: Pollution toxicity and remediation. *Chemosphere, 257*, 127199. https://doi.org/10.1016/j.chemosphere.2020.127199

Sinha, R., Singh, A. K., Bauddh, K., Sharma, T. R., & Sharma, P. (2021). Phytomining: A sustainable approach for recovery and extraction of valuable metals. In *Phytorestoration of abandoned mining and oil drilling sites* (pp. 487–506). https://doi.org/10.1016/B978-0-12-821200-4.00013-3

Srivastava, N. (2020). Phytoremediation of toxic metals/metalloids and pollutants by Brassicaceae plants. In *The plant family Brassicaceae* (pp. 409–435). https://doi.org/10.1007/978-981-15-6345-4_14

Tansel, B. (2017). From electronic consumer products to e-wastes: Global outlook, waste quantities, recycling challenges. *Environment International, 98*, 35–45. https://doi.org/10.1016/J.ENVINT.2016.10.002

Tognacchini, A., Rosenkranz, T., van der Ent, A., Machinet, G. E., Echevarria, G., & Puschenreiter, M. (2020). Nickel phytomining from industrial wastes: Growing nickel hyperaccumulator plants on galvanic sludges. *Journal of Environmental Management, 254*, 109798. https://doi.org/10.1016/J.JENVMAN.2019.109798

Tonietti, L., Esposito, M., Cascone, M., Barosa, B., Fiscale, S., Muscari Tomajoli, M. T., Sbaffi, T., Santomartino, R., Covone, G., Cordone, A., Giovannelli, D., & Rotundi, A. (2024). *Unveiling the bioleaching versatility of Acidithiobacillus ferrooxidans*. https://doi.org/10.31223/X51D64

Tributsch, H. (2001). Direct versus indirect bioleaching. *Hydrometallurgy, 59*(2–3), 177–185. https://doi.org/10.1016/S0304-386X(00)00181-X

Tsydenova, O., & Bengtsson, M. (2011). Chemical hazards associated with treatment of waste electrical and electronic equipment. *Waste Management, 31*(1), 45–58. https://doi.org/10.1016/J.WASMAN.2010.08.014

Vera, M., Schippers, A., Hedrich, S., & Sand, W. (2022). Progress in bioleaching: Fundamentals and mechanisms of microbial metal sulfide oxidation—Part a. *Applied Microbiology and Biotechnology, 106*(21), 6933. https://doi.org/10.1007/S00253-022-12168-7

Wang, J., Qin, W. q., Zhang, Y. s., Yang, C. r., Zhang, J. w., Nai, S. s., Shang, H., & Qiu, G. z. (2008). Bacterial leaching of chalcopyrite and bornite with native bioleaching microorganism. *Transactions of Nonferrous Metals Society of China, 18*(6), 1468–1472. https://doi.org/10.1016/S1003-6326(09)60027-3

Watling, H. R., Shiers, D. W., Li, J., Chapman, N. M., & Douglas, G. B. (2014). Effect of water quality on the leaching of a low-grade copper sulfide ore. *Minerals Engineering, 58*, 39–51. https://doi.org/10.1016/J.MINENG.2014.01.005

Wen, Y. M., Cheng, Y., Tang, C., & Chen, Z. L. (2013). Bioleaching of heavy metals from sewage sludge using indigenous iron-oxidizing microorganisms. *Journal of Soils and Sediments, 13*(1), 166–175. https://doi.org/10.1007/S11368-012-0580-3/METRICS

White, C., Sayer, J. A., & Gadd, G. M. (1997). Microbial solubilization and immobilization of toxic metals: Key biogeochemical processes for treatment of contamination. *FEMS Microbiology Reviews, 20*(3–4), 503–516. https://doi.org/10.1111/J.1574-6976.1997.TB00333.X

Widmer, R., Oswald-Krapf, H., Sinha-Khetriwal, D., Schnellmann, M., & Böni, H. (2005). Global perspectives on e-waste. *Environmental Impact Assessment Review, 25*(5), 436–458. https://doi.org/10.1016/J.EIAR.2005.04.001

Xavier, J. C., Costa, P. E. S., Hissa, D. C., Melo, V. M. M., Falcão, R. M., Balbino, V. Q., Mendonça, L. A. R., Lima, M. G. S., Coutinho, H. D. M., & Verde, L. C. L. (2019). Evaluation of the microbial diversity and heavy metal resistance genes of a microbial community on contaminated environment. *Applied Geochemistry, 105*, 1–6. https://doi.org/10.1016/J.APGEOCHEM.2019.04.012

Yaashikaa, P. R., Priyanka, B., Senthil Kumar, P., Karishma, S., Jeevanantham, S., & Indraganti, S. (2022). A review on recent advancements in recovery of valuable and toxic metals from e-waste using bioleaching approach. *Chemosphere, 287*, 132230. https://doi.org/10.1016/J.CHEMOSPHERE.2021.132230

Yachkula, A., Rozova, O., Abashina, T., Vainshtein, M., Grouzdev, D., & Bulaev, A. (2022). Attempts to stimulate leaching activity of Acidithiobacillus ferrooxidans strain TFBk. *Minerals, 12*(8), 1051. https://doi.org/10.3390/MIN12081051

Zeng, W. m., Cai, Y. x., Hou, C. w., Liu, A. j., Peng, T. j., Chen, M., Qiu, G. z., & Shen, L. (2020). Influence diversity of extracellular DNA on bioleaching chalcopyrite and pyrite by Sulfobacillus thermosulfidooxidans ST. *Journal of Central South University, 27*(5), 1466–1476. https://doi.org/10.1007/S11771-020-4382-2

Zhang, R., Hedrich, S., Römer, F., Goldmann, D., & Schippers, A. (2020). Bioleaching of cobalt from Cu/Co-rich sulfidic mine tailings from the polymetallic Rammelsberg mine, Germany. *Hydrometallurgy, 197*, 105443. https://doi.org/10.1016/J.HYDROMET.2020.105443

Zulkernain, N. H., Basant, N., Ng, C. C., Kriti, Salari, M., & Mallick, S. (2023). Recovery of precious metals from e-wastes through conventional and phytoremediation treatment methods: A review and prediction. *Journal of Material Cycles and Waste Management, 25*(5), 2726–2752. https://doi.org/10.1007/S10163-023-01717-5

Nanoparticles, Microorganisms, and E-Waste: An Integrated Approach to Environmental Sustainability

Palani Bharath, Sivaprakasam Vignesh, Balu Kamaraj, Sridharan Selvaraj, and D. Thirumal Kumar

Abstract In today's technologically advanced society, "electronic garbage" or "e-waste" presents an alarming concern to our green enivornment. Due to the transformation of our planet towards digitalization, people are bound to use electronic devices for their needs. Though there are pros and cons accompanying it, the darker side of digitalization needs to be taken care of more. The demand for smart electronic devices is increasing at a higher rate. In the end, these will generate a lot of electronic waste which needs to be processed or recycled. Traditional methods will fall short in this digital era as the quantity will be more and the amount of toxins they emit will also be high. Therefore, we need better approaches to clean up e-waste from our environment.

This chapter explores how e-waste is managed using nanoparticles and microorganisms. Nanoparticles are either synthesized from e-waste as a source or can be applied for remediation purposes. Nanoparticles can be synthesized using eco-friendly methods like green synthesis using plant extracts and microorganisms. On the other hand, microorganisms play a vast role in e-waste management. Through certain processes like bioleaching, biosorption, and bioaccumulation, microorganisms leach, adsorb, and accumulate metals by themselves, thereby helping in metal

P. Bharath · S. Vignesh
Department of Research, Meenakshi Academy of Higher Education and Research (Deemed to be University), Chennai, Tamil Nadu, India

B. Kamaraj
Department of Dental Education, College of Dentistry, Imam Abdulrahman Bin Faisal University, Dammam, Saudi Arabia

S. Selvaraj
School of Agricultural Sciences, Dhanalakshmi Srinivasan University, Perambalur, Tamil Nadu, India

D. Thirumal Kumar (✉)
Faculty of Allied Health Sciences, Meenakshi Academy of Higher Education and Research (Deemed to be University), Chennai, Tamil Nadu, India
e-mail: thirumal@maher.ac.in; thirumalkumard@gmail.com

A. P. Das, S. Priyadarsini (eds.), *Electronic Waste and Environmental Pollution*, Sustainable Environmental Waste Management Strategies,
https://doi.org/10.1007/978-3-031-90287-1_21

recovery. Apart from the individual contributions, microorganisms are also used as a consortium of the same species or different species of bacteria and fungi to extract metals from e-waste. Following this, the chapter discusses how nanoparticles and microorganisms can be used as a synergy to attain greater efficiencies. Recent advances in the field of nanobiotechnology to mitigate e-waste are also discussed. The chapter highlights the novel nanoparticle-microbe combinations and how they are applied in real-world cases. Due to some unsafe practices, toxic substances are being released into the environment posing a major threat to living organisms. The chapter then discusses on to the part where the safety and regulatory policies of different countries to manage e-waste are implemented. Finally, the emerging technologies and future prospects in the field of nanobiotechnology in e-waste management are addressed. Even though we have achieved greater heights in technological aspects, there are some gaps that need to be filled in this area to ensure the safety and health of all living organisms. Furthermore, effective strategies are needed in this hour to combat this huge task to make our world better place to live.

Keywords E-waste · Nanobiotechnology · Nanoparticles · Microorganisms · Bioleaching

1 Introduction

1.1 Background and Scope

Massive amounts of e-waste are generated and pose a major risk to the world's environment. Economically weak nations have problems in e-waste management because they lack good regulation policies and facilities to minimize or recycle e-waste (Mishra et al., 2023). This makes the situation more problematic and concerning. Due to these poor regulation policies, the environment is getting more contaminated and further affects due to the release of toxic substances from e-waste (Das & Mishra, 2008; Das & Singh, 2011; Paul et al., 2023). These toxic substances are exposed to various living organisms via different sources and routes (Grant et al., 2013; Heacock et al., 2016). The e-waste mostly ends up in municipal waste due to dumping of e-waste through landfilling or being exported to some countries for recycling purposes (Robinson, 2009). In 2019, it was estimated by the WHO that around 53.6 million tonnes of e-waste were generated, of which a small amount was collected for recycling (Forti et al., 2020). Though there are many conventional methods for e-waste remediation, somehow they lack efficiency in removing contaminants. Hence, novel techniques and approaches are needed for greater results (Rajput et al., 2022). Nanotechnology holds immense potential in many fields, such as agriculture, healthcare, environment, toxicology, etc., (Sthiannopkao & Wong,

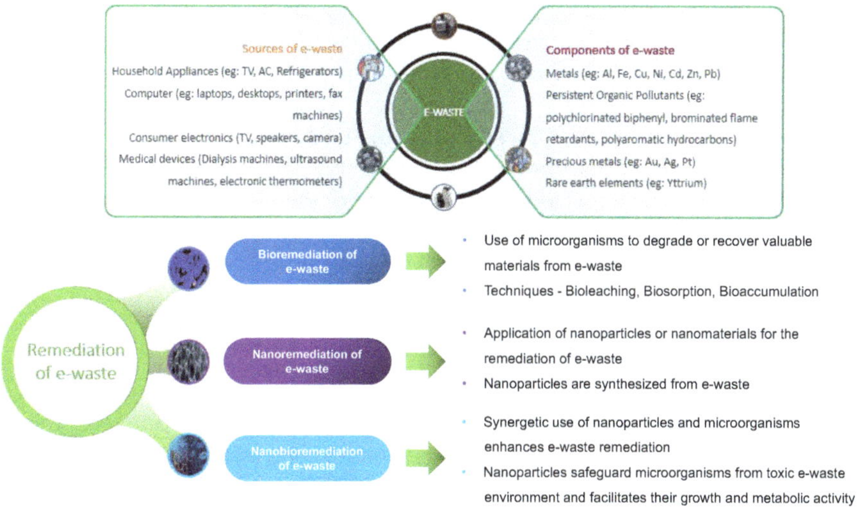

Fig. 1 Graphical abstract summarizing the integrated approach to environmental sustainability through the interaction of nanoparticles, microorganisms, and e-waste

2013). Due to various advantages like large surface area, high sensitivity and reactivity, unique electronic properties, and catalytic properties, nanomaterials or nanoparticles have been used widely for environmental applications (Corsi et al., 2018). Integrating nanoparticles with biological systems proves to be convincing, and there are big opportunities as these integrated systems are less explored (Rajput et al., 2022). In Figure 1, the schematic diagram highlights the integration of nanotechnology and bioremediation techniques for efficient recycling, pollutant degradation, and resource recovery from e-waste, emphasizing a sustainable approach to waste management and ecosystem restoration.

2 E-Waste: An Emerging Environmental Challenge

2.1 Definition and Types of E-Waste

E-waste is defined as the various forms of waste electrical and electronic equipment (WEEE) that have no more usage (Amankwaa et al., 2017). It is more of anything with a plug, electric cord, or battery. It consists of WEEE or products that are unfit for the purpose for which they were designed. Types of e-waste also includes, electric and electronic items such as large household appliances (refrigerators, dishwashers, vacuum cleaners, washing machines, cookers, electric radiators, air conditioners, microwaves, etc.), small household appliances (mixer-grinders, toasters, irons, clocks, etc.) (Kaya, 2016), computer and telecom appliances (fax machines, phones, Wi-Fi-modems, laptops/desktop computers, printers and scanners, radios, calculating devices, DVDs, network cables, etc.) (Robinson, 2009), consumer equipment (television, speakers, camcorders, digital cameras, etc.) (Shahabuddin et al., 2023), medical devices (dialysis machines, ultrasound machines, respiratory ventilators, electronic thermometers, cardiology equipment, etc.), and electrical and electronic tools (electrical saw, sewing machines, LED bulbs, low-pressure sodium lamps, smoke detectors, thermostats, etc.) (Amankwah-Amoah, 2016; Shahabuddin et al., 2023).

2.2 E-Waste Composition and Hazardous Components

E-waste contains different metals, non-metals, and chemicals with different properties. Composite materials consist of minerals such as polymers, glass fibres, ceramics alumina, mica, and silica (Chauhan et al., 2018). Inorganic metal substances include base metals (iron, aluminum, copper, and tin), precious metals (Au, Ag, and Pt), heavy metals (Pb, Cd, Ni, Hg, and Cr), and rare earth metals (yttrium) (Murali et al., 2022). Organic chemicals such as arsenic, lithium, chlorofluorocarbon (CFC), cobalt, barium, selenium (Forti et al., 2020; Priyadarsini & Das, 2024), polyaromatic hydrocarbons (PAHs), polychlorinated biphenyls (PCBs), brominated flame retardants (BFRs), and polychlorinated dibenzo-p-dioxin furans (PCDDs) are used (Wu et al., 2019). However, substances such as Cd, Cr, Hg, Pb, CFC, PCBs, PAHs, and PCDD are toxic, hazardous, and cause ozone depletion (Awasthi et al., 2016; Lin et al., 2022).

3 Nanobiotechnology in E-Waste Management

3.1 Nanoparticles for E-Waste Remediation

The application of nanotechnology for e-waste remediation has yet to be fully explored and only limited research articles published in this area. E-waste contains a higher concentration of rare earth elements, which can cause environmental contamination if it is not properly disposed of. Silica nanoparticles and plastic nanoparticles were employed in a previous study by Brewer et al. in 2022 to assess the retention of rare earth elements and were found to have an increase in retention capacity during the leaching process (Brewer et al., 2022). Gold is one of the precious metals used in many electronic components due to its good conducting properties, and some nanoparticles were used to recover it from the e-waste. Hollow polyaniline nanospheres (P(VAn-g-PANI)) were synthesized to recover gold using a novel approach. This nanosphere helped in reducing the gold ions as well as in the stabilization of polymer nanosphere-assisted gold nanoparticles

(Au@P(VAn-g-PANI)) formation. These polymer nanosphere-assisted gold nanoparticles can be combined with indium tin oxide (ITO) to extract metals from various electronic appliances (Tian et al., 2020). Nanopapers that were made of cellulose fibres (CF) and ionic covalent organic frameworks (COF) were used for the recovery of gold. Aquaregia was used initially to dissolve metals from the e-waste and then treated with CF-COF nanopapers. These nanopapers selectively capture gold, thereby helping it recover (Xu et al., 2023).

In another study by Hussein et al. in 2020, mesoporous silica nanospheres (MSNS) were employed to leach gold. This nanosphere was mobilized separately with two chromogenic probes, thiazole yellow G (TYG), and amacid yellow M (AYM). MSNS-TYG and MSNS-AYM recovered Au^{3+} ions at about 87.3% and 80.6%, respectively (Hussein et al., 2020). Nanoporous polymer microspheres functionalized with nitriles were employed for the application of the gold recovery. The results stated that these microspheres had more selectivity for gold and less selectivity for other metals, which helped in the recovery of gold more than other metals (Dogan et al., 2019). Zinc oxide nanoparticles (ZnO NPs) were used as nano adsorbents to remove toxic metals such as Cd, Pb, and Hg from the aqueous e-waste solution. In an acidic environment, these metals were removed at a higher rate, above 88% (Nayar, 2019). Polybrominated diphenyl ethers (PBDEs) are mostly found at e-waste recycling sites. These are released into the environment and pose several risks. PBDEs were degraded using a biochar-assisted nano zero-valent iron (BC-nZVI) and found to have better degradation under acidic conditions (Li et al., 2019).

3.2 *Microbial Bioremediation of E-Waste*

Bioremediation of e-waste involves the utilization of microorganisms such as bacteria, fungi, yeast, algae, or their enzymes and metabolites to degrade, detoxify, or mobilize e-waste components (Dixit et al., 2015). Microorganisms often used for bioremediation include chemolithotrophic bacteria (acidophiles), heterotrophic bacteria, and fungi. Acidophiles are classified into three types based on the growth temperature, namely mesophiles (28–37 °C), moderate thermophiles (40–60 °C), and extreme thermophiles (60–80 °C) (Yaashikaa et al., 2022). Other chemolithotrophs include iron-oxidizing, sulfur-oxidizing, and iodide-oxidizing bacteria, which oxidize iron, sulfur, and iodide (Becci et al., 2021). Iodide-oxidizing bacteria such as *Roseovarius mucosus* and *R. tolerans* were used to leach gold using three methods (one-step, two-step, and spent medium) in which the two-step method effectively yielded more gold. In another study, a novel thiourea bioleaching process was developed for gold to facilitate leaching. This offered a new alternative approach to iron-oxidizing bacteria/archaea facilitating gold bioleaching (Kudpeng et al., 2020). Various iron-oxidizing bacteria/archaea were studied, of which *Acidiplasma* species, Fv-Ap, a strain of acidophilic archaea, exhibited both PCB and thiourea tolerance (Rizki et al., 2019). *Pseudomonas balearica* was collected from an e-waste recycling facility and employed to leach Au and Ag. This strain possesses tolerance to e-waste toxicity (Kumar et al., 2018).

Acidithiobacillus thiooxidans is a mesophilic bacteria that produces sulfuric acid to leach copper from grounded e-waste via direct and indirect bioleaching (Hong & Valix, 2014; Priyadarsini & Das, 2025)). In addition to *A. thiooxidans, Chromobacterium violaceum,* and *Pseudomonas fluorescens* were employed to recover gold and copper via a novel bio-oxidation method. *Chromobacterium violaceum* and *Pseudomonas fluorescens* produce cyanides that dissolve gold and copper. Biooxidation was done before the bioleaching of e-waste by *Acidithiobacillus ferrooxidans* which helped to remove more than 80% of manganese (Das & Ghosh, 2023b, d). This approach was novel and more effective than applying *C. violaceum* and *P. fluorescens* without bio-oxidation (Pham & Ting, 2009). Another study by Natarajan & Ting in 2015 reported using the same bacteria, but instead of *P. fluorescens*, *P. aeruginosa* was used and made into different pure cultures. These bacteria also produce cyanides used for gold recovery via two-step bioleaching. Of these three, *C. violaceum* exhibited more cyanide production and gold recovery than others (Natarajan & Ting, 2015). *C. violaceum* was isolated from an abandoned gold mine and used to leach gold and copper from PCBs (Kumar et al., 2021). *F. casuarinae* and *Frankia* sp., a class of actinobacteria, were employed for the recovery of metals in which *F. casuarinae* and *Frankia* sp. leached out gold (75%) and Copper (94%) from the PCBs via two-step process (Marappa et al., 2020). Mostly used acidophiles in bioleaching of e-waste are *Acidiphilium acidophilum, A. thiooxidans,* and *Leptospirillum ferrooxidans* (Brandl et al., 2001; Jowkar et al., 2018; Sheel & Pant, 2018). In a study, *A. acidophilum* was grown in e-waste to leach Cu, Ni, Zn, and Pb. The leaching efficiency of these metals was found to be 79%, 39%, 29%, and 10% in 60 days (Priya & Hait, 2018). *A. thiooxidans* was used to recover indium and strontium, primarily found in electronics as a part of LCD screens and was taken from discarded laptops for leaching (Jowkar et al., 2018). *Lactobacillus acidophilus* with a chemical leachant called ammonium thiosulfate was employed to recover gold, and this novel leaching-sorption approach achieved a recovery of about 85% after 90 days (Sheel & Pant, 2018). *Paenibacillus* sp. was utilized for metal extraction using a two-step fermentation method from e-waste. With this approach, the bacteria leached more than 80% of manganese and zinc and 50% of copper, cadmium, iron, and palladium (Waghmode et al., 2021; Das et al., 2015; Das & Ghosh, 2018, 2023a; Ghosh et al., 2016).

In the case of fungal species, *Aspergillus niger* was exploited more than any other species (Mohanty et al., 2017). Many studies can be seen using *A. niger* for e-waste bioremediation. In a study, *A. niger* was isolated from an activated sludge and exploited in both one- and two-step bioleaching. The findings stated that two-step bioleaching was more effective. Organic acids are produced by *A. niger* which help in mobilizing the metals and detoxify the toxins produced by the metals (Narayanasamy et al., 2018). *A. niger* was applied to leach metals (Al, Pb, Cu, Zn, Fe, Li, Ni, and Mn) from old mobile phones and tantalum capacitors (Priyadarsini & Das, 2025). The fungi showed an efficiency of leaching about 75% of Al, 61% of Pb, 52% of Cu, 35% of Zn, 29% of Fe, 21% of Ni, and 5% of Mn (Brandl et al., 2001). *Penicillium simplicissimum* with *A. niger* was employed to mobilize Al, Zn,

Table 1 Microorganisms involved in biosorption and bioaccumulation of metals

Methods	Name of the organisms	Metals	References
Biosorption	*Bacillus sphaericus*	Chromium	Velásquez and Dussan (2009)
	Galdieria sulphuraria	Yttrium, cerium, europium, and terbium	Palmieri et al. (2022)
	Aspergillus carbonarius	Chromium	Lakshmi et al. (2020)
	Aspergillus niger	Lead, copper, and stannous	Hiremath et al. (2015)
	Leptolyngbya boryana	Arsenic	Zhu et al. (2020)
	Plectonema boryanum	Gold	Lengke et al. (2006)
Bioaccumulation	*Halomonas* sp.	Cadmium	Kavitha (2014)
	Penicillium expansum	Lanthanum and terbium	Di Piazza et al. (2017)
	Saccharomyces cerevisiae	Copper	Zhang and Min (2010)

Ni, and Pb by growing them in the same e-waste environment. They produced oxalate and citrate that efficiently leached more than 95% of Al, Zn, Ni, and Pb with more than 65% of Cu and Sn (Kudpeng et al., 2020). *P. simplicissimum* was used to extract Cu and Ni from mobile phone PCBs. The efficiency was observed to be 90% for Cu and 89% for Ni (Arshadi et al., 2019). Following this, the same bacteria were used to extract the same metals (Cu and Ni) from mobile phones in bubble column bioreactors (Nili et al., 2022). Multiple fungi species, such as *A. niger, Ganoderma lucidum, Trametes versicolor, Pleurotus florida*, and *Lentinus edodes* and two bacterial species (*Streptomyces* spp. and *Pseudomonas* spp.), were cultured separately to validate the efficiency of metal leaching in PCBs. These organisms produce laccase, an enzyme that catalyzes the bioleaching process. Among these organisms, *P. florida* and *Pseudomonas* spp. were found to have more biosorption and bioleaching abilities than others (Kaur et al., 2022). Rosa et al. in 2022 used *Trichoderma harzianum, Penicillium glandicola*, and *Aspergillus tubingensis* to investigate their potential to produce siderophores to mobilize copper, palladium, and yttrium from the e-waste (Rosa et al., 2022). Table 1 shows the list of microorganisms involved in the biosorption and bioaccumulation of metals, and Table 2 shows the bioleaching approaches for e-waste.

Table 2 Bioleaching approaches for e-waste

	One-step bioleaching	Two-step bioleaching	Spent bioleaching
Process	Microbial culture is inoculated into the culture medium, along with the e-waste in the bioreactor (Hosseini et al., 2022)	Microorganisms are cultured separately and introduced in a medium when they enter the logarithmic growth phase (Baniasadi et al., 2019)	Microorganisms are cultured separately for metabolite production, and cells are removed via centrifugation to get a cell-free supernatant medium to which e-waste is added (Vakilchap et al., 2016)
Steps	Single step: e-waste and microorganisms are added simultaneously in the reactor	Two-step: 1. Microorganisms grown separately 2. Leaching of metals	Single step: Metal leaching using metabolites (non-contact bioleaching)
Advantages	A simple process that follows direct bioleaching	Enhances the leaching ability of microorganisms and metabolite production (Liu et al., 2016)	The most suitable method for microorganisms in toxic environments and high pulp density (Reed et al., 2016)
Disadvantages	A toxic e-waste environment affects the bio-recovery process (Magoda & Mekuto, 2022)	Leached metals can accumulate in the microbial cells and hence reduce metal recovery (Ilyas et al., 2013; Natrajan & Ting, 2013)	Metabolites are produced only in the presence of metals (Natrajan & Ting, 2013)
	Reduces the growth of microorganisms and low metabolite production (Magoda & Mekuto, 2022)	Less log phase (Işıldar et al., 2016; Natrajan & Ting, 2013)	Lower lag phase (Ilyas et al., 2013)

4 Green Synthesis of Nanoparticles from E-Waste

4.1 Biological Methods of Nanoparticle Synthesis

Biological or bio-mediated processes are bottom-up methods that refer to synthesizing nanoparticles utilizing biological agents such as plants, microorganisms (bacteria, fungi, yeast, etc.), and their enzymes. These biological agents serve as both reducing and stabilizing agents, thus replacing the chemicals. They offer several advantages over physical and chemical routes, ensuring a clean, safe, non-toxic, inexpensive, and eco-friendly approach to producing nanoparticles (Jayaseelan et al., 2012; Malik et al., 2014; Mohanpuria et al., 2008). Plant materials such as leaves and roots are prepared as extracts and used to synthesize nanoparticles. They contain various phytochemicals that facilitate metal ion reduction to form nanoparticles (Hano & Abbasi, 2021). The nanoparticles can also be synthesized using microorganisms such as bacteria, fungi, algae, etc. Other microorganisms, such as actinobacteria, yeast, and viruses, are also used to synthesize nanoparticles (Fariq et al., 2017). Microorganisms produce enzymes that reduce metal ions to elemental form and help in nanoparticle synthesis. These nanoparticles are synthesized via intracellular or extracellular routes (Salem & Fouda, 2021).

4.2 Biogenic Nanoparticles for E-Waste Remediation

Biogenic nanoparticles are synthesized using microorganisms or plants or from their metabolites (Chauhan et al., 2023). In a study by Rozas et al. in 2017, Hyhel-1, a bacterial strain isolated from *Hymeniacidon heliophila* (marine sponge), was employed to leach metals from the printer motherboards. These bacteria have the potential to recover copper and also produce Cu nanoparticles. The production of lipopeptides by the bacteria catalyzes the leaching process. This process required no solvents or other acidification processes, thus offered a new bioleaching strategy (Rozas et al., 2017). *Pseudomonas* sp. and *Fusarium oxysporum* were used to leach copper from the integrated circuits (ICs) of electronic boards. They produced copper nanoparticles with a size range of approximately between 84–130 nm and 93–115 nm (Majumder, 2012). Lichen-associated bacteria such as *Parmotrema tintorum* and *P. recticulatum* were utilized to leach Cu and Fe from e-waste. They produced copper and iron nanoparticles in the 10–100 nm (Subbaiyan et al., 2023). Mixed cultures of *A. ferrooxidans, L. ferrooxidans*, and *A. thiooxidans* were employed to leach metals from PCBs through a two-step method. Cu was leached from the e-waste during the initial step and found to be over 90%. Besides that, there was a considerable amount of Au in the e-waste, so *P. putida* strains were employed to leach Au as they produce cyanides that help in the solubilization of Au. This was done during the second step. Then, *Fusarium oxysporum* and *Bacillus cereus* were added to the obtained leach solutions which facilitate the synthesis of Cu nanoparticles (Toshev et al., 2021).

5 Microorganisms in E-Waste Remediation

5.1 Role of Microbes in E-Waste Degradation

Microorganisms play a vital role in e-waste remediation, and they use different mechanisms like bioleaching, biosorption, and bioaccumulation and are exploited via in-situ or ex-situ bioremediation (Bala et al., 2022; Dixit et al., 2015). The bioleaching process refers to the utilization of microorganisms, to mobilize metal ions from insoluble ores or convert heavy metals to a soluble form (Das et al., 2011; Das & Ghosh, 2023c; Ghosh et al., 2023). These microorganisms use different mechanisms like acidolysis, redoxolysis, and complexolysis (Rohwerder et al., 2003; Vera et al., 2013). Based on the mode of action of microorganisms, bioleaching was classified into two types: direct (contact) and indirect (non-contact) bioleaching.

In direct bioleaching, the microorganisms attach to the metal's surface through direct physical contact with them leading to metal dissolution by oxidizing them enzymatically. Non-contact bioleaching does not involve direct contact with the metals; instead, it produces potent oxidizing agents that help in the reduction of metals (Bosecker, 1997; Ji et al., 2022; Rohwerder et al., 2003; Sand et al., 2001; Tributsch, 2001). Thiosulfate and polysulfide pathways are two pathways associated with non-contact bioleaching through which the dissolution of metal sulfides occurs (Schippers & Sand, 1999). Both iron- and sulfur-oxidizing bacteria play a significant role in mobilizing the metals from metal sulfides using these two pathways, where Fe^{2+} is oxidized into Fe^{3+}. Then, Fe^{3+} reacts with metal sulfides and is reduced into Fe^{2+} (Ji et al., 2022). Various microorganisms are used for bioleaching, predominantly acidophiles, thermophiles, and cyanogenic bacteria. Apart from these organisms, fungi and other species like algae, yeast, and archaea are also employed in bioleaching (Bosecker, 1997; Takatsugi et al., 2011). Biosorption and bioaccumulation are the other two most commonly used strategies in e-waste remediation using microorganisms. Biosorption is a passive process (i.e., the mechanism cannot be controlled metabolically) in which microorganisms uptake heavy metals, and these metals get adsorbed or bind to the surface of the microorganisms. This process involves mechanisms like adsorption, exchange of ions, surface complexation, and precipitation, which help in metal adsorption. The cell surface of microorganisms contains proteins, lipids, and polysaccharides and also possesses some functional groups like carboxyl, carbonyl, amine, amide, phosphodiester, phosphate, phenolic, etc. that attract metals to their surfaces (Fomina & Gadd, 2014; Kaur et al., 2022; Scott & Karanjkar, 1992). Bioaccumulation is an active process

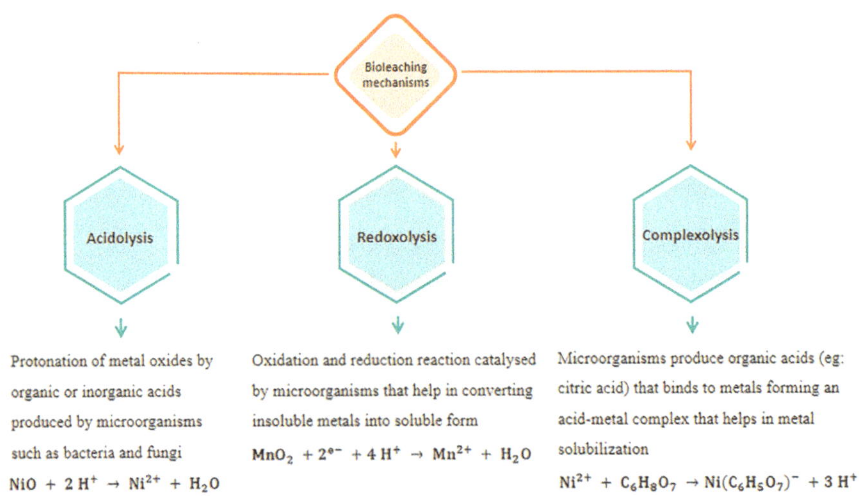

Fig. 2 Mechanism of bioleaching by microorganisms

that requires the metabolic activity of microorganisms. In this approach, microorganisms uptake heavy metals intracellularly. This depends on their physiochemical properties, structural properties, metal toxicity, and specificity (Blackwell et al., 1995; Davis et al., 2003). Figure 2 describes the mechanism of bioleaching used by the microorganism to extract metals from ores or secondary sources.

5.2 Microbial Consortia for Enhanced Remediation

Mixed cultures of cyanogenic bacteria (*C. violaceum* and *P. aeruginosa*, *C. violaceum* and *P. fluorescens*, *P. aeruginosa*, *P. fluorescens*, and *C. violaceum*, and *P. aeruginosa* and *P. fluorescens*) were utilized to recover gold from electronic waste. From these mixed cultures, the first mixed culture (*C. violaceum* and *P. aeruginosa*) recovered gold at the maximum level. They also continuously produced cyanide during the bioleaching process (Pham & Ting, 2009). A Frankia consortium (two *Frankia* sp. + two *Frankia casuarinae*) was used to recover gold and silver from PCBs. This consortium leached gold and silver at a rate of 90% and 75% (Narayanasamy et al., 2022). A novel method of metal solubilization using a biooxidation–leaching–separation cycle for the recovery of copper was developed. This approach utilized a consortium of *Leptospirillum ferriphilum* and *Sulfobacillus thermosulfidooxidans*, resulting in higher recovery of Cu and found to have a strong metal tolerance (Wu et al., 2018). In another study, a mixed culture of *Leptospirillum ferrooxidans*, *A. thiooxidans*, and *A. ferrooxidans* was used for leaching metals from spent mobile phones. Cu and Ni were recovered at a higher rate of 99% and 96%. Other metals such as Al, Zn, Fe, and Mn were obtained at the rate of 70%, 77%, 89%, and 43%. These acidophiles were also evaluated against tantalum capacitors. They showed great persistence in mobilizing Cu, Ni, Fe, Zn, Al, and Mn at an efficiency of 92%, 88%, 78%, 77%, 70%, and 57% (Sikander et al., 2022). A mixed consortium of *Thiobacillus ferrooxidans* and *Thiobacillus thiooxidans* (presently known as *A. thiooxidans*) was utilized to extract metals from electronic straps. As they were grown on electronic scrap, the initial growth phase was affected by the toxic environment but at medium concentrations, they mobilized more than 90% of Zn, Cu, Al, and Ni (Brandl et al., 2001). Consortia of two bacterial strains (*Cupriavidus necator* and *P. putida*) along with two fungal strains (*Neurospora sitophila* and *Boeremia exigua*) were utilized for bioleaching of Ag or Cu (Schneiter et al., 2023). A mixed fungal culture (*Purpureocillium lilacinum* and *A. niger*) was employed for bioleaching of certain low-grade metals (Zn, Cu, Al, Sn, and Pb) from PCBs in a stirred tank reactor (Xia et al., 2018). A combination of three fungi, *Sphingomonas* sp. MXB8, *Candida orthopsilosis* MXL20, and *A. niger* MX7 and MX5, was used to recover gold and silver from the PCB of mobile phones (Díaz-Martínez et al., 2019). Two strains of *A. niger* (MX7 and MXPE6) were collected from an e-waste-contaminated soil environment and used as a consortium to validate their efficiency in leaching gold from the PCBs. The results showed a higher gold recovery (Argumedo-Delira et al., 2019).

6 Integrated Approaches to E-Waste

6.1 Nanoparticle-Microorganism Synergy

Using nanoparticles with microorganisms to clear e-waste contaminants is a valuable approach, but few studies have related to this. This integrated approach of nanomaterials or nanoparticles with microorganisms can benefit more from e-waste nanobioremediation. E-waste contains heavy metals and other components that are toxic to microorganisms. These components inhibit the growth and activity of microorganisms. Nanoparticles can safeguard microorganisms and enhance their growth and metabolic activity (Sharma & Sood, 2023). Heavy metals can be reduced into metal ions by the action of bacterial enzymes, and these enzymes help in synthesizing nanoparticles from e-waste (Yadav et al., 2017). A nano-bio treatment method utilizing a nanoscale zero-valent iron and *Sphingomonas* sp. PH-07 strain was employed for the degradation of PBDEs. Initially, PBDEs were debrominated to low-brominated diphenyl ethers with the help of nano zero valent iron (nZVI). These low-brominated DEs were then biologically treated by *Sphingomonas* sp. PH-07, which effectively reduced them into bromophenols and other metabolites (Kim et al., 2012). A polyvinylpyrrolidone (PVP) coated iron oxide nanoparticles along with *Halomonas* sp. was used to remove lead and cadmium. This combined approach turned out to be more effective in metal removal in a shorter period (Raj & Das, 2023). The nanoparticles helped in reducing the toxicity of metals and improved bacterial growth. This combined approach can be implemented in e-waste to analyze whether it can remove toxic metals (Cao et al., 2020). Aroclor 1248 is one of the specific formulations of polychlorinated biphenyls that were widely used in electrical applications (like capacitors, transformers, etc.). These polychlorinated biphenyls are considered significant contaminants that can accumulate in many organisms. Using bimetallic Pd/Fe nanoparticles and *Burkholderia xenovorans* LB400, these compounds were degraded effectively (Le et al., 2015).

6.2 Challenges

Due to the toxic nature of e-waste, microbial growth can be reduced or inhibited impacting the leaching activity. This is considered one of the major challenges in bioleaching as they are the catalysts that help mobilize metals from e-waste. To overcome this, a study was reported on a subculturing strategy using an iron-oxidizing bacterial culture (*Leptospirillum ferriphilum*), which inhibited e-waste toxicity by improving microbial resistance (Anaya-Garzon et al., 2021). This will be the major challenge for an industrial application or commercial process. The two-step method can help to go through this. This method involves culturing microbes separately in the absence of e-waste. Following this, the metals in e-waste are mobilized with the help of metabolites produced by the microorganisms (Brandl et al., 2001). Even though bioremediation is an eco-friendly technique, it consumes more time for remediation than physical and chemical processes. Another big challenge is scaling up the nanobioremediation process from the lab scale to the industrial scale, which is quite difficult and requires lots of time, resources, and money. This translation process is one of the most difficult challenges many face (Jordahl et al., 1997; Panpatte & Jhala, 2021). Nanotoxicity will be a major concern when we use nanoparticles in any application, so there is a chance of accumulation and transfer of nanoparticles that can produce harmful effects on the host's health (Cedervall et al., 2012).

7 Case Studies and Success Stories

In a study by Vestola et al. in 2010), Cu and Ni were fully recovered from recycled electronic scraps and two other solid sample wastes using *Acidithiobacillus* sp. and *Leptospirillum* sp. (Vestola et al., 2010). About 100% copper recovery was achieved using *A. niger.* Following certain parameters like pH at 7.0, temperature at 30 °C for 5 days, aeration of 200 rpm, and a pulp density of 0.5%, full recovery was achieved (Sharada et al., 2021). *A. ferrooxidans* was investigated to carry out a process of displacement of copper ions from the solution through oxidation where Fe^{2+} is oxidized into Fe^{3+}. The extracted Cu was found to be 100% from the PCBs within 48 h (Benzal et al., 2020). Bioreactor systems were used for bioleaching of metals on a larger scale from Ni-Cd spent batteries. A maximum recovery of Cd (100%) and Ni (66.1%) was obtained in 50 days using indigenous *Thiobacilli*. They produced sulfuric acid in the bioreactors that helped leach Cd and Ni (N. Zhu et al., 2003). A study was conducted to recover In and Sr from the LCDs of discarded laptops using *A. thiooxidans*. This approach could be taken up on a large scale because it was more effective and could be used as an alternative to chemical leaching (Sheel & Pant, 2018).

8 Environmental Implications and Risk Assessment

8.1 Ecotoxicity and Environmental Impact

A study has been conducted in China at an e-waste site abandoned long ago. The results showed that compounds such as Cu and Cd were still present in the site and soil, and surrounding ecosystems (paddy fields and ponds) were also contaminated (Wu et al., 2015). In another recycling site in China, metals such as Cu, Cd, Sb, and Pb were found in the contaminated soil (Han et al., 2019). The water utilized for the acid wash was released into the streams, ultimately ending up in the nearby ponds. The unrecovered waste materials were also found to be dumped near the streams, posing ecological risks (Wu et al., 2015). The surroundings of e-waste recycling sites were found to be polluted with some poisonous compounds such as PBDDs, PCDDs, and Co-PCBs (Suzuki et al., 2016). Heavy metals are released into the atmosphere, posing several threats to the nearby areas (Gangwar et al., 2019). Similarly, during the e-waste recycling process, PDBE, PCDDs, and PAHs were released into the air resulting in polluting the air (Wang et al., 2022). The soil, water, and air have been polluted due to the release of hazardous substances from e-waste during processing or remediation, affecting the biotic components. These substances were transferred to humans through the food chain (Bai et al., 2011; Huang et al., 2010, 2011; Pradhan & Kumar, 2014). Besides the ecological risks, some studies have reported on the presence of e-waste components in higher-living organisms. High levels of PBDEs were found to be present in the brains of fish, the liver of mice, chickens, ducks and frogs, and eggs and muscles of mice, chickens, ducks and frogs (Qin et al., 2011; Zhao et al., 2016). Through the food chain, these toxic substances are transferred to humans and affect them (Huang et al., 2010). Organic pollutants such as PCBs and PBDEs have been detected in the hair of Chinese residents who were exposed to e-waste (Zhao et al., 2008). Dioxins, an e-waste constituent were detected in breast milk and affected the children due to breastfeeding by the mother (Chan et al., 2007; Li et al., 2017). During pregnancy, the fetus also gets exposed to the e-waste constituents causing respiratory problems (Li et al., 2017; Zeng et al., 2016; Zheng et al., 2008).

8.2 Safety and Regulatory Considerations

The safety and regulatory aspects of e-waste differ significantly across different countries in the world, which depends on:

- Level of development—Developed countries have more stringent regulations and better infrastructure than developing countries (Sthiannopkao & Wong, 2013).
- Environmental regulations—Countries with stricter natural controls.
- Economic factors—In emerging countries, e-waste management is hindered by economic barriers.

Countries have e-waste legislations like the Basel Convention for the control of transboundary movements of hazardous wastes (adopted by 187 countries) (Kempel, 1999), Product stewardship Act (2011) (For Australia, while New Zealand and Pacific Island nations following this law), Extended Producer Responsibility (EPR) regulations including a ban on e-waste import (Canada, Kenya, Ghana, Madagascar, and Nigeria), EPR principles (Bolivia, Costa Rica, Chile, Colombia, Ecuador, Peru, and Mexico), Circular Economy Promotion Law (China) (Widmer et al., 2005), and among Southern Asian countries, India is sole country with laws regarding e-waste (since 2011). E-waste rules (2011) for management and handling were first announced in 2016 and amended in 2018. These rules were enacted under the Environment (Protection) Act of 1986 (Rajya Sabha Secretariat, 2011). These rules set the general guidelines for electronic waste management and disposal in India. Other guidelines and regulations include Extended Producer Responsibility (EPR), state e-waste policies, BIS certificates, and pollution control boards. EPR is a policy where the manufacturers took the responsibility for their products (Manomaivibool, 2009).

9 Future Directions and Challenges

9.1 Emerging Technologies and Innovations

Silica nanotubes were designed as chemical nanosensors that were functionalized with azo-chromophores to detect palladium and cobalt ions in e-waste (Abou-Melha et al., 2021). Advancements in the field of nanotechnology can lead to these kinds of innovative nanosystems to monitor, detect, and extract metals from e-waste helping us clear the e-waste from our land. Bioinformatics and nanotechnology can also play a crucial role in cleaning up e-waste from our environment. Bioinformatics can play a role in this remediation process. Based on the omics (genomics, transcriptomics, proteomics, and metabolomics) data, we can predict the efficiency of bioremediation as it can correlate DNA sequences with mRNA, proteins, and metabolites. With the help of genomics, we can understand the microbes' genome employed in the bioremediation process. Metabolomics can play a significant role as it can provide information about the metabolic pathways of the microorganisms and the key metabolites involved in the degradation of e-waste components. It can also provide information about the environmental factors influencing microbial metabolism. Using CRISPR technology, we can edit the genes of the microorganisms involved in e-waste remediation and make them more resistant to toxins produced from e-waste. It can also enhance remediation efficiency. Besides CRISPR, there are other genetic engineering techniques like gene cloning and recombinant DNA technology, which also have the potential to play a part in e-waste remediation.

9.2 Research Gaps and Future Prospects

Research gaps exist in evaluating the toxicity and transportation of engineered nanoparticles and microbial byproducts that will be released during the remediation process. Considering these difficulties, we must develop innovative, safe, non-toxic, green nanoparticles that can overcome the hardships and enhance remediation efficiency. These can help in improving the effectiveness of nanobioremediation of e-waste. Future developments may pave the way for better and more integrated systems using microbes and nanoparticles to achieve greater sustainability. More research needs to be done to evaluate the nanoparticle-microorganism synergy for e-waste remediation. The correct combination of using specific nanoparticles with specific microbes can enhance the nanobioremediation process.

10 Conclusion

10.1 Key Findings and Takeaways

Nanobioremediation, an integrated approach combining the application of nanotechnology and biological remediation, tends to be more valuable than other existing approaches. Nanotechnology has expanded its wide application in many sectors, such as medicine, agriculture, and the environment. Nanotechnology can provide excellent solutions to our existing environmental issues. Nanoparticles or nanomaterials for remediation can enhance efficiency as they have desired physicochemical properties like high surface area, high chemical reactivity, unique optical properties, etc. We can tailor these properties of nanoparticles or nanomaterials according to our applications. Microorganisms can degrade toxic metals, but their abilities tend to decrease when the concentration increases and their population is also affected more. Nanoparticles can reduce the metal's toxicity, thereby allowing microorganisms to grow and improving their efficiency in the bioremediation process. Greener approaches using biological agents to synthesize or fabricate nanoparticles can reduce their toxicity and be an alternative approach for synthesizing nanoparticles. Nanoparticles and microorganisms can make an excellent team for removing hazardous components such as metals and other pollutants from e-waste.

10.2 Call to Action for Sustainable E-Waste Management

Due to rapid urbanization, industrialization, and technological advancements, the production of e-waste is increasing gradually across the globe, which is concerning, and we should find a way to minimize or recycle e-waste without harming our environment and ecosystems. Improper handling methods such as dumping or landfilling, incineration, acid baths, dismantling, and other unsafe methods increase the risk to all living organisms in ecosystems. Due to these dangerous activities, harmful substances are emitted or released into the ecosystems leading to soil, air, and water contaminations. These, in turn, affect whole biotic components from lower to higher living organisms via the food chain. New innovative, greener, and sustainable approaches must be discovered and implemented to manage e-waste effectively. Nanobioremediation techniques are primarily explored in different areas, such as soil, air, and wastewater, but e-waste management is still budding, and more research needs to be done. Nanobiotechnology offers promising solutions for sustainable development and remediation of e-waste. As this is an alarming threat to our planet, scientists, researchers, and others must work more on this and develop sustainable strategies to encounter e-waste.

References

Abou-Melha, K. S., Al-Hazmi, G. A. A., Habeebullah, T. M., Althagafi, I., Othman, A., El-Metwaly, N. M., Shaaban, F., & Shahat, A. (2021). Functionalized silica nanotubes with azo-chromophore for enhanced Pd2+ and Co2+ ions monitoring in E-wastes. *Journal of Molecular Liquids, 329*, 115585. https://doi.org/10.1016/j.molliq.2021.115585

Amankwaa, E. F., Adovor Tsikudo, K. A., & Bowman, J. A. (2017). 'Away' is a place: The impact of electronic waste recycling on blood lead levels in Ghana. *Science of the Total Environment, 601–602*, 1566–1574. https://doi.org/10.1016/j.scitotenv.2017.05.283

Amankwah-Amoah, J. (2016). Global business and emerging economies: Towards a new perspective on the effects of e-waste. *Technological Forecasting and Social Change, 105*, 20–26. https://doi.org/10.1016/j.techfore.2016.01.026

Anaya-Garzon, J., Hubau, A., Joulian, C., & Guezennec, A.-G. (2021). Bioleaching of E-waste: Influence of printed circuit boards on the activity of acidophilic iron-oxidizing bacteria. *Frontiers in Microbiology, 12*, 669738. https://doi.org/10.3389/fmicb.2021.669738

Argumedo-Delira, R., Gómez-Martínez, M. J., & Soto, B. J. (2019). Gold bioleaching from printed circuit boards of mobile phones by Aspergillus niger in a culture without agitation and with glucose as a carbon source. *Metals, 9*(5), 521. https://doi.org/10.3390/met9050521

Arshadi, M., Nili, S., & Yaghmaei, S. (2019). Ni and Cu recovery by bioleaching from the printed circuit boards of mobile phones in non-conventional medium. *Journal of Environmental Management, 250*, 109502. https://doi.org/10.1016/j.jenvman.2019.109502

Awasthi, A. K., Zeng, X., & Li, J. (2016). Environmental pollution of electronic waste recycling in India: A critical review. *Environmental Pollution, 211*, 259–270. https://doi.org/10.1016/j.envpol.2015.11.027

Bai, J., Xiao, R., Gong, A., Gao, H., & Huang, L. (2011). Assessment of heavy metal contamination of surface soils from typical paddy terrace wetlands on the Yunnan Plateau of China. *Physics and Chemistry of the Earth, Parts A/B/C, 36*(9–11), 447–450. https://doi.org/10.1016/j.pce.2010.03.025

Bala, S., Garg, D., Thirumalesh, B. V., Sharma, M., Sridhar, K., Inbaraj, B. S., & Tripathi, M. (2022). Recent strategies for bioremediation of emerging pollutants: A review for a green and sustainable environment. *Toxics, 10*(8), 484. https://doi.org/10.3390/toxics10080484

Baniasadi, M., Vakilchap, F., Bahaloo-Horeh, N., Mousavi, S. M., & Farnaud, S. (2019). Advances in bioleaching as a sustainable method for metal recovery from e-waste: A review. *Journal of Industrial and Engineering Chemistry, 76*, 75–90. https://doi.org/10.1016/j.jiec.2019.03.047

Becci, A., Amato, A., Rodríguez-Maroto, J. M., & Beolchini, F. (2021). Bioleaching of end-of-life printed circuit boards: Mathematical modeling and kinetic analysis. *Industrial & Engineering Chemistry Research, 60*(11), 4261–4268. https://doi.org/10.1021/acs.iecr.0c05566

Benzal, E., Solé, M., Lao, C., Gamisans, X., & Dorado, A. D. (2020). Elemental copper recovery from e-wastes mediated with a two-step bioleaching process. *Waste and Biomass Valorization, 11*(10), 5457–5465. https://doi.org/10.1007/s12649-020-01040-2

Blackwell, K. J., Singleton, I., & Tobin, J. M. (1995). Metal cation uptake by yeast: A review. *Applied Microbiology and Biotechnology, 43*(4), 579–584. https://doi.org/10.1007/BF00164757

Bosecker, K. (1997). Bioleaching: Metal solubilization by microorganisms. *FEMS Microbiology Reviews, 20*(3–4), 591–604. https://doi.org/10.1111/j.1574-6976.1997.tb00340.x

Brandl, H., Bosshard, R., & Wegmann, M. (2001). Computer-munching microbes: Metal leaching from electronic scrap by bacteria and fungi. *Hydrometallurgy, 59*(2–3), 319–326. https://doi.org/10.1016/S0304-386X(00)00188-2

Brewer, A., Dror, I., & Berkowitz, B. (2022). Electronic waste as a source of rare earth element pollution: Leaching, transport in porous media, and the effects of nanoparticles. *Chemosphere, 287*, 132217. https://doi.org/10.1016/j.chemosphere.2021.132217

Cao, X., Alabresm, A., Chen, Y. P., Decho, A. W., & Lead, J. (2020). Improved metal remediation using a combined bacterial and nanoscience approach. *Science of the Total Environment, 704*, 135378. https://doi.org/10.1016/j.scitotenv.2019.135378

Cedervall, T., Hansson, L.-A., Lard, M., Frohm, B., & Linse, S. (2012). Food chain transport of nanoparticles affects behaviour and fat metabolism in fish. *PLoS One, 7*(2), e32254. https://doi.org/10.1371/journal.pone.0032254

Chan, Xing, G. H., Xu, Y., Liang, Y., Chen, L. X., Wu, S. C., Wong, C. K. C., Leung, C. K. M., & Wong, M. H. (2007). Body loadings and health risk assessment of polychlorinated dibenzo-*p* -dioxins and dibenzofurans at an intensive electronic waste recycling site in China. *Environmental Science & Technology, 41*(22), 7668–7674. https://doi.org/10.1021/es071492j

Chauhan, G., Jadhao, P. R., Pant, K. K., & Nigam, K. D. P. (2018). Novel technologies and conventional processes for recovery of metals from waste electrical and electronic equipment: Challenges & opportunities—A review. *Journal of Environmental Chemical Engineering, 6*(1), 1288–1304. https://doi.org/10.1016/j.jece.2018.01.032

Chauhan, A., Anand, J., Parkash, V., & Rai, N. (2023). Biogenic synthesis: A sustainable approach for nanoparticles synthesis mediated by fungi. *Inorganic and Nano-Metal Chemistry, 53*(5), 460–473. https://doi.org/10.1080/24701556.2021.2025078

Corsi, I., Winther-Nielsen, M., Sethi, R., Punta, C., Della Torre, C., Libralato, G., Lofrano, G., Sabatini, L., Aiello, M., Fiordi, L., Cinuzzi, F., Caneschi, A., Pellegrini, D., & Buttino, I. (2018). Ecofriendly nanotechnologies and nanomaterials for environmental applications: Key issue and consensus recommendations for sustainable and ecosafe nanoremediation. *Ecotoxicology and Environmental Safety, 154*, 237–244. https://doi.org/10.1016/j.ecoenv.2018.02.037

Das, A. P., & Ghosh, S. (2018). Bioleaching of manganese from mining waste materials. *Materials Today: Proceedings, 5*(1), 2381–2390. https://doi.org/10.1016/J.MATPR.2017.11.459

Das, A. P., & Ghosh, S. (2023a). Manganese and its application. In *Manganese mining microorganisms* (pp. 29–40). https://doi.org/10.1016/B978-0-12-822147-1.00007-8

Das, A. P., & Ghosh, S. (2023b). Mechanism of manganese bioleaching. In *Manganese mining microorganisms* (pp. 87–98). https://doi.org/10.1016/B978-0-12-822147-1.00014-5

Das, A. P., & Ghosh, S. (2023c). Microorganisms in manganese biomining processes. In *Manganese mining microorganisms* (pp. 55–61). https://doi.org/10.1016/B978-0-12-822147-1.00013-3

Das, A. P., & Ghosh, S. (2023d). Mixed culture bioleaching: An insight into manganese biomining process and efficacy. In *Manganese Mining Microorganisms* (pp. 129–141). https://doi.org/10.1016/B978-0-12-822147-1.00001-7

Das, A. P., & Mishra, S. (2008). Hexavalent Chromium (VI): Health hazards and environmental pollutant. *Journal of Environmental Research and Development, 2*(3), 386–392.

Das, A. P., & Singh, S. (2011). Occupational health assessment of chromite toxicity among Indian miners. *Indian Journal Of Occupational And Environmental Medicine, 15*(1), 6. https://doi.org/10.4103/0019-5278.82998

Das, A. P., Sukla, L. B., Pradhan, N., & Nayak, S. (2011). Manganese biomining: A review. *Bioresource Technology, 102*(16), 7381–7387. https://doi.org/10.1016/J.BIORTECH.2011.05.018

Das, A. P., Ghosh, S., Mohanty, S., & Sukla, L. B. (2015). Consequences of manganese compounds: A review. *Toxicological & Environmental Chemistry, 96*(7), 981–997. https://doi.org/10.1080/02772248.2015.1005428

Davis, T. A., Volesky, B., & Mucci, A. (2003). A review of the biochemistry of heavy metal biosorption by brown algae. *Water Research, 37*(18), 4311–4330. https://doi.org/10.1016/S0043-1354(03)00293-8

Di Piazza, S., Cecchi, G., Cardinale, A. M., Carbone, C., Mariotti, M. G., Giovine, M., & Zotti, M. (2017). Penicillium expansum Link strain for a biometallurgical method to recover REEs from WEEE. *Waste Management, 60*, 596–600. https://doi.org/10.1016/j.wasman.2016.07.029

Díaz-Martínez, M. E., Argumedo-Delira, R., Sánchez-Viveros, G., Alarcón, A., & Mendoza-López, M. R. (2019). Microbial bioleaching of Ag, Au and Cu from printed circuit boards of mobile phones. *Current Microbiology, 76*(5), 536–544. https://doi.org/10.1007/s00284-019-01646-3

Dixit, R., Wasiullah, Malaviya, D., Pandiyan, K., Singh, U., Sahu, A., Shukla, R., Singh, B., Rai, J., Sharma, P., Lade, H., & Paul, D. (2015). Bioremediation of heavy metals from soil and aquatic environment: An overview of principles and criteria of fundamental processes. *Sustainability, 7*(2), 2189–2212. https://doi.org/10.3390/su7022189

Dogan, N. A., Hong, Y., Ozdemir, E., & Yavuz, C. T. (2019). Nanoporous polymer microspheres with nitrile and amidoxime functionalities for gas capture and precious metal recovery from E-waste. *ACS Sustainable Chemistry & Engineering, 7*(1), 123–128. https://doi.org/10.1021/acssuschemeng.8b05490

Ester, Rosa Simone, Di Piazza Grazia, Cecchi Michela, Mazzoccoli Micol, Zerbini Anna Maria, Cardinale Mirca, Zotti (2022) Applied tests to select the most suitable fungal strain for the recovery of critical raw materials from electronic waste powder recycling 7(5);72. https://doi.org/10.3390/recycling7050072

Fariq, A., Khan, T., & Yasmin, A. (2017). Microbial synthesis of nanoparticles and their potential applications in biomedicine. *Journal of Applied Biomedicine, 15*(4), 241–248. https://doi.org/10.1016/j.jab.2017.03.004

Fomina, M., & Gadd, G. M. (2014). Biosorption: Current perspectives on concept, definition and application. *Bioresource Technology, 160*, 3–14. https://doi.org/10.1016/j.biortech.2013.12.102

Forti, V., Balde, C. P., Kuehr, R., & Bel, G. (2020). The Global E-waste Monitor 2020: Quantities, flows and the circular economy potential.

Gangwar, C., Choudhari, R., Chauhan, A., Kumar, A., Singh, A., & Tripathi, A. (2019). Assessment of air pollution caused by illegal e-waste burning to evaluate the human health risk. *Environment International, 125*, 191–199. https://doi.org/10.1016/j.envint.2018.11.051

Ghosh, S., Mohanty, S., Akcil, A., Sukla, L. B., & Das, A. P. (2016). A greener approach for resource recycling: Manganese bioleaching. *Chemosphere, 154*, 628–639. https://doi.org/10.1016/J.CHEMOSPHERE.2016.04.028

Ghosh, S., Tripathy, B., Dey, S., & Das, A. P. (2023). Comparative investigation of fungal and bacterial manganese biomining mechanisms. In *Manganese mining microorganisms* (pp. 99–115). https://doi.org/10.1016/b978-0-12-822147-1.00016-9

Grant, K., Goldizen, F. C., Sly, P. D., Brune, M.-N., Neira, M., Van Den Berg, M., & Norman, R. E. (2013). Health consequences of exposure to e-waste: A systematic review. *The Lancet Global Health, 1*(6), e350–e361. https://doi.org/10.1016/S2214-109X(13)70101-3

Han, Y., Tang, Z., Sun, J., Xing, X., Zhang, M., & Cheng, J. (2019). Heavy metals in soil contaminated through e-waste processing activities in a recycling area: Implications for risk management. *Process Safety and Environmental Protection, 125*, 189–196. https://doi.org/10.1016/j.psep.2019.03.020

Hano, C., & Abbasi, B. H. (2021). Plant-based green synthesis of nanoparticles: Production, characterization and applications. *Biomolecules, 12*(1), 31. https://doi.org/10.3390/biom12010031

Heacock, M., Kelly, C. B., Asante, K. A., Birnbaum, L. S., Bergman, Å. L., Bruné, M.-N., Buka, I., Carpenter, D. O., Chen, A., Huo, X., Kamel, M., Landrigan, P. J., Magalini, F., Diaz-Barriga, F., Neira, M., Omar, M., Pascale, A., Ruchirawat, M., Sly, L., et al. (2016). E-waste and harm to vulnerable populations: A growing global problem. *Environmental Health Perspectives, 124*(5), 550–555. https://doi.org/10.1289/ehp.1509699

Hiremath, L., Kumar, V. G., & Sathyanarayana, B. S. (2015). Exploitation of live fungus (*Aspergillus niger* RV11) for bioremediation of heavy metals from e-waste. *International Journal of Chemical and Environmental Engineering, 6*(3), 184–189.

Hong, Y., & Valix, M. (2014). Bioleaching of electronic waste using acidophilic sulfur oxidising bacteria. *Journal of Cleaner Production, 65*, 465–472. https://doi.org/10.1016/j.jclepro.2013.08.043

Hosseini, S. M., Vakilchap, F., Baniasadi, M., Mousavi, S. M., Khodadadi Darban, A., & Farnaud, S. (2022). Green recovery of cerium and strontium from gold mine tailings using an adapted acidophilic bacterium in one-step bioleaching approach. *Journal of the Taiwan Institute of Chemical Engineers, 138*, 104482. https://doi.org/10.1016/j.jtice.2022.104482

Huang, H., Zhang, S., Christie, P., Wang, S., & Xie, M. (2010). Behavior of Decabromodiphenyl Ether (BDE-209) in the soil−plant system: Uptake, translocation, and metabolism in plants and dissipation in soil. *Environmental Science & Technology, 44*(2), 663–667. https://doi.org/10.1021/es901860r

Huang, H., Zhang, S., & Christie, P. (2011). Plant uptake and dissipation of PBDEs in the soils of electronic waste recycling sites. *Environmental Pollution, 159*(1), 238–243. https://doi.org/10.1016/j.envpol.2010.08.034

Hussein, M. A., Alamry, K. A., El Shishtawy, R. M., Elshehy, E. A., & El-Said, W. A. (2020). Nanoporous colorant sensors and captors for simultaneous recognition and recovery of gold from E-wastes. *Waste Management, 116*, 166–178. https://doi.org/10.1016/j.wasman.2020.07.030

Ilyas, S., Chi, R., & Lee, J. (2013). Fungal bioleaching of metals from mine tailing. *Mineral Processing and Extractive Metallurgy Review, 34*(3), 185–194. https://doi.org/10.1080/08827508.2011.623751

Işıldar, A., Van De Vossenberg, J., Rene, E. R., Van Hullebusch, E. D., & Lens, P. N. L. (2016). Two-step bioleaching of copper and gold from discarded printed circuit boards (PCB). *Waste Management, 57*, 149–157. https://doi.org/10.1016/j.wasman.2015.11.033

Jayaseelan, C., Rahuman, A. A., Kirthi, A. V., Marimuthu, S., Santhoshkumar, T., Bagavan, A., Gaurav, K., Karthik, L., & Rao, K. V. B. (2012). Novel microbial route to synthesize ZnO nanoparticles using Aeromonas hydrophila and their activity against pathogenic bacteria and fungi. *Spectrochimica Acta Part A: Molecular and Biomolecular Spectroscopy, 90*, 78–84. https://doi.org/10.1016/j.saa.2012.01.006

Ji, X., Yang, M., Wan, A., Yu, S., & Yao, Z. (2022). Bioleaching of typical electronic waste—printed circuit boards (WPCBs): A short review. *International Journal of Environmental Research and Public Health, 19*(12), 7508. https://doi.org/10.3390/ijerph19127508

Jordahl, J. L., Foster, L., Schnoor, J. L., & Alvarez, P. J. J. (1997). Effect of hybrid poplar trees on microbial populations important to hazardous waste bioremediation. *Environmental Toxicology and Chemistry, 16*(6), 1318–1321. https://doi.org/10.1002/etc.5620160630

Jowkar, M. J., Bahaloo-Horeh, N., Mousavi, S. M., & Pourhossein, F. (2018). Bioleaching of indium from discarded liquid crystal displays. *Journal of Cleaner Production, 180*, 417–429. https://doi.org/10.1016/j.jclepro.2018.01.136

Kaur, P., Sharma, S., Albarakaty, F. M., Kalia, A., Hassan, M. M., & Abd-Elsalam, K. A. (2022). Biosorption and bioleaching of heavy metals from electronic waste varied with microbial genera. *Sustainability, 14*(2), 935. https://doi.org/10.3390/su14020935

Kavitha, A. V. (2014). Extraction of precious metals from e-waste. *Journal of Chemical and Pharmaceutical Sciences, Special Issue, 3*, 147–149.

Kaya, M. (2016). Recovery of metals and nonmetals from electronic waste by physical and chemical recycling processes. *Waste Management, 57*, 64–90. https://doi.org/10.1016/j.wasman.2016.08.004

Kempel. (1999). The negotiations on the Basel convention on the transboundary movement of hazardous wastes and their disposal: A national delegation perspective. *International Negotiation, 4*(3), 413–434. https://doi.org/10.1163/15718069920848552

Kim, Y., Murugesan, K., Chang, Y., Kim, E., & Chang, Y. (2012). Degradation of polybrominated diphenyl ethers by a sequential treatment with nanoscale zero valent iron and aerobic biodegradation. *Journal of Chemical Technology & Biotechnology, 87*(2), 216–224. https://doi.org/10.1002/jctb.2699

Kudpeng, K., Bohu, T., Morris, C., Thiravetyan, P., & Kaksonen, A. H. (2020). Bioleaching of gold from sulfidic gold ore concentrate and electronic waste by Roseovarius tolerans and Roseovarius mucosus. *Microorganisms, 8*(11), 1783. https://doi.org/10.3390/microorganisms8111783

Kumar, A., Saini, H. S., & Kumar, S. (2018). Bioleaching of gold and silver from waste printed circuit boards by Pseudomonas balearica SAE1 isolated from an e-waste recycling facility. *Current Microbiology, 75*(2), 194–201. https://doi.org/10.1007/s00284-017-1365-0

Kumar, A., Saini, H. S., Şengör, S., Sani, R. K., & Kumar, S. (2021). Bioleaching of metals from waste printed circuit boards using bacterial isolates native to abandoned gold mine. *Biometals, 34*(5), 1043–1058. https://doi.org/10.1007/s10534-021-00326-9

Lakshmi, S., Suvedha, K., Sruthi, R., Lavanya, J., Varjani, S., & Nakkeeran, E. (2020). Hexavalent chromium sequestration from electronic waste by biomass of *Aspergillus carbonarius. Bioengineered, 11*(1), 708–717. https://doi.org/10.1080/21655979.2020.1780828

Le, T. T., Nguyen, K.-H., Jeon, J.-R., Francis, A. J., & Chang, Y.-S. (2015). Nano/bio treatment of polychlorinated biphenyls with evaluation of comparative toxicity. *Journal of Hazardous Materials, 287*, 335–341. https://doi.org/10.1016/j.jhazmat.2015.02.001

Lengke, M. F., Ravel, B., Fleet, M. E., Wanger, G., Gordon, R. A., & Southam, G. (2006). Mechanisms of gold bioaccumulation by filamentous cyanobacteria from gold(III)−chloride complex. *Environmental Science & Technology, 40*(20), 6304–6309. https://doi.org/10.1021/es061040r

Li, X., Tian, Y., Zhang, Y., Ben, Y., & Lv, Q. (2017). Accumulation of polybrominated diphenyl ethers in breast milk of women from an e-waste recycling center in China. *Journal of Environmental Sciences, 52*, 305–313. https://doi.org/10.1016/j.jes.2016.10.008

Li, H., Zhu, F., & He, S. (2019). The degradation of decabromodiphenyl ether in the e-waste site by biochar supported nanoscale zero-valent iron/persulfate. *Ecotoxicology and Environmental Safety, 183*, 109540. https://doi.org/10.1016/j.ecoenv.2019.109540

Lin, S., Ali, M. U., Zheng, C., Cai, Z., & Wong, M. H. (2022). Toxic chemicals from uncontrolled e-waste recycling: Exposure, body burden, health impact. *Journal of Hazardous Materials, 426*, 127792. https://doi.org/10.1016/j.jhazmat.2021.127792

Liu, R., Li, J., & Ge, Z. (2016). Review on Chromobacterium Violaceum for gold bioleaching from E-waste. *Procedia Environmental Sciences, 31*, 947–953. https://doi.org/10.1016/j.proenv.2016.02.119

Magoda, K., & Mekuto, L. (2022). Biohydrometallurgical recovery of metals from waste electronic equipment: Current status and proposed process. *Recycling, 7*(5), 67. https://doi.org/10.3390/recycling7050067

Majumder, D. R. (2012). Bioremediation: Copper nanoparticles from electronic-waste. *International Journal of Engineering Science and Technology, 4*(10), 4380–4389.

Malik, P., Shankar, R., Malik, V., Sharma, N., & Mukherjee, T. K. (2014). Green chemistry based benign routes for nanoparticle synthesis. *Journal of Nanoparticles, 2014*, 1–14. https://doi.org/10.1155/2014/302429

Manomaivibool, P. (2009). Extended producer responsibility in a non-OECD context: The management of waste electrical and electronic equipment in India. *Resources, Conservation and Recycling, 53*(3), 136–144. https://doi.org/10.1016/j.resconrec.2008.10.003

Marappa, N., Ramachandran, L., Dharumadurai, D., & Nooruddin, T. (2020). Recovery of gold and other precious metal resources from environmental polluted E-waste printed circuit board by bioleaching Frankia. *International Journal of Environmental Research, 14*(2), 165–176. https://doi.org/10.1007/s41742-020-00254-5

Mishra, S., Ghosh, S., van Hullebusch, E. D., Singh, S., & Das, A. P. (2023). A critical review on the recovery of base and critical elements from electronic waste-contaminated streams using microbial biotechnology. *Applied Biochemistry and Biotechnology*, 2023, 1–30. https://doi.org/10.1007/S12010-023-04440-X

Mohanpuria, P., Rana, N. K., & Yadav, S. K. (2008). Biosynthesis of nanoparticles: Technological concepts and future applications. *Journal of Nanoparticle Research, 10*(3), 507–517. https://doi.org/10.1007/s11051-007-9275-x

Mohanty, S., Ghosh, S., Nayak, S., & Das, A. P. (2017). Bioleaching of manganese by Aspergillus sp. isolated from mining deposits. *Chemosphere, 172*, 302–309. https://doi.org/10.1016/J.CHEMOSPHERE.2016.12.136

Murali, A., Sarswat, P. K., Benedict, J., Plummer, M. J., Shine, A. E., & Free, M. L. (2022). Determination of metallic and polymeric contents in electronic waste materials and evaluation of their hydrometallurgical recovery potential. *International journal of Environmental Science and Technology, 19*(4), 2295–2308. https://doi.org/10.1007/s13762-021-03285-3

Narayanasamy, M., Dhanasekaran, D., Vinothini, G., & Thajuddin, N. (2018). Extraction and recovery of precious metals from electronic waste printed circuit boards by bioleaching acidophilic fungi. *International journal of Environmental Science and Technology, 15*(1), 119–132. https://doi.org/10.1007/s13762-017-1372-5

Narayanasamy, M., Dhanasekaran, D., & Thajuddin, N. (2022). Frankia consortium extracts high-value metals from e-waste. *Environmental Technology & Innovation, 28*, 102564. https://doi.org/10.1016/j.eti.2022.102564

Natarajan, G., & Ting, Y.-P. (2015). Gold biorecovery from e-waste: An improved strategy through spent medium leaching with pH modification. *Chemosphere, 136*, 232–238. https://doi.org/10.1016/j.chemosphere.2015.05.046

Natrajan, G., & Ting, Y. P. (2013). Two-step bioleaching and spent medium leaching of gold from electronic scrap material using *Chromobacterium violaceum. Advanced Materials Research, 825*, 270–273. https://doi.org/10.4028/www.scientific.net/AMR.825.270

Nayar, R. (2019). Removal of toxic cation from aqueous e-waste material using zno as a nanoadsorbent by adsorption technique. J Curr Chem Pharm Sc, 9(1), 119.

Nili, S., Arshadi, M., & Yaghmaei, S. (2022). Fungal bioleaching of e-waste utilizing molasses as the carbon source in a bubble column bioreactor. *Journal of Environmental Management, 307*, 114524. https://doi.org/10.1016/j.jenvman.2022.114524

Palmieri, M., Iovinella, M., Davis, S. J., Di Cicco, M. R., Lubritto, C., Race, M., Papa, S., Fabbricino, M., & Ciniglia, C. (2022). Galdieria sulphuraria ACUF427 freeze-dried biomass as novel biosorbent for rare earth elements. *Microorganisms, 10*(11), 2138. https://doi.org/10.3390/microorganisms10112138

Panpatte, D. G., & Jhala, Y. K. (Eds.). (2021). *Microbial rejuvenation of polluted environment: Volume 1* (Vol. 25). Springer Singapore. https://doi.org/10.1007/978-981-15-7447-4

Paul, A., Dey, S., Ram, D. K., & Das, A. P. (2023). Hexavalent chromium pollution and its sustainable management through bioremediation. *Geomicrobiology Journal*, 1–11. https://doi.org/10.1080/01490451.2023.2218377

Pham, V. A., & Ting, Y. P. (2009). Gold bioleaching of electronic waste by cyanogenic bacteria and its enhancement with bio-oxidation. *Advanced Materials Research, 71–73*, 661–664. https://doi.org/10.4028/www.scientific.net/AMR.71-73.661

Pradhan, J. K., & Kumar, S. (2014). Informal e-waste recycling: Environmental risk assessment of heavy metal contamination in Mandoli industrial area, Delhi, India. *Environmental Science and Pollution Research, 21*(13), 7913–7928. https://doi.org/10.1007/s11356-014-2713-2

Priya, A., & Hait, S. (2018). Feasibility of bioleaching of selected metals from electronic waste by Acidiphilium acidophilum. *Waste and Biomass Valorization, 9*(5), 871–877. https://doi.org/10.1007/s12649-017-9833-0

Priyadarsini, S., & Das, A. P. (2024). Analytical and structural characterization of waste lithium-ion batteries for their effective recycling strategy. *Environmental Chemistry and Ecotoxicology*. https://doi.org/10.1016/j.enceco.2024.12.004

Priyadarsini, S., & Das, A. P. (2025). Lithium bioleaching: A review on microbial-assisted sustainable technology for lithium bio-circularity. *Journal of Water Process Engineering, 69*, 106744. https://doi.org/10.1016/j.jwpe.2024.106744

Qin, X., Qin, Z., Li, Y., Zhao, Y., Xia, X., Yan, S., Tian, M., Zhao, X., Xu, X., & Yang, Y. (2011). Polybrominated diphenyl ethers in chicken tissues and eggs from an electronic waste recycling area in Southeast China. *Journal of Environmental Sciences, 23*(1), 133–138. https://doi.org/10.1016/S1001-0742(10)60384-1

Raj, K., & Das, A. P. (2023). Lead pollution: Impact on environment and human health and approach for a sustainable solution. *Environmental Chemistry and Ecotoxicology, 5*, 79–85. https://doi.org/10.1016/J.ENCECO.2023.02.001

Rajput, V. D., Minkina, T., Upadhyay, S. K., Kumari, A., Ranjan, A., Mandzhieva, S., Sushkova, S., Singh, R. K., & Verma, K. K. (2022). Nanotechnology in the restoration of polluted soil. *Nanomaterials, 12*(5), 769. https://doi.org/10.3390/nano12050769

Rajya Sabha Secretariat. (2011). *E-Waste in India.*

Reed, D. W., Fujita, Y., Daubaras, D. L., Jiao, Y., & Thompson, V. S. (2016). Bioleaching of rare earth elements from waste phosphors and cracking catalysts. *Hydrometallurgy, 166*, 34–40. https://doi.org/10.1016/j.hydromet.2016.08.006

Rizki, I. N., Tanaka, Y., & Okibe, N. (2019). Thiourea bioleaching for gold recycling from e-waste. *Waste Management, 84*, 158–165. https://doi.org/10.1016/j.wasman.2018.11.021

Robinson, B. H. (2009). E-waste: An assessment of global production and environmental impacts. *Science of the Total Environment, 408*(2), 183–191. https://doi.org/10.1016/j.scitotenv.2009.09.044

Rohwerder, T., Gehrke, T., Kinzler, K., & Sand, W. (2003). Bioleaching review part A. *Applied Microbiology and Biotechnology, 63*(3), 239–248. https://doi.org/10.1007/s00253-003-1448-7

Rozas, E. E., Mendes, M. A., Nascimento, C. A. O., Espinosa, D. C. R., Oliveira, R., Oliveira, G., & Custodio, M. R. (2017). Bioleaching of electronic waste using bacteria isolated from the marine sponge Hymeniacidon heliophila (Porifera). *Journal of Hazardous Materials, 329*, 120–130. https://doi.org/10.1016/j.jhazmat.2017.01.037

Salem, S. S., & Fouda, A. (2021). Green synthesis of metallic nanoparticles and their prospective biotechnological applications: An overview. *Biological Trace Element Research, 199*(1), 344–370. https://doi.org/10.1007/s12011-020-02138-3

Sand, W., Gehrke, T., Jozsa, P.-G., & Schippers, A. (2001). (Bio)chemistry of bacterial leaching—Direct vs. Indirect bioleaching. *Hydrometallurgy, 59*(2–3), 159–175. https://doi.org/10.1016/S0304-386X(00)00180-8

Schippers, A., & Sand, W. (1999). Bacterial leaching of metal sulfides proceeds by two indirect mechanisms via thiosulfate or via polysulfides and sulfur. *Applied and Environmental Microbiology, 65*(1), 319–321. https://doi.org/10.1128/AEM.65.1.319-321.1999

Schneiter, A., Rüegg, K., Kim, J., Biselli, S., Vaideeswaran, K., Sereda, O., Briand, D., & Bindschedler, S. (2023). *Fungal biodegradation unravels potential low-tech pathway for paper electronics waste valorization*. In Review. https://doi.org/10.21203/rs.3.rs-3443595/v1.

Scott, J. A., & Karanjkar, A. M. (1992). Repeated cadmium biosorption by regenerated Enterobacter aerogenes biofilm attached to activated carbon. *Biotechnology Letters, 14*(8), 737–740. https://doi.org/10.1007/BF01021653

Shahabuddin, M., Uddin, M. N., Chowdhury, J. I., Ahmed, S. F., Uddin, M. N., Mofijur, M., & Uddin, M. A. (2023). A review of the recent development, challenges, and opportunities of electronic waste (e-waste). *International journal of Environmental Science and Technology, 20*(4), 4513–4520. https://doi.org/10.1007/s13762-022-04274-w

Sharada, H. M., Abdel-Halim, S. A., Hafez, M. A., Elbarbary, T. A., Abdel-Fatah, Y., & Ibrahim, I. A. (2021). Bioleaching of copper from electronic waste using Aspergillus niger. *Biointerface Research in Applied Chemistry, 12*(6), 8406–8425. https://doi.org/10.33263/BRIAC126.84068425

Sharma, M., & Sood, K. (2023). Nano-bioremediation: An emerging weapon for emerging pollutants. In N. George, V. Dwibedi, S. K. Rath, & P. S. Chauhan (Eds.), *Management and miti-*

gation of emerging pollutants (pp. 273–291). Springer International Publishing. https://doi.org/10.1007/978-3-031-41005-5_10

Sheel, A., & Pant, D. (2018). Recovery of gold from electronic waste using chemical assisted microbial biosorption (hybrid) technique. *Bioresource Technology, 247*, 1189–1192. https://doi.org/10.1016/j.biortech.2017.08.212

Sikander, A., Kelly, S., Kuchta, K., Sievers, A., Willner, T., & Hursthouse, A. S. (2022). Chemical and microbial leaching of valuable metals from PCBs and tantalum capacitors of spent mobile phones. *International Journal of Environmental Research and Public Health, 19*(16), 10006. https://doi.org/10.3390/ijerph191610006

Sthiannopkao, S., & Wong, M. H. (2013). Handling e-waste in developed and developing countries: Initiatives, practices, and consequences. *Science of the Total Environment, 463–464*, 1147–1153. https://doi.org/10.1016/j.scitotenv.2012.06.088

Subbaiyan, R., Ganesan, A., Sasikumar, B., Rajendran, S., & Ramasubramanian, B. (2023). Synthesis and characterization of ferrous and copper nanoparticles from E-waste using biological reduction by lichen-associated bacteria and their application in antifouling activity. *Applied Biochemistry and Biotechnology, 195*(5), 3142–3155. https://doi.org/10.1007/s12010-022-04293-w

Suzuki, G., Someya, M., Matsukami, H., Tue, N. M., Uchida, N., Tuyen, L. H., Viet, P. H., Takahashi, S., Tanabe, S., Brouwer, A., & Takigami, H. (2016). Comprehensive evaluation of dioxins and dioxin-like compounds in surface soils and river sediments from e-waste-processing sites in a village in northern Vietnam: Heading towards the environmentally sound management of e-waste. *Emerging Contaminants, 2*(2), 98–108. https://doi.org/10.1016/j.emcon.2016.03.001

Takatsugi, K., Sasaki, K., & Hirajima, T. (2011). Mechanism of the enhancement of bioleaching of copper from enargite by thermophilic iron-oxidizing archaea with the concomitant precipitation of arsenic. *Hydrometallurgy, 109*(1–2), 90–96. https://doi.org/10.1016/j.hydromet.2011.05.013

Tian, X., Zhang, B., Hou, J., Gu, M., & Chen, Y. (2020). In situ preparation and unique electrical behaviors of gold@hollow polyaniline nanospheres through recovery of gold from simulated e-waste. *Bulletin of the Chemical Society of Japan, 93*(3), 373–378. https://doi.org/10.1246/bcsj.20190286

Toshev, S., Loukanov, A., Emin, S., & Nakabayashi, S. (2021). Biotechnological recycling and recovery of metals from secondary raw materials through biogenic synthesis of nanoparticles. *Sustainable Extraction and Processing of Raw Materials Journal, 2*, 68–73.

Tributsch, H. (2001). Direct versus indirect bioleaching. *Hydrometallurgy, 59*(2–3), 177–185. https://doi.org/10.1016/S0304-386X(00)00181-X

Vakilchap, F., Mousavi, S. M., & Shojaosadati, S. A. (2016). Role of Aspergillus niger in recovery enhancement of valuable metals from produced red mud in Bayer process. *Bioresource Technology, 218*, 991–998. https://doi.org/10.1016/j.biortech.2016.07.059

Velásquez, L., & Dussan, J. (2009). Biosorption and bioaccumulation of heavy metals on dead and living biomass of Bacillus sphaericus. *Journal of Hazardous Materials, 167*(1–3), 713–716. https://doi.org/10.1016/j.jhazmat.2009.01.044

Vera, M., Schippers, A., & Sand, W. (2013). Progress in bioleaching: Fundamentals and mechanisms of bacterial metal sulfide oxidation—Part A. *Applied Microbiology and Biotechnology, 97*(17), 7529–7541. https://doi.org/10.1007/s00253-013-4954-2

Vestola, E. A., Kuusenaho, M. K., Närhi, H. M., Tuovinen, O. H., Puhakka, J. A., Plumb, J. J., & Kaksonen, A. H. (2010). Acid bioleaching of solid waste materials from copper, steel and recycling industries. *Hydrometallurgy, 103*(1–4), 74–79. https://doi.org/10.1016/j.hydromet.2010.02.017

Waghmode, M. S., Gunjal, A. B., & Patil, N. N. (2021). Bioleaching of electronic waste. *Pollution, 7*(1). https://doi.org/10.22059/poll.2020.308119.872

Wang, H., Hao, R., Nie, L., Zhang, X., & Zhang, Y. (2022). Pollution characteristics and risk assessment of air multi-pollutants from typical e-waste dismantling activities. *Environmental Pollution, 294*, 118630. https://doi.org/10.1016/j.envpol.2021.118630

Widmer, R., Oswald-Krapf, H., Sinha-Khetriwal, D., Schnellmann, M., & Böni, H. (2005). Global perspectives on e-waste. *Environmental Impact Assessment Review, 25*(5), 436–458. https://doi.org/10.1016/j.eiar.2005.04.001

Wu, Q., Leung, J. Y. S., Geng, X., Chen, S., Huang, X., Li, H., Huang, Z., Zhu, L., Chen, J., & Lu, Y. (2015). Heavy metal contamination of soil and water in the vicinity of an abandoned e-waste recycling site: Implications for dissemination of heavy metals. *Science of the Total Environment, 506–507*, 217–225. https://doi.org/10.1016/j.scitotenv.2014.10.121

Wu, W., Liu, X., Zhang, X., Zhu, M., & Tan, W. (2018). Bioleaching of copper from waste printed circuit boards by bacteria-free cultural supernatant of iron–sulfur-oxidizing bacteria. *Bioresources and Bioprocessing, 5*(1), 10. https://doi.org/10.1186/s40643-018-0196-6

Wu, Z., Han, W., Xie, M., Han, M., Li, Y., & Wang, Y. (2019). Occurrence and distribution of polybrominated diphenyl ethers in soils from an e-waste recycling area in northern China. *Ecotoxicology and Environmental Safety, 167*, 467–475. https://doi.org/10.1016/j.ecoenv.2018.10.029

Xia, M., Bao, P., Liu, A., Wang, M., Shen, L., Yu, R., Liu, Y., Chen, M., Li, J., Wu, X., Qiu, G., & Zeng, W. (2018). Bioleaching of low-grade waste printed circuit boards by mixed fungal culture and its community structure analysis. *Resources, Conservation and Recycling, 136*, 267–275. https://doi.org/10.1016/j.resconrec.2018.05.001

Xu, Q., Du, X.-H., Luo, D., Strømme, M., Zhang, Q.-F., & Xu, C. (2023). Gold recovery from E-waste using freestanding nanopapers of cellulose and ionic covalent organic frameworks. *Chemical Engineering Journal, 458*, 141498. https://doi.org/10.1016/j.cej.2023.141498

Yaashikaa, P. R., Priyanka, B., Senthil Kumar, P., Karishma, S., Jeevanantham, S., & Indraganti, S. (2022). A review on recent advancements in recovery of valuable and toxic metals from e-waste using bioleaching approach. *Chemosphere, 287*, 132230. https://doi.org/10.1016/j.chemosphere.2021.132230

Yadav, K. K., Singh, J. K., Gupta, N., & Kumar, V. (2017). A review of nanobioremediation technologies for environmental cleanup: A novel biological approach. *Journal of Materials and Environmental Sciences, 8*(2), 740–757.

Zeng, X., Xu, X., Zheng, X., Reponen, T., Chen, A., & Huo, X. (2016). Heavy metals in PM 2.5 and in blood, and children's respiratory symptoms and asthma from an e-waste recycling area. *Environmental Pollution, 210*, 346–353. https://doi.org/10.1016/j.envpol.2016.01.025

Zhang, J., & Min, H. (2010). Characterization of a multimetal resistant Burkholderia fungorum isolated from an e-waste recycling area for its potential in Cd sequestration. *World Journal of Microbiology and Biotechnology, 26*(2), 371–374. https://doi.org/10.1007/s11274-009-0163-7

Zhao, G., Wang, Z., Dong, M. H., Rao, K., Luo, J., Wang, D., Zha, J., Huang, S., Xu, Y., & Ma, M. (2008). PBBs, PBDEs, and PCBs levels in hair of residents around e-waste disassembly sites in Zhejiang Province, China, and their potential sources. *Science of the Total Environment, 397*(1–3), 46–57. https://doi.org/10.1016/j.scitotenv.2008.03.010

Zhao, Y., Li, Y., Qin, X., Lou, Q., & Qin, Z. (2016). Accumulation of polybrominated diphenyl ethers in the brain compared with the levels in other tissues among different vertebrates from an e-waste recycling site. *Environmental Pollution, 218*, 1334–1341. https://doi.org/10.1016/j.envpol.2016.08.091

Zheng, L., Wu, K., Li, Y., Qi, Z., Han, D., Zhang, B., Gu, C., Chen, G., Liu, J., Chen, S., Xu, X., & Huo, X. (2008). Blood lead and cadmium levels and relevant factors among children from an e-waste recycling town in China. *Environmental Research, 108*(1), 15–20. https://doi.org/10.1016/j.envres.2008.04.002

Zhu, N., Zhang, L., Li, C., & Cai, C. (2003). Recycling of spent nickel–cadmium batteries based on bioleaching process. *Waste Management, 23*(8), 703–708. https://doi.org/10.1016/S0956-053X(03)00068-0

Zhu, F., Yang, M., Luo, Z., Yu, R., Hu, G., & Yan, Y. (2020). Bioaccumulation and biotransformation of arsenic in Leptolyngbya boryana. *Environmental Science and Pollution Research, 27*(24), 29993–30000. https://doi.org/10.1007/s11356-020-09294-y

www.ingramcontent.com/pod-product-compliance
Lightning Source LLC
Chambersburg PA
CBHW060153100226
39428CB00003B/52
* 9 7 8 3 0 3 1 9 0 2 8 6 4 *